AF324050

2D Inorganic Materials beyond Graphene

2D Inorganic Materials beyond Graphene

Editors

C N R Rao

Jawaharlal Nehru Centre for Advanced Scientific Research, India
Indian Institute of Science, India

U V Waghmare

Jawaharlal Nehru Centre for Advanced Scientific Research, India

World Scientific

NEW JERSEY · LONDON · SINGAPORE · BEIJING · SHANGHAI · HONG KONG · TAIPEI · CHENNAI · TOKYO

Published by

World Scientific Publishing Europe Ltd.

57 Shelton Street, Covent Garden, London WC2H 9HE

Head office: 5 Toh Tuck Link, Singapore 596224

USA office: 27 Warren Street, Suite 401-402, Hackensack, NJ 07601

Library of Congress Cataloging-in-Publication Data
Names: Rao, C. N. R. (Chintamani Nagesa Ramachandra), 1934– editor. |
 Waghmare, U. V. (Umesh V.), editor.
Title: 2D inorganic materials beyond graphene / edited by C.N.R. Rao (Jawaharlal Nehru Centre
 for Advanced Scientific Research, India & Indian Institute of Science, India),
 U.V. Waghmare (Jawaharlal Nehru Centre for Advanced Scientific Research, India).
Description: New Jersey : World Scientific, 2017. | Includes bibliographical references.
Identifiers: LCCN 2016054153 | ISBN 9781786342690 (hc : alk. paper)
Subjects: LCSH: Nanostructured materials. | Graphene. | Layer structure (Solids)
Classification: LCC TA418.9.N35 T877 2017 | DDC 620.1/15--dc23
LC record available at https://lccn.loc.gov/2016054153

British Library Cataloguing-in-Publication Data
A catalogue record for this book is available from the British Library.

Desk Editors: Kalpana Bharanikumar/Mary Simpson

Typeset by Stallion Press
Email: enquiries@stallionpress.com

Printed in Singapore

Preface

The discovery of graphene created a great sensation in physical and materials sciences. The extraordinary properties of graphene are not only remarkable but also most unexpected. Encouraged by this discovery, there has been much interest in exploring and understanding the properties of nanosheets of other 2D materials, molybdenum disulphide currently being the most representative of such materials. A variety of transition metal dichalcogenides possess layered structures besides the materials like boron nitride (BN). Many new 2D materials have come to the fore, phospherene and borocarbonitrides ($B_xC_yN_z$) being two important ones. These materials exhibit not only interesting properties but also with several potential applications. We considered it more appropriate to edit a book covering the current status and recent developments in this area. This book is a result of such a desire on our part to present the varied aspects of the 2D materials.

The book contains 10 chapters, of which Chapter 1 deals with the essential aspects of synthesis, structure, characterisation, properties and applications of 2D materials including transition metal dichalcogenides, BN and borocarbonitrides. Some of the 2D materials including graphene, h-BN, TMX_2 exhibit honeycomb structures and associated topologically non-trivial valleys in the valence and conduction band structures. Theoretical treatment of this aspect and its generalisation to 3D Weyl semimetals are discussed in Chapter 2. With a brief introduction to the oxygen reduction reaction relevant to fuel cells, Chapter 3 presents the properties of 2D silicene, chalcogenides, carbides, borides, borocarbonitrides and layered double hydroxides relevant to catalysis of this reaction. Phosphorene, a recent addition to 2D materials, exhibits a remarkable anisotropy and interesting properties. Its atomic and electronic structure and defects, carrier mobility and transport, electronic topological transition, tunability of its properties with strain, and properties relevant to gas sensing, photocatalytic applications, field-effect transistor are discussed in Chapter 4.

An emerging area of research today concerns heterostructures based on 2D materials. A detailed description of the fabrication, structure, functionality and devices based on 2D van der Waals hybrids is presented in Chapter 5, highlighting how clever combinations of 2D materials can result in hybrids with unusual properties. Synthesis, structure and thermoelectric properties of chalcogenides, such as Bi_2Te_3 are discussed in Chapter 6. Commercial applications of 2D materials require scalability in their synthesis or deposition on substrates, and associated techniques for their diagnostics. The present status and challenges in plasma-assisted chemical and physical deposition methods for the growth of graphene, 2D transition metal dichalcogenides and BN ultra-thin films are discussed in Chapter 7. The structure and properties of an interface of a metallic electrode with a nanomaterial like MoS_2 are key to its use in nanoelectronic devices. A comprehensive understanding of contacts to MoS_2 with different mechanisms of Fermi level pinning and strategies to develop low-resistance contacts are discussed in Chapter 8. 2D materials bend readily and can be subjected to mechanical strain. The strain-dependent electronic, optical and vibrational properties of 2D TMX_2 materials and a possibility of engineering their transport properties are discussed in Chapter 9. Most 2D materials synthesised on a relatively large-scale contain defects and grain boundaries, which have a strong impact on their properties and technological applications. Theoretical analysis of the structure, electronic and vibrational signatures of various point defects, grain boundaries and structural faults in graphene, h-BN and TMX_2 is presented in Chapter 10.

We are quite pleased with the contents of the book and hope that it will satisfy the needs of beginning researchers as well as practitioners. Clearly, this area will expand further in the years to come when it will be necessary to come out with a new version of the book.

C. N. R. Rao
Umesh V. Waghmare
Bangalore (India)
2016

About the Authors

C. N. R. Rao is the Linus Pauling Research Professor at the Jawaharlal Nehru Centre for Advanced Scientific Research and Honorary Professor at the Indian Institute of Science. His research interests are mainly in the chemistry of materials. He is a Fellow of the Royal Society, London, foreign associate of the US National Academy of Sciences and member of Japan, French, Russian as well as several other science academies. He is the recipient of the Einstein Gold Medal of UNESCO, the Hughes and Royal Medals of the Royal Society, the August Wilhelm von Hofmann medal of the German Chemical Society, the Dan David Prize and Trieste Science Prize for materials research and the first India Science Prize.

Umesh V. Waghmare is a Professor in the Theoretical Sciences Unit and Dean, Academic Affairs at Jawaharlal Nehru Centre for Advanced Scientific Research. His research is in the area of first-principles theory, multi-scale modeling and simulations of materials. He is a Fellow of the three Science academies in India. He is the recipient of the Institute Silver Medal of Indian Institute of Technology, Mumbai (1990), IBM Faculty grant award (2009), SS Bhatnagar Prize in Physical Sciences (2010) and the Infosys Prize (2015).

Contents

Chapter 2. Topological Valleytronics   67

Motohiko Ezawa

**Chapter 5.   2D van der Waals Hybrid:
Structures, Properties and Devices**   **169**

*Md. Ali Aamir, Tanweer Ahmed, Kimberly Hsieh,
Saurav Islam, Paritosh Karnatak, Ranjit Kashid,
Phanibhusan Singha Mahapatra, Jayanta Mishra,
Tathagata Paul, Avradip Pradhan, Kallol Roy,
Anindita Sahoo and Arindam Ghosh*

Chapter 6.   Thermoelectric Energy Conversion in Layered Metal Chalcogenides   239

Satya N. Guin, Ananya Banik and Kanishka Biswas

Chapter 7.   Plasma Chemical and Physical Vapour Deposition Methods and Diagnostics for 2D Materials   **275**

Majed A. Alrefae, Nicholas R. Glavin,
Andrey A. Voevodin and Timothy S. Fisher

Chapter 8. Metal Contacts to MoS$_2$ **317**

Naveen Kaushik, Sameer Grover,
Mandar M. Deshmukh and Saurabh Lodha

Chapter 9.   Strain Dependent Properties of 2D MX_2 (M = Mo and W; X = S, Se and Te)   349

Tribhuwan Pandey, Swastibrata Bhattacharyya and Abhishek K. Singh

Chapter 10.   Point Defects, Grain Boundaries and Planar Faults in 2D h-BN and TMX$_2$: Theory and Simulations   389

Anjali Singh and Umesh V. Waghmare

Chapter 1

Transition Metal Dichalcogenides and Other Layered Materials

Manoj K. Jana and C. N. R. Rao*

*New Chemistry Unit, International Centre for
Materials Science and Sheikh Saqr Laboratory,
Jawaharlal Nehru Centre for Advanced Scientific Research
(JNCASR), Bangalore 560064, India
cnrrao@jncasr.ac.in

Abstract. The importance of two-dimensional (2D) materials needs to be recognised in a big way due to the discovery of graphene. It is noteworthy that research on 2D inorganic layered materials in the last few years has shown that these materials are equally fascinating, exhibiting several novel properties and phenomena. Especially, noteworthy is the variety of fascinating properties of single and few layers of MoS_2 and related dichalcogenides. WTe_2 exhibits the largest magnetoresistance recorded in materials. Then there are topological insulators such as Bi_2Se_3. Besides chalcogenides, there are materials such as boron nitrides and Borocarbonitrides with several interesting features. This chapter presents a brief review of the various 2D materials of interest today.

1.1 Introduction

The sensational discovery of graphene [1, 2], one-atom thick two-dimensional (2D) layer of sp^2 carbon atoms, and its amazing properties have prompted the scientific community to investigate other 2D materials such as layered transition metal dichalcogenides (TMDs), boron nitride (BN) and borocarbonitrides (BCN). These 2D

"

inorganic analogues of graphene comprise a novel class of electronic materials providing exciting avenues in diverse areas such as spintronics, electronics, optoelectronics, photovoltaics, sensing, catalysis and energy storage [3–6]. Interestingly, modification of interlayer interaction, changes in symmetry elements and quantum confinement of carriers cause significant changes in the electronic structure and properties of single- and few-layer 2D inorganic materials such as TMDs, compared to their bulk counterparts [4]. Several TMDs and other 2D materials have been synthesised in single- and few-layer forms by various methods, characterised and investigated for a variety of properties of both fundamental and practical interest. In this context, MoS_2 has gained great attention owing to certain unique properties with potential applications [7]. The ability to dynamically control the valley-spin polarisation in monolayer MoS_2 by optical pumping has demonstrated the possibilities in valleytronics [8]. Isolated atomic layers of the 2D materials can be configured into complex heterostructures bound by van der Waals interactions, with a precise control over the sequence. The van der Waals heterostructures are found to exhibit unusual properties and phenomena [9–12]. Furthermore, hybrids of graphene and 2D inorganic analogues are reported to exhibit novel and enhanced properties owing to synergy [13]. In this chapter, we present the synthesis, characterisation, properties and applications of 2D layered inorganic materials which include TMDs, BN, BCN and related materials. Some of the latest discoveries pertinent to distorted layered TMDs such as extremely large magnetoresistance and pressure-driven superconductivity are also detailed.

1.2 Synthesis and Characterisation of 2D Graphene Analogues

1.2.1 Transition metal dichalcogenides

Bulk layered TMDs with the general formula MX_2 (M = transition metal, X = chalcogen) are solids comprising covalently bonded X−M−X sandwiches, stacked along the crystallographic c-axis via weak van der Waals interactions. By virtue of anisotropic bonding, single layers of TMDs can be obtained by physical and chemical methods.

1.2.1.1 *Physical methods of synthesis*

Electronic grade monolayers of TMDs are obtained by mechanical exfoliation using the scotch-tape technique as in the case of graphene, but the yields are quite low [14, 15]. Sonication-assisted liquid-phase exfoliation (LPE) of bulk samples in a solvent medium produces single and few layers of layered materials in high yields, for applications in electrochemical storage and catalysis. LPE is widely used to prepare single and few layers of TMDs such as MoX_2 (X = S, Se, Te), WS_2, $TaSe_2$, $NbSe_2$ and $NiTe_2$, as well as h-BN, Bi_2Te_3 and other layered materials [16]. The LPE method relies on the dispersing solvent or surfactant to overcome the cohesive energy between the layers of the solute. Hence, the surface energies of the dispersing solvent and the exfoliated TMD material should match [17]. The Hansen solubility parameters represent contributions from hydrogen (δ_H), dispersion (δ_D) and polar bonding (δ_P) to the total cohesive energy density. The energy of exfoliation is represented by the enthalpy of mixing, given by $\Delta H_{\mathrm{mix}} = \phi(1 - \phi)[(\delta_{D,A} - \delta_{D,B})^2 + (\delta_{P,A} - \delta_{P,B})^2 + (\delta_{H,A} - \delta_{H,B})^2)]$ where A, B and φ represent solute, solvent and volume fractions of the solute, respectively [16]. For an efficient exfoliation, all the three Hansen solubility parameters of the solvent should match those of the solute. After screening a wide range of solvents, Coleman *et al.* [16] have shown solvents such as isopropyl alcohol (IPA), dimethylformadimide (DMF), and *N*-methyl-2-pyrrolidone (NMP) to be most suitable media for dispersing a variety of TMDs, and BN. The insets in Figures 1(a)–1(c) show photographs of the dispersions of MoS_2 and WS_2 in NMP, and dispersion of BN in IPA, obtained by LPE route. The respective transmission electron microscope (TEM) images (Figures 1(a)–1(c)) show transparent ultrathin layers with large lateral dimensions. As seen from the high-resolution TEM (HRTEM) images (Figures 1(d) and 1(e)), single layer TMDs prepared by LPE method retain their bulk 2H-structure discussed below. In order to prepare monolayers of MoS_2 and WS_2 in good yields, Zhou *et al.* [18] have used mixtures

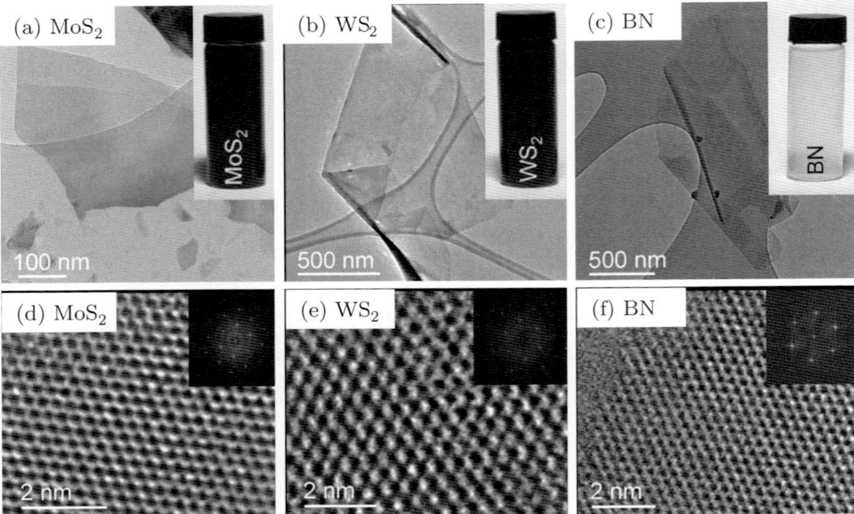

Figure 1. (a)–(c) Low-resolution TEM images of (a) MoS$_2$, (b) WS$_2$ and (c) BN flakes obtained by the LPE route. The insets in (a)–(c) show photographs of the dispersions of MoS$_2$ and WS$_2$ in NMP and BN in IPA; (d)–(f) HRTEM images of (d) MoS$_2$, (e) WS$_2$ and (f) BN monolayers, respectively. The insets show respective Fast Fourier transforms (FFT) of the images. (Adapted from Ref. [16].)

of low-boiling solvents like ethanol and water in appropriate ratios as determined by the Hansen solubility parameters. A variety of layered TMDs, transition metal oxides and h-BN have been exfoliated in an aqueous medium containing sodium cholate as the surfactant to prevent reaggregation of nanosheets. Yao *et al.* [19] have employed low-energy ball milling in combination with ultrasonication for efficient exfoliation of various TMDs, by means of additional shear forces. Sonication has the tendency to fragment the nanosheets, reducing their lateral dimensions. However, by carefully optimising the initial bulk mass, power and duration of ultrasonication, and centrifuge conditions, the proportion of moderately large ($>1\,\mu$m) and thin nanosheets can be maximised.

Dispersions of free-standing nanosheets of VS$_2$ [20] and VSe$_2$ [21] have been prepared by refluxing the bulk materials in formamide whereby the solvent molecules intercalate in between the layers and cause swelling of bulk VS_2 and VSe_2 until exfoliation. Laser thinning of multilayered MoS_2 is another small-scale approach to preparing monolayer MoS_2 with a defined shape and size [22]. Single and few layer flakes have also been prepared by irradiating DMF suspensions of bulk TMDs such as (Mo, W)S_2 and (Mo, W)Se_2 with a KrF excimer laser [23].

1.2.1.2 *Chemical methods of synthesis*

One of the effective methods to produce TMD nanosheets is the lithium intercalation-exfoliation route in which the bulk material is soaked in hexane solutions of n-butyl lithium (typically 1–2 weeks at room temperature or 3–4 days at $100°C$ for MoS_2) followed by an exfoliation of lithiated material in ultrasonicated water [24]. Single layers of MoS_2 [25], $MoSe_2$ [26], WS_2 [27], TaS_2 [28] and TiS_2 [29] have been prepared by using this route with nearly 100% yield. Interestingly, the stable 2H-phase of TMDs such as MoS_2 and WS_2 transforms into the metastable 1T-phase upon Li-intercalation [4]. The 2H-phase with trigonal prismatic coordination (D_{3h}) of the transition metal is semiconducting with a band gap between the empty $d_{x^2-y^2,xy}$ (e) and the filled d_{z^2} (a_1) bands of the transition metal. On the other hand, the 1T-phase with octahedral coordination (D_{3d}) of the transition metal is metallic with partially filled degenerate $d_{xy,yz,xz}$ (t_{2g}) band of the transition metal [4]. Both the 2H- and 1T-phases are composed of X$-$M$-$X sandwich structures (where M $=$ transition metal and X $=$ chalcogen). The letters, H and T, respectively, denote hexagonal and trigonal, whereas the digits indicate the number of X$-$M$-$X stacks in a single unit cell. Figure 2(c) illustrates the structures of 2H- and 1T-MoS_2 viewed down the crystallographic c-axis. 2H-MoS_2 has a trigonal prismatic coordination of Mo and S atoms (Figure 2(c)) where the S atoms of the top plane lie directly above the S atoms in the bottom plane giving rise to the AbA BaB type of stacking sequence (where the upper and lower case letters denote X and M, respectively). On the other hand, 1T-phase has a stacking sequence of the AbC AbC type; the S atoms in the top and bottom planes are off-set from one another by $30°$ and the Mo atoms lie within the octahedral voids of

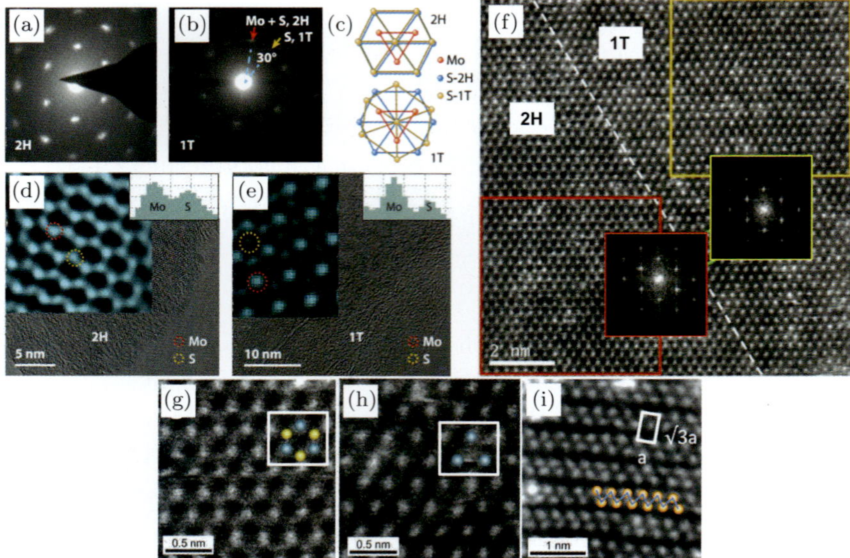

Figure 2. Electron microscopy of 2H- and 1T-MoS$_2$ monolayers prepared by Li-intercalation and exfoliation. (a) and (b) SAED patterns from monolayers of (a) 2H- and (b) 1T-MoS$_2$; (c) 2H- and 1T-MoS$_2$ structures viewed down c-axis; (d) and (e) HRTEM images of (d) 2H- and (e) 1T-MoS$_2$ monolayers. The left insets in (d) and (e) show magnified images of Mo- (red circle) and S-atomic arrangements (yellow circle), and the right insets in (d) and (e) show intensity line scans along Mo and S atoms; (f) HRTEM image of the heterojunction of 2H- and 1T-MoS$_2$-phases. The insets are corresponding SAED patterns. The white dashed line demarcates 1T- and 2H-phases; (g) and (h) Dark-field STEM images of (g) 2H- and (h) 1T-regions of MoS$_2$ monolayer. Yellow and blue balls indicate S and Mo atoms, respectively; (i) STEM image of 1T$'$-WS$_2$ monolayer showing zigzag clustering of W atoms. $\sqrt{3}$a $\times$ a unit cell of the superstructure is indicated by white box, where a is the lattice constant. ((a)–(e) adapted from Ref. [30]; (f)–(h) adapted from Ref. [31] and (i) adapted from Ref. [4]).

the sulphur sub-lattice (Figure 2(c)). Accordingly, the selected area electron diffraction (SAED) pattern of 2H-MoS$_2$ shows a hexagonal pattern of spots (Figure 2(a)), whereas that of 1T-MoS$_2$ reveals additional spots at 30° angular spacing in between the hexagonal spots of the 2H structure (see Figure 2(b)). HRTEM images of 2H- and 1T-monolayers are shown in Figures 2(d) and 2(e), respectively. As seen from the magnified HRTEM images (viewed down the c-axis), each Mo is surrounded by three S atoms in monolayer 2H-MoS$_2$

(Figure 2(d), left inset), whereas six S atoms surround a Mo atom in monolayer 1T-MoS$_2$ (Figure 2(e), left inset) [30]. A local atomic rearrangement is induced through charge transfer from the valence s orbital of the alkali metal to the d orbital of the transition metal centre leading to the transformation from semiconducting 2H- to the metallic 1T-phase.

Chhowalla and co-workers [4, 31, 32] have identified through a high-resolution scanning transmission electron microscope (STEM) imaging, the coherent heterojunctions of 2H- and 1T-phases of MoS$_2$ (Figure 2(f)) in samples prepared by Li-intercalation and exfoliation. In addition to regions of 2H- and 1T-MoS$_2$, they have observed regions of zigzag superlattices with $\sqrt{3}a \times a$ (or $2a \times a$) arrangement (where a is the lattice constant) [31]. These superlattices, identified as distorted 1T′-phases, are formed by chain-clustering of metal atoms away from their equilibrium positions. Interestingly, these distorted phases are metastable even after removing the Li-intercalant, and their properties are expected to be significantly different from their regular counterparts. The 2H- and 1T-phases being semiconducting and metallic, respectively, their novel heterojunctions are interesting for molecular electronic devices. In spite of high yield (nearly 100%) of single layer TMDs, the Li-intercalation and exfoliation route is limited by long lithiation-time, difficulty over controlling the degree of lithium-insertion and submicron size of flakes.

Zheng *et al.* [33] have reported a two-step expansion and intercalation strategy to prepare single layer MoS$_2$ flakes with dimensions as large as about $400\,\mu\mathrm{m}^2$. Bulk MoS$_2$ is first treated with hydrazine (N$_2$H$_4$) at 130°C for 48 h during which N$_2$H$_4$ intercalates and partially transforms to N$_2$H$_5^+$. The latter species are thermally unstable and decomposes to N$_2$, NH$_3$ and H$_2$ at high temperatures, expanding the MoS$_2$ sheets significantly ($>$100 times) compared to the initial volume. In the second step, the expanded MoS$_2$ flakes are intercalated by alkali naphthalenide and exfoliated in water with mild sonication to prevent fragmentation of the nanosheets. Figure 3 shows atomic-force microscopy (AFM) and TEM images of the as-exfoliated monolayers of MoS$_2$ and WS$_2$. Sodium naphthalenide is found to be most effective for obtaining flakes with larger dimensions.

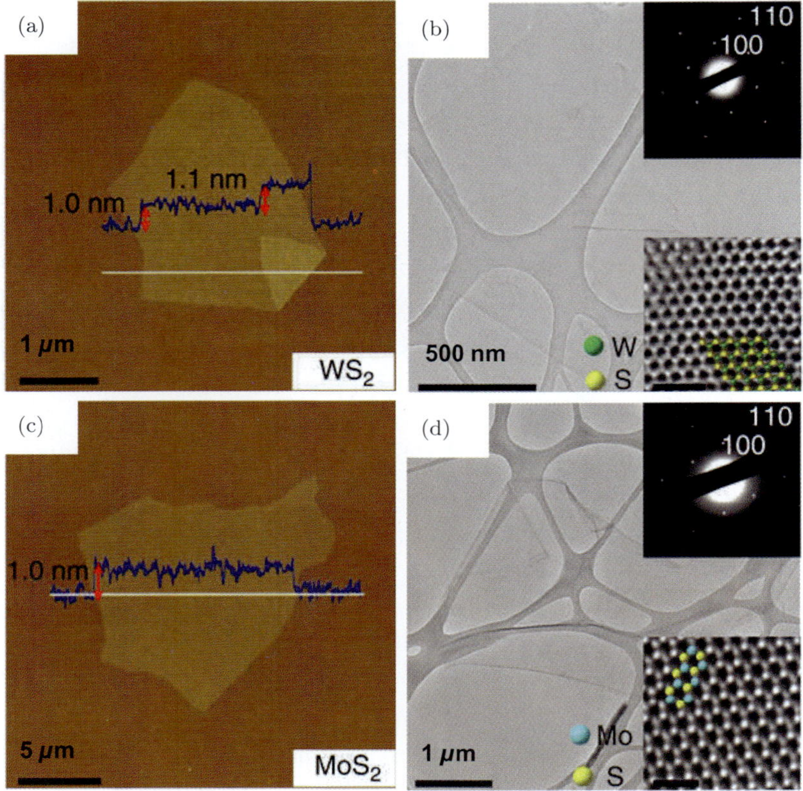

Figure 3. (a) AFM and (b) TEM images of Na-exfoliated WS_2 monolayer; (c) AFM and (d) TEM images of Na-exfoliated MoS_2 monolayers. The upper and lower insets are corresponding SAED and aberration-corrected HRTEM images, respectively. (Adapted from Ref. [33].)

Monolayers of TiS_2, TaS_2 and NbS_2 as well as few layers of $TiSe_2$, $NbSe_2$ and $MoSe_2$ have been prepared by this route in high yields.

A tandem molecular intercalation (TMI) strategy (Figure 4(a)) has been reported by Jeong $et\ al.$ [34], wherein a short alkylamine intercalates, expands the interlayer gap of multilayer TMDs and facilitates the intercalation of another long primary alkylamine. Owing to different chain lengths, a bilayer configuration with empty spaces between the intercalates is adopted in order to minimise the van der Waals force between them. Finally, spontaneous exfoliation

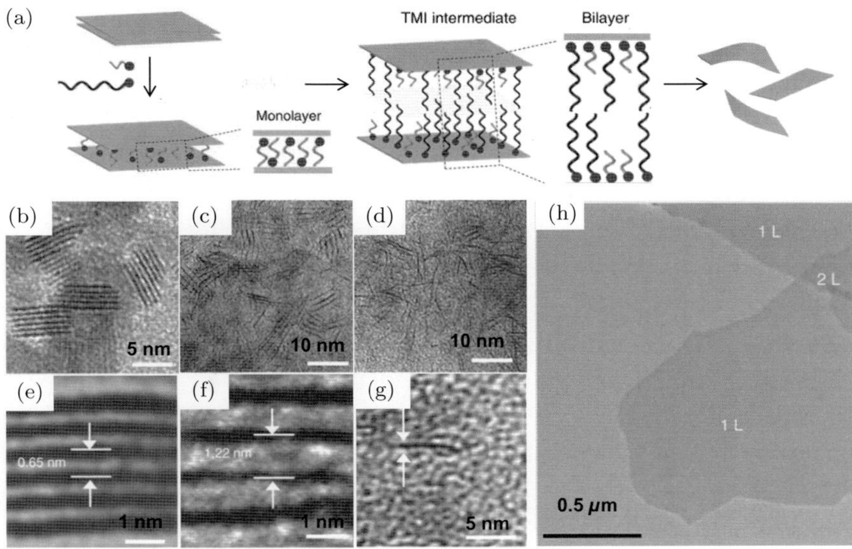

Figure 4. (a) Schematic of TMI intercalation process; (b)–(d) TEM and (e)–(g) magnified TEM images of starting multilayer WSe_2 flakes (b) and (e), WSe_2 intercalated with ethoxide in bilayer arrangement (c) and (f) and monolayer WSe_2 nanosheets (d) and (g); (h) TEM image of as-exfoliated single layer MoS_2. (Adapted from Ref. [34].)

occurs to generate single layer TMDs (see Figure 4(a)). After stirring multilayer WSe_2 in a solution of two intercalates *viz.* sodium ethoxide and sodium hexanolate in DMSO for 7 h at room temperature, single layer nanosheets of WSe_2 can be isolated by centrifugation. This process can be scaled up to sub-gram scale. During the course of intercalation, the interlayer spacing of WSe_2 expands from 6.5 Å (Figure 4(e)) to 12.2 Å (Figure 4(f)) suggesting that intercalation of the ethoxide proceeds to form a bilayer arrangement between the layers. Figures 4(b)–4(g) show TEM images illustrating the step-wise TMI process during exfoliation of multilayer WSe_2 into single layers. Similarly, micron-sized single layer nanosheets of MoS_2 can be obtained by using the same pair of intercalates with stirring at room temperature for 48 h (Figure 4(h)). The intercalate acts as a Lewis base, while TMD acts as a Lewis acid. Therefore, smaller the energy-difference between highest occupied and lowest unoccupied molecular orbitals of intercalate and TMD, respectively, greater the efficiency of intercalation. Whereas a weak Lewis base such as alkylamine is effective for groups IV and V TMDs, a strong Lewis base such as an alkoxide is needed for the exfoliation of group VI TMDs like MoS_2 and WSe_2. By using suitable intercalates, single layer nanostructures of groups IV (TiS_2, ZrS_2), V (NbS_2) and VI (MoS_2, WSe_2) TMDs have been successfully generated by the TMI route which has an advantage of being a mild single-step process at room temperature without requiring harsh conditions like ultrasonication or profuse H_2 evolution.

Zeng *et al.* [35] have synthesised nanosheets of various TMDs such as (Mo, W)S_2, (Ti, Ta)S_2 and ZrS_2 by electrochemical Li-intercalation and subsequent exfoliation in water. Li-insertion into TMD is carried out by a controlled galvanostatic discharge at a current density of 0.05 mA in a standard battery workstation with the Li-foil as the anode and the bulk-layered TMD as the cathode. During the electrochemical discharge, the Li-anode is oxidised giving out electrons that flow towards cathode; Li-ions migrate simultaneously through the electrolyte ($LiPF_6$) from anode to cathode where they intercalate in between the layers and get reduced to Li-metal by the incoming electrons. Upon sonicating the lithiated TMD in water, the intercalated Li-metal reacts with water to liberate H_2 gas which expands the interlayer distance of TMD causing exfoliation. This method yields dispersions containing mainly single layers. Interestingly, unlike the case of Li-intercalation in solution, the transformation from the 2H form to the 1T form was not observed after electrochemical Li-intercalation of MoS_2 [35]. It is noted that the insufficient Li-insertion causes incomplete exfoliation, whereas excess Li-insertion causes decomposition of TMD and hence a careful optimisation is crucial. Zeng *et al.* [36] have been able to optimise the cut-off voltage and discharge current required for optimal Li-insertion to obtain few-layer BN, $NbSe_2$, WSe_2, Sb_2Se_3 and Bi_2Te_3.

Chemical vapour deposition (CVD) is a widely employed chemical method to fabricate highly uniform and electronic grade TMD monolayers with large dimensions. CVD of TMDs usually employs four approaches, *viz.* sulphurisation or selenisation of pre-deposited thin films of metal or metal oxides, a vapour-phase reaction of metal-based precursors and chalcogen, thermal decomposition of single-source precursors and vapour-transport of TMD powders. A thin film of Mo, pre-deposited on SiO_2 substrate, can be sulphurised to an MoS_2 thin film by annealing in the presence of elemental sulphur [37]. However, owing to the difficulty in growing a uniform thin film of metal, both single and few-layer MoS_2 are formed on the substrates. Alternatively, single layer MoS_2 can be grown if sulphur is preloaded on a Cu surface such that it is available only to form a single MoS_2 layer [38]. By layer-by-layer sulphurisation of an MoO_3 thin layer, the wafer-scale uniform MoS_2 thin films with controlled thickness have been prepared. A thin film of MoO_3 with a desired thickness is initially grown on sapphire substrate by thermal-evaporation, reduced to MoO_2 or other Mo-based forms under H_2/Ar atmosphere at $500°C$, and annealed in a sulphur-rich environment at $1000°C$. The as-grown MoS_2 thin film can be transferred onto arbitrary substrates (Figures 5(a)–5(c)) [39, 40]. Large-area (about 1 cm^2) nanosheets of WS_2 with tunable thickness, have been grown similarly by sulphurisation of thermally evaporated thin films of WO_x [41].

Continuous films of single- and few-layer MoS_2 have been grown on SiO_2/Si substrates coated with graphene oxide and perylene-3,4,9,10-tetracarboxylate of potassium, by the vapour-phase reaction of MoO_3 and S powders. The organic coating favours the growth of MoS_2 layers by wetting the substrate and lowering the free energy of nucleation [42]. Li *et al.* [43] have grown ternary $MoS_{2x}Se_{2(1-x)}$ nanosheets with tunable band edge emission through a similar vapour-phase reaction between MoO_3, sulphur and selenium powders (Figures 5(f) and 5(g)). Few-layer MoS_2 is grown on a graphitic-surface through thermal-reduction of amorphous MoS_3 in the presence of reduced graphite oxide at high temperatures ($\sim 1000°C$) under high-vacuum [44]. It was, however, not possible to achieve a complete coverage of the substrate through this method. Three-layer MoS_2 sheets can be grown on various insulating substrates through thermal decomposition of single source precursor, ammonium thiomolybdate $((NH_4)_2MoS_4)$. The substrates were first dip-coated in $(NH_4)_2MoS_4$ solution, and annealed under Ar/H_2 flow

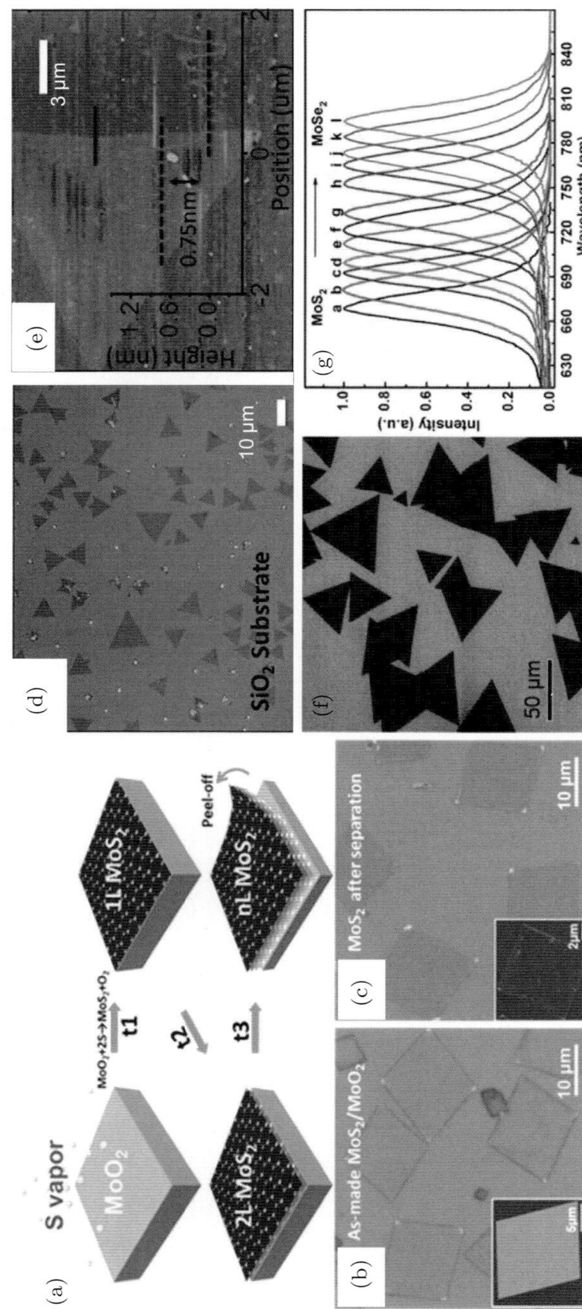

Figure 5. (a) Schematic showing the synthesis and transfer of MoS_2 thin films synthesised by sulphurisation of MoO_2; (b) and (c) are respectively, the optical images of as-synthesised MoS_2/MoO_2 and the transferred MoS_2 flake. Corresponding AFM images are given as insets; (d) optical microscope image of MoS_2 flakes grown on SiO_2/Si by physical vapour transport of MoS_2 powders and (e) AFM height profile of a typical MoS_2 monolayer; (f) SEM image of the ternary $MoS_{2x}Se_{2(1-x)}$ nanosheets obtained by vapour-phase reaction of MoO_3 with sulphur and selenium powders and (g) photoluminescence (PL) spectra of a series of $MoS_{2x}Se_{2(1-x)}$ nanosheets, respectively. ((a)–(c) adapted from Ref. [46] and (f) and (g) adapted from Ref. [40]; (d) and (e) adapted from Ref. [43].)

at 500°C and then at 1000°C [45]. The chemical reaction involved here is $(NH_4)_2MoS_4 + H_2 \rightarrow 2NH_3 + 2H_2S + MoS_2$. Wu *et al.* [46] have employed a vapour–solid (VS) growth technique for obtaining monolayer MoS_2 on insulating substrates such as sapphire and Si-wafer, though physical vapour transport of MoS_2 powder under a low pressure of 20 Torr (Figures 5(d) and 5(e)). The hot zone was maintained at 900°C and substrates were placed in a cold zone at 650°C for the deposition of MoS_2. Feng *et al.* [47] have grown uniformly distributed monolayer $MoS_{2(1-x)}Se_{2x}$ alloys on SiO_2/Si substrates through direct evaporation of MoS_2 and $MoSe_2$ powders at a high temperature of about 950°C and subsequent downstream alloying at 650°C. It is noted that reaction conditions have to be carefully optimised at several stages to obtain uniform and defect-free films.

Other chemical methods of synthesis include solid-state thermal decomposition of precursors and solution-phase reactions. High yields of few-layer $(Mo,W)S_2$ and $(Mo,W)Se_2$ are obtained by heating a mixture of molybdic or tungstic acid and excess thiourea or selenourea (1:48 ratio) at 773 K for 3 h [48, 49]. X-ray diffraction patterns of the few-layer samples (Figure 6(a)) reveal vanishing (002) reflections indicating that the nanosheets are ultrathin. Figures 6(b) and 6(c) show, respectively, the TEM image of few layer MoS_2 and the Fourier filtered HRTEM image of a single layer region of $MoSe_2$ layer. Few layer MoS_2 is also prepared hydrothermally by the reaction of MoO_3 and KSCN in an aqueous medium at 453 K while few-layer $MoSe_2$ can be prepared by the reaction of the molybdic acid and selenium metal in aqueous $NaBH_4$ solution at 453 K [48, 49]. Few layers of MoS_2 and $MoSe_2$ are obtained using the same precursors under microwave conditions with ethylene glycol as the solvent [23]. Free-standing nanosheets of MoS_2 and WS_2 can be synthesised under solvothermal conditions by decomposing $(NH_4)_2MoS_4$ or $(NH_4)_2WS_4$, in an oleylamine solvent at 360°C [50]. Mahler *et al.* [51] have recently synthesised monolayers of 2H-WS_2 and distorted 1T-WS_2 in oleylamine using WCl_6 and CS_2 as precursors. When hexamethyldisilazane (HMDS) is added to oleylamine, 2H-WS_2 is formed instead of 1T-WS_2. Figure 7(a) shows

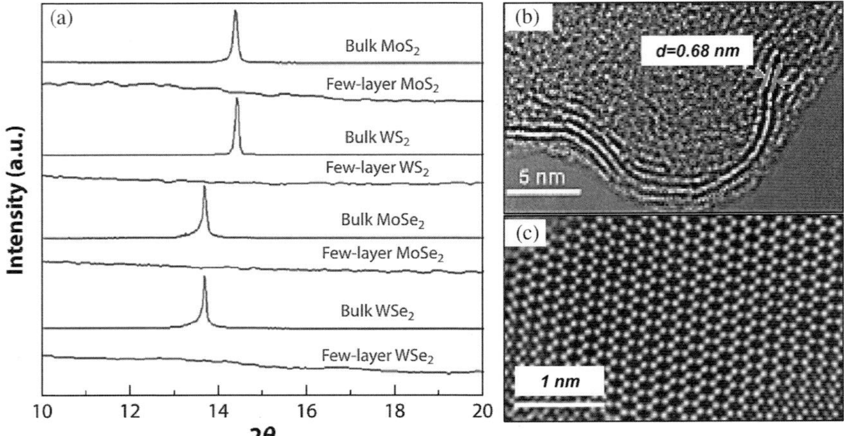

Figure 6. (a) X-ray diffraction patterns of bulk and few-layer $Mo(S, Se)_2$ and $W(S, Se)_2$ synthesised by thermal decomposition of molybdic or tungstic acid in excess thiourea or selenourea; (b) TEM image of few-layer MoS_2 showing the interlayer d-spacing and (c) Fourier filtered HRTEM image of single layer region of $MoSe_2$, synthesised by the above route. ((a) and (c) adapted from Ref. [6] and (b) adapted from Ref. [25].)

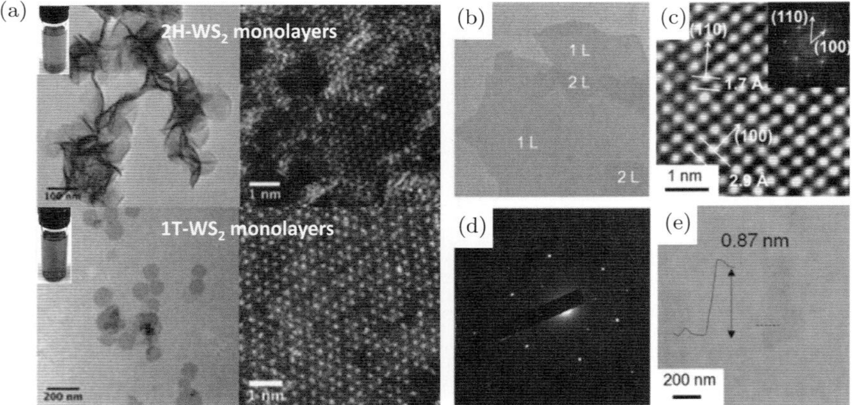

Figure 7. (a) TEM and HRTEM images of colloidal 2H- and 1T-WS_2 nanosheets synthesised solvothermally; (b) low-magnification TEM image, (c) HRTEM image along with FFT pattern as inset, (d) SAED pattern and (e) AFM image of single layer TiS_2 nanosheets synthesised by DCCI method. (Adapted from Ref. [51].)

low- and high-resolution TEM images of 2H- and 1T-WS_2 monolayers obtained in the presence and absence of HMDS, respectively. Yoo *et al.* [52] have developed a protocol called diluted chalcogen continuous influx (DCCI) for the solution-phase synthesis of single layer sheets of group IV metal disulphides. The continuous influx of dilute H_2S gas throughout the growth-stage is crucial to obtaining nanosheets with large lateral dimensions. 1-dodecanethiol was used for the slow *in situ* formation of H_2S which reacts with the metal chlorides (MCl_4, M = Ti, Zr, Hf) to form large single layers of the disulphides. Figures 7(b)–7(e) show TEM and AFM images of single layers of TiS_2 prepared by DCCI route.

1.2.2 Boron nitride

BN is a wide band-gap insulator and exhibits useful thermal, mechanical and dielectric properties [53]. Being isostructural to carbon, BN exists in different crystalline polymorphs: hexagonal, rhombohedral and cubic forms. The most commonly occurring hexagonal BN (h-BN, also called white graphene) is a white slippery compound with a layered structure analogous to graphite whereas the cubic BN (c-BN) is similar to diamond. There is also a rare wurtzite form of BN which is similar to lonsdaleite. BN is not found in nature but prepared synthetically. The first report by Balmain on the synthesis of BN using the reaction between molten H_3BO_3 and potassium cyanide (KCN) dates back to 1842 [54].

Mechanical peeling of bulk-layered BN using scotch tape can produce nanosheets of BN with large lateral dimensions [55], but this is not as effective method as in the case of graphene for preparing single or few-layer BN nanosheets because of the relatively strong interaction between BN basal planes via formation of bridging chemical bonds between atoms of adjacent layers. These interactions contribute to a metastable energy minimum by decreasing the number of dangling bonds at the edges or tips and energetically stabilise the multilayer form [56, 57]. Nanosheets of BN can be obtained via gentle shear forces using a mild wet-ball milling of bulk BN powder in a milling agent such as benzyl benzoate under N_2

atmosphere [58]. These nanosheets have a high density of defects [59]. Sonication-assisted LPE in solvents such as ethylene glycol, DMF, NMP, *N,N*-dimethylacetamide and IPA produces nanosheets of BN (see Figures 1(c) and 1(f)) [16, 60, 61]. When water is used as the dispersing solvent, in addition to few-layered nanosheets, monolayers and nanoribbons of BN are formed. Most nanosheets have reduced lateral dimensions due to the fragmentation of parent h-BN sheets induced by sonication-assisted hydrolysis, as evidenced from ammonia test and spectroscopy [62]. Water-soluble nanosheets of BN are obtained by ultrasonication of BN powder functionalised with a Lewis base such as amine molecules with long lipophilic or hydrophilic chains. The amines form complexes or adducts with the electron-deficient B atoms on BN upon heating the mixture for 4–6 days under N_2 flow, and facilitate the exfoliation of bulk BN [63]. When the bulk BN powder is ultrasonicated in methylsulphonic acid (MSA), protonation of BN nanosheet edges and surfaces by MSA causes the desired repulsion between adjacent layers and leads to exfoliation [64].

A finely ground mixture of hydroxides such as NaOH or KOH and bulk BN powders when heated at 180°C for 2 h, produces nanosheets and nanoscrolls of BN. The layer edges initially curl upon adsorption of cations (Na^+ or K^+) on the outermost BN surface. Anions and cations intercalate between the layers, and the subsequent adsorption of anions (OH^-) on the positively charged curved surface drives continuous curling of the BN layer, leading to peeling of the layers off the parent material. The as-exfoliated nanosheets of BN can be transferred to any substrate by redispersion in common solvents such as water and ethanol. This method has the advantages of being simple, one-step and low-cost [65].

CVD growth of h-BN thin films employs chemical precursors such as BF_3-NH_3, BCl_3-NH_3, $B_2H_6-NH_3$ or the decomposition of a single source precursor such as borazine ($B_3N_3H_6$), [66] trichloroborazine ($B_3N_3H_3Cl_3$) [67] or hexachloroborazine ($B_3N_3Cl_6$) [68]. The first h-BN monolayer film was grown epitaxially by decomposition of $B_3N_3H_6$ on transition metals such as Pt(111) and Ru(0001) surfaces as substrates [69]. Epitaxial h-BN monolayer has been successfully grown on a number of transition metal surfaces, such as Ni(111), Cu(111), Pt(111), Pd(111), Pd(110), Fe(110), Mo(110), Cr(110), Rh(111) and Ru(001) [70–72]. Among them, the h-BN/Ni(111) and h-BN/Cu(111) interfaces have 1×1 commensurate structure where h-BN is buckled due to a slight mismatch between h-BN and Ni(111) or Cu(111) surfaces as revealed by low-energy electron diffraction (LEED) results [71].

A low-pressure CVD (LPCVD) method was used to grow monolayer BN (Figure 8(a)) on Cu substrates from NH_3-BH_3 precursor [73]. CVD growth of BN on Ru(0001) at a low $B_3N_3H_6$ pressure leads to high-quality BN monolayers [74]. When an atomically clean Rh(111) surface was exposed to $B_3H_6N_3$ vapour at $800°C$ inside an ultrahigh vacuum chamber with subsequent cooling to room temperature, few-layer BN with an unusual nanomesh-like morphology was formed [75]. STM images of the product revealed an ordered nanostructure with a periodicity of ~3 nm (Figures 8(d)–8(f)). The formation of pores is attributed to the large lattice mismatch between Rh(111) surface and BN. On changing the precursor to trichloroborazine, islands and superstructures of BN are formed [67]. Large-area BN film with a uniform thickness of about 1 nm, were grown on pre-annealed Cu foils with NH_3-BH_3 precursor at $1000°C$. The corresponding AFM image and height profile are shown in Figures 8(b) and 8(c) [76].

Various systems have been developed for the non-epitaxial CVD growth of 2D BN nanosheets. Vertically-aligned BN nanosheets are grown on a Si substrate at $800°C$ from a gaseous mixture of $BF_3-N_2-H_2$ via a microwave plasma CVD technique [77]. The growth of protruding BN nanosheets instead of a uniform granular film is ascribed to the strong etching effect of fluorine, and the electrical field generated in the plasma sheath. Morphology and thickness of the BN nanosheets can be altered by varying the ratio and flow rates of BF_3 and H_2. Alternatively, Pakdel *et al.* [78] have developed a thermal CVD technique in which solid precursors (B, MgO and FeO powders) were heated up to $1000–1300°C$ under NH_3 flow to grow vertically-aligned BN nanosheets on Si/SiO_2 substrates. Mechanically exfoliated graphene and pyrolytic graphite have also

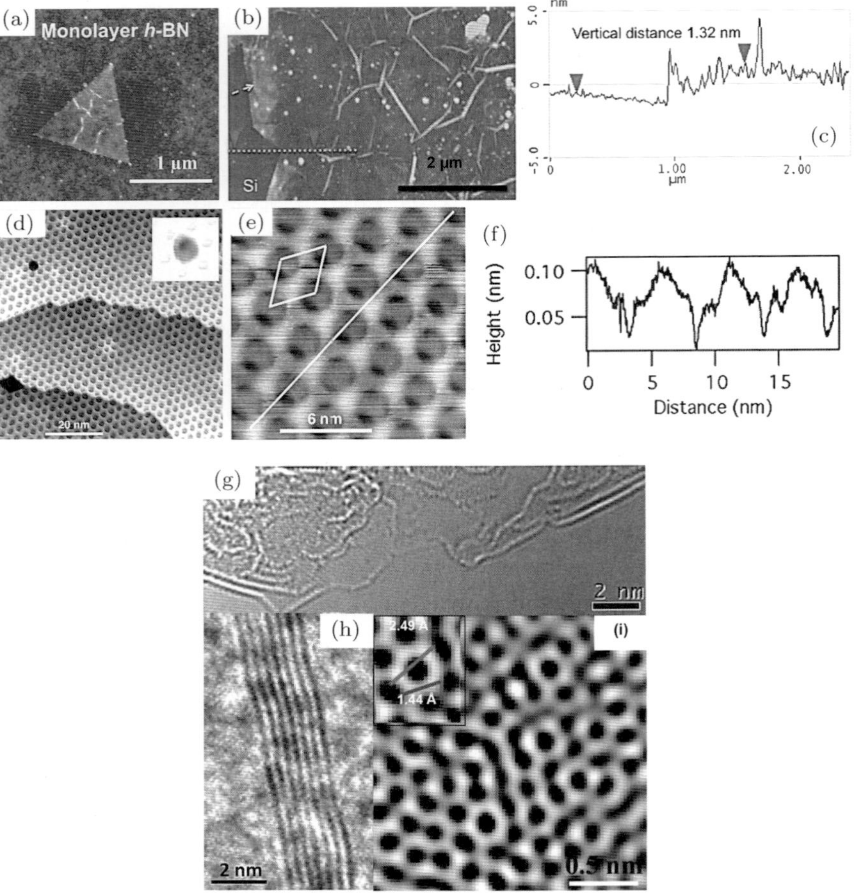

Figure 8. (a) AFM image of a triangular BN monolayer grown by LPCVD route; (b) AFM image and (c) height profile of a large h-BN film with a uniform thickness of about 1 nm, synthesised by CVD of NH_3-BH_3; (d) constant-current STM image of a BN nanomesh formed by high-temperature decomposition of borazine on Rh(111) surface; (e) high-resolution STM image showing two layers of BN mesh that are offset. The mesh unit cell is indicated by a white box; (f) cross-sectional profile along the diagonal white line in (e); (g, h) TEM images of few-layer BN prepared by decomposition of a mixture of boric acid and urea in (g) 1:48 and (h) 1:12 ratios; (i) HRTEM image of few-layer BN (1:48 ratio of boric acid and urea) showing a B–N distance of 1.44 Å and nearest B–B or N–N distance of 2.49 Å. ((a) adapted from Ref. [73]; (b) and (c) adapted from Ref. [76]; (d)–(f) adapted from Ref. [75] and (g)–(i) adapted from Ref. [85].)

been used as a substrates to grow BN films by CVD from NH_3-BH_3 mixture [79]. Heating a mixture of graphene and B_2O_3 at 1650°C yields BN and BN–C films, with graphene serving as a sacrificial template [80].

A substrate- and catalyst-free 'chemical blowing' route has been developed for the preparation of large area BN nanosheets with a high throughput [81]. In this method, a multi-step heating of BH_3-NH_3 leads to the formation of densely packed bubbles with atomically thin $B-N-H$ walls, due to dehydrogenation. After annealing at high temperatures, the product is dispersed in ethanol and transferred to substrates, upon which these thin bubbles collapse into BN atomic sheets. The introduction of ethanol during heating process leads to C_x-BN nanosheets. High-purity, pinhole-free micron-size bilayer BN films can be grown by magnetron sputtering of B under N_2-Ar flow for use as thin-film dielectrics [82]. Ar-plasma etching of a composite of BN nanotubes and poly(methyl methacrylate) film leads to unwrapping of BN nanotubes to form half-opened nanoribbons [83].

Few-layer BN has been prepared in high yields by the thermal decomposition of precursors. A two-step route involving combustion of boric acid with sodamide, ammonium carbonate and ammonium bromide followed by annealing to 1000°C yields h-BN nanoplates [84]. Alternatively, decomposition of a mixture of boric acid and urea at 900°C in N_2 atmosphere produces few-layer BN nanosheets in good yields [85]. The number of layers and the surface area can be tuned by varying the ratio of boric acid and urea. The number of layers is found to decrease with the decreasing molar ratio of boric acid and urea. TEM images (Figures 8(g) and 8(h)) of few-layer BN nanosheets prepared using 1:48 and 1:12 molar ratios of boric acid and urea reveal samples with 1–2, and 8–14 layers, respectively. HRTEM image of a thin BN layer (Figure 8(i)) clearly reveals the B–N bond distance of 1.44 Å and the nearest $B-B$ (or $N-N$) distance of 2.49 Å. The as-prepared BN nanosheets can be readily solubilised in Lewis bases.

1.2.3   BCN, $B_xC_yN_z$

2D BCNs can be considered to possess features somewhere between BN and graphene. In fact, they generally contain BN and graphene domains and also random B−C−N networks as revealed by spectroscopic and theoretical studies [86]. The carbon-content of the material can be varied in order to bring about changes in properties [86]. BCN can be generated by depositing BN and graphene sheets alternately or in some other proportion by employing CVD with appropriate precursors. $NH_3 \cdot BH_3$ has been used as the precursor of BN and CH_4 $(+H_2)$, as a precursor for graphene [87, 88] to obtain excellent 2-layered structures of BCN. Another way to make bulk BCN compositions is by the reaction of boric acid, carbon and urea (or NH_3) at high temperatures [86, 89]. The activated charcoal and graphene have been used as the source of carbon, the use of graphene giving rise to large sheets of BCN. Gas-phase reaction of BCl_3 (BBr_3) with NH_3 and CH_4 also yields BCN [90]. BCNs possess fascinating properties [13, 86], such as high surface areas, good uptake of CO_2 and application in energy devices. An important feature of these materials is that by varying the carbon content, one can vary the electronic structure as well as properties. Thus, BCN and BC_5N have band gaps of 3.60 eV and 3.0 eV, respectively [91].

1.2.4   Other layered chalcogenides

Besides TMDs, there are other layered metal chalcogenides of interest. Bi_2Se_3, Bi_2Te_3 and Sb_2Te_3 are known to be topological insulators (TIs) in which metallic surface states are embedded within an insulating bulk band gap. Layered Bi- and Sb-based TIs are also known to be the state-of-art thermoelectric materials [92]. These are anisotropic materials containing quintuple layers (QL), periodically stacked along the crystallographic c-axis by weak van der Waals interactions. Each QL, for example in Bi_2Se_3, is about 1 nm thickness, being composed of five covalently bonded atomic planes $[Se_2−Bi−Se_1−Bi−Se_2]$. Free-standing five-atom-thick Bi_2Se_3 single layers have been prepared through Li-intercalation and exfoliation of the bulk Bi_2Se_3 [93]. Ultrathin nanodiscs of Bi_2Se_3 have been synthesised by heating $Bi(NO_3)_3$ and Na_2SeO_3 in ethylene glycol containing small amounts of hydroxlamine [94]. Free-standing few-layer Bi_2Se_3 nanosheets can be obtained through a green ionothermal synthesis by heating a mixture of bismuth acetate and selenourea in a room temperature ionic liquid, 1-ethyl-3-methylimidazolium tetrafluoroborate ([EMIM][BF$_4$]) at 180°C for 4 h. Ionothermal decomposition of the single source precursor, $Bi(Se-C_6H_6N)_3$ in the same ionic liquid yields nanodiscs of few-layer Bi_2Se_3 [95]. Few-layer Bi_2Te_3 nanosheets have been prepared by employing various methods such as mechanical exfoliation [96], chemical vapour transport [97] and vapour-phase deposition [98]. Nanosheets of Sb_2Te_3 are prepared by microwave-heating of $SbCl_3$, Na_2TeO_3 and hydrazine hydrate in ethylene glycol solvent [99]. Single-crystalline Sb_2Te_3 nanosheets and nanobelts have also been synthesised by the reaction of $SbCl_3$ and TeO_2 with hydrazine hydrate employed as the reducing agent [100]. Nanosheets of SnS and SnS_2 with important applications in optoelectronics and Li-ion batteries, have been prepared by thermal decomposition of single-source precursors [101] and solvothermal [102] or ionothermal synthesis [103].

1.2.5   Metal oxides

Ultrathin layers of metal oxides are important for use as thin dielectrics in transistors [104]. Nanosheets of various layered metal oxides have been prepared [6]. Layered TiO_2 has been extensively studied for its dielectric properties and photocatalytic activity [104, 105]. Rutile form of TiO_2 has one of the highest dielectric constants (80–100) among the binary oxides [104]. Nanosheets of TiO_2 have been used in Li-ion batteries [106]. Sasaki *et al.* [107] have prepared nanosheets from layered titanates through intercalation induced exfoliation in an aqueous medium containing intercalants such as tetrabutylammonium ions or long-chain amines. Figures 9(a) and 9(b) show, respectively, the structural model of a $Ti_{1-\delta}O_2$ monolayer and the AFM image of a $Ti_{1-\delta}O_2$ nanosheet with a thickness of $\sim$1 nm, prepared via exfoliation in tetrabutylammonium hydroxide solution [104]. MoO_3 is another well-studied layered oxide. Figure 9(c) shows

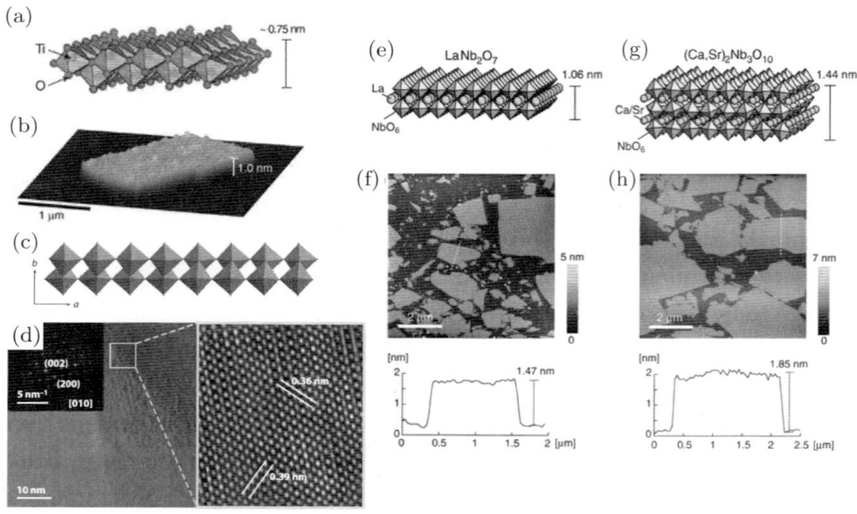

Figure 9. (a) Structural model of $Ti_{1-\delta}O_2$ and (b) AFM image of a $Ti_{1-\delta}O_2$ nanosheet on a SiO_2/Si substrate; (c) structural model of α-MoO_3 single layer; (d) TEM image (left), and HRTEM image (right) of few-layer MoO_3 prepared by air-annealing of few-layer MoS_2. Top-left inset shows the electron diffraction pattern; (e, g) structural models of (e) $LaNb_2O_7$ and (g) $(Ca, Sr)_2Nb_3O_{10}$ perovskites; (f, h) AFM images and height profiles of (f) $LaNb_2O_7$ and (h) $Ca_2Nb_3O_{10}$ nanosheets prepared by delamination in TBAH solution. ((a),(b) and (e)–(h) adapted from Ref. [104] and (d) adapted from Ref. [108].)

the structure of a monolayer α-MoO_3 composed of stacked MoO_6 octahedra. Several synthesis methods have been employed for the preparation of MoO_3 nanosheets *viz.* Li-intercalation/exfoliation, LPE, thermal annealing of few-layer MoS_2 and template-assisted synthesis using graphene sheets [108]. Figure 9(d) shows the TEM and HRTEM images of few-layer MoO_3 prepared by annealing few-layer MoS_2 in the air [108].

Many of the layered perovskite oxides have high dielectric constants and ferroelectric properties. Ultrathin layers of perovskite oxides such as $Ca_2Nb_3O_{10}$ and $LaNb_2O_7$ have been prepared by delamination of corresponding protonic precursors in tetrabutylammonium hydroxide (TBAH) solution [104]. Figures 9(e) and 9(g) show the structural models of $LaNb_2O_7$ and $(Ca,Sr)_2Nb_3O_{10}$, respectively. AFM images and height profiles of $LaNb_2O_7$ and

$Ca_2Nb_3O_{10}$ perovskite nanosheets obtained by delamination route are shown in Figures 9(f) and 9(h). These oxides have been used as high-κ dielectric materials. Members of the Aurivillius family with the general formula $(Bi_2O_2)(A_{m-1}B_mO_{3m+1})$, as well as of the Ruddlesden–Popper family with general formula $A_{m+1}M_mO_{3m+1}$, possess interesting dielectric (ferroelectric) properties, as exemplified by $Bi_4Ti_3O_{12}$. Nanosheets of several oxides from the Aurivillius family ($Bi_2A_{n-1}B_nO_{3n+3}$, where A = Bi^{3+}, Ba^{2+}, etc. and B = Ti^{4+}, Fe^{3+}, etc.) have been prepared by Li-intercalation and subsequent exfoliation in water [109]. Oxides from the Ruddlesden–Popper family such as $Sr_3Ti_2O_7$ and $Sr_4Ti_3O_{10}$ exhibit novel electronic and magnetic properties besides dielectric properties. Interestingly, the magnetic and electronic properties of manganese oxides of this family vary with the dimensionality [110, 111]. Nickel oxides, particularly $Ln_{m+1}Ni_mO_{3m+1}$ (Ln = rare earth), also exhibit electronic properties that vary with m or dimensionality.

1.3   Novel Properties of 2D Graphene Analogues

1.3.1   Electronic structure and properties

2H-MoS_2 is centro-symmetric (P63/mmc) with an indirect band gap in its bulk form. The calculated electronic structures of bulk 2H-MoS_2 reveal an indirect band gap ($\sim$1.2 eV) between the valence band maximum (VBM) at the Γ point and the conduction band minimum (CBM) near the midpoint along Γ-K path in the Brillouin zone. In addition, a direct band gap of $\sim$1.8 eV is present between CBM and VBM at the K point in the Brillouin zone (inset of Figure 10(a)) [112–115]. The indirect band gap along the Γ–K path increases with the decreasing number of layers and exceeds the direct band gap in the monolayer regime (Figure 10(d)), making monolayer MoS_2 a direct band gap semiconductor [113–115]. However, the direct band gap at the K point does not depend on the number of layers as the conduction band states at the K point are principally composed of strongly localised d-orbitals of Mo atoms. The latter have minimal interlayer interaction since Mo atoms are sandwiched between two S-planes in a single layer. On the other hand, states associated

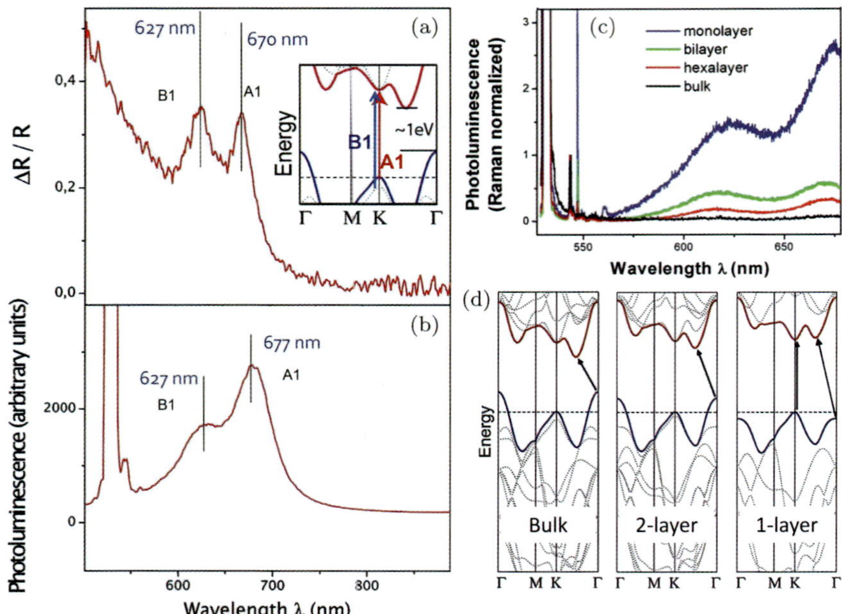

Figure 10. (a) Reflection and (b) PL spectra of ultrathin MoS_2 showing two peaks at 670 nm and 627 nm which correspond to the A1 and B1 direct excitonic transitions at the K point in Brillouin zone. The inset in (a) shows the electronic structure of bulk MoS_2 with an indirect band gap of $\sim 1\,eV$. The higher energy A1 and B1 direct excitonic transitions at the K point are indicated by red and blue arrows, respectively; (c) PL spectra (normalised by Raman intensity) of MoS_2 layers with varied thickness, show a dramatic increase in the luminescence-efficiency in MoS_2 monolayer; (d) electronic band structure of bulk, bilayer and monolayer MoS_2 showing the evolution of indirect band gap along Γ-K path. The arrows indicate the lowest energy transitions. (Adapted from Ref. [113].)

with the indirect band gap along the Γ–K path arise from a linear combination of anti-bonding p_z-orbitals on S atoms and d-orbitals on Mo atoms and therefore exhibit strong energy dependence on the number of layers due to strong interlayer coupling [113–115]. Carrier confinement due to suppressed interlayer hopping in ultrathin regime leads to an increasing indirect band gap with decreasing number of layers. The indirect-to-direct gap transition from bulk to monolayer is also evident from the rapid rise in photoconductivity of MoS_2 monolayer around 1.8 eV [115].

Monolayer MoS_2 with the space group P-6m2 has time-reversal symmetry $[E\uparrow(\vec{k}) = E\downarrow(-\vec{k})]$ but no inversion symmetry $[E\uparrow(\vec{k}) = E\uparrow(-\vec{k})]$. As a result, spin-orbit coupling (SOC) splits the otherwise two-fold degenerate valence bands at the K point, into two bands with the spin-up and spin-down character [116, 117]. Therefore, two direct excitonic transitions *viz.* A1 and B1, are allowed at the K point (Figure 10(a), inset). Experimental PL and reflectance spectra of monolayer MoS_2 exhibit two peaks at 670 nm and 627 nm due to A1 and B1 direct excitonic transitions, respectively (Figures 10(a) and 10(b)) [113, 118, 119]. Owing to indirect-to-direct band-gap transition, PL is drastically enhanced in MoS_2 monolayer compared to its multilayered counterparts (Figure 10(c)) and the quantum efficiency of PL in a monolayer MoS_2 is about 10^4 times that of the bulk MoS_2 [113, 115]. The enhanced PL from MoS_2 monolayer has been ascribed to the significant lowering of intraband relaxation rate [113, 115]. On the other hand, PL emission is absent in the metallic 1T-MoS_2 phase obtained by Li-intercalation. Upon annealing 1T-MoS_2 at 300°C, the excitonic features emerge due to phase-transformation to 2H-phase [32].

In addition to layer thickness, the band gap in group VI TMDs can be tuned by electric fields or a mechanical strain. The indirect band gap of bilayer TMDs can be driven to zero by applying an electric field of 2–3 Vnm^{-1} perpendicular to the layers [120]. Under strain, the band gaps of mono and bilayer MoS_2 decrease to zero leading to an insulator-to-metal transition [121–124]. Chemical stimuli can also tune the band gap in MoS_2. MoS_2 being p-type does not interact with electron-accepting tetracyanoethylene (TCE) but interacts with electron-donating tetrathiafulvalene (TTF) which forms a radical cation by donating an electron. First-principles calculations reveal a large band gap reduction in MoS_2 upon interaction with TTF [125]. Unlike graphene which can be doped with both electrons and holes, only electron-doping can be induced in a MoS_2 field-effect transistor (FET) by gate voltage [126].

BN nanosheets show a sharp absorption band in the deep UV range of 210–220 nm [127], and hence referred to as white graphenes in the literature. Experimental wide band gaps of pure

BN nanoribbons and nanosheets are about $5.3\,\text{eV}$ [128] and $5.7\,\text{eV}$, respectively [88, 127]. Incorporation of C into BN matrix results in smaller and variable band gaps [88, 127, 129]. BN nanosheets exhibit strong cathodoluminescence (CL) in the deep ultraviolet range [77, 127] and are promising for compact UV laser devices.

1.3.2   Raman spectroscopy

Raman spectroscopy is an excellent tool to determine the thickness and stacking of 2D TMDs and also to directly probe and monitor charge-doping and to study their mechanical and thermal properties. Group theory predicts four first order Raman modes in 2H-MoS_2, namely E_{2g}^2, E_{1g}, E_{2g}^1 and A_{1g} with frequencies at $32\,\text{cm}^{-1}$, $286\,\text{cm}^{-1}$, $383\,\text{cm}^{-1}$ and $408\,\text{cm}^{-1}$, respectively [6, 130]. E_{2g}^1 and A_{1g} are the only intense modes corresponding to the in-plane and out-of-plane vibrations, respectively (inset of Figure 11(b)). As the number of layers decreases, the E_{2g}^1 mode stiffens while the A_{1g} mode softens (Figures 11(a) and 11(b)) [131, 132]. However, in the frame of coupled harmonic oscillators, both the E_{2g}^1 and A_{1g} modes should stiffen with increasing number of layers as the interlayer van der Waals interactions increase the effective restoring forces acting on the atoms [131]. The shift in the A_{1g} mode agrees with the model but the shift in E_{2g}^1 mode is anomalous and is attributed to the dielectric screening of long-range Coulomb interactions which increase with number of layers [131, 132]. For 2H-$MoSe_2$, the A_{1g} mode is the most intense vibration, occurring at a frequency lower than that of the E_{2g}^1 vibration [133]. As the number of layers is reduced, the A_{1g} mode softens while the E_{2g}^1 mode stiffens [133, 134]. The A_{1g} and E_{2g}^1 modes are degenerate in bulk 2H-WSe_2, giving rise to only a single Raman band. As the dimensionality is decreased, this peak splits into two with the higher intensity E_{2g}^1 band occurring at a frequency lower than that of the A_{1g} band [135].

The Raman spectrum of mechanically exfoliated monolayer TMDs exhibits significant temperature-dependence. The temperature coefficients of the A_{1g} and E_{2g}^1 modes of MoS_2 monolayer were respectively found to be $-0.0123\,\text{cm}^{-1}\text{K}^{-1}$ and $-0.0132\,\text{cm}^{-1}\text{K}^{-1}$

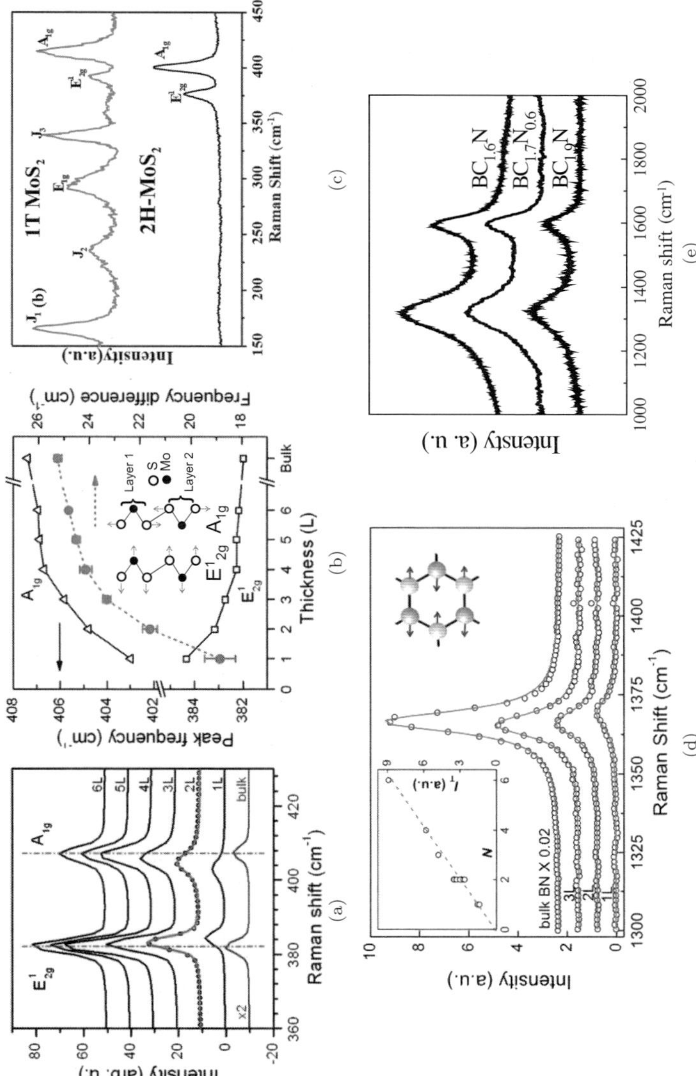

Figure 11. (a) Raman spectra of bulk MoS$_2$ and mechanically exfoliated thin layers (nL) of MoS$_2$; (b) Frequencies of E^1_{2g} and A_{1g} Raman bands (left vertical axis) and variation of their separation (right vertical axis) as a function of number of layers; (c) Raman spectra of Li-exfoliated monolayer 1T- and 2H-MoS$_2$; (d) Raman spectra of bulk as well as atomically thin BN. The left inset shows the variation of integrated intensity (I_T) with the number of layers, N. The right inset shows a schematic of the associated phonon mode; (e) Raman spectra of B$_x$C$_y$N$_z$ samples prepared by high temperature urea-route. ((a) and (b) adapted from Ref. [132]; (c) adapted from Ref. [26]; (d) adapted from Ref. [142] and (e) adapted from Ref. [86]).

[136–138]. In MoSe$_2$ monolayer, the temperature coefficient of the A_{1g} vibration is much smaller ($-0.0045 \times 10^{-2}\,\mathrm{cm^{-1}K^{-1}}$) than that of MoS$_2$ monolayer owing to the difference in the strain-phonon modes as suggested by the first-principles calculations [133]. When exposed to shockwaves, few-layer MoS$_2$ and related TMDs undergo reduction in interlayer (002) separation resulting in softening of the E_{2g}^1 and A_{1g} bands [139]. Raman spectra of monolayer 1T-MoS$_2$ and 1T-MoSe$_2$ generated by Li-intercalation and exfoliation are different compared with their 2H-counterparts owing to changes in symmetry [26]. Figure 11(c) compares the Raman spectra of 2H-MoS$_2$ and 1T-MoS$_2$. Raman spectrum of 1T-MoS$_2$ exhibits new Raman bands of J1, J2 and J3 in addition to E_{2g}^1 and A_{1g} modes which are blue-shifted with respect to 2H-MoS$_2$.

Td$-$WTe$_2$, a group VI TMD, has an orthorhombic crystal structure and occurs as a distorted variant of 1T structure with a $\sqrt{3} \times 1$ superstructure of dimerised zigzag W$-$W chains running along crystallographic a-axis (see Figure 21(a)). In contrast with other TMDs, such as MoS$_2$ which exhibit a strong dependence of the electronic structure on the number of layers, first-principles calculations reveal that bulk as well as monolayer Td$-$WTe$_2$ have similar electronic band structures, both being semimetallic [140]. Td$-$WTe$_2$ has 33 zone-centre Raman active modes of which five distinct Raman bands are experimentally observed around 112 cm^{-1}, 118 cm^{-1}, 134 cm^{-1}, 165 cm^{-1} and 212 cm^{-1} in bulk Td$-$WTe$_2$ (Figure 12(a)). Based on the symmetry analysis and the Raman tensor, the intense bands at 165 cm^{-1} and 212 cm^{-1} have been assigned to the A_1' and the A_1'' modes, respectively [140]. In contrast with the 2H-polytypes of group VI TMDs such as MoS$_2$, the low crystal-symmetry of Td$-$WTe$_2$ results in mixing of the out-of-plane and in-plane components of atomic displacements. The A_1' mode is associated with in-plane displacements of W atoms and out-of-plane (z-direction) displacements of Te atoms whereas their motion is vice versa in the A_1'' mode. From the thickness-dependent measurements, the A_1'' band is found to be most sensitive to the number of layers, stiffening to about 4 cm^{-1} in a 3-layer flake compared with bulk Td$-$WTe$_2$ whereas the A_1' band hardly changes with the number of layers

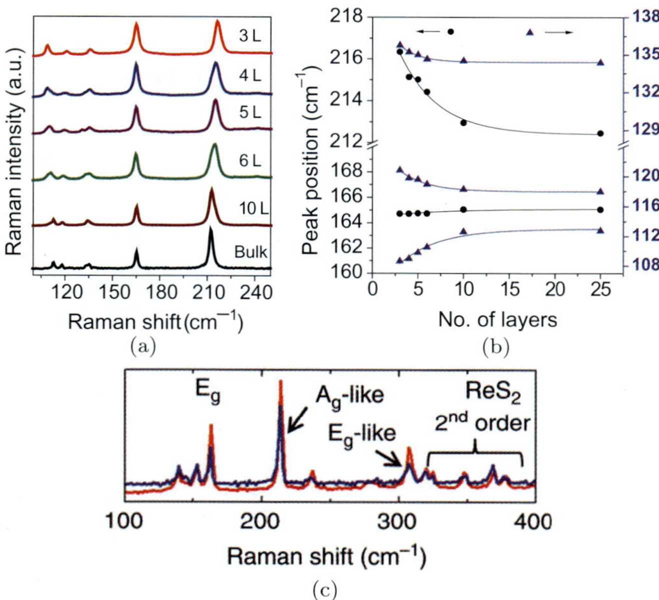

Figure 12. (a) Raman spectra of mechanically exfoliated Td$-$WTe$_2$ flakes of different thicknesses and (b) variation of peak frequencies as a function of number of layers (solid lines are guides to eye); (c) Raman spectra from bulk (blue) and monolayer (red) ReS$_2$. ((a) and (b) adapted from Ref. [140] and (c) adapted from Ref. [141].)

although both belong to the same A_1 symmetry (Figure 12(b)). The Te atoms of the same plane vibrate in-phase in the A_1' vibration, whereas their motion is out-of-phase in the A_1'' vibration. The A_1' mode is, therefore, more localised to a layer of Td$-$WTe$_2$ and exhibits weaker or no dependence on the number of layers [140].

Bulk ReS$_2$, a group VII TMD, occurs in a distorted 1T (1T$'$) structure with dimerised zigzag Re$-$Re chains and behaves as electronically and vibrationally decoupled monolayers stacked together [141]. Bulk and monolayer ReS$_2$ have nearly identical electronic band structures with a direct band gap of about 1.5 eV. As a result, the Raman spectrum of ReS$_2$ does not depend on the number of layers (Figure 12(c)). Optical absorption is also insensitive to modulation of interlayer distance by hydrostatic pressure ($dE_{\text{gap}}/dP \sim 0.02$ eV per GPa) [141]. First-principles calculations reveal that Peierls distortion of the 1T structure of ReS_2 prevents ordered stacking and minimises the interlayer overlap of wave functions causing the decoupling of individual layers of ReS_2. Besides, weak intralayer polarisation owing to slight charge differences between neighbouring planes of Re and S also results in very weak interlayer van der Waals interactions [141].

Hexagonal BN has D_{6h}^4 space group symmetry and the irreducible representation of zone-centre optical phonons is given by $\Gamma = 2E_{2g} + 2B_{1g} + A_{2u} + E_{1u}$. While the E_{2g} modes are Raman-active, the A_{2u} and E_{1u} modes are IR-active, and the B_{1g} mode is optically inactive. The low-frequency E_{2g} mode is characterised by entire planes sliding past each other whereas the high-frequency E_{2g} mode is due to B and N atoms moving against each other within a plane [71]. The Raman spectrum of BN nanosheets exhibits a dominant high-frequency intralayer E_{2g} band in the range of 1363–1374 cm^{-1} [77]. BN monolayer exhibits stiffening of E_{2g} band (around 1370 cm^{-1}) by up to 4 cm^{-1} relative to bulk BN, which was attributed to a slightly shorter B–N bond in the isolated monolayer (Figure 11(d)) [142]. $B_xC_yN_z$, which is considered as a hybrid of graphene and BN, shows two Raman bands at 1300 cm^{-1} and 1600 cm^{-1} (Figure 11(e)). The Raman band around 1300 cm^{-1} can be deconvoluted into two bands due to the B–N vibrational mode and the D-band of graphene. The Raman band around 1600 cm^{-1} is similar to the G-band of graphene [86].

1.3.3 Magnetic properties

TMDs have been investigated for layer-dependent magnetism following the discovery of defect-induced room-temperature ferromagnetism in the non-magnetic graphene. In fact, MoS_2 shows commonalities with graphene such as ferromagnetic ordering along zigzag-edges and novel exchange-bias phenomenon which are attractive for spintronics. Electron paramagnetic resonance (EPR) investigations by Panich *et al.* [143] revealed the presence of a large density of dangling bonds carrying unpaired electrons in inorganic fullerene-like MoS_2 nanoparticles. MoS_2 and WS_2 clusters (Mo_6S_{12} and W_6S_{12}) are rendered magnetic due to unsaturated metal centres with partially filled d-orbitals [144]. Zhang *et al.* [145] reported ferromagnetism in MoS_2 near room-temperature and ascribed it to a high density of prismatic edges containing unsaturated Mo and S. Lithium-intercalated MoS_2 prepared by using n-butyl lithium also shows ferromagnetism near room-temperature [146]. Room-temperature magnetic ordering is induced in MoS_2 upon irradiating with 2 MeV proton beam and the temperature-dependent magnetisation exhibits ferrimagnetic behaviour with a Curie temperature of 895 K. Proton irradiation can induce isolated vacancies, vacancy-clusters, edge states, and lattice-reconstructions which may result in room-temperature magnetism in MoS_2 [147]. Graphene-like MoS_2 is ferromagnetic near room temperature with the zero-field-cooled (ZFC) and the field-cooled (FC) magnetisation diverging from about 300 K [23, 148]. Such divergence of the ZFC and FC curves is seen in magnetically frustrated systems like spin-glasses with randomly distributed ferromagnetic and antiferromagnetic domains. The temperature-dependent magnetisation of few-layer MoS_2 is shown in Figure 13(a). It is ferromagnetic at room temperature exhibiting

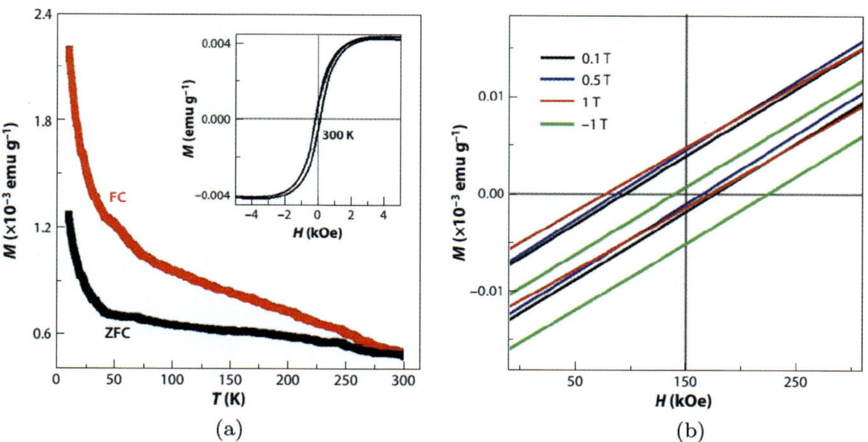

Figure 13. (a) Temperature-dependent magnetisation (ZFC and FC) of few-layer MoS_2 at an applied field of 100-Oe. Hysteresis behaviour of few-layer MoS_2 at 300 K is shown in the inset; (b) Field-dependent magnetisation in few-layer MoS_2 at 2 K, shows the dependence of exchange bias on the applied field. (Adapted from Ref. [23].)

hysteresis in magnetisation vs. applied magnetic field curve (inset of Figure 13(a)) [23]. The coexistence of antiferromagnetic and ferromagnetic interactions was also evident by the exchange bias behaviour in few-layer MoS_2 samples at different applied magnetic fields (Figure 13(b)). A negative exchange bias of 15.5 Oe at 1T field, decreases with decreasing applied field and increases in the opposite direction to -15.5 Oe at $-1T$ (Figure 13(b)) [23]. The existence of ferromagnetism in MoS_2 is attributed partly to the zigzag edges in the magnetic ground state [149]. First-principles calculations reveal that MoS_2 nanoribbon with armchair-terminated edges is semiconducting with a spin-unpolarised non-magnetic ground state, whereas that with zigzag-terminated edges is metallic with a spin-polarised magnetic ground state [150]. MoS_2 nanosheets become non-magnetic beyond a certain thickness as shown by thickness-dependent magnetic force microscopy (MFM) of few-layer MoS_2 nanosheets [151].

Magnetic properties of BN and BCN layers have been investigated. Theoretical studies indicate that vacancy defects or adatoms can induce magnetic properties in BN monolayers [152, 153]. It has been shown that N-vacancies have no influence on the magnetic properties, while B-vacancies induce spin-polarisation in the nearest three N atoms. Room-temperature ferromagnetism is observed in graphene-like BN and has been ascribed to the free spins that become available either from the defect-induced unpaired electrons or due to conversion of sp^3 to sp^3–sp^2 hybridisation [154].

Magnetic properties of $BC_{1.9}N$ with high BET surface area of 1990 m^2g^{-1}, prepared by the high temperature urea-route, have been studied [86]. Figure 14(a) shows the zero field-cooled (ZFC) and field-cooled (FC) magnetisation of $BC_{1.9}N$ at 500 Oe field. The divergence between the ZFC and FC curves is seen up to room-temperature and the curves overlap at a higher field of 1 Tesla as shown in the inset of Figure 14(a). This behaviour is similar to that of magnetically frustrated systems. Figure 14(b) shows the room-temperature hysteresis with a saturation magnetisation (M_S) of about 0.014 emu g^{-1} and a coercive field (H_C) value of about 120 Oe.

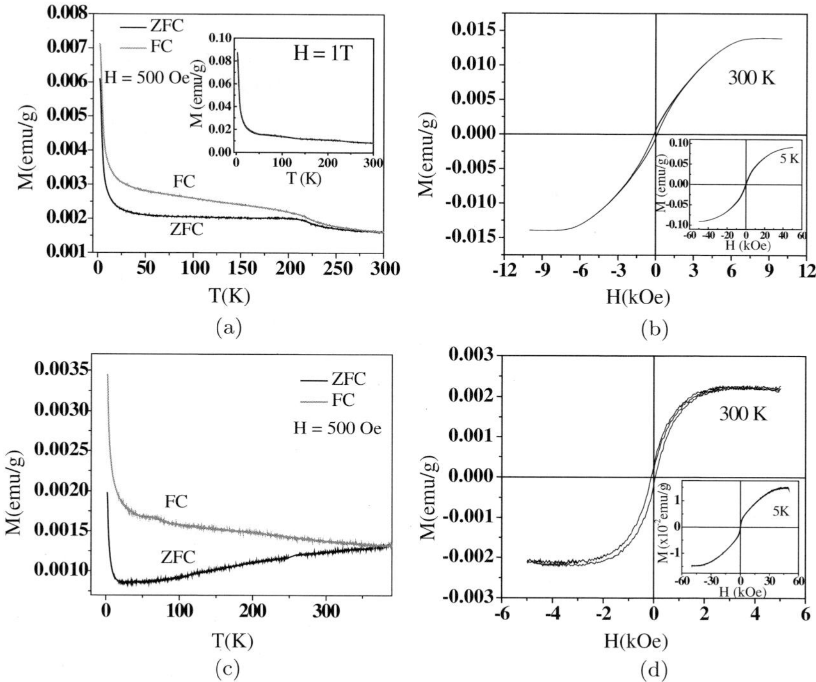

Figure 14. (a) ZFC–FC magnetisation curves at 500 Oe field for the high surface area $BC_{1.9}N$ sample prepared by urea-route. The inset show ZFC-FC curve at an applied field of 1 T; (b) magnetic hysteresis loop of the same $BC_{1.9}N$ at room temperature and at 5 K (inset); (c) ZFC–FC magnetisation curves at 500 Oe for the $BC_{1.6}N$ sample synthesised by gas-phase reaction; (d) magnetic hysteresis loop of the same $BC_{1.6}N$ at room temperature and at 5 K (inset). (Adapted from Ref. [86].)

The inset of Figure 14(b) shows the hysteresis at 5 K with M_S and H_C values of about $0.092\,\mathrm{emu\ g^{-1}}$ and 300 Oe, respectively. $BC_{1.6}N$ prepared by gas-phase reaction of BBr_3, methane and ammonia at $950^\circ C$, also exhibits a similar magnetic behaviour but with a lower magnetisation. We show the ZFC and FC curves of this sample in Figure 14(c). Room temperature hysteresis gives M_S and H_C values of about $2.2 \times 10^{-3}\ \mathrm{emu\ g^{-1}}$ and 65 Oe respectively for this sample (Figure 14(d)) [86]. The hysteresis loop at 5 K is shown in the inset of Figure 14(d).

1.3.4 Valley-spin coupling and trions

Electronic and spintronic devices exploit the charge and spin of electrons, respectively, whereas valleytronics is based on the condition where the valence or conduction bands of some materials have two or more degenerate valleys separated in the momentum space [8]. For realising a valleytronic device, it is essential to create a valley-polarisation by manipulating the population of carriers in these valleys. Monolayer MoS_2 is an ideal system for valleytronics with conduction and valence band edges having two degenerate valleys at the K and K′(-K) points of the 2D hexagonal Brillouin zone (Figure 15(a)). The direct band gaps of about 1.9 eV at these two valley points (K and K′) are suitable for valley-polarisation through optical pumping [8, 155, 156]. Given the lack of inversion symmetry in monolayer MoS_2, the spin-orbit interaction splits the otherwise degenerate valence bands by ∼160 meV into two bands of spin-up ($E\uparrow$) and spin-down ($E\downarrow$) character [156–158]. Thus, the spin projection, S_z, is well-defined along the c-axis of the crystal. In monolayer MoS_2, this broken spin degeneracy along with time-reversal symmetry results in coupling of the valley and spin of the valence bands giving rise to a valley-dependent optical selection rule [8, 155]. As a consequence, the inter-band transitions near K or K′ valleys exclusively couple to left or right circularly polarised light, respectively, (see Figure 15(b)) and the polarisation of circular component of the band-edge luminescence is same as that of the circularly polarised excitation (Figure 15(c)). Valley and spin are decoupled in bilayer MoS_2 where the inversion symmetry as well as time reversal symmetry restore the spin-degeneracy by splitting the 4-fold degenerate valence bands into two spin-degenerate valence bands. Valley-dependent selection rule is, therefore, not allowed and negligible circular polarisation is exhibited by bilayer MoS_2 under the same conditions (Figure 15(d)) [155].

Recently, tightly bound negatively charged trions have been spectroscopically identified in monolayer MoS_2 FET [159]. Trions are quasi-particles composed of a hole and two electrons, which can be created optically with spin and valley-polarised holes. The

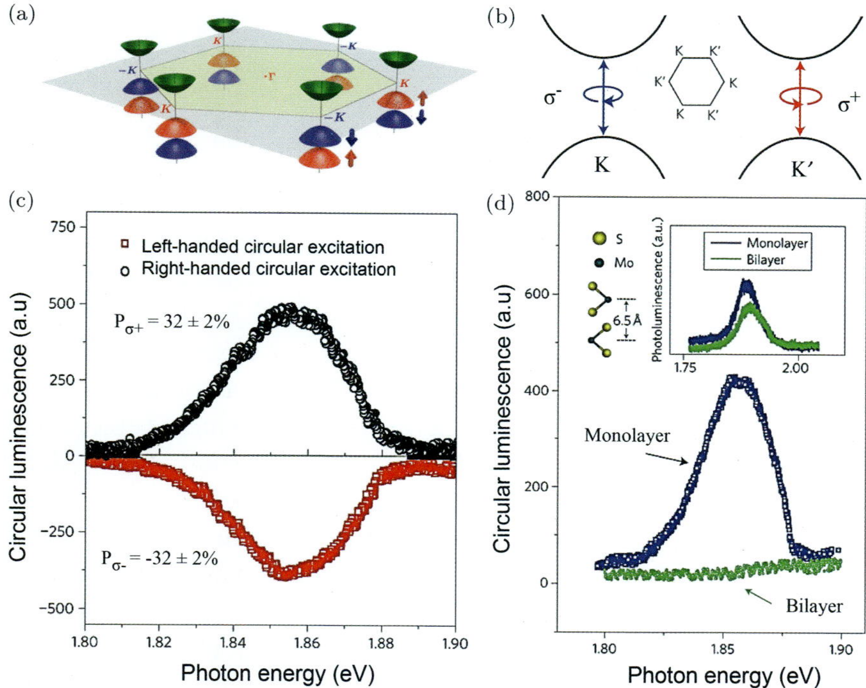

Figure 15. (a) Schematic of the MoS_2 electronic structure showing opposite spin-orbit splitting of valence band maxima at the K and K$'$ points of hexagonal Brillouin zone. Green surfaces represent conduction band valleys while red and blue surfaces represent spin-up and spin-down valence band valleys, respectively; (b) valley-dependent optical selection rules at K and K$'$ points in k-space; (c) polarisation-resolved luminescence upon circularly polarised excitation from a HeNe laser (1.96 eV) at 10 K; (d) circularly polarised luminescence from MoS_2 monolayers (blue) and bilayers (green) under circular excitation (1.96 eV) at 10 K. Circular polarisation is negligible in bilayer MoS_2. Top-left and top-right insets show respectively, schematic of a MoS_2 bilayer unit cell, and the PL spectra of monolayer (blue) and bilayer (green) MoS_2. ((a) adapted from Ref. [156] and (c) and (d) adapted from Ref. [155].)

trions possess a relatively large binding energy of about 20 meV, which is an order of magnitude higher than that found in conventional quantum wells. Such high binding energies originate from significantly enhanced Coulomb interactions, arising from reduced dielectric screening in monolayer MoS_2. The existence of tightly bound trions opens up possibilities of novel many-body phenomena and may influence the development of new optoelectronic and photonic devices.

1.4   Applications of 2D Graphene Analogues

1.4.1   Field effect transistors (FETs) and sensors

2D materials are attractive for nanoelectronics owing to the relative ease of fabricating complex device structures. Graphene-based FETs exhibiting a charge carrier mobility as high as $10^6\,cm^2\,V^{-1}s^{-1}$ have been fabricated [160]. However, graphene has a vanishing band gap and a finite band gap is required for various transistor applications. Single layer MoS_2 has a band gap being a direct band gap semiconductor. An n-type FET based on single layer MoS_2 with hafnium oxide as the top gate dielectric exhibits high current on/off ratios greater than 10^8 and high FET mobility of about $200\,cm^2\,V^{-1}s^{-1}$ (Figures 16(a) and 16(b)), comparable to those of thin silicon films and graphene nanoribbons [161]. MoS_2 FET with high-k Al_2O_3 as the top-gate dielectric showed an electron-mobility as high as $517\,cm^2\,V^{-1}s^{-1}$ and current on/off ratio exceeding 10^8 [162]. Single layer MoS_2 FET top-gated with high-k Al_2O_3 exhibited a mobility of $\sim170\,cm^2\,V^{-1}s^{-1}$, much greater than that of bilayer ($\sim25\,cm^2\,V^{-1}s^{-1}$) and triplelayer ($\sim15\,cm^2\,V^{-1}s^{-1}$) MoS_2 FETs. The degrading performance with increasing MoS_2 thickness was attributed to an increasing dielectric constant with thickness [163]. Flexible FETs based on thin-film MoS_2 with current on/off ratio of 10^5 were fabricated using ion gel gate-dielectrics [164]. Chakraborty *et al.* [126] have fabricated top-gated transistors based on single layer MoS_2 which exhibited a high current on/off ratio of about 10^5 and a mobility of $50\,cm^2\,V^{-1}s^{-1}$. Though MoS_2 is p-type, transistors fabricated out of 15 nm-thick sheets showed ambipolar behaviour with Hall mobility of holes ($86\,cm^2\,V^{-1}s^{-1}$) being greater than that of electrons ($44\,cm^2\,V^{-1}\,s^{-1}$) implying that p-type operation is more favoured [165]. A FET based on MoS_2 nanosheets prepared by the LPE route exhibits n-type characteristics with low current on/off ratios of 3–4 and a low FET mobility of $0.117\,cm^2\,V^{-1}\,s^{-1}$ [166]. Temperature-dependent measurements of electrical transport

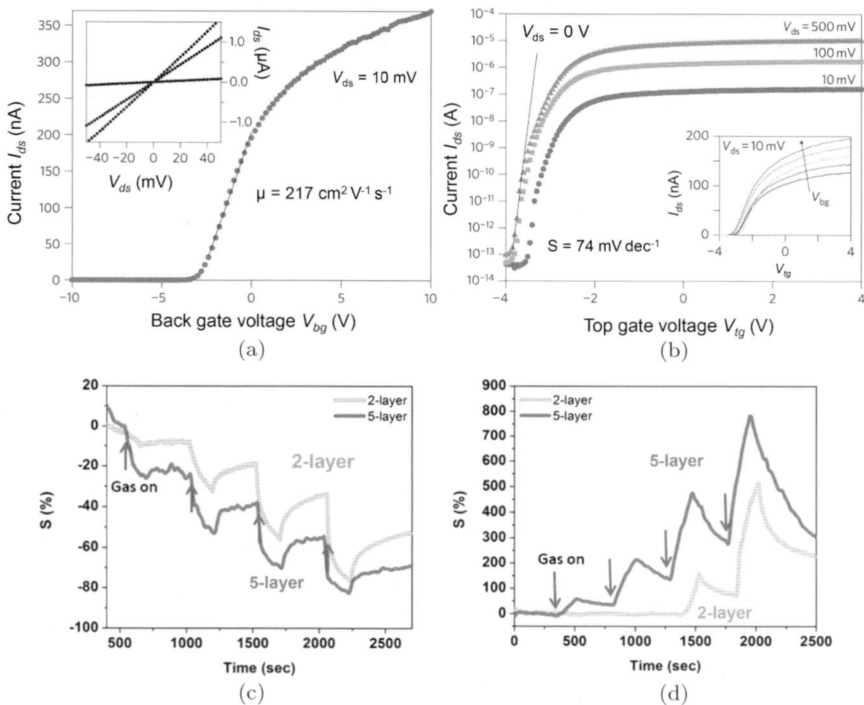

Figure 16. (a) Room-temperature transfer characteristics of the single layer MoS_2 FET at an applied bias voltage, V_{ds} of $10\,mV$. The inset shows the $I_{ds} - V_{ds}$ curve obtained for different V_{bg} values (0, 1 and 5 V); (b) $I_{ds} - V_{tg}$ curve acquired with bias voltages of 10, 100 and $500\,mV$. [bg: back gate, tg: top gate, ds: drain and source, μ: mobility, and S: sub-threshold swing with unit mV dec^{-1}, where a dec (decade) represents a 10-fold increase in the drain current, I_d]; (c) and (d) cyclic sensing-behaviour of two- and five-layer MoS_2 flakes under intermittent exposure to (c) NH_3 and (d) NO_2 gases in increasing concentrations (100, 200, 500 and 1000 ppm). ((a) and (b) adapted from Ref. [161] and (c) and (d) adapted from Ref. [178].)

indicated variable range hopping at lower temperatures which was explained in terms of Coulomb potential of randomly distributed charges at the $MoS_2 - SiO_2$ interface [167]. Integrated small-signal analogue amplifiers have been fabricated by connecting two top-gated single layer MoS_2 transistors in series [168]. A FET comprising semiconducting few-layer MoS_2 channel in-between single layer graphene and a thin film of metal, exhibited a current on/off ratio greater than 10^3, and a high current density of up to $5000\,\mathrm{A\,cm^{-2}}$ [169].

Late *et al.* [170] have reported hysteresis behaviour in back-gated FETs based on mechanically exfoliated single layer MoS_2. The increase in hysteresis of MoS_2 FET with humidity and under illumination is attributed respectively to the trap-states induced by water molecules absorbed on the surface, and the photo-sensitivity. Passivation of the device with a $30\,\mathrm{nm}$-thick Si_3N_4 layer increases the conductivity of the FET by more than $100\,$times, and suppresses the hysteresis. The FET mobility also increases due to an improved contact and suppressed Coulomb scattering. Hao *et al.* [171] have investigated the influence of chemisorbed oxygen and water molecules on the performance of back-gated MoS_2 transistors.

Back-gated FETs based on mechanically exfoliated ultrathin $MoSe_2$, exhibit n-type characteristics with a current on/off ratio exceeding 10^6 and a FET mobility of $50\,\mathrm{cm^2\,V^{-1}\,s^{-1}}$ [172]. A strong phonon-dominated scattering is suggested from the temperature-dependence of FET mobilities in these devices. single layer WSe_2 FETs with the high-κ gate dielectric and chemically doped source and drain contacts showed ambipolar characteristics, with large hole mobilities close to about $250\,\mathrm{cm^2\,V^{-1}\,s^{-1}}$ and a high current on/off ratios exceeding 10^6 at room temperature [173, 174].

Transistors based on mechanically exfoliated nanosheets of MoS_2 with 1–4 layers have been used to detect down to 2-ppm levels of NO with an 80% decrease in the channel conductance [175]. Flexible transistors with MoS_2 as the thin-film channel and reduced graphene oxide (RGO) as the electrode, have been used for sensing NO_2. Instead, when Pt-decorated MoS_2 is used as the channel, the sensitivity of FET enhances by about three times with a detection limit as low as 2 ppb [176]. Devices based on electrochemically reduced MoS_2 thin films, exhibiting high conductivity and rapid electron-transfer, have been used for the selective detection of glucose and dopamine [177]. Few-layer MoS_2 transistors are shown to sense humidity and gases such as NO_2 and NH_3 (Figures 16(c) and 16(d)); the sensing is based on the changes in resistivity of the device in the presence of these gases as a result of charge transfer to MoS_2 at different applied fields [178]. Phototransistors of single and bilayer MoS_2 can detect green light whereas those of triplelayer MoS_2 can detect red light [179]. MoS_2 with 10–60 nm thickness was shown to detect UV-to-near-IR radiation [179]. FETs of few-layer WS_2 prepared by sonication-assisted exfoliation in IPA show ambipolar characteristics, with a high current on/off ratio of about 10^5 and good photosensitivity to visible light [180].

BN is a wide band-gap insulator ($E_g \sim 5$–$6\,eV$) and exhibits good dielectric properties. BN was employed as an efficient dielectric-substrate for devices based on graphene [181]. BN has remarkable chemical and thermal stability. In addition, the absence of dangling bonds and surface charge traps in BN leads to chemical stability and enhanced mobility in graphene transistors with BN as a dielectric layer [182, 183]. Heterostructures of multilayered BN nanosheets and graphene of the type, $BN-C-BN$, have exhibited remarkable charge mobility, as high as $5,00,000\,cm^2\,V^{-1}\,s^{-1}$ [183]. Single- and few-layer BN were employed as thin-film dielectrics for graphene and MoS_2 FETs to suppress the charge-trapping at the interface between the active material and SiO_2 [184, 185]. Thin nanosheets of BN have also been used in field-effect tunnelling transistor devices where BN acts as a tunnel barrier between two layers of graphene; the tunnel current in these devices is found to be exponentially dependent on the BN thickness as expected for quantum tunnelling [186].

1.4.2 Supercapacitors

MoS_2 samples with a high density of nanowalls show an electrochemical double layer capacitance comparable to that of carbon nanotube arrays [187]. Nanoporous films of edge-oriented MoS_2 obtained by electrochemical anodisation of Mo metal in the presence of sulphur vapour, exhibit a capacitance of up to $14.5\,mF\,cm^{-2}$ from galvanostatic charge-discharge measurements at a current density of $1\,mA\,cm^{-2}$ and $12.5\,mF\,cm^{-2}$ from cyclic voltammetry (CV) at a scan rate of $50\,mV\,s^{-1}$. Interestingly, the capacitance of these films increases from $2.2\,mF\,cm^{-2}$ to $10.5\,mF\,cm^{-2}$ after 10,000 cycles at a current density of $10\,\mathrm{mA\,cm^{-2}}$ [188]. Composites of MoS_2 and graphene prepared by cysteine-assisted solution-phase synthesis showed a maximum specific capacitance of about $243\,\mathrm{F\,g^{-1}}$, which is considerably higher compared with the individual components. The electrodes demonstrate long-term cyclability, with only about 7.7% decrease in specific capacitance even after 1000 cycles [189]. Composites of MoS_2 and reduced graphene oxide (RGO) in different proportions have been investigated for supercapacitor applications [190] and the best results are obtained with MoS_2/RGO (1:2) composite as shown in Figure 17. The quasi-rectangular CV curves for

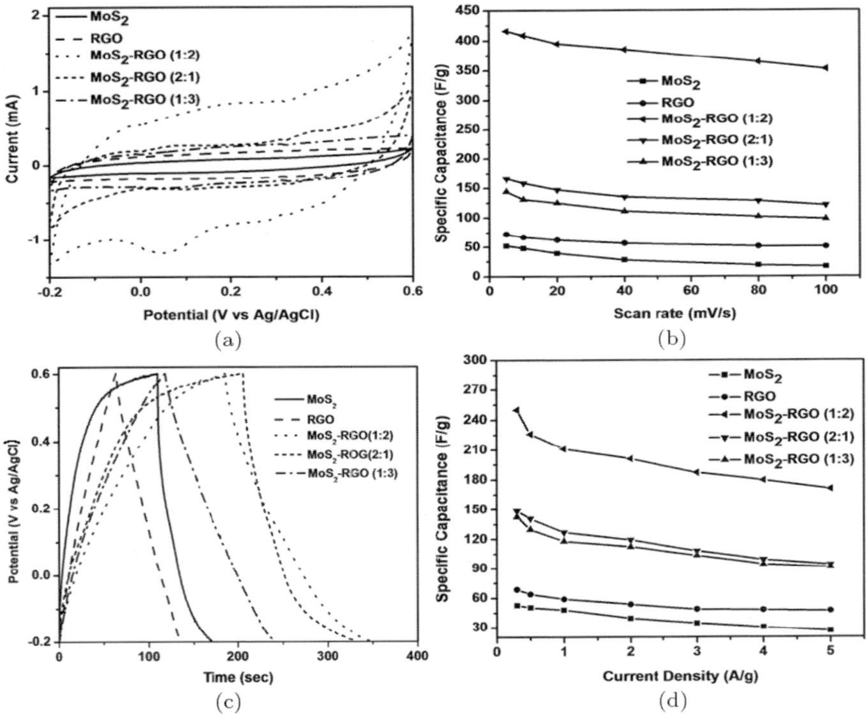

Figure 17. (a) Cyclic voltammograms of MoS_2, RGO and MoS_2/RGO electrodes acquired at a scan rate of $100\,\mathrm{mV/s}$; (b) Specific capacitance of MoS_2 RGO, and MoS_2/RGO electrodes obtained at different scan rates; (c) Galvanostatic charge–discharge curves of MoS_2, RGO and MoS_2/RGO electrodes at a current density of 1 A/g; (d) Specific capacitance of MoS_2, RGO and MoS_2/RGO electrodes at different current densities. (Adapted from Ref. [190]).

MoS_2/RGO electrodes are akin to those of an ideal supercapacitor, giving a maximum capacitance of 416 F g^{-1} with MoS_2/RGO (1:2) at a scan rate of 5 mV s^{-1}. As shown in Figure 17(b), the current increases and the capacitance decreases with an increasing scan rate. The charge–discharge curves obtained at a current density of 1 A g^{-1} are shown in Figure 17(c). The discharge time of MoS_2/RGO composite (1:2) is considerably longer than that of other composites, and the nearly symmetrical curve is indicative of its remarkable charge-storing ability. The calculated specific capacitance of this composite is about 249 F g^{-1} at a current density of 0.3 A g^{-1}. Figure 17(d) shows the specific capacitance of electrodes composed of MoS_2, RGO and MoS_2/RGO, obtained at various current densities. The high electronic conductivity of RGO and surface properties of both MoS_2 and RGO result in an improved performance of MoS_2/RGO composites, the calculated synergistic effect being 118% in the case of MoS_2/RGO (1:2) composite. Similarly, composites of MoS_2 and polyaniline (PANI) in different ratios showed an excellent supercapacitor performance [191]. The MoS_2/PANI composites in 1:1 and 1:6 ratios showed capacitance values of 417 F g^{-1} and 567 F g^{-1}, respectively, with improved cyclic stability relative to PANI alone. MoS_2/polypyrrole nanocomposites synthesised by *in situ* oxidative polymerisation exhibit a specific capacitance of about 700 F g^{-1} at a scan rate of 10 mV s^{-1} [192].

Composites of BN and graphene ($BN_{1-x}G_x$) have been prepared by covalent cross-linking using the carbodiimide coupling reaction, and examined for supercapacitor application. The specific capacitance values of $BN_{0.25}G_{0.75}$, $BN_{0.5}G_{0.5}$ and $BN_{0.75}G_{0.25}$ composites are 238 F g^{-1}, 204 F g^{-1} and 116 F g^{-1}, respectively, at a current density of 0.3 A g^{-1}. The high specific capacitance of $BN_{1-x}G_x$ has been ascribed to the enhanced surface area together with the presence of slit-like micropores in the composite induced by covalent cross-linking [193].

High surface-area BCN, ($BC_{4.5}N$ and BC_3N) have been prepared by the urea route and investigated for supercapacitor applications [194]. $BC_{4.5}N$ and BC_3N showed high BET surface areas of 1280 m^2 g^{-1} and 1300 m^2 g^{-1}, respectively. Electrodes of $BC_{4.5}N$ and

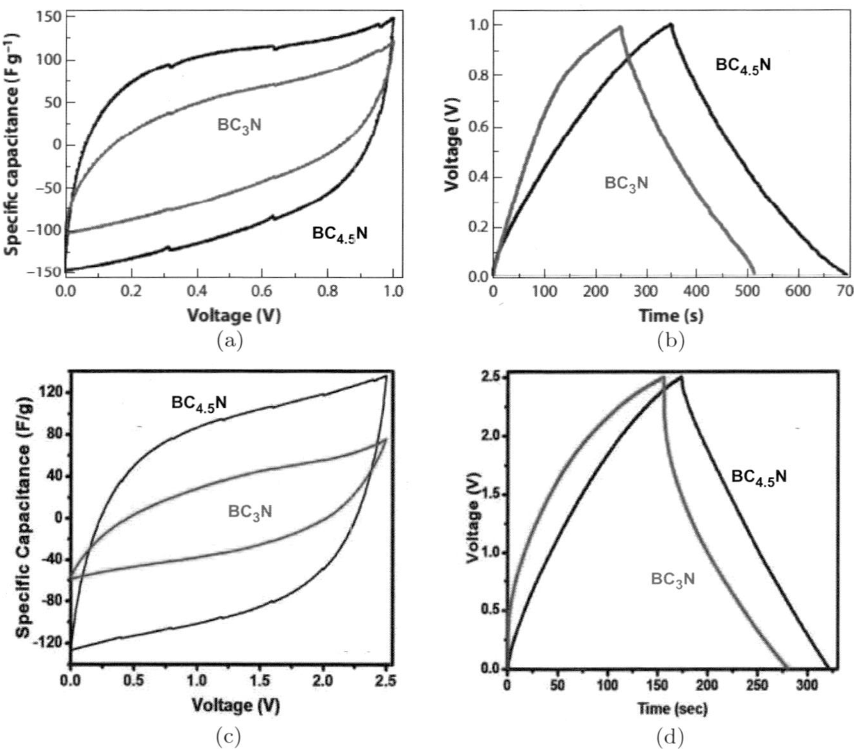

Figure 18. Supercapacitor performance of BC_3N and $BC_{4.5}N$. Cyclic voltammograms recorded at a scan rate of $100\,mV\,s^{-1}$ in (a) 6 M KOH aqueous electrolyte and (c) $TEABF_4$ ionic liquid; Galvanostatic charge/discharge curves in (b) 6 M KOH aqueous electrolyte at $100\,mA\,g^{-1}$ current density and (d) $TEABF_4$ ionic liquid at 1 A g^{-1} current density. $TEABF_4$: tetraethylammonium tetrafluoroborate. (Adapted from Ref. [194]).

BC_3N showed specific capacitance values of about 144 and 102 F g^{-1}, respectively, at a scan rate of $100\,mV\,s^{-1}$ and the current density of $100\,mA$ g^{-1} (Figures 18(a) and 18(b)), with an excellent capacity retention even after 1000 cycles. In an ionic liquid medium, $BC_{4.5}N$ and BC_3N electrodes exhibited specific capacitance values of $\sim$234 and 203 $F\,g^{-1}$, respectively, at a current density of $1\,A\,g^{-1}$ (Figures 18(c) and 18(d)) and show a maximum energy density of $32.5\,W\,h\,kg^{-1}$ [194]. Interestingly, composites few-layer MoO_3 and BCN exhibited a high capacitance of $471\,F\,g^{-1}$ at a scan rate of

$10\,\mathrm{mV\,s^{-1}}$ and a current density of $1\,\mathrm{A\,g^{-1}}$, which is significantly enhanced relative to pristine BCN alone [108].

1.4.3   Lithium ion batteries

Layered TMDs, MX_2 (M = Mo, W, Ti, V and X = S, Se), undergo rapid intercalation and de-intercalation of guest ions (e.g., Li^+ and Na^+ ions) between the layers and thus act as good host materials for battery applications. Haering *et al.* [195] demonstrated the first lithium-ion battery employing MoS_2 as the anode material. Chemically exfoliated and restacked MoS_2 nanosheets exhibit high surface area and a high charge capacity of $800\,\mathrm{mA\,h^{-1}g^{-1}}$ in the first cycle which is retained up to $750\,\mathrm{mA\,h^{-1}g^{-1}}$ even after 20 cycles at a current density of $50\,\mathrm{mA\,g^{-1}}$ [196]. Hydrothermally synthesised MoS_2 nanoflakes and nanoplates showed high specific capacities of $1000\,\mathrm{mA\,h^{-1}g^{-1}}$ [197] and $1062\,\mathrm{mA\,h^{-1}g^{-1}}$ [198], respectively. Nanosheets and nanoflakes of WS_2 also exhibited good reversible capacity [199–201]. Mesoporous WS_2, prepared using SBA-15 as the template, exhibits high surface area and a good storage capacity of $805\,\mathrm{mA\,h^{-1}g^{-1}}$ at $100\,\mathrm{mA\,g^{-1}}$ current density [202]. Graphene has been widely used as a commercial anode material due to its high surface area and chemical stability [203, 204]. Composites of graphene and TMDs have been studied extensively for Li-ion storage. MoS_2/graphene hybrid composites showed a high reversible capacity of $\sim1290\,\mathrm{mA\,h^{-1}g^{-1}}$ which is retained up to 50 cycles at a current density of $100\,\mathrm{mA\,g^{-1}}$; a high specific capacity of $1040\,\mathrm{mA\,h^{-1}g^{-1}}$ was retained even at much higher discharge rates of $1\,\mathrm{A\,g^{-1}}$ [205]. Replacing graphene with N-doped graphene in MoS_2/graphene composites further enhances the capacity and cyclability as shown in Figures 19(a) and 19(b) [206]. An increase in capacity is observed with the increasing number of cycles and may be caused by increasing vacancies and defects in N-doped graphene. The latter appears to promote more Li-ion insertion in the course of battery operation (Figure 19(a)) [206]. WS_2/graphene composites also show improved specific capacity at high current densities and good cyclability compared to the bare WS_2 [207]. David *et al.* [208] have reported Na-ion batteries with

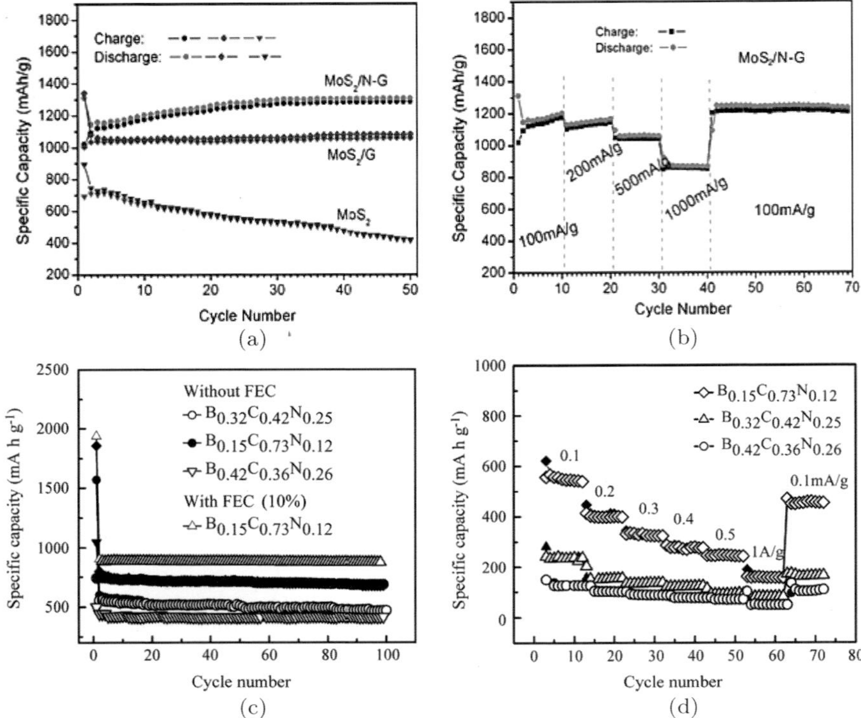

Figure 19. (a) Specific capacity vs. number of cycles for MoS_2, MoS_2/graphene and MoS_2/N-graphene electrodes at a current density of $100\,\text{mA h}^{-1}\text{g}^{-1}$ and (b) current rate capabilities of MoS_2/N-graphene electrode at various current densities; (c) Specific capacity vs. number of cycles for three different compositions of $B_xC_yN_z$ samples at $0.05\,\text{Ag}^{-1}$ current density and (d) current rate capabilities of the three $B_xC_yN_z$ samples in the range of 0.1–$1\,\text{A g}^{-1}$ current density. ((a) and (b) adapted from Ref. [206]; (c) and (d) adapted from Ref. [209].)

MoS_2/graphene composites as anode materials. These composites showed a capacity of $230\,\text{mA h}^{-1}\text{g}^{-1}$ with retention up to several cycles even at high current densities of $200\,\text{mA g}^{-1}$. Graphene or carbon in composites improves the overall electrochemical performance and stability by (a) absorbing the mechanical stress during volume changes associated with Li/Na insertion and de-insertion cycles, (b) improving the electrical conductivity and (c) preventing the undesired diffusion of polysulphides into the electrolyte.

$B_x C_y N_z$ samples prepared by high temperature urea-route exhibit high lithium cyclability and current rate capability when used as anode materials in Li-ion batteries [209]. The samples possess high surface areas and are likely to contain BCN rings in addition to domains of graphene and BN. The electrochemical performance of $B_x C_y N_z$ samples was found to be strongly dependent on the composition and surface area. Among the samples, the carbon-rich $B_{0.15}C_{0.73}N_{0.12}$ with the highest surface area shows specific capacities of about $710 \, \text{mA h g}^{-1}$ at $0.05 \, \text{A g}^{-1}$ and $150 \, \text{mA h g}^{-1}$ at $1 \, \text{A g}^{-1}$ with an excellent stability even after 100 cycles (Figures 19(c) and 19(d)). With fluoroethylene carbonate as an additive, the specific capacity of this sample increases to nearly $900 \, \text{mA h g}^{-1}$ at $0.05 \, \text{A g}^{-1}$. Solid–electrolyte layer at the interface of $B_x C_y N_z$ and the electrolyte is believed to play a crucial role in the performance of the $B_x C_y N_z$ [209].

1.4.4 Gas adsorption and storage

Graphene-like $B_x C_y N_z$ samples prepared by the high-temperature reaction of boric acid, urea and activated charcoal, have been explored for storage of greenhouse gases such as CO_2 and methane [89]. Samples with compositions close to BC_2N exhibit high BET surface areas in the range of $1500–1990 \, \text{m}^2\text{g}^{-1}$. The sample with the highest surface area of $1990 \, \text{m}^2\text{g}^{-1}$ shows remarkable CO_2-uptake of 128 wt.% at 195 K and 0.1 MPa, and 64 wt.% at 298 K and 5 MPa. CH_4-uptake increases exponentially with increasing surface area, ranging from 5 to 15 wt.% at 298 K and 5 MPa for different samples [89]. Adsorption of CO_2 on $BC_{1.7}N_{0.6}$ is found to be nearly 10 times more selective over N_2 at 1 atm and 293 K (Figure 20(a)) [86]. On $BC_{1.7}N_{0.6}$, selectivities of CO_2 adsorption over H_2, N_2 and CH_4 were found to be 28:1, 8:1 and 4:1, respectively at 293 K and 5000 kPa while a remarkable selectivity of 68:1 is observed for CH_4 over H_2 at 293 K and 4000 kPa (Figure 20(b)) [86]. First-principles calculations indicate that the $B_x C_y N_z$ surface has strongly adsorbing sites for CO_2 and CH_4 than does graphene or BN. Defect sites (e.g., Stone–Wales defects) specially favour adsorption of CO_2 and CH_4 [89].

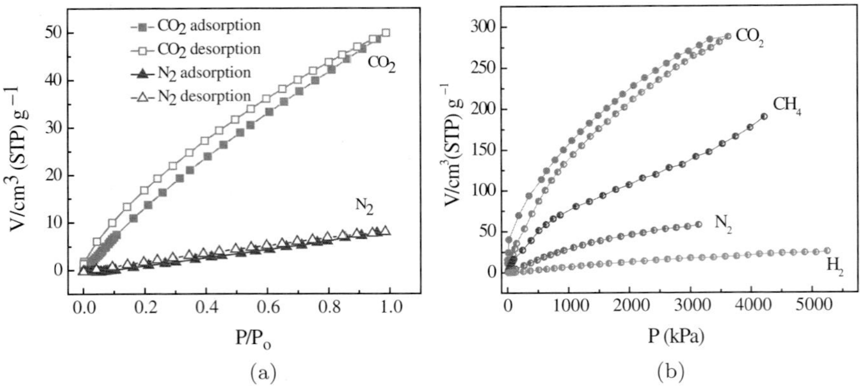

Figure 20. (a) Comparison of CO_2 and N_2 uptakes of $BC_{1.7}N_{0.6}$ at 293 K and 1 atm; (b) High-pressure adsorption isotherms of CO_2, CH_4, N_2 and H_2 at 293 K for $BC_{1.7}N_{0.6}$. (Adapted from Ref. [86]).

1.4.5   Catalysis

MoS_2 is well-known as a hydrodesulphurisation catalyst [210–212]. A high conversion-efficiency of about 98% at 375°C has been achieved for hydrodesulphurisation of thiophene to butane with few-layer MoS_2 decorated with Co and Ni nanoparticles as a catalyst [213]. Ni-Fe/MoS_2 composites have shown good catalytic activity for the oxidation of hydrazine [214]. Hinnemann *et al.* [215] have demonstrated MoS_2 nanoparticles to act as good catalysts for the hydrogen evolution reaction (HER) with a low overpotential of 0.1–0.2 V. Based on theoretical studies, the edge sites [216] and vacancies [217] in MoS_2 are shown to be catalytically active sites for HER. Increasing the density of active edge sites and surface area of MoS_2 catalyst improves the photocatalytic and electrochemical HER activities [218–222]. Electrocatalytic HER activity is further enhanced in MoS_2/graphene composites [223] owing to a better electrical coupling to the electrode facilitated by graphene [224]. MoS_2 has been used as a co-catalyst along with light-absorbing semiconductors such as CdS [225, 226], CdSe [227] and TiO_2 [228], and dyes such as [Ru(bpy)$_3$] and eosin [229], for photocatalytic HER in aqueous media. When a small weight percent of graphene is incorporated in the composites, photocatalytic

HER activity is enhanced as graphene acts as an electron-sink, promoting better charge-separation [230]. Charge-separation is further enhanced by forming a p–n junction between p-type MoS_2 and n-type N-doped graphene. These composites of MoS_2 and N-doped graphene exhibited improved photocatalytic [30] and electrocatalytic HER activity [231] compared with MoS_2/graphene composites.

In metallic $1T$-MoS_2 and $1T$-$MoSe_2$ containing t_{2g} (d_{yx}, d_{zx}, d_{zy}) orbitals of Mo partly filled with two electrons, an extra electron would result in a stable half-filled configuration of t_{2g} orbitals (Figure 21(a)). $1T$-MoS_2 and $1T$-$MoSe_2$ can therefore readily accept an electron from a dye like eosin and catalyse the HER as shown in the schematic in Figure 21(b). $1T$-MoS_2 prepared by Li-intercalation and exfoliation route indeed exhibits an activity, 600 times higher than that of the few-layer $2H$-MoS_2 (see inset of Figure 21(c)) [30]. It is also found that $1T$-$MoSe_2$ shows even higher HER activity $(\sim 62 \pm 5\,\mathrm{mmol\,g^{-1}h^{-1}})$ compared with $1T$-MoS_2 (Figure 21(c)) [26]. Based on theoretical studies, a lower work function of $MoSe_2$, and weaker binding of hydrogen to Se than to S edge sites are considered responsible for the higher activity of $1T$-$MoSe_2$. $1T$-WS_2 nanosheets obtained by Li-intercalation and exfoliation show high activity for electrocatalytic HER [27, 232]. High-angle annular dark field (HAADF) imaging in an aberration-corrected STEM has revealed a high concentration of the zigzag-like superlattices in the exfoliated WS_2 nanosheets, identified as strained metallic $1T$-phase ($1T'$). The enhanced electrocatalytic HER activity of WS_2 nanosheets has been attributed to the presence of tensile strain induced by zigzag-like distortion in the $1T$-phase [27]. Density functional theory calculations show that the tensile strain increases the density of states near the Fermi level and facilitates hydrogen-binding [27]. $1T$-MoS_2 also shows better electrocatalytic HER activity than $2H$-MoS_2. It was thought earlier that the edge sites are the active catalytic sites in MoS_2, but a recent study involving oxidation of MoS_2 edges shows that, in addition to the edges, the basal plane of $1T$-MoS_2 nanosheets is also catalytically active for electrocatalytic HER (Figure 21(d)) [233].

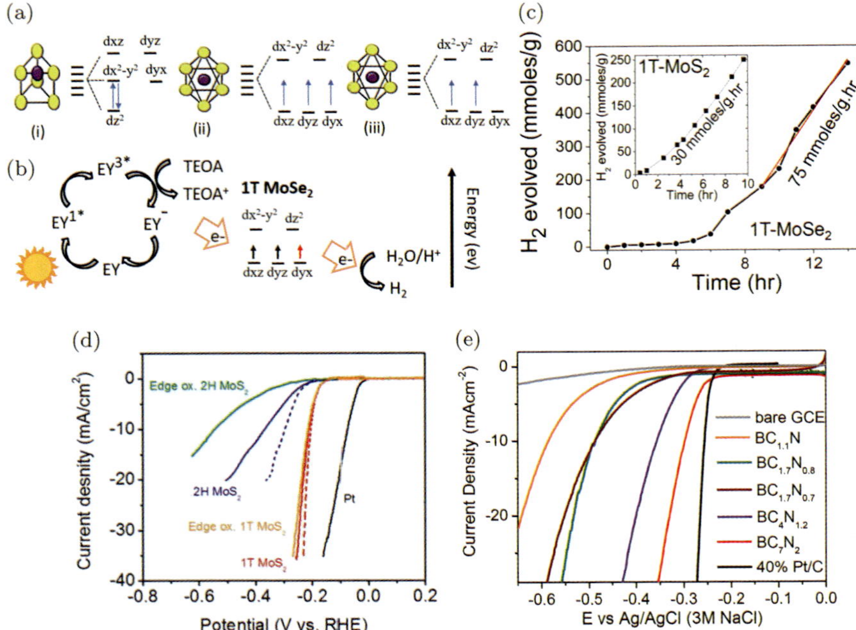

Figure 21. (a) The crystal field induced electronic configuration of (i) 2H-MoSe$_2$, (ii) Li-intercalated MoSe$_2$ and (iii) 1T-MoSe$_2$; (b) possible mechanism of photocatalytic HER with 1T-MoSe$_2$ as co-catalyst. EY-eosin, EY^1-singlet-state EY, EY^3-triplet-state EY and TEOA-triethanolamine; (c) hydrogen evolution as a function of time with 1T-MoSe$_2$ and that of 1T-MoS$_2$ co-catalysts (inset); (d) polarisation curves for electrocatalytic HER before and after edge-oxidation of 1T and 2H-MoS$_2$ nanosheets (dashed lines are *iR*-corrected polarisation curves); (e) composition-dependent electrocatalytic HER performance of B$_x$C$_x$N$_z$ samples. ((a)–(c) adapted from Ref. [26]; (d) adapted from Ref. [233] and (e) adapted from Ref. [236].)

B$_x$C$_y$N$_z$ with tunable B/N codoping levels, synthesised by annealing GO in the presence of boric acid and ammonia, is an excellent electrocatalyst for ORR [234]. ORR activity in few-layer B$_x$C$_y$N$_z$ has been found to be composition-dependent [235]. The best activity is found in the carbon-rich B$_{0.15}$C$_{0.72}$N$_{0.12}$ sample whose onset potential of 0.198 V vs. normal hydrogen electrode (NHE) is more positively shifted relative to that of B- or N-doped graphene and is close to that of commercial Pt/C [235]. A mechanistic study reveals a four-electron pathway with comparable kinetic parameters and enhanced methanol-tolerance with respect to the Pt/C electrode [235]. Recently, a superior activity for electrocatalytic HER was reported in metal-free $B_xC_yN_z$ catalysts [236]. Of the samples with different compositions, carbon-rich BC_7N_2 sample exhibited the best activity at par with Pt-catalyst (Figure 21(e)) with a highly positive onset potential of $-56\,$mV vs. reversible hydrogen electrode (RHE) and a high current density of $10\,$mA cm^{-2} at an over potential of $70\,$mV [236]. The high performance of this composition has been ascribed to low proportion of insulating BN domains, high percentage of B–C bonds and larger proportion of pyridinic nitrogen. Theoretical studies further indicate a favourable alignment of the conduction band in BC_7N_2 with respect to the redox level for proton reduction.

1.5 Distorted Layered TMDs and Their Novel Properties

Distorted layered ditellurides of W and Mo with zigzag metal-metal chains (Figures 22(a) and 22(b) have gained significant attention due to their fascinating properties such as extremely large ($>10^5\%$) and non-saturating magnetoresistance (XMR), and pressure-driven superconductivity. Unidirectional (along with the W$-$W chains) positive XMR of 452,700% has been reported recently in crystals of semimetallic Td$-$WTe$_2$ at 4.5 K in a magnetic field of 14.7 T, reaching 13 million% at 0.53 K in a field of 60 T (Figure 22(c)) [237]. Td$-$WTe$_2$ occurs as a distorted variant of a 1T structure with a $\sqrt{3}\times1$ superstructure of zigzag W$-$W chains along crystallographic a-axis (Figure 22(a)). The single crystals of Td$-$WTe$_2$ are grown by chemical vapour transport of polycrystalline Td$-$WTe$_2$ using Br$_2$ as the transport-agent. Even at very high applied magnetic fields (60 Tesla), XMR in Td$-$WTe$_2$ does not saturate due to perfectly balanced electron–hole compensation in Td$-$WTe$_2$ particularly at low temperatures. This observation has been reinforced by a temperature-dependent study of angle-resolved photoemission [238]. The study reveals a uniaxial Fermi surface composed of a tiny electron and hole pockets in the momentum

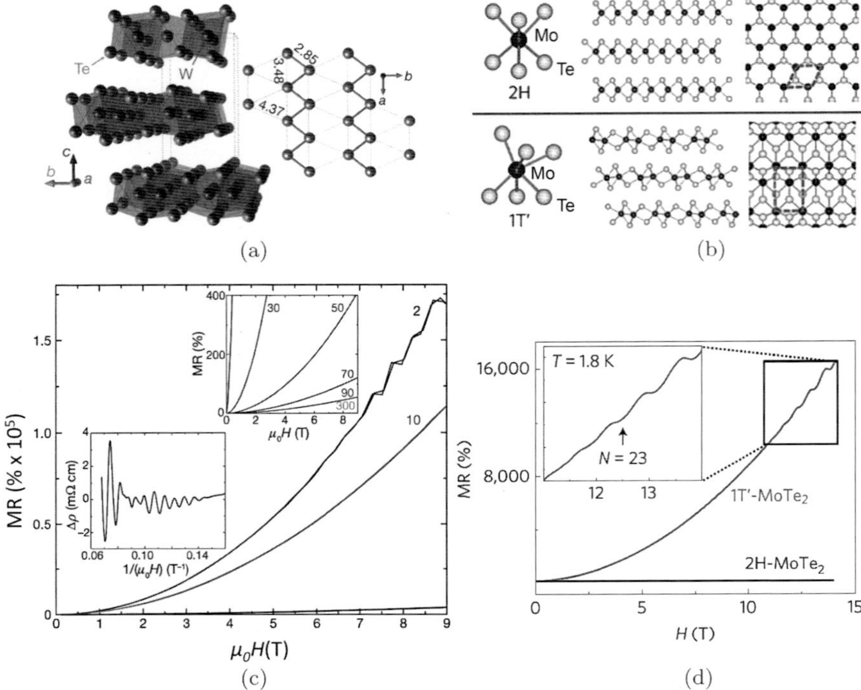

Figure 22. (a) Crystal structures of (a) Td$-$WTe$_2$ and (b) 2H- and 1T$'$-MoTe$_2$; (c) Field-dependent MR in Td$-$WTe$_2$ at different temperatures (numbers in kelvins). The upper inset shows the MR in Td$-$WTe$_2$ at higher temperatures while the lower inset shows the SdH oscillations at 4.5 K; (d) Comparison of field-dependent MR of 1T$'$- and 2H-MoTe$_2$ at T = 1.8 K. The inset shows the SdH oscillations in 1T$'$-MoTe$_2$ at 1.8 K. ((a) and (c) adapted from Ref. [237] and (b) and (d) adapted from Ref. [244]).

space along W$-$W chain direction. The relative size of the pockets is temperature dependent approaching a perfect compensation in the limit of zero temperature. At high temperatures, the disruption of the balance between these pockets leads to suppression of XMR.

A recent study of the pressure-dependent study of MR in Td$-$WTe$_2$ crystals has shown that with increasing pressure, the Fermi surface topology changes drastically and the XMR is strongly suppressed [239]. Through an analysis of Shubnikov–de Haas oscillations (SdH), four Fermi surfaces have been identified under ambient pressure which are likely two pairs of hole and electron pockets along the $\Gamma-X$ direction (i.e., along $W-W$ chains) [239]. While one pair of hole and electron pockets (γ and δ) disappears with increasing pressure, the other pair of hole and electron pockets (α and β) persists to high pressure but the relative size of the α and β pockets decreases with increasing pressure, accompanied by a strong suppression of the XMR. These results also show that a perfect electron-hole compensation forms the electronic basis for the non-saturating XMR in WTe_2. Pan *et al.* [240] have reported a dome-shaped superconductivity in $Td-WTe_2$, that appears at a pressure of 2.5 GPa, reaching a maximum critical temperature (Tc) of 7 K at around 16.8 GPa, followed by a monotonic decrease in T with increasing pressure. WTe_2 being a semimetal has quite small density of states at the Fermi level ($N(E_F)$) and therefore does not exhibit superconductivity even down to 0.3 K at ambient pressure. Applied pressure rapidly increases $N(E_F)$, an essential condition for the onset of superconductivity. A simultaneous suppression of MR with increasing pressure is due to increasing difference between the electron and hole Fermi pockets. Kang *et al.* [241] have also reported superconductivity in $Td-WTe_2$ arising from a suppressed MR state under strain. With increasing pressure, MR is found to gradually decrease and turn off at a critical pressure of 10.5 GPa, where superconductivity begins to emerge.

In contrast to $MoS(Se)_2$, $MoTe_2$ shows a thermally induced structural transition around $900°C$ from the semiconducting 2H-phase to semimetallic $1T'$-phase [242] $1T'$-$MoTe_2$ with monoclinic crystal structure is a distorted variant of 1T-phase with in-plane $Mo-Mo$ bonds resulting in pseudo-hexagonal layers with zigzag Mo-chains (Figure 22(b)). First-principles calculations show that 2H-$MoTe_2$ is the most stable form at low temperatures. However, selective synthesis of $1T'$- and 2H-$MoTe_2$ is feasible given the small energy difference between 2H- and $1T'$-$MoTe_2$ (~ 35 meV) [243, 244]. Centimetre-scale, uniform 2H- and $1T'$-$MoTe_2$ thin films were grown at $650°C$ by tellurisation of Mo thin films deposited on Si/SiO_2 substrates. A thin film of $1T'$-$MoTe_2$ is formed initially which converts gradually to the stable 2H-$MoTe_2$ phase over a prolonged growth time under excess

Te atmosphere. The phase transition is driven by the relaxation of tensile strain in the presence of an excess Te. Annealing 2H-MoTe$_2$ under a lower partial pressure of Te at the same temperature followed by a rapid quenching leads to a reverse transition from 2H-MoTe$_2$ to 1T′-MoTe$_2$ [245].

Laser-induced phase patterning has been employed to fabricate an ohmic heterojunction between semiconducting 2H- and metallic 1T′-MoTe$_2$ that is stable up to 300°C [243]. These heterojunctions increase the carrier mobility of the MoTe$_2$ transistor by a factor of ∼50, while maintaining a high on/off current ratio of 10^6. The phase transition from 2H- to 1T′-phase originates from the Te-vacancies induced by laser-irradiation at elevated temperatures. Density-functional calculations reveal that Te-vacancies in excess of 3% cause the 1T′-phase to be more stable than the 2H-phase [243]. Owing to strong SOC and a narrow band gap of ∼1 eV, 2H−MoTe$_2$ is useful in valleytronics [246] and semiconductor-electronics [247–250]. A mobility as high as ∼40 cm^2 V^{-1}s^{-1} has been reported for FETs based on few-layer 2H-MoTe$_2$ [244]. On the other hand, bulk semimetallic 1T′-MoTe$_2$ exhibits a giant MR of 16,000% in a magnetic field of 14 T at 1.8 K (Figure 22(d)) [244]. Pressure-driven superconductivity has been reported in single crystals of orthorhombic Td−MoTe$_2$ (isostructural to Td−WTe$_2$), a low-temperature (<120 K) polymorph of 1T′-MoTe$_2$, with a maximum critical temperature of 8.2 K at 11.7 GPa [251].

1.6   Concluding Remarks

The earlier sections clearly bring out how 2D inorganic layered compounds have emerged to become materials of vital importance rivalling graphene. The recent discovery of extraordinary magnetoresistance is noteworthy. It appears that there will be more surprises in the properties of this group of 2D materials. It also seems likely that many of the properties of these layered materials will be exploited in the near future.

References

[1] A. K. Geim and K. S. Novoselov, *Nat. Mater.* **6**, 183 (2007).

[2] C. N. R. Rao, A. K. Sood, R. Voggu and K. S. Subrahmanyam, *J. Phys. Chem. Lett.* **1**, 572 (2010).

[3] Q. H. Wang, K. Kalantar-Zadeh, A. Kis, J. N. Coleman and M. S. Strano, *Nat Nanotechnol.* **7**, 699 (2012).

[4] M. Chhowalla, H. S. Shin, G. Eda, L.-J. Li, K. P. Loh and H. Zhang, *Nat. Chem.* **5**, 263 (2013).

[5] C. N. R. Rao and U. Maitra, *Ann. Rev. Mater. Res.* **45**, 29 (2015).

[6] C. N. R. Rao, H. S. S. Ramakrishna Matte and U. Maitra, *Angew. Chem. Int. Ed.* **52**, 13162 (2013).

[7] C. N. R. Rao, U. Maitra and U. V. Waghmare, *Chem. Phys. Lett.* **609**, 172 (2014).

[8] K. F. Mak, K. He, J. Shan and T. F. Heinz, *Nat. Nanotehnol.* **7**, 494 (2012).

[9] A. K. Geim and I. V. Grigorieva, *Nature* **499**, 419 (2013).

[10] A. Azizi, S. Eichfeld, G. Geschwind, K. Zhang, B. Jiang, D. Mukherjee, L. Hossain, A. F. Piasecki, B. Kabius, J. A. Robinson and N. Alem, *ACS Nano.* **9**, 4882 (2015).

[11] C.-H. Lee, G.-H. Lee, A. M. van der Zande, W. Chen, Y. Li, M. Han, X. Cui, G. Arefe, C. Nuckolls, T. F. Heinz, J. Guo, J. Hone and P. Kim, *Nat. Nanotechnol.* **9**, 676 (2014).

[12] H. Fang, C. Battaglia, C. Carraro, S. Nemsak, B. Ozdol, J. S. Kang, H. A. Bechtel, S. B. Desai, F. Kronast, A. A. Unal, G. Conti, C. Conlon, G. K. Palsson, M. C. Martin, A. M. Minor, C. S. Fadley, E. Yablonovitch, R. Maboudian and A. Javey, *PNAS* **111**, 6198 (2014).

[13] C. N. R. Rao, K. Gopalakrishnan and U. Maitra, *ACS Appl. Mater. Interfaces* **7**, 7809 (2015).

[14] C. N. R. Rao and A. Nag, *Eur. J. Inorg. Chem.* **2010**, 4244 (2010).

[15] M. B. Ruben, G. N. Cristina, G. H. Julio and Z. Felix, *Nanoscale* **3**, 20 (2011).

[16] J. N. Coleman, M. Lotya, A. O'Neill, S. D. Bergin, P. J. King, U. Khan, K. Young, A. Gaucher, S. De, R. J. Smith, I. V. Shvets, S. K. Arora, G. Stanton, H.-Y. Kim, K. Lee, G. T. Kim, G. S. Duesberg, T. Hallam, J. J. Boland, J. J. Wang, J. F. Donegan, J. C. Grunlan, G. Moriarty, A. Shmeliov, R. J. Nicholls, J. M. Perkins, E. M. Grieveson, K. Theuwissen, D. W. McComb, P. D. Nellist and V. Nicolosi, *Science* **331**, 568 (2011).

[17] R. J. Smith, P. J. King, M. Lotya, C. Wirtz, U. Khan, S. De, A. O'Neill, G. S. Duesberg, J. C. Grunlan, G. Moriarty, J. Chen,

J. Wang, A. I. Minett, V. Nicolosi and J. N. Coleman, *Adv. Mater.* **23**, 3944 (2011).

[18] K.-G. Zhou, N.-N. Mao, H.-X. Wang, Y. Peng and H.-L. Zhang, *Angew. Chem. Int. Ed.* **50**, 10839 (2011).

[19] Y. Yao, Z. Lin, Z. Li, X. Song, K.-S. Moon and C.-p. Wong, *J. Mater. Chem.* **22**, 13494 (2012).

[20] J. Feng, L. Peng, C. Wu, X. Sun, S. Hu, C. Lin, J. Dai, J. Yang and Y. Xie, *Adv. Mater.* **24**, 1969 (2012).

[21] K. Xu, P. Chen, X. Li, C. Wu, Y. Guo, J. Zhao, X. Wu and Y. Xie, *Angew. Chem. Int. Ed.* **52**, 10477 (2013).

[22] A. Castellanos-Gomez, M. Barkelid, A. M. Goossens, V. E. Calado, H. S. J. van der Zant and G. A. Steele, *Nano Lett.* **12**, 3187 (2012).

[23] H. S. S. R. Matte, U. Maitra, P. Kumar, B. Govinda Rao, K. Pramoda and C. N. R. Rao, *Z. Anorg. Allg. Chem.* **638**, 2617 (2012).

[24] P. Joensen, R. F. Frindt and S. R. Morrison, *Mater. Res. Bull.* **21**, 457 (1986).

[25] H. S. S. R. Matte, A. Gomathi, A. K. Manna, D. J. Late, R. Datta, S. K. Pati and C. N. R. Rao, *Angew. Chem. Int. Ed.* **49**, 4059 (2010).

[26] U. Gupta, B. S. Naidu, U. Maitra, A. Singh, S. N. Shirodkar, U. V. Waghmare and C. N. R. Rao, *APL Mat.* **2**, 092802 (2014).

[27] D. Voiry, H. Yamaguchi, J. Li, R. Silva, D. C. B. Alves, T. Fujita, M. Chen, T. Asefa, V. B. Shenoy, G. Eda and M. Chhowalla, *Nat. Mater.* **12**, 850 (2013).

[28] U. Gupta, B. G. Rao, U. Maitra, B. E. Prasad and C. N. R. Rao, *Chem. Asian J.* **9**, 1311 (2014).

[29] P. Ganal, W. Olberding, T. Butz and G. Ouvrard, *Solid State Ionics* **59**, 313 (1993).

[30] U. Maitra, U. Gupta, M. De, R. Datta, A. Govindaraj and C. N. R. Rao, *Angew. Chem. Int. Ed.* **52**, 13057 (2013).

[31] G. Eda, T. Fujita, H. Yamaguchi, D. Voiry, M. Chen and M. Chhowalla, *ACS Nano.* **6**, 7311 (2012).

[32] G. Eda, H. Yamaguchi, D. Voiry, T. Fujita, M. Chen and M. Chhowalla, *Nano Lett.* **11**, 5111 (2011).

[33] J. Zheng, H. Zhang, S. Dong, Y. Liu, C. Tai Nai, H. Suk Shin, H. Young Jeong, B. Liu and K. Ping Loh, *Nat. Commun.* **5**, 2995 (2013).

[34] S. Jeong, D. Yoo, M. Ahn, P. Miró, T. Heine and J. Cheon, *Nat. Commun.* **6**, 5763 (2014).

[35] Z. Zeng, Z. Yin, X. Huang, H. Li, Q. He, G. Lu, F. Boey and H. Zhang, *Angew. Chem. Int. Ed.* **50**, 11093 (2011).

[36] Z. Zeng, T. Sun, J. Zhu, X. Huang, Z. Yin, G. Lu, Z. Fan, Q. Yan, H. H. Hng and H. Zhang, *Angew. Chem. Int. Ed.* **51**, 9052 (2012).

[37] Y. Zhan, Z. Liu, S. Najmaei, P. M. Ajayan and J. Lou, *Small* **8**, 966 (2012).

[38] D. Kim, D. Sun, W. Lu, Z. Cheng, Y. Zhu, D. Le, T. S. Rahman and L. Bartels, *Langmuir.* **27**, 11650 (2011).

[39] Y.-C. Lin, W. Zhang, J.-K. Huang, K.-K. Liu, Y.-H. Lee, C.-T. Liang, C.-W. Chu and L.-J. Li, *Nanoscale* **4**, 6637 (2012).

[40] X. Wang, H. Feng, Y. Wu and L. Jiao, *J. Am. Chem. Soc.* **135**, 5304 (2013).

[41] A. L. Elías, N. Perea-López, A. Castro-Beltrán, A. Berkdemir, R. Lv, S. Feng, A. D. Long, T. Hayashi, Y. A. Kim, M. Endo, H. R. Gutiérrez, N. R. Pradhan, L. Balicas, T. E. Mallouk, F. López-Urías, H. Terrones and M. Terrones, *ACS Nano.* **7**, 5235 (2013).

[42] Y.-H. Lee, X.-Q. Zhang, W. Zhang, M.-T. Chang, C.-T. Lin, K.-D. Chang, Y.-C. Yu, J. T.-W. Wang, C.-S. Chang, L.-J. Li and T.-W. Lin, *Adv. Mater.* **24**, 2320 (2012).

[43] H. Li, X. Duan, X. Wu, X. Zhuang, H. Zhou, Q. Zhang, X. Zhu, W. Hu, P. Ren, P. Guo, L. Ma, X. Fan, X. Wang, J. Xu, A. Pan and X. Duan, *J. Am. Chem. Soc.* **136**, 3756 (2014).

[44] V. O. Koroteev, L. G. Bulusheva, A. V. Okotrub, N. F. Yudanov and D. V. Vyalikh, *Phys. Status Solid. B* **248**, 2740 (2011).

[45] K.-K. Liu, W. Zhang, Y.-H. Lee, Y.-C. Lin, M.-T. Chang, C.-Y. Su, C.-S. Chang, H. Li, Y. Shi, H. Zhang, C.-S. Lai and L.-J. Li, *Nano Lett.* **12**, 1538 (2012).

[46] S. Wu, C. Huang, G. Aivazian, J. S. Ross, D. H. Cobden and X. Xu, *ACS Nano.* **7**, 2768 (2013).

[47] Q. Feng, Y. Zhu, J. Hong, M. Zhang, W. Duan, N. Mao, J. Wu, H. Xu, F. Dong, F. Lin, C. Jin, C. Wang, J. Zhang and L. Xie, *Adv. Mater.* **26**, 2648 (2014).

[48] H. S. S. Ramakrishna Matte, A. Gomathi, A. K. Manna, D. J. Late, R. Datta, S. K. Pati and C. N. R. Rao, *Angew. Chem. Int. Ed.* **49**, 4059 (2010).

[49] H. S. S. R. Matte, B. Plowman, R. Datta and C. N. R. Rao, *Dalton Trans.* **40**, 10322 (2011).

[50] C. Altavilla, M. Sarno and P. Ciambelli, *Chem. Mater.* **23**, 3879 (2011).

[51] B. Mahler, V. Hoepfner, K. Liao and G. A. Ozin, *J. Am. Chem. Soc.* **136**, 14121 (2014).

[52] D. Yoo, M. Kim, S. Jeong, J. Han and J. Cheon, *J. Am. Chem. Soc.* **136**, 14670 (2014).

[53] A. Pakdel, Y. Bando and D. Golberg, *Chem. Soc. Rev.* **43**, 934 (2014).

[54] W. H. Balmain, *Philos. Mag. Series 3* **21**, 270 (1842).

[55] N. Alem, R. Erni, C. Kisielowski, M. D. Rossell, W. Gannett and A. Zettl, *Phys. Rev. B* **80**, 155425 (2009).

[56] X. Blase, A. De Vita, J. C. Charlier and R. Car, *Phys. Rev. Lett.* **80**, 1666 (1998).

[57] M. Buongiorno Nardelli, C. Brabec, A. Maiti, C. Roland and J. Bernholc, *Phys. Rev. Lett.* **80**, 313 (1998).

[58] L. H. Li, Y. Chen, G. Behan, H. Zhang, M. Petravic and A. M. Glushenkov, *J. Mater. Chem.* **21**, 11862 (2011).

[59] J. Y. Huang, H. Yasuda and H. Mori, *J. Am. Ceram. Soc.* **83**, 403 (2000).

[60] C. Zhi, Y. Bando, C. Tang, H. Kuwahara and D. Golberg, *Adv. Mater.* **21**, 2889 (2009).

[61] W.-Q. Han, L. Wu, Y. Zhu, K. Watanabe and T. Taniguchi, *Appl. Phys. Lett.* **93**, 223103 (2008).

[62] Y. Lin, T. V. Williams, T.-B. Xu, W. Cao, H. E. Elsayed-Ali and J. W. Connell, *J. Phys. Chem. C* **115**, 2679 (2011).

[63] Y. Lin, T. V. Williams and J. W. Connell, *J. Phys. Chem. Lett.* **1**, 277 (2010).

[64] Y. Wang, Z. Shi and J. Yin, *J. Mater. Chem.* **21**, 11371 (2011).

[65] X. Li, X. Hao, M. Zhao, Y. Wu, J. Yang, Y. Tian and G. Qian, *Adv. Mater.* **25**, 2200 (2013).

[66] A. C. Adams, *J. Electrochem. Soc.* **128**, 1378 (1981).

[67] W. Auwärter, H. U. Suter, H. Sachdev and T. Greber, *Chem. Mater.* **16**, 343 (2004).

[68] G. Constant and R. Feurer, *J. Less Common Met.* **82**, 113 (1981).

[69] M. T. Paffett, R. J. Simonson, P. Papin and R. T. Paine, *Surface Science* **232**, 286 (1990).

[70] E. Rokuta, Y. Hasegawa, K. Suzuki, Y. Gamou, C. Oshima and A. Nagashima, *Phys. Rev. Lett.* **79**, 4609 (1997).

[71] A. Pakde l, Y. Bando and D. Golberg, *Chem. Soc. Rev.* **43**, 934 (2014).

[72] A. B. Preobrajenski, A. S. Vinogradov and N. Mårtensson, *Surf. Sci.* **582**, 21 (2005).

[73] K. K. Kim, A. Hsu, X. Jia, S. M. Kim, Y. Shi, M. Hofmann, D. Nezich, J. F. Rodriguez-Nieva, M. Dresselhaus, T. Palacios and J. Kong *Nano Lett.* **12**, 161 (2012).

[74] P. Sutter, J. Lahiri, P. Albrecht and E. Sutter, *ACS Nano.* **5**, 7303 (2011).

[75] M. Corso, W. Auwärter, M. Muntwiler, A. Tamai, T. Greber and J. Osterwalder, *Science* **303**, 217 (2004).

[76] L. Song, L. Ci, H. Lu, P. B. Sorokin, C. Jin, J. Ni, A. G. Kvashnin, D. G. Kvashnin, J. Lou, B. I. Yakobson and P. M. Ajayan, *Nano. Lett.* **10**, 3209 (2010).

[77] J. Yu, L. Qin, Y. Hao, S. Kuang, X. Bai, Y.-M. Chong, W. Zhang and E. Wang, *ACS Nano.* **4**, 414 (2010).

[78] A. Pakdel, C. Zhi, Y. Bando, T. Nakayama and D. Golberg, *ACS Nano.* **5**, 6507 (2011).

[79] Z. Liu, L. Song, S. Zhao, J. Huang, L. Ma, J. Zhang, J. Lou and P. M. Ajayan, *Nano Lett.* **11**, 2032 (2011).

[80] W.-Q. Han, H.-G. Yu and Z. Liu, *Appl. Phys. Lett.* **98**, 203112 (2011).

[81] X. Wang, C. Zhi, L. Li, H. Zeng, C. Li, M. Mitome, D. Golberg and Y. Bando, *Adv. Mater.* **23**, 4072 (2011).

[82] P. Sutter, J. Lahiri, P. Zahl, B. Wang and E. Sutter, *Nano Lett.* **13**, 276 (2013).

[83] H. Zeng, C. Zhi, Z. Zhang, X. Wei, X. Wang, W. Guo, Y. Bando and D. Golberg, *Nano Lett.* **10**, 5049 (2010).

[84] Z. Zhao, Z. Yang, Y. Wen and Y. Wang, *J. Am. Ceram. Soc.* **94**, 4496 (2011).

[85] A. Nag, K. Raidongia, K. P. S. S. Hembram, R. Datta, U. V. Waghmare and C. N. R. Rao, *ACS Nano.* **4**, 1539 (2010).

[86] N. Kumar, K. Moses, K. Pramoda, S. N. Shirodkar, A. K. Mishra, U. V. Waghmare, A. Sundaresan and C. N. R. Rao, *J. Mater. Chem. A* **1**, 5806 (2013).

[87] M. P. Levendorf, C.-J. Kim, L. Brown, P. Y. Huang, R. W. Havener, D. A. Muller and J. Park, *Nature* **488**, 627 (2012).

[88] L. Ci, L. Song, C. Jin, D. Jariwala, D. Wu, Y. Li, A. Srivastava, Z. F. Wang, K. Storr, L. Balicas, F. Liu and P. M. Ajayan, *Nat. Mater.* **9**, 430 (2010).

[89] N. Kumar, K. S. Subrahmanyam, P. Chaturbedy, K. Raidongia, A. Govindaraj, K. P. S. S. Hembram, A. K. Mishra, U. V. Waghmare and C. N. R. Rao, *ChemSusChem* **4**, 1662 (2011).

[90] N. Kumar, K. Raidongia, A. K. Mishra, U. V. Waghmare, A. Sundaresan and C. N. R. Rao, *J. Solid State Chem.* **184**, 2902 (2011).

[91] M. Kota, N. S. Sharmila, U. V. Waghmare and C. N. R. Rao, *Mater. Res. Express* **1**, 025603 (2014).

[92] W. Liu, K. C. Lukas, K. McEnaney, S. Lee, Q. Zhang, C. P. Opeil, G. Chen and Z. Ren, *Energy Environ. Sci.* **6**, 552 (2013).

[93] Y. Sun, H. Cheng, S. Gao, Q. Liu, Z. Sun, C. Xiao, C. Wu, S. Wei and Y. Xie, *J. Am. Chem. Soc.* **134**, 20294 (2012).

[94] Y. Min, G. D. Moon, B. S. Kim, B. Lim, J.-S. Kim, C. Y. Kang and U. Jeong, *J. Am. Chem. Soc.* **134**, 2872 (2012).

[95] M. K. Jana, K. Biswas and C. N. R. Rao, *Chem. Eur. J.* **19**, 9110 (2013).

[96] D. Teweldebrhan, V. Goyal and A. A. Balandin, *Nano Lett.* **10**, 1209 (2010).

[97] Y. Zhao, R. W. Hughes, Z. Su, W. Zhou and D. H. Gregory, *Angew. Chem. Int. Ed.* **50**, 10397 (2011).

[98] G. Hao, X. Qi, Y. Liu, Z. Huang, H. Li, K. Huang, J. Li, L. Yang and J. Zhong, *J. App. Phys.* **111**, 114312 (2012).

[99] G.-H. Dong, Y.-J. Zhu and L.-D. Chen, *J. Mater. Chem.* **20**, 1976 (2010).

[100] G.-H. Dong, Y.-J. Zhu, G.-F. Cheng and Y.-J. Ruan, *J. Alloys Compd.* **550**, 164 (2013).

[101] Y. Zhang, J. Lu, S. Shen, H. Xu and Q. Wang, *Chem. Commun.* **47**, 5226 (2011).

[102] C. Zhai, N. Du and H. Z. D. Yang, *Chem. Commun.* **47**, 1270 (2011).

[103] M. K. Jana, H. B. Rajendra, A. J. Bhattacharyya and K. Biswas, *CrystEngComm* **16**, 3994 (2014).

[104] M. Osada and T. Sasaki, *Adv. Mater.* **24**, 210 (2012).

[105] W.-S. Wang, D.-H. Wang, W.-G. Qu, L.-Q. Lu and A.-W. Xu, *J. Phys. Chem. C*, **116**, 19893 (2012).

[106] J. S. Chen and X. W. Lou, *Mater. Today* **15**, 246 (2012).

[107] T. Sasaki and M. Watanabe, *J. Am. Chem. Soc.* **120**, 4682 (1998).

[108] M. B. Sreedhara, H. S. S. R. Matte, A. Govindaraj and C. N. R. Rao, *Chem. Asian J.* **8**, 2430 (2013).

[109] M. B. Sreedhara, B. E. Prasad, M. Moirangthem, R. Murugavel and C. N. R. Rao, *J. Solid State Chem.* **224**, 21 (2015).

[110] R. A. Mohan Ram, P. Ganguly and C. N. R. Rao, *J. Solid State Chem.* **70**, 82 (1987).

[111] C. N. R. Rao, P. Ganguly, K. K. Singh and R. A. M. Ram, *J. Solid State Chem.* **72**, 14 (1988).

[112] J. A. Wilson and A. D. Yoffe, *Adv. Phys.* **18**, 193 (1969).

[113] A. Splendiani, L. Sun, Y. Zhang, T. Li, J. Kim, C.-Y. Chim, G. Galli and F. Wang, *Nano Lett.* **10**, 1271 (2010).

[114] T. Korn, S. Heydrich, M. Hirmer, J. Schmutzler and C. Schüller, *Appl. Phys. Lett.* **99**, 102109 (2011).

[115] K. F. Mak, C. Lee, J. Hone, J. Shan and T. F. Heinz, *Phy. Rev. Lett.* **105**, 136805 (2010).

[116] E. S. Kadantsev and P. Hawrylak, *Solid State Commun.* **152**, 909 (2012).

[117] Z. Y. Zhu, Y. C. Cheng and U. Schwingenschlögl, *Phys. Rev. B* **84**, 153402 (2011).

[118] R. Coehoorn, C. Haas, J. Dijkstra, C. J. F. Flipse, R. A. de Groot and A. Wold, *Phy. Rev. B* **35**, 6195 (1987).

[119] R. Coehoorn, C. Haas and R. A. de Groot, *Phy. Rev. B* **35**, 6203 (1987).

[120] A. Ramasubramaniam, D. Naveh and E. Towe, *Phy. Rev. B* **84**, 205325 (2011).

[121] E. Scalise, M. Houssa, G. Pourtois, V. Afanas'ev and A. Stesmans, *Nano Res.* **5**, 43 (2012).

[122] S. Bhattacharyya and A. K. Singh, *Phy. Rev. B* **86**, 075454 (2012).

[123] W. S. Yun, S. W. Han, S. C. Hong, I. G. Kim and J. D. Lee, *Phy. Rev. B* **85**, 033305 (2012).

[124] A. P. Nayak, S. Bhattacharyya, J. Zhu, J. Liu, X. Wu, T. Pandey, C. Jin, A. K. Singh, D. Akinwande and J.-F. Lin, *Nat Commun.* **5**, 3731 (2014).

[125] S. Dey, H. S. S. R. Matte, S. N. Shirodkar, U. V. Waghmare and C. N. R. Rao, *Chemistry–An Asian Journal* **8**, 1780 (2013).

[126] B. Chakraborty, A. Bera, D. V. S. Muthu, S. Bhowmick, U. V. Waghmare and A. K. Sood, *Phy. Rev. B* **85**, 161403 (2012).

[127] A. Pakdel, X. Wang, C. Zhi, Y. Bando, K. Watanabe, T. Sekiguchi, T. Nakayama and D. Golberg, *J. Mater. Chem.* **22**, 4818 (2012).

[128] Z.-G. Chen, J. Zou, G. Liu, F. Li, Y. Wang, L. Wang, X.-L. Yuan, T. Sekiguchi, H.-M. Cheng and G. Q. Lu, *ACS Nano.* **2**, 2183 (2008).

[129] L. Qin, J. Yu, S. Kuang, C. Xiao and X. Bai, *Nanoscale* **4**, 120 (2012).

[130] T. J. Wieting and J. L. Verble, *Phys. Rev. B* **3**, 4286 (1971).

[131] A. Molina-Sánchez and L. Wirtz, *Phy. Rev. B* **84**, 155413 (2011).

[132] C. Lee, H. Yan, L. E. Brus, T. F. Heinz, J. Hone and S. Ryu, *ACS Nano.* **4**, 2695 (2010).

[133] D. J. Late, S. N. Shirodkar, U. V. Waghmare, V. P. Dravid and C. N. R. Rao, *ChemPhysChem* **15**, 1592 (2014).

[134] S. Tongay, J. Zhou, C. Ataca, K. Lo, T. S. Matthews, J. Li, J. C. Grossman and J. Wu, *Nano Lett.* **12**, 5576 (2012).

[135] H. Sahin, S. Tongay, S. Horzum, W. Fan, J. Zhou, J. Li, J. Wu and F. M. Peeters, *Phy. Rev. B* **87**, 165409 (2013).

[136] S. Sahoo, A. P. S. Gaur, M. Ahmadi, M. J. F. Guinel and R. S. Katiyar, *J. Phys. Chem. C* **117**, 9042 (2013).

[137] N. A. Lanzillo, A. Glen Birdwell, M. Amani, F. J. Crowne, P. B. Shah, S. Najmaei, Z. Liu, P. M. Ajayan, J. Lou, M. Dubey, S. K. Nayak, apos and T. P. Regan, *Appl. Phys. Lett.* **103**, 093102 (2013).

[138] S. Najmaei, Z. Liu, P. M. Ajayan and J. Lou, *Appl. Phys. Lett.* **100**, 013106 (2012).

[139] K. Vasu, H. S. S. R. Matte, S. N. Shirodkar, V. Jayaram, K. P. J. Reddy, U. V. Waghmare and C. N. R. Rao, *Chem. Phys. Lett.* **582**, 105 (2013).

[140] M. K. Jana, A. Singh, D. J. Late, C. R. Rajamathi, K. Biswas, C. Felser, U. V. Waghmare and C. N. R. Rao, *J. Phys. Condens. Matter.* **27**, 285401 (2015).

[141] S. Tongay, H. Sahin, C. LKo, A. uce, W. Fan, K. Liu, J. Zhou, Y.-S. Huang, C.-H. Ho, J. Yan, D. F. Ogletree, S. Aloni, J. Ji, S. Li, J. Li, F. M. Peeters and J. Wu, *Nat Commun.* **5**, 3252 (2014).

[142] R. V. Gorbachev, I. Riaz, R. R. Nair, R. Jalil, L. Britnell, B. D. Belle, E. W. Hill, K. S. Novoselov, K. Watanabe, T. Taniguchi, A. K. Geim and P. Blake, *Small*, **7**, 465 (2011).

[143] A. M. Panich, A. I. Shames, R. Rosentsveig and R. Tenne, *J. Phys. Condens. Matter.* **21**, 395301 (2009).

[144] P. Murugan, V. Kumar, Y. Kawazoe and N. Ota, *Phys. Rev. A* **71**, 063203 (2005).

[145] J. Zhang, J. M. Soon, K. P. Loh, J. Yin, J. Ding, M. B. Sullivian and P. Wu, *Nano Lett.* **7**, 2370 (2007) .

[146] D. Li, C. Zhang, G. Du, R. Zeng, S. Wang, Z. Guo, Z. Chen and H. Liu, *J. Chin. Chem. Soc.* **59**, 1196 (2012).

[147] S. Mathew, K. Gopinadhan, T. K. Chan, X. J. Yu, D. Zhan, L. Cao, A. Rusydi, M. B. H. Breese, S. Dhar, Z. X. Shen, T. Venkatesan and J. T. L. Thong, *Appl. Phys. Lett.* **101**, 102103 (2012).

[148] C. N. R. Rao, H. S. S. R. Matte, K. S. Subrahmanyam and U. Maitra, *Chem. Sci.* **3**, 45 (2012).

[149] S. Tongay, S. S. Varnoosfaderani, B. R. Appleton, J. Wu and A. F. Hebard, *Appl. Phys. Lett.* **101**, 123105 (2012).

[150] Y. Li, Z. Zhou, S. Zhang and Z. Chen, *J. Am. Chem. Soc.* **130**, 16739 (2008).

[151] H. Li, X. Qi, J. Wu, Z. Zeng, J. Wei and H. Zhang, *ACS Nano.* **7**, 2842 (2013).

[152] M. S. Si and D. S. Xue, *Phys. Rev. B* **75**, 193409 (2007).

[153] P. Dev, Y. Xue and P. Zhang, *Phys. Rev. Lett.* **100**, 117204 (2008).

[154] B. Song, J. C. Han, J. K. Jian, H. Li, Y. C. Wang, H. Q. Bao, W. Y. Wang, H. B. Zuo, X. H. Zhang, S. H. Meng and X. L. Chen, *Phys. Rev. B* **80**, 153203 (2009).

[155] H. Zeng, J. Dai, W. Yao, D. Xiao and X. Cui, *Nat. Nanotechnol.* **7**, 490 (2012).

[156] D. Xiao, G.-B. Liu, W. Feng, X. Xu and W. Yao, *Phys. Rev. Lett.* **108**, 196802 (2012).

[157] Z. Y. Zhu, Y. C. Cheng and U. Schwingenschlögl, *Phys. Rev. B* **84**, 153402 (2011).

[158] T. Cheiwchanchamnangij and W. R. L. Lambrecht, *Phys. Rev. B* **85**, 205302 (2012).

[159] K. F. Mak, K. He, C. Lee, G. H. Lee, J. Hone, T. F. Heinz and J. Shan, *Nat. Mater.* **12**, 207 (2013).

[160] F. Schwierz, *Nat. Nanotechnol.* **5**, 487 (2010).

[161] B. Radisavljevic, A. Radenovic, J. Brivio, V. Giacometti and A. Kis, *Nat. Nanotechnol.* **6**, 147 (2011).

[162] H. Liu and P. D. Ye. arXiv:1112.4397 [cond.mat.mtrl-sci]. (2011).

[163] S.-W. Min, H. S. Lee, H. J. Choi, M. K. Park, T. Nam, H. Kim, S. Ryu and S. Im, *Nanoscale* **5**, 548 (2013).

[164] J. Pu, Y. Yomogida, K.-K. Liu, L.-J. Li, Y. Iwasa and T. Takenobu *Nano Lett.* **12**, 4013 (2012).

[165] Y. Zhang, J. Ye, Y. Matsuhashi and Y. Iwasa, *Nano Lett.* **12**, 1136 (2012).

[166] K. Lee, H.-Y. Kim, M. Lotya, J. N. Coleman, G.-T. Kim and G. S. Duesberg, *Adv. Mater.* **23**, 4178 (2011).

[167] S. Ghatak, A. N. Pal and A. Ghosh, *ACS Nano.* **5**, 7707 (2011).

[168] B. Radisavljevic, M. B. Whitwick and A. Kis, *Appl. Phys. Lett.* **101**, 043103 (2012).

[169] W. J. Yu, Z. Li, H. Zhou, Y. Chen, Y. Wang, Y. Huang and X. Duan *Nat Mater.* **12**, 246 (2013).

[170] D. J. Late, B. Liu, H. S. S. R. Matte, V. P. Dravid, C. N. R. and Rao, *ACS Nano.* **6**, 5635 (2012).

[171] Q. Hao, P. Lijia, Y. Zongni, L. Junjie, S. Yi and W. Xinran, *Appl. Phys. Lett.* **100**, 123104 (2012).

[172] S. Larentis, B. Fallahazad and E. Tutuc, *Appl. Phys. Lett.* **101**, 223104 (2012).

[173] V. Podzorov, M. E. Gershenson, C. Kloc, R. Zeis and E. Bucher *Appl. Phys. Lett.* **84**, 3301 (2004).

[174] H. Fang, S. Chuang, T. C. Chang, K. Takei, T. Takahashi and A. Javey, *Nano Lett.* **12**, 3788 (2012).

[175] H. Li, J. Wu, Z. Yin and H. Zhang, *Acc. Chem. Res.* **47**, 1067 (2014).

[176] Q. He, Z. Zeng, Z. Yin, H. Li, S. Wu, X. Huang and H. Zhang *Small* **8**, 2994 (2012).

[177] S. Wu, Z. Zeng, Q. He, Z. Wang, S. J. Wang, Y. Du, Z. Yin, X. Sun, W. Chen and H. Zhang, *Small* **8**, 2264 (2012).

[178] D. J. Late, Y.-K. Huang, B. Liu, J. Acharya, S. N. Shirodkar, J. Luo, A. Yan, D. Charles, U. V. Waghmare, V. P. Dravid and C. N. R. Rao, *ACS Nano.* **7**, 4879 (2013).

[179] W. Choi, M. Y. Cho, A. Konar, J. H. Lee, G.-B. Cha, S. C. Hong, S. Kim, J. Kim, D. Jena, J. Joo and S. Kim, *Adv. Mater.* **24**, 5832 (2012).

[180] W. Sik Hwang, M. Remskar, R. Yan, V. Protasenko, K. Tahy, S. Doo Chae, P. Zhao, A. Konar, H. Xing, A. Seabaugh and D. Jena, *Appl. Phys. Lett.* **101**, 013107 (2012).

[181] C. R. Dean, A. F. Young, I. Meric, C. Lee, L. Wang, S. Sorgenfrei, K. Watanabe, T. Taniguchi, P. Kim, K. L. Shepard and J. Hone, *Nat. Nanotechnol.* **5**, 722 (2010).

[182] J. Xue, J. Sanchez-Yamagishi, D. Bulmash, P. Jacquod, A. Deshpande, K. Watanabe, T. Taniguchi, P. Jarillo-Herrero and B. J. LeRoy *Nat. Mater.* **10**, 282 (2011).

[183] A. S. Mayorov, R. V. Gorbachev, S. V. Morozov, L. Britnell, R. Jalil, L. A. Ponomarenko, P. Blake, K. S. Novoselov, K. Watanabe, T. Taniguchi and A. K. Geim, *Nano Lett.* **11**, 2396.

[184] H. Xu, J. Wu, Y. Chen, H. Zhang and J. Zhang, *Chem. Asian J.* **8**, 2446 (2013).

[185] G.-H. Lee, Y.-J. Yu, X. Cui, N. Petrone, C.-H. Lee, M. S. Choi, D.-Y. Lee, C. Lee, W. J. Yoo, K. Watanabe, T. Taniguchi, C. Nuckolls, P. Kim and J. Hone, *ACS Nano.* **7**, 7931 (2013).

[186] L. Britnell, R. V. Gorbachev, R. Jalil, B. D. Belle, F. Schedin, M. I. Katsnelson, L. Eaves, S. V. Morozov, A. S. Mayorov, N. M. R. Peres, A. H. Castro Neto, J. Leist, A. K. Geim, L. A. Ponomarenko and K. S. Novoselov, *Nano Lett.* **12**, 1707 (2012).

[187] J. M. Soon and K. P. Loh, *Electrochem. Solid-State Lett.* **10**, A250 (2007).

[188] H. Tang, J. Wang, H. Yin, H. Zhao, D. Wang and Z. Tang, *Adv. Mater.* **27**, 1117 (2014).

[189] K.-J. Huang, L. Wang, Y.-J. Liu, Y.-M. Liu, H.-B. Wang, T. Gan and L.-L. Wang, *Int. J. Hydrog. Energy* **38**, 14027 (2013).

[190] K. Gopalakrishnan, K. Pramoda, U. Maitra, U. Mahima, M. A. Shah and C. N. R. Rao, *Nanomater. Energy* **4**, 9 (2014).

[191] K. Gopalakrishnan, S. Sultan, A. Govindaraj and C. N. R. Rao, *Nano Energy* **12**, 52 (2015).

[192] H. Tang, J. Wang, H. Yin, H. Zhao, D. Wang and Z. Tang, *Advanced Materials* **27**, 1117 (2014).

[193] R. Kumar, K. Gopalakrishnan, I. Ahmad and C. N. R. Rao, *Adv. Funct. Mater.* **25**, 5910 (2015).

[194] K. Gopalakrishnan, K. Moses, A. Govindaraj and C. N. R. Rao, *Solid State Commun.* **175-176**, 43 (2013).

[195] R. R. Haering, J. A. R. Stiles and K. Brandt. (Ed.: US Patent 4224390) (1980).

[196] G. Du, Z. Guo, S. Wang, R. Zeng, Z. Chen and H. Liu, *Chem. Commun.* **46**, 1106 (2010).

[197] C. Feng, J. Ma, H. Li, R. Zeng, Z. Guo and H. Liu, *Mater. Res. Bull.* **44**, 1811 (2009).

[198] H. Hwang, H. Kim and J. Cho, *Nano Lett.* **11**, 4826 (2011).

[199] C. Feng, L. Huang, Z. Guo and H. Liu, *Electrochem. Commun.* **9**, 119 (2007).

[200] J.-w. Seo, Y.-w. Jun, S.-w. Park, H. Nah, T. Moon, B. Park, J.-G. Kim, Y. J. Kim and J. Cheon, *Angew. Chem. Int. Ed.* **46**, 8828 (2007).

[201] R. Bhandavat, L. David and G. Singh, *J. Phys. Chem. Lett.* **3**, 1523 (2012).

[202] H. Liu, D. Su, G. Wang and S. Z. Qiao, *J. Mater. Chem.* **22**, 17437 (2012).

[203] A. Ghosh, K. S. Subrahmanyam, K. S. Krishna, S. Datta, A. Govindaraj, S. K. Pati and C. N. R. Rao, *J. Phys. Chem. C* **112**, 15704 (2008).

[204] M. Winter and R. J. Brodd, *Chem. Rev.* **104**, 4245 (2004).

[205] K. Chang and W. Chen, *Chem. Commun.* **47**, 4252 (2011).

[206] K. Chang, D. Geng, X. Li, J. Yang, Y. Tang, M. Cai, R. Li and X. Sun, *Adv. Energy Mater.* **3**, 839 (2013).

[207] K. Shiva, H. S. S. Ramakrishna Matte, H. B. Rajendra, A. J. Bhattacharyya and C. N. R. Rao, *Nano Energy* **2**, 787 (2013).

[208] L. David, R. Bhandavat and G. Singh, *ACS Nano.* **8**, 1759 (2014).

[209] S. Sen, K. Moses, A. J. Bhattacharyya and C. N. R. Rao, *Chem. Asian J.* **9**, 100 (2013).

[210] R. Prins, *Adv. Catal.* **46**, 399 (2001).

[211] J. V. Lauritsen, M. Nyberg, J. K. Nørskov, B. S. Clausen, H. Topsøe, E. Lægsgaard and F. Besenbacher, *J. Catal.* **224**, 94 (2004).

[212] Y. Tsverin, R. Popovitz-Biro, Y. Feldman, R. Tenne, M. R. Komarneni. Z. Yu, A. Chakradhar, A. Sand and U. Burghaus, *Mater. Res. Bull.* **47**, 1653 (2012).

[213] B. G. Rao, H. S. S. R. Matte, P. Chaturbedy and C. N. R. Rao, *ChemPlusChem* **78**, 419 (2013).

[214] X. Zhong, H. Yang, S. Guo, S. Li, G. Gou, Z. Niu, Z. Dong, Y. Lei, J. Jin, R. Li and J. Ma, *J. Mater. Chem.* **22**, 13925 (2012).

[215] B. Hinnemann, P. G. Moses, J. Bonde, K. P. Jørgensen, J. H. Nielsen, S. Horch, I. Chorkendorff and J. K. Nørskov, *J. Am. Chem. Soc.* **127**, 5308 (2005).

[216] T. F. Jaramillo, K. P. Jørgensen, J. Bonde, J. H. Nielsen, S. Horch and I. Chorkendorff, *Science* **317**, 100 (2007).

[217] C. Ataca and S. Ciraci, *Phys. Rev. B* **85**, 195410 (2012).

[218] D. Merki and X. Hu, *Energy Environ. Sci.* **4**, 3878 (2011).

[219] A. B. Laursen, S. Kegnaes, S. Dahl and I. Chorkendorff, *Energy Environ. Sci.* **5**, 5577 (2012).

[220] D. Merki, S. Fierro, H. Vrubel and X. Hu, *Chem. Sci.* **2**, 1262 (2011).

[221] J. Kibsgaard, Z. Chen, B. N. Reinecke and T. F. Jaramillo, *Nat. Mater.* **11**, 963 (2012).

[222] V. W.-H. Lau, A. F. Masters, A. M. Bond and T. Maschmeyer, *Chem. Eur. J.* **18**, 8230 (2012).

[223] Y. Li, H. Wang, L. Xie, Y. Liang, G. Hong and H. Dai, *J. Am. Chem. Soc.* **133**, 7296 (2011).

[224] E. G. S. Firmiano, M. A. L. Cordeiro, A. C. Rabelo, C. J. Dalmaschio. A. N. Pinheiro, E. C. Pereira and E. R. Leite, *Chem. Commun.* **48**, 7687 (2012).

[225] X. Zong, G. Wu, H. Yan, G. Ma, J. Shi, F. Wen, L. Wang and C. Li, *J. Phys. Chem. C* **114**, 1963 (2010).

[226] X. Zong, H. Yan, G. Wu, G. Ma, F. Wen, L. Wang and C. Li, *J. Am. Chem. Soc.* **130**, 7176 (2008).

[227] F. A. Frame and F. E. Osterloh, *J. Phys. Chem. C* **114**, 10628 (2010).

[228] W. Zhou, Z. Yin, Y. Du, X. Huang, Z. Zeng, Z. Fan, H. Liu, J. Wang and H. Zhang. (2012) *Small* **9**, 140.

[229] S. Min and G. Lu, *J. Phys. Chem. C* **116**, 25415 (2012).

[230] Q. Xiang, J. Yu and M. Jaroniec, *J. Am. Chem. Soc.* **134**, 6575 (2012).

[231] F. Meng, J. Li, S. K. Cushing, M. Zhi and N. Wu, *J. Am. Chem. Soc.* **135**, 10286 (2013).

[232] M. A. Lukowski, A. S. Daniel, C. R. English, F. Meng, A. Forticaux, R. J. Hamers and S. Jin, *Energy Environ. Sci.* **7**, 2608 (2014).

[233] D. Voiry, M. Salehi, R. Silva, T. Fujita, M. Chen, T. Asefa, V. B. Shenoy, G. Eda and M. Chhowalla, *Nano Lett.* **13**, 6222 (2013).

[234] S. Wang, L. Zhang, Z. Xia, A. Roy, D. W. Chang, J.-B. Baek and L. Dai, *Angew. Chem. Int. Ed.* **51**, 4209 (2012).

[235] K. Moses, V. Kiran, S. Sampath and C. N. R. Rao, *Chem. Asian J.* **9**, 838 (2014).

[236] M. Chhetri, S. Maitra, H. Chakraborty, U. V. Waghmare and C. N. R. Rao, *Energy Environ. Sci.* **9**, 95 (2016).

[237] M. N. Ali, J. Xiong, S. Flynn, J. Tao, Q. D. Gibson, L. M. Schoop, T. Liang, N. Haldolaarachchige, M. Hirschberger, N. P. Ong and R. J. Cava, *Nature* **514**, 205 (2014).

[238] I. Pletikosić, M. N. Ali, A. V. Fedorov, R. J. Cava and T. Valla, *Phys Rev. Lett.* **113**, 216601.

[239] P. L. Cai, J. Hu, L. P. He, J. Pan, X. C. Hong, Z. Zhang, J. Zhang, J. Wei, Z. Q. Mao and S. Y. Li, *Phys. Rev. Lett.* **115**, 057202 (2015).

[240] X.-C. Pan, X. Chen, H. Liu, Y. Feng, Z. Wei, Y. Zhou, Z. Chi, L. Pi, F. Yen, F. Song, X. Wan, Z. Yang, B. Wang, G. Wang and Y. Zhang, *Nat Commun.* **6**, 7805 (2015).

[241] D. Kang, Y. Zhou, W. Yi, C. Yang, J. Guo, Y. Shi, S. Zhang, Z. Wang, C. Zhang, S. Jiang, A. Li, K. Yang, Q. Wu, G. Zhang, L. Sun and Z. Zhao, *Nat Commun.* **6**, 7804 (2015).

[242] B. Brown, *Acta Cryst.* **20**, 268 (1966).

[243] S. Cho, S. Kim, J. H. Kim, J. Zhao, J. Seok, D. H. Keum, J. Baik, D.-H. Choe, K. J. Chang, K. Suenaga, S. W. Kim, Y. H. Lee and H. Yang, *Science* **349**, 625 (2015).

[244] D. H. Keum, S. Cho, J. H. Kim, D.-H. Choe, H.-J. Sung, M. Kan, H. Kang, J.-Y. Hwang, S. W. Kim, H. Yang, K. J. Chang and Y. H. Lee, *Nat. Phys.* **11**, 482 (2015).

[245] J. C. Park, S. J. Yun, H. Kim, J.-H. Park, S. H. Chae, S.-J. An, J.-G. Kim, S. M. Kim, K. K. Kim and Y. H. Lee, *ACS Nano.* **9**, 6548 (2015).

[246] X. Zhang, X.-F. Qiao, W. Shi, J.-B. Wu, D.-S. Jiang and P.-H. Tan, *Chem. Soc. Rev.* **44**, 2757 (2015).

[247] S. Fathipour, N. Ma, W. S. Hwang, V. Protasenko, S. Vishwanath, H. G. Xing, H. Xu, D. Jena, J. Appenzeller and A. Seabaugh, *Appl. Phys. Lett.* **105**, 192101 (2014).

[248] Y.-F. Lin, Y. Xu, S.-T. Wang, S.-L. Li, M. Yamamoto, A. Aparecido-Ferreira, W. Li, H. Sun, S. Nakaharai, W.-B. Jian, K. Ueno and K. Tsukagoshi, *Adv. Mater.* **26**, 3263 (2014).

[249] S. Nakaharai, M. Yamamoto, K. Ueno, Y.-F. Lin, S.-L. Li and K. Tsukagoshi, *ACS Nano.* **9**, 5976 (2015).

[250] N. R. Pradhan, D. Rhodes, S. Feng, Y. Xin, S. Memaran, B.-H. Moon, H. Terrones, M. Terrones and L. Balicas, *ACS Nano.* **8**, 5911 (2014).

[251] Y. Qi, P. G. Naumov, M. N. Ali, C. R. Rajamathi, O. Barkalov, Y. Sun, C. Shekhar, S.-C. Wu, V. Süß, M. Schmidt, E. Pippel, P. Werner, R. Hillebrand, T. Förster, E. Kampertt, W. Schnelle, S. Parkin, R. J. Cava, C. Felser, B. Yan and S. A. Medvedev. arXiv:1508.03502 [cond-mat.mtrl-sci] (2015).

Chapter 2

Topological Valleytronics

Motohiko Ezawa

Department of Applied Physics,
University of Tokyo, Hongo 7-3-1, Tokyo, Japan
ezawa@ap.t.u-tokyo.ac.jp

Abstract. In this chapter, we review the topological aspect of valleytronics in the two- and three-dimensions. We first take an example of the honeycomb lattice, where K and K' points act as the valley degree of freedom. The valley-selective optical absorption known as the circular dichroism occurs when we apply a circularly polarised light into the honeycomb system. We show that the K and K' points have the opposite Chern numbers. The K and K' points shift and annihilate for highly anisotropic honeycomb lattices. We also review the basic of the Weyl semimetal, where the Weyl and anti-Weyl points act as the valley-degrees of freedom. The Weyl and anti-Weyl points have the opposite monopole charges in the momentum space. Accordingly, the Weyl and anti-Weyl points will never be gapped out without a pair annihilation of them.

2.1 Introduction

The valley represents an isolated region, which can accommodate electrons in the momentum space. Valley-polarised states are realised in the surface state of silicon [1–3], AlAs [4], MoS_2 [5–8], diamond [9, 10] and Bismuth [11]. The notion of valleytronics was introduced [12, 13] and developed [14, 15] in the study of graphene. A recent main target of valleytronics is transition-metal dichalcogenides [7, 8]. The valley-selective pumping due to the circular dichroism is realised in these systems [15, 16, 23]. Valleytronics has been extended to

"

spin-valleytronics by incorporating the spin degree of freedom [17]. It will be realised in silicene, germanene and stanene, which are monolayer silicon, germanium and tin, respectively [18].

Graphene, the two-dimensional (2D) honeycomb carbon, has opened a new field of condensed matter physics. Its low-energy theory is described by the Dirac fermion at the two inequivalent K and K' points. A set of the K and K' points is regarded as a valley-degree of freedom. As in the case of spintronics, the valleytronics extends the idea of electronics by including the valley-degree of freedom.

In this chapter, we first describe the valleytronics in the honeycomb system, where two valleys are analytically formulated. The low-energy theory is well described in terms of the Dirac fermions. We show that the Chern number is quantised to be one-half for each valley with the opposite values in the presence of the staggered potential.

The first step of the valleytronics is how to make valley-polarised states. Such state can be constructed by applying a circularly polarised light to the honeycomb system. Namely, only electrons at the K (K') point are excited by the right-handed (left-handed) circularly polarised light. This valley-selective optical absorption is due to the fact that the two Dirac fermions at the K and the K' points have the opposite chiralities. This is known as the circular dichroism since the optical absorption is selectively occurs at one valley. Recent experiments on transition-metal dichalcogenides have demonstrated this circular dichroism.

Next, we investigate the anisotropic honeycomb lattice, where the positions of the K and K' points shift. At the critical anisotropy point, two Dirac cones at the K and K' points merge. When the anisotropy is larger than this critical point, the system becomes gapped. This is viewed as a pair annihilation of the K and K' points.

We show further examples of lattice models equipped with valleys, which include the Kagome lattice and the multiorbital honeycomb lattice. Interestingly, the band structures of them and the honeycomb lattice are exactly identical except for the perfect flat bands. Accordingly, the same physics and valleytronics are applicable to such systems.

Finally, we summarise recent developments of the Weyl semimetal, which is the three-dimensional (3D) semimetal with a linear dispersion. The Weyl semimetal is described by the 2×2 matrices. It can never be gapped out by adding the mass term, which only shifts the Weyl point. Only exceptional case is a pair annihilation of the Weyl and anti-Weyl points. We interpret it by defining the monopole charge in the momentum space, which characterises the topological nature of the Weyl semimetal. It is shown that the monopole charge is onehalf with the opposite signs for the Weyl and anti-Weyl points.

The honeycomb system and the Weyl semimetal share similar physics although the dimension is different. Namely, they have one pair of valleys which have the opposite chiralities. They are characterised by the topological indices such as the Berry phase in the 2D semimetal, the Chern number in the 2D insulator and the monopole charge in the Weyl semimetal. The position of the valleys can be tuned by controlling the model parameters. The two valleys merge and disappear in the process, which can be interpreted as the annihilation of a pair of valleys. In this process, the topological charge is conserved. In this sense, these phenomena demonstrate a topological nature of the valley. Hence, we name it as "topological valleytronics".

2.2 Honeycomb Lattice

The simplest lattice which has a valley-degrees of freedom is a honeycomb lattice. It consists of two triangular sublattices made of inequivalent lattice sites A and B (Figure 1(a)). The reciprocal lattice is also a honeycomb lattice in the momentum space (Figure 1(b)), which constitutes the Brillouin zone. The tight-binding model for the honeycomb lattice is [19, 20],

$$\hat{H} = -t \sum_{\langle i,j \rangle s} c_{is}^{\dagger} c_{js}, \tag{2.1}$$

where $c_{is}^{\dagger}$ creates an electron with spin polarisation $s = \uparrow \downarrow$ at site i, $\langle i, j \rangle$ runs over all the nearest neighbour hopping sites and t is the

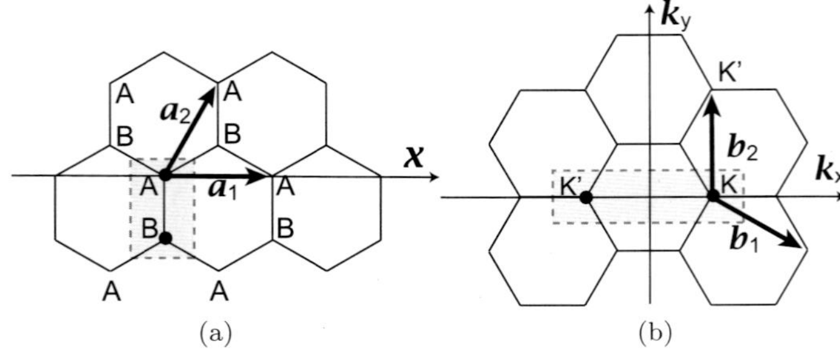

Figure 1. (a) The honeycomb structure, made of two fundamental vectors $\boldsymbol{a}_1$ and $\boldsymbol{a}_2$, consists of two sublattices made of A and B sites. A dotted rectangular represents a unit cell. (b) The reciprocal lattice is also a honeycomb lattice. A dotted rectangular represents a unit cell, which contains two inequivalent points K and K'.

transfer energy. In the momentum representation, the Hamiltonian (2.1) is rewritten as:

$$\hat{H} = \sum_s \int d^2k' \left(c_A^\dagger, c_B^\dagger \right) H \begin{pmatrix} c_A \\ c_B \end{pmatrix}, \tag{2.2}$$

with

$$H = \begin{pmatrix} 0 & f\left(\boldsymbol{k}'\right) \\ f^*\left(\boldsymbol{k}'\right) & 0 \end{pmatrix}, \tag{2.3}$$

where

$$f\left(\boldsymbol{k}'\right) = t \left(e^{-iak_y'/\sqrt{3}} + 2e^{iak_y'/2\sqrt{3}} \cos \frac{ak_x'}{2} \right). \tag{2.4}$$

The energy spectrum is obtained as:

$$E\left(\boldsymbol{k}'\right) = |f\left(\boldsymbol{k}'\right)| = t\sqrt{1 + 4\cos \frac{ak_x'}{2} \cos \frac{\sqrt{3}ak_y'}{2} + 4\cos^2 \frac{ak_x'}{2}}. \tag{2.5}$$

We show the band structure in Figure 2(a). It consists of valleys or cones near the Fermi surface. The cones touch the Fermi surface at two inequivalent points in the Brillouin zone (Figure 2(b)). Namely,

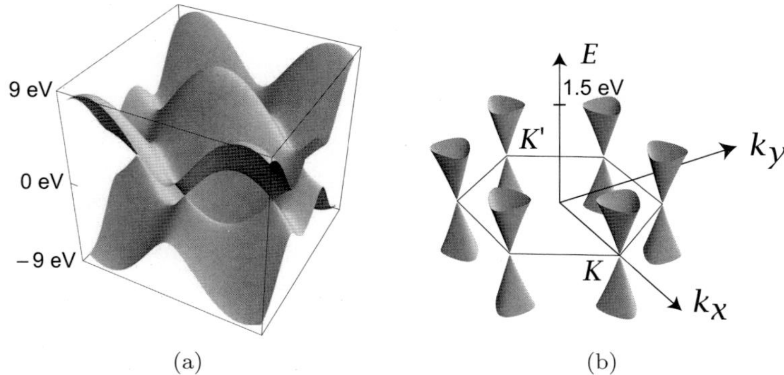

Figure 2. Band structure of graphene. (a) Six valleys are seen in this figure, among which only two valleys are inequivalent. (b) The gap is closed at the K and K' points, where the band structure looks like a cone. It is called the Dirac cone because the dispersion is linear.

the gap closes at $\boldsymbol{k}' = \boldsymbol{K}_\xi$ with $\xi = \pm$, where,

$$\boldsymbol{K}_\xi = \frac{1}{a}\left(\xi\frac{4\pi}{3}, 0\right). \tag{2.6}$$

These points are termed as the K and K' points.

We are interested in physics near the Fermi energy. We make the Taylor expansion of Eq. (2.4) around $\boldsymbol{k}' = \boldsymbol{K}_\xi$. By setting $\boldsymbol{k}' = \boldsymbol{K}_\xi + \boldsymbol{k}$, we obtain

$$f\left(\boldsymbol{k} + \boldsymbol{K}_\xi\right) = \xi k_x - i k_y \quad \text{for } |\boldsymbol{k}| \ll a^{-1}, \tag{2.7}$$

and the dispersion is linear,

$$E\left(\boldsymbol{k} + \boldsymbol{K}_\xi\right) = \pm\hbar v_{\mathrm{F}}\sqrt{k_x^2 + k_y^2} \equiv \pm\hbar v_{\mathrm{F}} k. \tag{2.8}$$

Hence, the low-energy physics near the Fermi energy is described by the Dirac theory,

$$H^\xi = \begin{pmatrix} 0 & \hbar v_{\mathrm{F}}(\xi k_x - i k_y) \\ \hbar v_{\mathrm{F}}(\xi k_x + i k_y) & 0 \end{pmatrix} = \hbar v_{\mathrm{F}}(\xi k_x \tau_x + k_y \tau_y),$$

$$\tag{2.9}$$

where $v_{\mathrm{F}} = \frac{\sqrt{3}}{2\hbar}at$ is the Fermi velocity.

We refer to the K and K' points also as the K_ξ points or the Dirac points. The cone-shaped parts of the energy spectrum are referred to as the Dirac cones (Figure 2). We note that the number of the Dirac cones is always even in the tight-binding theory, as is consistent with the Nielsen–Ninomiya theorem [21].

2.2.1 Berry connection, Berry phase and Berry curvature

We start with the Berry connection, which is the "gauge potential" in the momentum space defined for each band index n by

$$a_\mu^n(\boldsymbol{k}) = -i\langle\psi_n(\boldsymbol{k})|\,\partial_{k_\mu}\psi_n(\boldsymbol{k})\rangle. \tag{2.10}$$

The Berry phase is defined as a loop integration of the Berry connection,

$$\Gamma_n = \sum_k \oint a_\mu^n dk_\mu. \tag{2.11}$$

The Berry curvature is defined by

$$F_n(\boldsymbol{k}) = \partial_{k_x} a_y^n(\boldsymbol{k}) - \partial_{k_y} a_x^n(\boldsymbol{k}), \tag{2.12}$$

which is the "magnetic field". The Chern number is defined by

$$C_n = \frac{1}{2\pi}\int d^2k\, F_n(\boldsymbol{k}). \tag{2.13}$$

All these quantities are defined for each band.

The Hamiltonian (2.9) of the honeycomb system is rewritten as:

$$H^\xi = \xi\hbar v_\mathrm{F} k \begin{pmatrix} 0 & e^{-i\xi\phi} \\ e^{i\xi\phi} & 0 \end{pmatrix}, \tag{2.14}$$

in the polar coordinate $k_x = k\cos\phi, k_y = k\sin\phi$. The eigenvalues are $\xi\hbar v_\mathrm{F} k$ with the eigenstates,

$$\psi_\xi(\boldsymbol{k}) = \frac{1}{\sqrt{2}}\begin{pmatrix} \xi e^{i\xi\phi} \\ 1 \end{pmatrix}. \tag{2.15}$$

There exist a single valence band. The Berry connection is a pure gauge, $a_\mu \propto \partial_\mu \phi$. The Berry phase of this band is calculated as:

$$\Gamma_{\mathrm{B}} = \oint a_\phi d\phi = 2\pi \left(-\frac{\xi}{2}\right) = -\xi\pi,\tag{2.16}$$

which is opposite between the K and K' points. The Berry curvature is singular and the Chern number is ill defined. This is because the system is semimetalic.

The topological insulator is characterised by the Chern number. We calculate the Chern number by including a staggered potential to the honeycomb lattice. The Hamiltonian (2.3) is modified as:

$$H = \begin{pmatrix} m & f\left(\boldsymbol{k}'\right) \\ f^*\left(\boldsymbol{k}'\right) & -m \end{pmatrix}.\tag{2.17}$$

We illustrate the Berry curvature with the use of the tight-binding Hamiltonian (2.2) together with (2.17) in Figure 3(a). The Berry curvature is strictly localised at the K and K' points. Consequently, we may employ the Dirac Hamiltonian to make the topological analysis at each valley, which is indexed by the valley index ξ. Namely, it is possible to assign the Chern number $\mathcal{C}^\xi$ to each valley.

The Dirac Hamiltonian is given by

$$H^\xi = \begin{pmatrix} m & \hbar v_{\mathrm{F}}(\xi k_x - i k_y) \\ \hbar v_{\mathrm{F}}(\xi k_x + i k_y) & -m \end{pmatrix}$$
$$= \hbar v_{\mathrm{F}}(\xi k_x \tau_x + k_y \tau_y) + m\tau_z.\tag{2.18}$$

At the K_ξ point, the wave function of the valence band is given by

$$\psi_v\left(\boldsymbol{k}\right) = \frac{1}{c}\begin{pmatrix} \hbar v_{\mathrm{F}} k(\xi k_x + i k_y) \\ -m + \sqrt{(\hbar v_{\mathrm{F}} k)^2 + m^2} \end{pmatrix},\tag{2.19}$$

with $c = \sqrt{(\hbar v_{\mathrm{F}} k)^2 + \left(-m + \sqrt{(\hbar v_{\mathrm{F}} k)^2 + m^2}\right)^2}$. The Berry connection is calculated as:

$$a_x\left(\boldsymbol{k}\right) = -\xi \frac{k_y}{2k^2}\left(1 + g(k)\right), \quad a_y\left(\boldsymbol{k}\right) = \frac{k_x}{2k^2}\left(1 + g(k)\right),\tag{2.20}$$

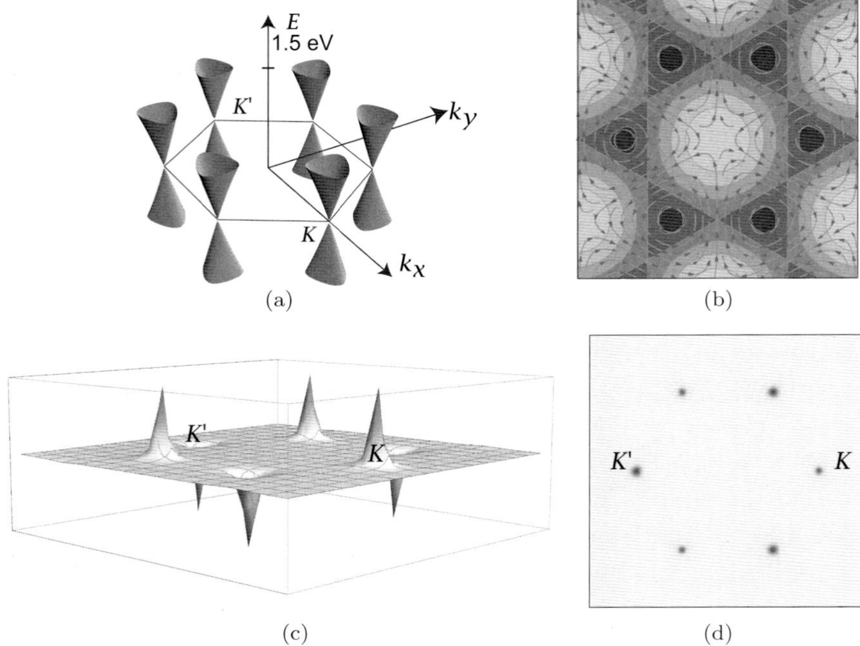

Figure 3. (a) The band structure of the honeycomb lattice. (b) The Berry curvature of the honeycomb lattice. There are vortex structures at the K and K' points, whose directions are opposite. (c) The bird's eye's view and (d) density plot of the Berry curvature of the honeycomb lattice. It is strictly localised at the K and K' points.

with $g(k) = m/\sqrt{(\hbar v_{\mathrm{F}} k)^2 + m^2}$. Substituting (2.20) into (2.12), we obtain

$$F^\xi\left(\boldsymbol{k}\right) = -\frac{\xi m}{2\left(\left(\hbar v_{\mathrm{F}} k\right)^2 + m^2\right)^{3/2}}. \tag{2.21}$$

Consequently, we may calculate the Chern number (2.13) explicitly as

$$\mathcal{C}^\xi = \frac{1}{2\pi} \int_0^\infty 2\pi k\, dk\, F^\xi\left(\boldsymbol{k}\right) = -\frac{\xi}{2}\mathrm{sgn}(m). \tag{2.22}$$

The Chern number is quantised as $\mathcal{C}^\xi = \pm\frac{1}{2}$ at each Dirac point. It is insensitive to a deformation of the band structure provided the gap is open. Since there are two Dirac points, the total Chern number is an integer.

It is well known [22] that the Hall conductance is proportional to the Chern number. It is given by $\sigma_{xy} = \sum_{\xi} \sigma_{xy}^{\xi}$ with

$$\sigma_{xy}^{\xi} = \begin{cases} -\dfrac{\xi}{2}\dfrac{e^2}{2\pi h}\mathrm{sgn}(m) & \text{for} \quad |\mu| < |m| \\[2em] -\dfrac{\xi}{2}\dfrac{|m|}{|\mu|}\dfrac{e^2}{2\pi h}\mathrm{sgn}(m) & \text{for} \quad |\mu| > |m|, \end{cases} \tag{2.23}$$

as a function of the chemical potential. It is quantised between the bulk band gap $|\mu| < |m|$, which is known as the quantum anomalous Hall effect. On the other hand, the Hall conductance is anti-proportional to the chemical potential outside the band gap $|\mu| > |m|$.

2.2.2 Pontryagin number

We present another formula to calculate the Chern number valid for the 2×2 Hamiltonian of the form $H = \boldsymbol{\tau} \cdot \boldsymbol{d}\,(\boldsymbol{k})$. We parametrise $\boldsymbol{d}\,(\boldsymbol{k})$ as

$$\hat{\boldsymbol{d}}\,(\boldsymbol{k}) = \boldsymbol{d}/|\boldsymbol{d}| = (\sin\theta\cos\phi, \sin\theta\sin\phi, \cos\theta). \tag{2.24}$$

The eigen function of the Hamiltonian is unique and given by

$$\psi_v\,(\boldsymbol{k}) = \begin{pmatrix} -e^{-i\phi}\sin\frac{\theta}{2} \\ \cos\frac{\theta}{2} \end{pmatrix}. \tag{2.25}$$

The Berry connection (2.10) and the Berry curvature (2.12) are calculated as:

$$a_{\mu}\,(\boldsymbol{k}) = \frac{1}{2}\,(1 - \cos\theta)\,\partial_{k_{\mu}}\phi, \tag{2.26}$$

$$F\,(\boldsymbol{k}) = \frac{1}{2}\sin\theta\,(\partial_{k_x}\theta\partial_{k_y}\phi - \partial_{k_y}\theta\partial_{k_x}\phi) = \left(\frac{\partial\hat{\boldsymbol{d}}}{\partial k_x} \times \frac{\partial\hat{\boldsymbol{d}}}{\partial k_y}\right) \cdot \hat{\boldsymbol{d}}. \tag{2.27}$$

Hence, the Chern number (2.13) is equivalent to the Pontryagin number,

$$C^{\xi} = \frac{1}{4\pi} \int d^2k \left(\frac{\partial \hat{\boldsymbol{d}}}{\partial k_x} \times \frac{\partial \hat{\boldsymbol{d}}}{\partial k_y} \right) \cdot \hat{\boldsymbol{d}}. \tag{2.28}$$

The Pontryagin number counts how many times the vector $\hat{\boldsymbol{d}}(\boldsymbol{k})$ wraps a sphere as k_μ moves over the whole 2D space $(-\infty < k_\mu < \infty)$. It is interesting that the Chern number is calculable solely by using the vector $\boldsymbol{d}(\boldsymbol{k})$ in the Hamiltonian $H = \boldsymbol{\tau} \cdot \boldsymbol{d}(\boldsymbol{k})$.

We apply the above formula to the present Hamiltonian given by (2.18), which is expressed as $H_s^{\eta} = \boldsymbol{\tau} \cdot \boldsymbol{d}$ with

$$\hat{\boldsymbol{d}}(\boldsymbol{k}) = \frac{1}{\sqrt{(\hbar v_{\mathrm{F}} k)^2 + m^2}} (\xi \hbar v_{\mathrm{F}} k_x, \hbar v_{\mathrm{F}} k_y, m), \tag{2.29}$$

or

$$\hat{d}_x \pm i\hat{d}_y = \sqrt{1 - \sigma^2(k)}\, e^{i\xi\phi}, \quad \hat{d}_z = \sigma(k) \tag{2.30}$$

in the polar coordinate $\boldsymbol{k} = (k, \phi)$. It agrees with the parametrisation (2.24) with $\sigma(k) = \cos\theta$.

Then, the Pontryagin number (2.28) is rewritten as:

$$C^{\xi} = \frac{\xi}{4\pi} \int d^2k \; \varepsilon_{ij} \partial_i \sigma \partial_j \theta = -\frac{\xi}{2} \int_0^1 d\sigma = -\frac{\xi}{2} \mathrm{sgn}(m), \tag{2.31}$$

which agrees with (2.22). The pseudospin texture $\hat{\boldsymbol{d}}(\boldsymbol{k})$ forms a meron structure in the momentum space as k moves from $k = 0$ to $k = \infty$ and ϕ moves from $\phi = 0$ to $\phi = 2\pi$, as shown in Figure 4. A meron is a topological structure which has a half integer Pontryagin number.

2.3   Valley-selective Optical Absorption

One of the key steps of the valleytronics is exciting only one valley. A simplest and efficient way is a valley optical excitation, where only electrons at one valley are excited by injecting circularly polarised light. We explore optical inter-band transitions from the state $|u_{\mathrm{v}}(\widetilde{\boldsymbol{k}})\rangle$ in the valence band to the state $|u_{\mathrm{c}}(\widetilde{\boldsymbol{k}})\rangle$ in the conduction band. The fundamental transition is a transition from the highest occupied

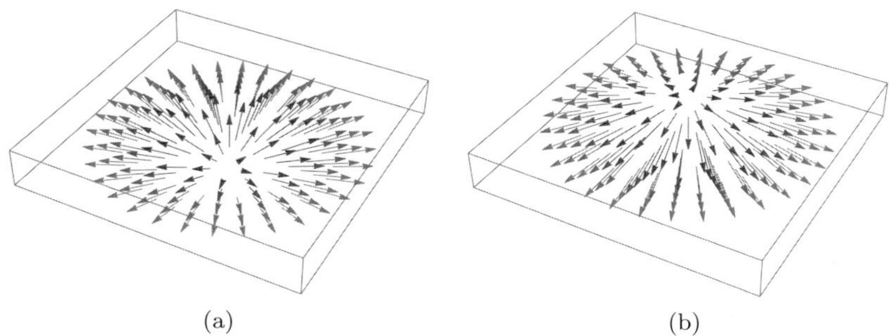

Figure 4. Illustration of a meron structure in momentum space. A meron with the Pontryagin number (a) $1/2$, whose core spin is pointing up direction, (b) $-1/2$, whose core spin is pointing down direction.

band to the lowest unoccupied band. We inject a beam of elliptical polarised light onto the honeycomb system. The corresponding electromagnetic potential is given by $\boldsymbol{A}(t) = (A_x \sin \omega t, A_y \cos \omega t)$. The electromagnetic potential is introduced into the Hamiltonian by way of the minimal substitution, that is, by replacing the momentum $\widetilde{k}_i$ with the covariant momentum $P_i \equiv \widetilde{k}_i + eA_i$. The resultant Hamiltonian simply reads $H(A) = H + \mathcal{P}_x A_x + \mathcal{P}_y A_y$, with

$$\mathcal{P}_x = \frac{\partial H}{\partial \widetilde{k}_x}, \quad \mathcal{P}_y = \frac{\partial H}{\partial \widetilde{k}_y}, \tag{2.32}$$

in the linear response theory. The coupling strength with optical fields of the right$(+)$ or left$(-)$ circular polarisation is given by $\mathcal{P}_{\pm}(k) = \mathcal{P}_x(k) \pm i\mathcal{P}_y(k)$. The optical matrix element between the initial and final states in the photoemission process is given by [15, 24–27]

$$P_i(\widetilde{\boldsymbol{k}}) \equiv m_0 \left\langle u_{\mathrm{c}}(\widetilde{\boldsymbol{k}}) \left| \frac{\partial H}{\partial \widetilde{k}_i} \right| u_{\mathrm{v}}(\widetilde{\boldsymbol{k}}) \right\rangle, \tag{2.33}$$

where m_0 is the free electron mass. Here, $|P_{\pm}^{\eta}(k)|$ is called the optical absorption.

The imaginary part of the dielectric function arises due to the inter-band absorption and is given by

$$\varepsilon_{\pm}(\omega) = \frac{\pi e^2}{\varepsilon_0 m_e^2 \omega^2} \sum_i \int_{BZ} \frac{d\widetilde{\boldsymbol{k}}}{(2\pi)^2} f(\widetilde{\boldsymbol{k}}) \left| P_{\pm}(\widetilde{\boldsymbol{k}}) \right|^2 \delta[E_{\mathrm{c}}(\widetilde{\boldsymbol{k}}) - E_{\mathrm{v}}(\widetilde{\boldsymbol{k}}) - \omega],$$

where $E_c(\widetilde{\boldsymbol{k}})$ and $E_v(\widetilde{\boldsymbol{k}})$ are the energies of the conduction and valence bands and $f(\widetilde{\boldsymbol{k}})$ is the Fermi distribution function. By adjusting the energy of light to the band edge, $\omega = E_c(\widetilde{\boldsymbol{k}}) - E_v(\widetilde{\boldsymbol{k}}) = 2|\tilde{m}|$, we find,

$$\varepsilon_\pm \left(\omega = 2|\tilde{m}|\right) = \frac{\pi e^2}{4\varepsilon_0 m_e^2 \tilde{m}^2} \left|P_\pm(0)\right|^2, \qquad (2.34)$$

The wave functions $|u_v(k)\rangle$ and $|u_c(k)\rangle$ are obtained explicitly by diagonalising the Hamiltonian (2.17). We have calculated numerically the optical absorption $|P_\pm^\xi(k)|$ as a function of k based on the tight-binding model, which we show in Figure 5.

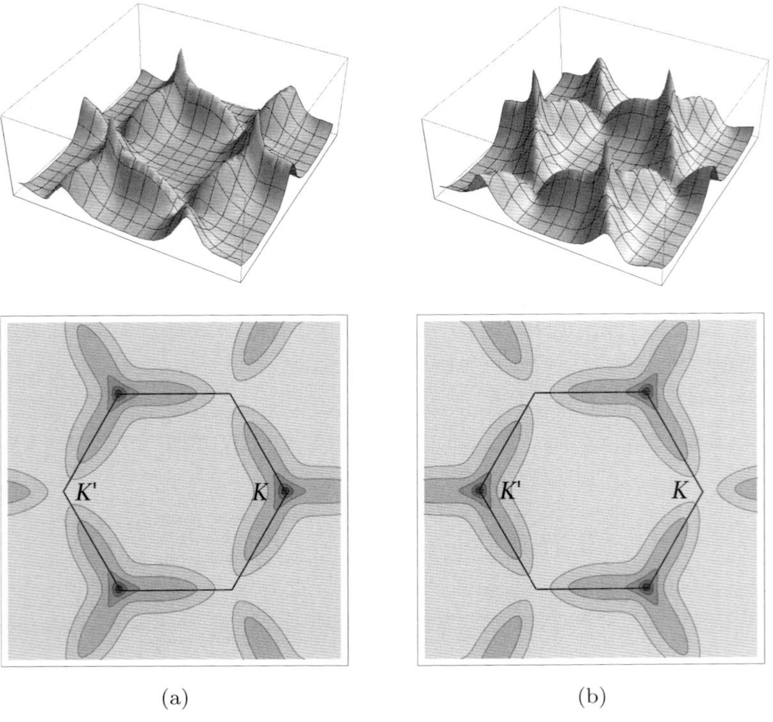

(a)   (b)

Figure 5. Optical absorption calculated from the tight-binding model. The vertical axis is $|P_+(k)|$ for (a) and $|P_-(k)|$ for (b). The optical absorption shows a peak structure at the $K(K')$ point in $|P_+(k)|$ $(|P_-(k)|)$, while there is no optical absorption at the $K'(K)$ point in $|P_+(k)|$ $(|P_-(k)|)$.

We may derive analytical results as follows. We start with

$$\mathcal{P}_x^{\xi} = \frac{1}{\hbar}\frac{\partial H_{\xi}}{\partial k_x} = v_{\mathrm{F}}\xi\tau_x, \qquad \mathcal{P}_y^{\xi} = \frac{1}{\hbar}\frac{\partial H_{\xi}}{\partial k_y} = v_{\mathrm{F}}\tau_y. \tag{2.35}$$

They are written in terms of the ladder operator of spins and pseudospins.

$$\mathcal{P}_{\pm}^{K} = v_{\mathrm{F}}\tau_{\pm}, \qquad \mathcal{P}_{\pm}^{K'} = -v_{\mathrm{F}}\tau_{\mp}, \tag{2.36}$$

where $s_{\pm} = s_x \pm is_y$ and $\tau_{\pm} = \tau_x \pm i\tau_y$. Then we are able to obtain an analytic formula for the transitions near the K_{ξ} point,

$$|P_{\pm}^{\xi}(k)|^2 = m_0^2 v_{\mathrm{F}}^2 \left(1 \pm \xi\frac{m}{\sqrt{m^2 + 4a^2t^2k^2}}\right)^2. \tag{2.37}$$

There exist the relations, $P_{+}^{K}(k) = P_{-}^{K'}(k)$, $P_{-}^{K}(k) = P_{+}^{K'}(k)$, reflecting the time-reversal symmetry. The right-handed circular polarisation at the K point and the left-handed circular polarisation at the K' point are equal. Especially, the optical absorption is simplified as,

$$|P_{\pm}^{\xi}(0)|^2 = m_0^2 v_{\mathrm{F}}^2 (1 \pm \xi\mathrm{sgn}\,m)^2 \tag{2.38}$$

for the condition $\omega = 2|\tilde{m}|$. We find,

$$|P_{\pm}^{\xi}(0)|^2 = 0, \tag{2.39}$$

when $\xi\mathrm{sgn}\,m = \mp 1$. For example, there is no optical absorption at the $K'(K)$ point when we apply right-handed (left-handed) circular polarised light for the materials with $m > 0$. As a result, we can selectively excite electron at one valley.

The k-resolved optical polarisation $\eta(\widetilde{\boldsymbol{k}})$ is given by [15, 23–25],

$$\eta(\widetilde{\boldsymbol{k}}) = \frac{|P_{+}(\widetilde{\boldsymbol{k}})|^2 - |P_{-}(\widetilde{\boldsymbol{k}})|^2}{|P_{+}(\widetilde{\boldsymbol{k}})|^2 + |P_{-}(\widetilde{\boldsymbol{k}})|^2}. \tag{2.40}$$

This quantity is the difference between the absorption of the left- and right-handed lights ($\pm$), normalised by the total absorption, around the K and K' point. Optical polarisations are perfectly polarised at the K and K' points ($\widetilde{\boldsymbol{k}} = 0$). Namely, the selection rule holds exactly at the K and K' points.

We obtain an analytic formula of the optical polarisation,

$$\eta^{\xi}(k) = \frac{2\xi m \sqrt{4v_{\mathrm{F}}^2 k^2 + m2}}{4v_{\mathrm{F}}^2 k^2 + 2m^2}.$$ (2.41)

The optical polarisation changes the sign at the topological phase transition $m = 0$.

2.4 Anisotropic Honeycomb Lattice

We investigate the honeycomb system where the hoppings are anisotropic. The Hamiltonian is modified as:

$$\hat{H} = \sum_s \int d^2 k' \left(c_{\mathrm{A}}^{\dagger}, c_{\mathrm{B}}^{\dagger} \right) \begin{pmatrix} 0 & f\left(\boldsymbol{k}'\right) \\ f^*\left(\boldsymbol{k}'\right) & 0 \end{pmatrix} \begin{pmatrix} c_{\mathrm{A}} \\ c_{\mathrm{B}} \end{pmatrix},$$ (2.42)

with

$$f\left(\boldsymbol{k}'\right) = t_1 e^{-iak_y'/\sqrt{3}} + 2t_2 e^{iak_y'/2\sqrt{3}} \cos \frac{ak_x'}{2}.$$ (2.43)

The energy spectrum reads,

$$E\left(\boldsymbol{k}'\right) = |f\left(\boldsymbol{k}'\right)| = \sqrt{t_1^2 + 4t_1 t_2 \cos \frac{ak_x'}{2} \cos \frac{\sqrt{3}ak_y'}{2} + 4t_2^2 \cos^2 \frac{ak_x'}{2}}.$$ (2.44)

We show the band structure in Figure 6. The gap closes at

$$\boldsymbol{K}_{\xi} = \frac{1}{a}\left(\frac{2}{\pi} \arctan\left(-\frac{t_1}{2t_2}, \xi \frac{\sqrt{4t_2^2 - t_1^2}}{2t_2}\right), 0\right),$$ (2.45)

when $4t_2^2 - t_1^2 < 0$. These two points merge at $4t_2^2 - t_1^2 = 0$. On the other hand, there are no solutions for $4t_2^2 - t_1^2 < 0$, which means that the system is gapped. We make the Taylor expansion of Eq. (2.4) around $\boldsymbol{M}$ point. By setting $\boldsymbol{k}' = \boldsymbol{M} + \boldsymbol{k}$, we obtain

$$f\left(\boldsymbol{k} + \boldsymbol{M}\right) = m + uk_x^2 - ivk_y \quad \text{for } |\boldsymbol{k}| \ll a^{-1},$$ (2.46)

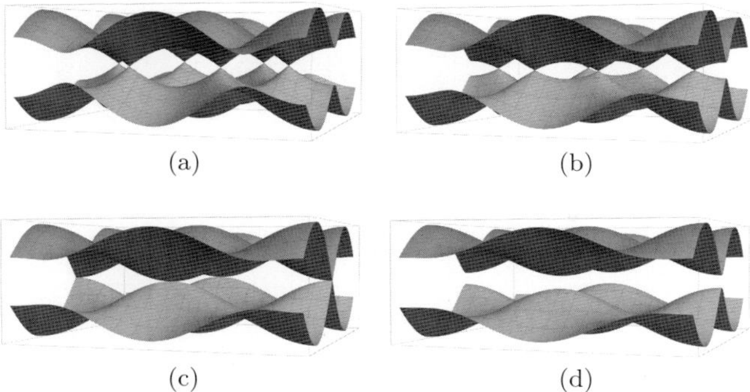

Figure 6. The band structure of the anisotropic honeycomb lattice for (a) $t_1 = 1$, (b) $t_1 = 1.5$, (c) $t_1 = 2$ and (d) $t_1 = 2.5$. We have set $t_2 = 1$. Two inequivalent Dirac cones merge at $t_1 = 2t_2$ and the system becomes gapped for $t_1 > 2t_2$.

with

$$m = t_1 - 2t_2, \quad u = \frac{\pi^2}{4}t_2, \quad v = \frac{\pi}{\sqrt{3}}\left(t_1 + t_2\right). \qquad (2.47)$$

Hence, the low-energy physics near the Fermi energy is described by,

$$H = \left(m + uk_x^2\right)\tau_x + vk_y\tau_y. \qquad (2.48)$$

The dispersion is linear along the k_y-axis, while it is parabolic along the k_x-axis:

$$E(\boldsymbol{k} + \boldsymbol{K}_\xi) = \pm\sqrt{(m + uk_x^2)^2 + v^2k_y^2}. \qquad (2.49)$$

The gap closes at $m = 0$. This type of Hamiltonian and energy dispersion are naturally realised in phosphorene, monolayer black phosphorus [26].

2.4.1 Band structure of anisotropic honeycomb nanoribbons

We investigate the change of the band structure of nanoribbon by changing the parameter t_2 continuously with t_1 being fixed [26]. We show the band structure with (a) the zigzag–zigzag edges, (b) the

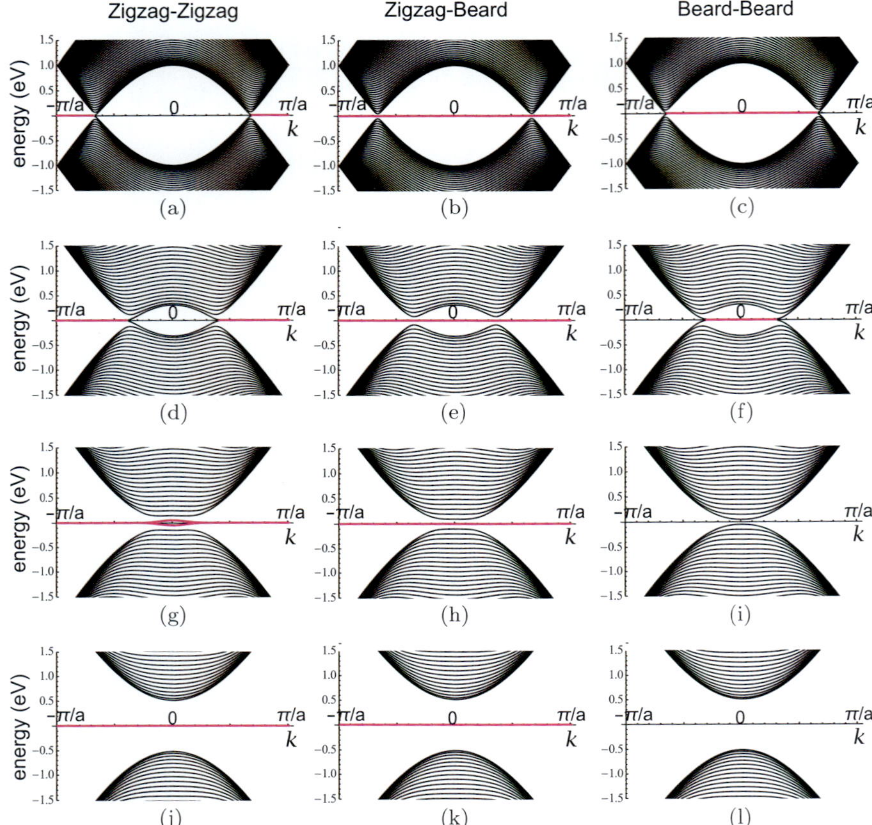

Figure 7. Band structure of anisotropic honeycomb nanoribbons. The flat edge states are marked in magenta. We have also set (a)–(c) $t_1 = 1$, (d)–(f) $t_1 = 1.7$, (g)–(i) $t_1 = 2$, (j)–(l) $t_1 = 2.5$. The unit cell contains 144 atoms. We have set $t_2 = 1$.

zigzag-beard edges, and (c) the beard–beard edges in Figure 7 for typical values of t_1 and t_2.

(i) We start with the case $t_2 = |t_1|$, where the energy spectrum (2.44) becomes that of graphene with two Dirac cones at the K and K' points. The perfect flat band emerges and connects the K and K' points, that is, it lies for (a) $-\pi \leq ak \leq -\frac{2}{3}\pi$ and $\frac{2}{3}\pi \leq ak \leq \pi$; (b) $-\pi \leq ak \leq \pi$; (c) $-\frac{2}{3}\pi \leq ak \leq \frac{2}{3}\pi$. It is attached to the bulk band. See Figures 7(a)–7(c).

(ii) As we increase t_2 but keeping t_1 fixed, the two Dirac points move towards $k = 0$, as is clear from (2.45). The flat band keeps to be present and connect the two Dirac points. See Figures 7(d)–7(f).

(iii) At $|t_2| = 2|t_1|$, the two Dirac points merge into one Dirac point at $k = 0$, as implied by (2.45). The flat band touches the bulk band at $k = 0$ for the zigzag–zigzag nanoribbon and the zigzag-beard nanoribbon, but disappears from the beard–beard nanoribbon. See Figures 7(g)–7(i).

(iv) For $|t_2| > 2|t_1|$, the bulk band shifts away from the Fermi level. The flat band is disconnected from the bulk band for the zigzag–zigzag nanoribbon and the zigzag-beard nanoribbon, where it extends over all region $-\pi \leq k \leq \pi$. On the other hand, the edge band becomes a part of the bulk band and disappears from the Fermi level for the beard–beard nanoribbon. See Figures 7(j)–7(l).

2.4.2 Topological origin of flat bands

The topological origin of the flat band has been discussed in graphene [27]. It is straightforward to apply the reasoning to the anisotropic honeycomb-lattice model (2.42). We consider one-dimensional (1D) Hamiltonian $H_k(k_x)$ in the k_x space, which is given by the 2-band model (2.42) at a fixed value of $k \equiv k_y$. We analyse the topological property of this 1D Hamiltonian. Because the k_x space is a circle due to the periodic condition, the homotopy class is $\pi_1(S^1) = \mathbb{Z}$.

We write the Hamiltonian as:

$$H_k(k_x) = \begin{pmatrix} 0 & F_k(k_x) \\ F_k^*(k_x) & 0 \end{pmatrix}. \tag{2.50}$$

It is important to remark that the gauge degree of freedom is present in this Hamiltonian. Indeed, the phase of the term $F_k(k_x)$ is irrelevant for the energy spectrum of the bulk system. However, this is not the case for the analysis of a nanoribbon since the way of taking the unit cell is inherent to the type of nanoribbon, as illustrated in Figure 8. It is necessary to make a gauge fixing, so that the hopping between the two atoms in the unit cell becomes real, namely, t_1 for the zigzag edge and t_2 for the beard edge.

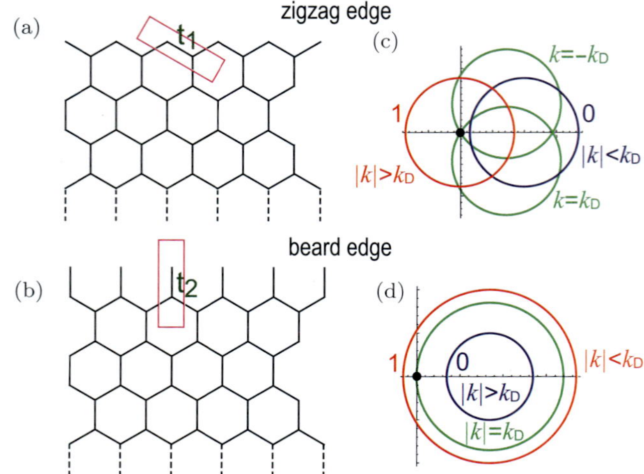

Figure 8. Unit cells and winding numbers for the zigzag and beard edges. It is necessary to make a gauge fixing in the Hamiltonian so that the hopping between the two atoms in the unit cell becomes real, namely, (a) t_1 for the zigzag edge and (b) t_2 for the beard edge. (c, d) The winding number reads $N_{\text{wind}}(k) = 1$ if the loop encircles the origin (red circle) and $N_{\text{wind}}(k) = 0$ if not (blue circle). The origin is represented by a dot, where the Hamiltonian is ill defined.

For the zigzag edge, we thus make the gauge fixing such that

$$F_k(k_x) = e^{-i\left(\frac{1}{2\sqrt{3}} a_x k_x + \frac{1}{2} ak\right)} (f_1 + f_2)$$

$$= t_1 \left(1 + e^{-iak}\right) + t_2 e^{-i\frac{1}{2}\left(\sqrt{3} a_x k_x - ak\right)},$$

where the parameter k corresponds to the momentum of zigzag nanoribbons. Here, t_1 for the link in the unit cell and $e^{-iak} t_1$ for the neighbouring link (Figure 8(a)). The topological number of the 1D system is given by,

$$N_{\text{wind}}(k) = \frac{1}{2\pi i} \oint dk_x \, \partial_{k_x} \log F_k(k_x). \tag{2.51}$$

By an explicit evaluation, $N_{\text{wind}}(k)$ is found to take only two values; $N_{\text{wind}} = 1$ for $|k| < k_D$ and $N_{\text{wind}} = 0$ for $\pi \geq |k| > k_D$. We may interpret this as the winding number as follows. We consider the complex plane for $F_k(k_x) = |F_k(k_x)| \exp[i\Theta_k(k_x)]$. The quantity $N_{\text{wind}}(k)$ counts how many times the complex number $F_k(k_x)$ winds around the origin as k_x moves from 0 to $2\pi/a$ for a fixed value of k. Note that the origin implies $F_k(k_x) = 0$, where the Hamiltonian is ill defined. We have shown such a loop for typical values of k in Figure 8(b), where the horizontal axis is for $\mathrm{Re}[F_k(k_x)]$ and the vertical axis is for $\mathrm{Im}[F_k(k_x)]$,

$$\mathrm{Re}[F_k(k_x)] = t_1(1 + \cos ak) + t_2 \cos \frac{1}{2}(\sqrt{3}a_x k_x - ak), \qquad (2.52\mathrm{a})$$

$$\mathrm{Im}[F_k(k_x)] = t_1 \sin ak + t_2 \sin \frac{1}{2}(\sqrt{3}a_x k_x - ak). \qquad (2.52\mathrm{b})$$

The loop surrounds the origin and $N_{\mathrm{wind}}(k) = 1$ when $-\pi \leq k < -k_{\mathrm{D}}$ and $k_{\mathrm{D}} < k \leq \pi$, while it does not and $N_{\mathrm{wind}}(k) = 0$ when $|k| < k_{\mathrm{D}}$: The loop touches the origin when $k = \pm k_{\mathrm{D}}$. We have demonstrated that the system is topological for $|k| > k_{\mathrm{D}}$ and trivial for $|k| < k_{\mathrm{D}}$.

We next appeal to the bulk-edge correspondence to the topological system. When we cut the bulk along the y-direction, the 1D system has an edge. Since the edge separates a topological insulator and the trivial state (i.e., the vacuum state), the gap must close at the edge, namely, there must appear a gapless edge mode. The gapless edge mode appears for all $|k| < k_{\mathrm{D}}$, implying the emergence of a flat band connecting the two Dirac points given by (2.45).

For the beard edge, we make the gauge fixing such that

$$F_k(k_x) = e^{i \frac{1}{\sqrt{3}} a_x k_x}(f_1 + f_2), \qquad (2.53)$$

and carry out an analogues argument (see Figure 8). We reach at the conclusion that a flat band appears for $|k| > k_{\mathrm{D}}$ and connects the two Dirac points given by (2.45) but in an opposite way to the case of the zigzag edge.

2.4.3 Wave function and energy spectrum of edge states

We have explained how the flat band appears in the anisotropic honeycomb-lattice model. The flat band corresponds to the quasi-flat band in the original Hamiltonian (2.42).

We construct an analytic form of the wave function at the zero-energy state in the anisotropic honeycomb-lattice model (2.42) as follows. We label the wave function of the atom on the outer most cite as ψ_1 and that of the atom next to it as ψ_2, and as so on. The total wave function is $\psi = \{\psi_1, \psi_2, \ldots, \psi_N\}$ if there are N atoms across the nanoribbon. The Hamiltonian is explicitly written as:

$$H = \begin{pmatrix} 0 & t_1 g & 0 & 0 & \cdots \\ t_1 g^* & 0 & t_2 & 0 & \cdots \\ 0 & t_2 & 0 & t_1 g & \cdots \\ 0 & 0 & t_1 g^* & 0 & \cdots \\ \cdots & \cdots & \cdots & \cdots & \cdots \end{pmatrix}, \qquad (2.54)$$

with $g = 1 + e^{-iak}$. The eigenvalue problem $H\psi = 0$ is easily solved [28], yielding $\psi_{2n} = 0$ and $\psi_{2n+1} = [t_1 (1 + e^{iak}) / t_2]^n \psi_1$. By solving the Hamiltonian matrix recursively from the outer most cite, we obtain the analytic form of the local density of states of the wave function for odd cite j,

$$|\psi(j)| = \alpha^j \sqrt{1 - \alpha^2}, \qquad (2.55)$$

with $\alpha = 2|t_1|(\cos \frac{ak}{2})/|t_2|$. The wave function is zero for an even cite. It is perfectly localised at the outer most cite when $ak = \pi$ and describes the flat band.

2.5 Weyl Semimetal

2.5.1 Continuum model of the Weyl semimetal

A Weyl semimetal is a 3D semimetal described by the 2-band theory [29, 30]. The Hamiltonian is given in the continuum theory by

$$H_{\mathrm{W}} = v_x k_x \sigma_x + v_y k_y \sigma_y + v_z k_z \sigma_z. \qquad (2.56)$$

The energy spectrum is linear,

$$E = \pm \sqrt{v_x^2 k_x^2 + v_y^2 k_y^2 + v_z^2 k_z^2}, \qquad (2.57)$$

and the band gap closes only at one point, $k_x = k_y = k_z = 0$. We call the point the Weyl or anti-Weyl point according to $\mathrm{sgn}(v_x v_y v_z) \gtrless 0$, where the monopole charge is $\mp\frac{1}{2}$, as shown in the next subsection.

Weyl semimetals are realised experimentally in TaAs [31–33] and NbAs [34] system.

An interesting feature of the Weyl semimetal is that it can never be gapped by introducing the mass term. There are three types of the "mass" terms, $\boldsymbol{m} = (m_x, m_y, m_z)$, depending on the three Pauli matrices,

$$H_m = \boldsymbol{m} \cdot \boldsymbol{\sigma} = m_x \sigma_x + m_y \sigma_y + m_z \sigma_z. \tag{2.58}$$

However, these "mass" terms only shift the Weyl point since they are absorbed into the momentum by shifting it as

$$\begin{aligned} H_{\mathrm{W}} + H_m &= [(v_x k_x + m_x)\,\sigma_x + (v_y k_y + m_y)\,\sigma_y + (v_z k_z + m_z)\,\sigma_z] \\ &= (v_x k'_x \sigma_x + v_y k'_y \sigma_y + v_z k'_z \sigma_z), \end{aligned} \tag{2.59}$$

where we have defined the shifted momentums

$$k'_x = k_x + \frac{m_x}{v_x}, \quad k'_y = k_y + \frac{m_y}{v_y}, \quad k'_z = k_z + \frac{m_z}{v_z}. \tag{2.60}$$

Only one way to gap out the Weyl semimetal is a "pair annihilation" of the Weyl and anti-Weyl points, as we discuss later.

2.5.2   Monopole charge of Weyl semimetal

The wave function for the valence band is given by

$$|\psi\rangle = \frac{1}{\sqrt{2vk\,(v_z k_z - vk)}} \begin{pmatrix} v_z k_z - vk \\ v_x k_x + i v_y k_y \end{pmatrix}, \tag{2.61}$$

where $vk = \sqrt{v_x^2 k_x^2 + v_y^2 k_y^2 + v_z^2 k_z^2}$. The Berry connection (2.10) is defined also in the 3D space and calculated as

$$A_x = \frac{-v_x v_y k_y}{2vk\,(v_z k_z - vk)}, \quad A_y = \frac{v_x v_y k_x}{2vk\,(v_z k_z - vk)}, \quad A_z = 0. \tag{2.62}$$

The Berry curvature (2.12) is the magnetic field in the momentum space,

$$\boldsymbol{B} = -\frac{v_x v_y v_z}{2v^3}\frac{\boldsymbol{k}}{k^3}, \tag{2.63}$$

which has an anti-hedgehog or hedgehog structure for $\mathrm{sgn}(v_x v_y v_z) \gtrless 0$. The divergence of $\boldsymbol{B}$ is zero at all points except the origin of $\boldsymbol{k}$,

$$\mathrm{div}\,\boldsymbol{B} = -2\pi\mathrm{sgn}\left(v_x v_y v_z\right)\delta\left(\boldsymbol{k}\right), \qquad (2.64)$$

where the coefficient of the delta function is determined by using the Gauss's law,

$$\iiint dV\,\mathrm{div}\,\boldsymbol{B}$$

$$= \iint d\boldsymbol{S}\cdot\boldsymbol{B} = 4\pi k^2 \frac{-\mathrm{sgn}\left(v_x v_y v_z\right)k}{2k^3} = -2\pi\mathrm{sgn}\left(v_x v_y v_z\right). \quad (2.65)$$

Consequently, the Weyl (anti-Weyl) semimetal has an anti-monopole (monopole) structure in the 3D momentum space [35].

2.5.3 Weyl semimetal and Chern number

We may associate the Chern number to the Weyl and anti-Weyl points. Since it is defined only on the 2D space, we treat k_z as a parameter and analyse the system of 2D massive Dirac fermions on the k_x–k_y space for each k_z. We are able to evaluate the Chern number of the Hamiltonian (2.56) as a function of k_z as

$$C\left(k_z\right) = \frac{1}{2\pi}\iint dk_x dk_y\,B_z = -\frac{1}{2}\frac{v_x v_y v_z k_z}{\left|v_x v_y v_z k_z\right|} = -\frac{1}{2}\mathrm{sgn}\left(v_x v_y v_z k_z\right).$$

$$(2.66)$$

The sign changes at $k_z = 0$, where a Weyl or anti-Weyl point exists. Namely, we can detect the position of the Weyl point by the change of the Chern number as a function of k_z.

2.6 Weyl and Anti-Weyl Points

It is an intriguing property that the lattice system contains an equal number of Weyl and anti-Weyl points. It is a reminiscence of the K and K' points in the honeycomb system. See Section 2.7.2 for an explicit example of a tight-binding model containing 4 pairs of Weyl and anti-Weyl points. Although a lattice system containing a single pair is yet unknown, it is possible to construct a continuous theory corresponding to it. A minimum model for a pair of Weyl and anti-Weyl points is

$$H = v\left(k_x\sigma_x + k_y\sigma_y + \left(k_z^2 - m\right)\sigma_z\right), \tag{2.67}$$

with a real mass m. There exist a Weyl point at $(k_x, k_y, k_z) = (0, 0, \sqrt{m})$ and an anti-Weyl point at $(k_x, k_y, k_z) = (0, 0, -\sqrt{m})$ when $m > 0$. Indeed, expanding the Hamiltonian around these points, we obtain

$$H_\pm = v\left(k_x\sigma_x + k_y\sigma_y \pm 2\sqrt{m}(k_z \mp \sqrt{m})\sigma_z\right), \tag{2.68}$$

where $\pm$ for the Weyl (anti-Weyl) point. Here, we note that the Weyl and anti-Weyl points merge as $m \to 0$, where the dispersion becomes parabolic along the k_z direction. For $m < 0$, there are none of these points. It represents the pair annihilation (creation) process of the Weyl and anti-Weyl points as the sign of m changes from positive (negative) to negative (positive). We show the band structure in Figure 9.

We calculate the Chern number of the Hamiltonian (2.67). The Berry curvature of the Hamiltonian (2.67) is given by

$$B_x = \frac{-k_x k_z}{\left(k_x^2 + k_y^2 + (k_z^2 - m)^2\right)^{3/2}}, \quad B_x = \frac{-k_y k_z}{\left(k_x^2 + k_y^2 + (k_z^2 - m)^2\right)^{3/2}},$$

$$B_z = \frac{m - k_z^2}{2\left(k_x^2 + k_y^2 + (k_z^2 - m)^2\right)^{3/2}}. \tag{2.69}$$

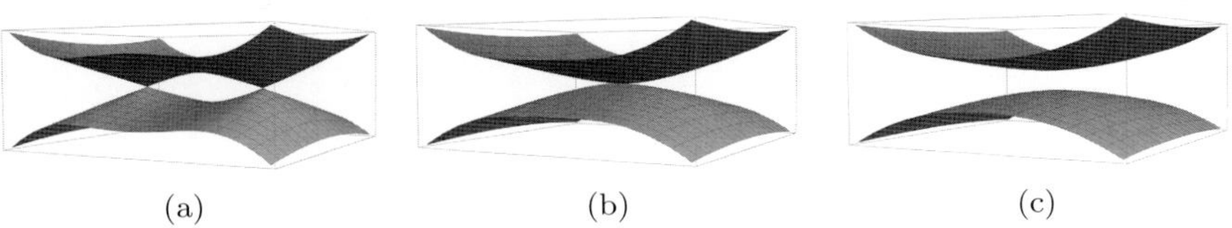

(a)   (b)   (c)

Figure 9. The band structure of the Hamiltonian (2.67) in the k_x–k_z plane with $k_y = 0$ for (a) $m = 1$, (b) $m = 0$ and (c) $m = -1$. There are Weyl and anti-Weyl points for $m > 0$. They merge at $m = 0$ and disappear for $m < 0$.

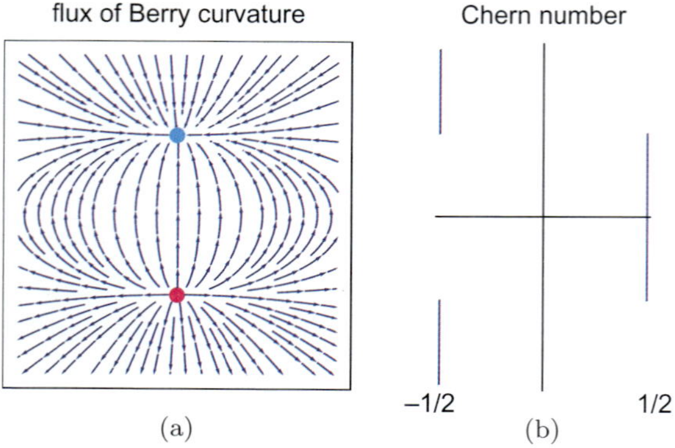

Figure 10. (a) The stream plot of the Berry curvature in the k_x–k_z plane. There are monopole and anti-monopole structures at the Weyl and anti-Weyl points. Cyan (Magenta) disk represents the Weyl (anti-Weyl) point. (b) The Chern number along the k_z axis. The sign is inverted at the Weyl and anti-Weyl points.

We show the Berry curvature in Figure 10. The Chern number is calculated as:

$$C\left(k_z\right) = \frac{1}{2\pi} \iint dk_x dk_y B_z = \frac{1}{2} \frac{m - k_z^2}{\left|m - k_z^2\right|} = \frac{1}{2}\mathrm{sgn}\left(m - k_z^2\right). \quad (2.70)$$

It is $-\frac{1}{2}$ for $|k_z| > \sqrt{m}$ and $\frac{1}{2}$ for $|k_z| < \sqrt{m}$, as illustrated in Figure 10.

The total Hall conductance on the k_x–k_y plane is given by the integral of the Chern number along the k_z axis,

$$C = \int dk_z C\left(k_z\right) = 2\sqrt{m} - \frac{1}{2}\int dk_z, \quad (2.71)$$

which is proportional to the distance between the Weyl and anti-Weyl points except for the constant $\frac{1}{2}\int dk_z$. Accordingly, the distance between the Weyl and anti-Weyl points can be detected by measuring the Hall conductance.

2.7 Chiral Anomaly

The system containing a pair of Weyl and anti-Weyl points is described by a set of 2-band Hamiltonians $H_\pm$ given by (2.68). They are combined as a 4-band Hamiltonian,

$$H = \begin{pmatrix} \boldsymbol{\sigma} \cdot (\boldsymbol{k} + \boldsymbol{b}) - b_0 & 0 \\ 0 & -\boldsymbol{\sigma} \cdot (\boldsymbol{k} - \boldsymbol{b}) + b_0 \end{pmatrix}, \qquad (2.72)$$

where $\boldsymbol{b} = (0, 0, -\sqrt{m})$ and we have assumed the symmetric velocity $v_x = v_y = v_z = 1$. We have also introduced the energy separation b_0 between the Weyl and anti-Weyl points as a possible freedom. It is convenient to introduce the chiral representation,

$$\gamma^0 = \begin{pmatrix} \boldsymbol{0} & \boldsymbol{1} \\ \boldsymbol{1} & \boldsymbol{0} \end{pmatrix}, \quad \gamma^j = \begin{pmatrix} \boldsymbol{0} & \sigma^j \\ -\sigma^j & \boldsymbol{0} \end{pmatrix}, \quad \gamma^5 = \begin{pmatrix} -\boldsymbol{1} & \boldsymbol{0} \\ \boldsymbol{0} & \boldsymbol{1} \end{pmatrix}. \qquad (2.73)$$

The second-quantised Hamiltonian follows from (2.72) as:

$$H = \int d^3 x \psi^\dagger \left(-i\gamma^0 \gamma^j \partial_j + b_\mu \gamma^0 \gamma^\mu \gamma^5 \right) \psi, \qquad (2.74)$$

with $\mu = 0, 1, 2, 3$ and $b_\mu = (b_0, \boldsymbol{b})$ and γ^μ are the Dirac matrices obeying the Clifford algebra,

$$\gamma^\mu \gamma^\nu + \gamma^\nu \gamma^\mu = g^{\mu\nu} = \text{diag.} \, (-1, 1, 1, 1) \, . \qquad (2.75)$$

The action is given by

$$S(\bar{\psi}, \psi) = \int d^4 x \bar{\psi} i \gamma^\mu \left(\partial_\mu + ieA_\mu + ib_\mu \gamma^5 \right) \psi \qquad (2.76)$$

with $\bar{\psi} = \psi^\dagger \gamma^0$ in the presence of the external electromagnetic field A_μ.

We first analyse this system classically. It is notable that the b_μ field is gauged away by making the chiral gauge transformation

$$\psi \to \psi' \equiv e^{-i\theta(x)\gamma^5/2}\psi, \quad \bar{\psi} \to \bar{\psi}' \equiv \bar{\psi}e^{-i\theta(x)\gamma^5/2}, \qquad (2.77)$$

where $\theta(x) = 2b_\mu x^\mu = 2\boldsymbol{b} \cdot \boldsymbol{x} - 2b_0 x^0$. Indeed, it is trivial to see

$$S(\bar{\psi}, \psi) \to S_0(\bar{\psi}', \psi') \equiv \int d^4 x \bar{\psi}' i \gamma^\mu \left(\partial_\mu + ieA_\mu \right) \psi'. \qquad (2.78)$$

Namely, a pair of the Weyl and anti-Weyl points with the separation $2\boldsymbol{b}$ is physically equivalent to the pair with no separation. However, this classical result is invalidated due to the chiral anomaly in the quantum theory. We may derive the chiral anomaly based on the

Fujikawa method [36–38]. The chiral anomaly leads to anomalous transport phenomena such as the chiral magnetic effect.

The quantum theory is formulated in terms of the path integration,

$$I = \int [d\bar{\psi}][d\psi] \exp[iS(\bar{\psi}, \psi)]. \tag{2.79}$$

A non-trivial Jacobian is produced when we apply the chiral gauge transformation (2.77) to eliminate the b_μ term. To derive the Jacobian, it is necessary to fix what the symbol $[d\bar{\psi}][d\psi]$ means. We introduce the eigenfunctions $\phi_n(x)$ of the Dirac operator,

$$\slashed{D}\phi_n(x) = \varepsilon_n \phi_n(x), \tag{2.80}$$

where we may choose any Dirac operator $\slashed{D}$. They form an orthonormal complete set, in terms of which we expand the field variables as:

$$\psi(x) = \sum_n \phi_n(x)c_n, \quad \overline{\psi}(x) = \sum_n \phi_n^*(x)\bar{c}_n, \tag{2.81}$$

$$\psi'(x) = \sum_n \phi_n(x)c_n', \quad \overline{\psi}'(x) = \sum_n \phi_n^*(x)\bar{c}_n', \tag{2.82}$$

together with

$$\int d^4x \phi_n^*(x)\phi_m(x) = \delta_{nm}, \tag{2.83}$$

where c_n, $\bar{c}_n$, c_n', $\bar{c}_n'$ are the Grassmann variables. We thus find,

$$[d\bar{\psi}][d\psi] = \prod_{mn} d\bar{c}_m dc_n, \quad [d\bar{\psi}'][d\psi'] = \prod_{mn} d\bar{c}_m' dc_n'. \tag{2.84}$$

We now study the chiral transformation (2.77) explicitly,

$$\psi'(x) = e^{-i\theta(x)\gamma^5/2}\psi(x) = \sum_m e^{-i\theta(x)\gamma^5/2}\phi_m(x)c_m = \sum_n \phi_n(x)c_n', \tag{2.85}$$

$$\overline{\psi}'(x) = \overline{\psi}(x)e^{-i\theta(x)\gamma^5/2} = \sum_n \phi_n^*(x)\bar{c}_n e^{-i\theta(x)\gamma^5/2} = \sum_n \phi_n^*(x)\bar{c}_n'. \tag{2.86}$$

Multiplying $\phi_n^\dagger(x)$ to $\psi'(x)$ and $\phi_n(x)$ to $\overline{\psi}'(x)$ and using the normalisation condition (2.83), we obtain,

$$c_n' = \sum_m \int d^4x\, \phi_n^\dagger(x) e^{-i\theta(x)\gamma^5/2} \phi_m c_m \equiv \sum_m U_{nm} c_m, \qquad (2.87)$$

$$\overline{c}_n' = \sum_m \overline{c}_m \int d^4x\, \phi_m^*(x) e^{-i\theta(x)\gamma^5/2} \phi_n(x) \equiv \sum_m \overline{c}_m U_{nm}, \qquad (2.88)$$

where

$$U_{nm} = \int d^4x\, \phi_n^\dagger(x) e^{-i\theta(x)\gamma^5/2} \phi_m. \qquad (2.89)$$

It follows that

$$\prod_n dc_n' = \det[U]^{-1} \prod_n dc_n, \quad \prod_m d\overline{c}_m' = \det[U]^{-1} \prod_n d\overline{c}_m, \qquad (2.90)$$

where the inverse of the determinant appears due to the Grassmannian property of the variables. We thus find:

$$\sum_n d\overline{c}_n dc_n = \det[U]^2 \prod_n d\overline{c}_n' dc_n' = \exp[2\mathrm{tr}\ln U] \prod_n d\overline{c}_n' dc_n'. \qquad (2.91)$$

Consequently, after the chiral transformation the path-integration formula (2.79) reads,

$$I = \int [d\overline{\psi}][d\psi] \exp[iS(\overline{\psi},\psi)] = \int [d\overline{\psi}'][d\psi'] \exp[iS_0(\overline{\psi}',\psi') + iS_\theta], \qquad (2.92)$$

with $S_\theta = -2i\mathrm{tr}\ln U$.

The extra action S_θ is obtained explicitly for small $\theta(x)$ as follows. We expand the exponential into a power series up to the order of $\theta(x)$,

$$U_{nm} = \delta_{nm} - \frac{i}{2} \int d^4x\, \theta(x) I_{nm}(x), \qquad (2.93)$$

with

$$I_{nm}(x) = \phi_n^*(x)\gamma^5 \phi_m(x). \qquad (2.94)$$

Hence,

$$\operatorname{tr}\ln U = -\frac{i}{2}\sum_n \int d^4x\,\theta(x) I_{nn}(x) = -\frac{i}{2}\int d^4x\,\theta(x) I(x), \qquad (2.95)$$

with

$$I(x) = \sum_n I_{nn}(x) = \sum_n \phi_n^*(x)\gamma^5\phi_m(x). \qquad (2.96)$$

We obtain,

$$S_\theta = -2i\operatorname{tr}\ln U = -\int d^4x\,\theta(x) I(x). \qquad (2.97)$$

The quantity $I(x)$ is ill-defined conditionally convergent. We evaluate it by regularising the large eigenvalues,

$$I(x) = \lim_{M\to\infty}\sum_n \phi_n^*(x)\gamma^5 \exp[-(\slashed{D})^2/M^2]\phi_n(x). \qquad (2.98)$$

It is most convenient to choose the simplest Dirac operator for $\slashed{D}$,

$$\slashed{D} = \gamma^\mu(\partial_\mu + ieA_\mu). \qquad (2.99)$$

The square of the Dirac operator is explicitly given by:

$$\slashed{D}^2 = D_\mu D^\mu + \frac{1}{4}[\gamma^\mu,\gamma^\nu][D_\mu,D_\nu] = -D_\mu D_\mu + \frac{ie}{4}[\gamma^\mu,\gamma^\nu]F_{\mu\nu}, \qquad (2.100)$$

where we have used

$$[D_\mu,D_\nu] = ie\,(\partial_\mu A_\nu - \partial_\nu A_\mu) = ieF_{\mu\nu}. \qquad (2.101)$$

Substituting this into (2.98) and using the completeness condition (2.83), we obtain,

$$I(x) = \lim_{M\to\infty}\int \frac{d^4k}{(2\pi)^2}\operatorname{Tr}\gamma^5 e^{-ikx}e^{-\slashed{D}^2/M^2}e^{ikx},$$

$$= \lim_{M\to\infty}\int \frac{d^4k}{(2\pi)^2}\operatorname{Tr}\gamma^5 \exp\left(\frac{(ik_\mu + D_\mu)^2}{M^2} - \frac{ie}{4M^2}[\gamma^\mu,\gamma^\nu]F_{\mu\nu}\right). \qquad (2.102)$$

By rescaling the momentum integration variables, $k_\mu \to M k_\mu$ and expanding it in a series of $1/M$, we may extract a well-defined term,

$$I(x) = -\frac{e^2}{32\pi^2} \mathrm{Tr}\gamma^5 \left[\gamma^\mu, \gamma^\nu\right]\left[\gamma^\alpha, \gamma^\beta\right] F_{\mu\nu} F_{\alpha\beta} = \frac{e^2}{32\pi^2}\varepsilon^{\mu\nu\alpha\beta} F_{\mu\nu} F_{\alpha\beta},$$

(2.103)

where we have used

$$\mathrm{Tr}\left[\gamma^5\gamma^\mu\gamma^\nu\gamma^\alpha\gamma^\beta\right] = -4i\varepsilon^{\mu\nu\alpha\beta}.$$

(2.104)

Substituting this into (2.97), we obtain,

$$S_\theta = -\frac{e^2}{32\pi^2}\varepsilon^{\mu\nu\alpha\beta}\int d^4x\theta(x)F_{\mu\nu}F_{\alpha\beta}$$

$$= -\frac{e^2}{4\pi^2}\int d^4x\theta(x)\boldsymbol{E}(x)\cdot\boldsymbol{B}(x).$$

(2.105)

It is rewritten into the Chern–Simons form,

$$S_\theta = +\frac{e^2}{8\pi^2}\varepsilon^{\mu\nu\alpha\beta}\int dtdr\partial_\mu\theta A_\nu\partial_\alpha A_\beta.$$

(2.106)

In the path-integration formula (2.79), the action reads $S_0(\bar{\psi}', \psi') + S_\theta(A)$ in the new gauge.

The current is given by differentiating the action by the gauge field. Due to the Chern–Simons action, the new term emerges,

$$j_\nu = \frac{\partial S_\theta}{\partial A_\nu} = \frac{e^2}{2\pi^2\hbar^2}\left(b_i\varepsilon^{i\nu\alpha\beta}\partial_\alpha A_\beta - b_0\varepsilon^{0\nu\alpha\beta}\partial_\alpha A_\beta\right).$$

(2.107)

It is rewritten as

$$\boldsymbol{j} = \frac{e^2}{2\pi^2\hbar^2}\boldsymbol{b}\times\boldsymbol{E} + \frac{e^2}{2\pi^2\hbar^2}b_0\boldsymbol{B}.$$

(2.108)

The first term represents the anomalous Hall effect and the second term represents the chiral magnetic effect. In the new gauge, the field b_μ appears only in the extra action $S_\theta(A)$ and hence $\boldsymbol{j}^\theta$ is the sole current that involves the field b_μ in the total current in the system: It is physical and must be observable experimentally.

2.7.1 Weyl semimetal from Dirac semimetal with magnetic field

The 3D Dirac fermions are described by the following standard Hamiltonian,

$$H = \gamma^0 \gamma_j k^j + \gamma^0 m. \tag{2.109}$$

We include the Zeeman term,

$$H = \gamma^0 \gamma_j k^j + \gamma^0 m + h\sigma_z \otimes \tau_0. \tag{2.110}$$

In the chiral representation of γ^μ, the Dirac Hamiltonian is reduced to:

$$H\left(k_x, k_y, k_z\right) = (\boldsymbol{k} \cdot \sigma)\tau_x + m\tau_z + h\sigma_z$$

$$= \begin{pmatrix} k_z + h & k_- & m & 0 \\ k_+ & -k_z - h & 0 & m \\ m & 0 & -k_z + h & k_- \\ 0 & m & k_+ & k_z - h \end{pmatrix}. \tag{2.111}$$

There are four eigenvalues,

$$E = \pm\sqrt{k_x^2 + k_y^2 + (\sqrt{k_z^2 + m^2} \pm h)^2}. \tag{2.112}$$

(i)   When $h < m$, the gap opens, with the gap being $\sqrt{m^2 - h^2}$.
(ii)   When $h = m$, the band touches parabolically.
(iii) When $h > m$, the gap closes at

$$k_x = 0, \quad k_y = 0 \quad \text{and} \quad k_z = \pm\sqrt{h^2 - m^2}, \tag{2.113}$$

where there emerge a pair of Weyl and anti-Weyl points. The two Dirac cones shift apart as h increases.

We explicitly derive the Weyl Hamiltonian. We first diagonalise the Hamiltonian with $k_x = k_y = 0$ by the unitary transformation U,

$$U^{-1} H\left(0, 0, k_z\right) U$$

$$= \text{diag.} \left(h - \sqrt{k_z^2 + m^2}, -h + \sqrt{k_z^2 + m^2}, h + \sqrt{k_z^2 + m^2},\right.$$

$$\left.-h - \sqrt{k_z^2 + m^2}\right). \tag{2.114}$$

The Hamiltonian with finite k_x and k_y is transformed by the same unitary transformation U,

$$U^{-1} H\left(k_x, k_y, k_z\right) U$$

$$= \begin{pmatrix} h - \sqrt{k_z^2 + m^2} & -k_- & 0 & 0 \\ -k_+ & -h + \sqrt{k_z^2 + m^2} & 0 & 0 \\ 0 & 0 & h + \sqrt{k_z^2 + m^2} & -k_- \\ 0 & 0 & -k_+ & -h - \sqrt{k_z^2 + m^2} \end{pmatrix}.$$

$$\tag{2.115}$$

When $h > 0$, the low energy theory is described by

$$PU^{-1} H\left(k_x, k_y, k_z\right) UP = \begin{pmatrix} h - \sqrt{k_z^2 + m^2} & -k_- \\ -k_+ & -h + \sqrt{k_z^2 + m^2} \end{pmatrix}. \tag{2.116}$$

There are two zeros at $k_z = \pm\sqrt{h^2 - m^2}$. Expanding it around $k_z = \pm\sqrt{h^2 - m^2}$, we derive the Weyl Hamiltonian,

$$H = -k_x \sigma_z - k_y \sigma_y \mp \sqrt{1 - \frac{m^2}{h^2}} \left(k_z \mp \sqrt{h^2 - m^2}\right) \sigma_z. \tag{2.117}$$

When $h < 0$, the low energy theory is described by

$$PU^{-1}H\left(k_x, k_y, k_z\right)UP = \begin{pmatrix} h + \sqrt{k_z^2 + m^2} & -k_- \\ -k_+ & -h - \sqrt{k_z^2 + m^2} \end{pmatrix}.$$

$$(2.118)$$

In the same way, we obtain the Weyl Hamiltonian,

$$H = -k_x\sigma_z - k_y\sigma_y \pm \sqrt{1 - \frac{m^2}{h^2}}\left(k_z \mp \sqrt{h^2 - m^2}\right)\sigma_z. \qquad (2.119)$$

The chiral current is defined by

$$j_5^\mu = \gamma^0\gamma^\mu\gamma_5, \qquad (2.120)$$

which read

$$j_5^0 = \gamma^5 = \sigma_0 \otimes \tau_x, \qquad j_5^i = \sigma_i \otimes \tau_0. \qquad (2.121)$$

where j_5^0 is the chiral charge and j_5^i describes spin. We show the vector plot of the chiral current in Figure 11, where we have plotted the band structure, the spin direction and the Berry magnetic field for each case (i), (ii) and (iii). It is clearly seen that a pair of monopole and anti-monopole emerges as the system becomes a Weyl semimetal by controlling the external magnetic field h.

2.7.2   Wilson–Dirac Hamiltonian

It is convenient to construct a tight-binding model realising a Weyl semimetal. The simplest method is to replace $\boldsymbol{k}$ by $\sin\boldsymbol{k}$ in the Hamiltonian (2.56),

$$H = v\boldsymbol{\sigma}\cdot\sin\boldsymbol{k} = v\left(\sigma_x\sin k_x + \sigma_y\sin k_y + \sigma_z\sin k_z\right). \qquad (2.122)$$

This is called the Wilson–Dirac Hamiltonian and is defined on the cubic lattice.

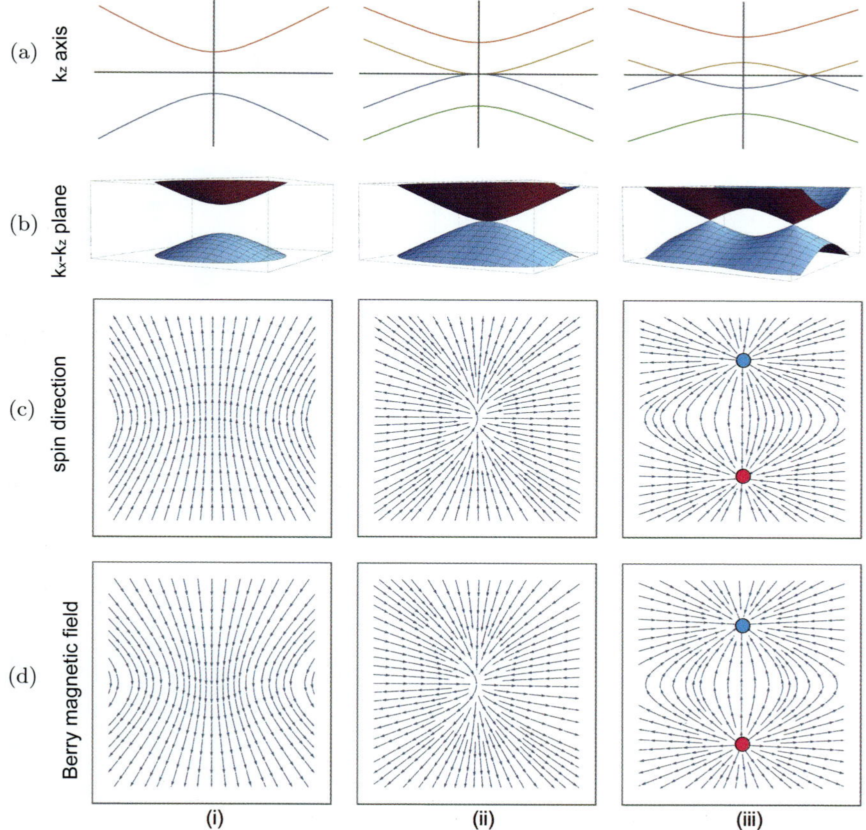

Figure 11. (a) Band structures along the k_z axis with $k_x = k_y = 0$. (b) Band structures along the k_x–k_z plane with $k_y = 0$. (c) Spin directions along k_x–k_z plane with $k_y = 0$. (d) Berry curvatures along k_x–k_z plane with $k_y = 0$. Cyan (Magenta) disk represents the Weyl (anti-Weyl) point.

The Weyl (anti-Weyl) points emerge at $(n_x\pi, n_y\pi, n_z\pi)$ when $n_x + n_y + n_z$ is even (odd). In the first Brillouin zone, there are inequivalent 4 Weyl and 4 anti-Weyl points. Each Weyl and anti-Weyl points are adjacent with the same distance. We show the Berry curvature and the Fermi surface in Figure 12.

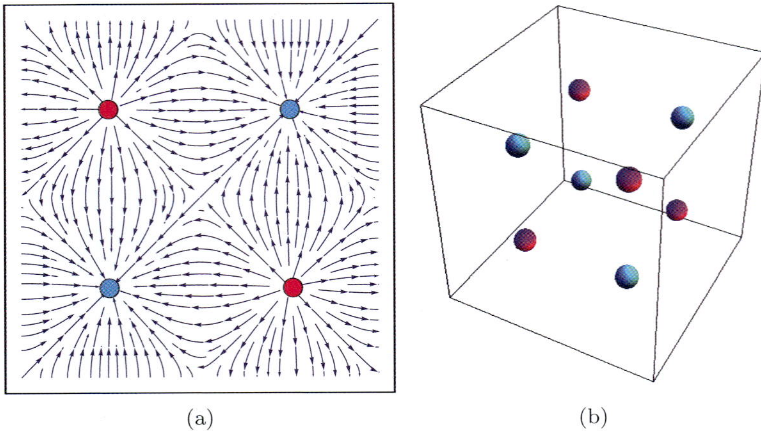

(a)   (b)

Figure 12. (a) The stream plot of the Berry curvature of the Wilson–Dirac model along the k_x–k_y plane with $k_z = 0$. (b) The position of the Weyl and anti-Weyl points in the first Brillouin zone. Cyan (Magenta) disk represents the Weyl (anti-Weyl) point.

2.8 Conclusion

We have explored the topological physics of valleytronics. Important observation is that two valleys have opposite chiralities, which are characterised by the topological indices such as the Berry phase in the 2D semimetal, the Chern number in the 2D insulator and the monopole charge in the Weyl semimetal. They can be pair annihilated by controlling the model parameters. During this process, the topological charge is conserved.

References

[1] K. Takashina *et al.*, *Phys. Rev. Lett.* **96**, 236801 (2006).

[2] D. Culcer, A. L. Saraiva, B. Koiller, X. Hu and S. Das Sarma, *Phys. Rev. Lett.* **108**, 126804 (2012).

[3] C. H. Yang *et al.*, *Nat. Comm.* **4**, 2069 (2013).

[4] N. C. Bishop *et al.*, *Phys. Rev. Lett.* **98**, 266404 (2007).

[5] K. F. Mak *et al.*, *Phys. Rev. Lett.* **105**, 136805 (2010).

[6] A. Splendiani *et al.*, *Nano Lett.* **10**, 1271 (2010).

[7] H. Zeng *et al.*, *Nat. Nanotechnol.* **7**, 490 (2012).

[8] T. Cao *et al.*, *Nat. Commun.* **3**, 887 (2012).

[9] J. Isberg *et al.*, *Nat. Mater.* **12**, 760 (2013).

[10] C. E. Nebel, *Nat. Mater.* **12**, 690 (2013).

[11] Z. Zhu *et al.*, *Nat. Phys.* **8**, 89 (2012).

[12] A. Rycerz *et al.*, *Nat. Phys.* **3**, 172 (2007).

[13] A. R. Akhmerov and C. W. J. Beenakker, *Phys. Rev. Lett.* **98**, 157003 (2007).

[14] D. Xiao *et al.*, *Phys. Rev. Lett.* **99**, 236809 (2007).

[15] W. Yao, D. Xiao and Q. Niu, *Phys. Rev. B* **77**, 235406 (2008).

[16] X. Xu *et al.*, *Nat. Phys.* **10**, 343 (2014).

[17] M. Ezawa, *Phys. Rev. B* **87**, 155415 (2013).

[18] M. Ezawa, *J. Phys. Soc. Jpn.* **84**, 121003 (2015).

[19] R. Saito, G. Dresselhaus and M. S. Dresselhaus, *Physical Properties of Carbon Nanotubes*, Imperial College Press, London (1998).

[20] M. I. Katsnelson, *Graphene: Carbon in Two Dimensions*, Cambridge University Press, Cambridge (2012).

[21] H. B. Nielsen and M. Ninomiya, *Nucl. Phys. B* **185**, 20 (1981).

[22] D. J. Thouless, M. Kohmoto, M. P. Nightingale and M. den Nijs, *Phys. Rev. Lett.* **49**, 405 (1982).

[23] D. Xiao, G.-B. Liu, W. Feng, X. Xu and W. Yao, *Phys. Rev. Lett.* **108**, 196802 (2012).

[24] J. Li, A. F. Morpurgo, M. Bütiker and I. Martin, *Phys. Rev. B* **82**, 245404 (2010).

[25] M. Ezawa, *Phys. Rev. B* **86**, 161407(R) (2012).

[26] M. Ezawa, *New J. Phys.* **16**, 115004 (2014).

[27] R. Shinsei and Y. Hatsugai, *Phys. Rev. Lett.* **89**, 077002 (2002).

[28] M. Kohmoto and Y. Hasegawa, *Phys. Rev. B* **76**, 205402 (2007).

[29] P. Hosur and X. L. Qi, *C. R. Physique* **14**, 857 (2013).

[30] A. M. Turner and A. Vishwanath, cond-mat/arXiv:1301.0330.

[31] S.-M. Huang *et al.*, *Nat. Commun.* **6**, 7373 (2015).

[32] S.-Y. Xu *et al.*, *Science* **349**, 613 (2015), S.-Y. Xu, *Science Advances* **10**, 1501092 (2015).

[33] B. Q. Lv *et al.*, *Phys. Rev. X* **5**, 031013 (2015).

[34] S.-Y. Xu, *Nat. Phys.* **11**, 748 (2015).

[35] S. Murakami, *New J. Phys.* **9**, 356 (2007).

[36] K. Fujikawa, *Phys. Rev. D*, **21**, 2848 (1980).

[37] K. Fujikawa and H. Suzuki, *Path Integrals and Quantum Anomalies* (International Series of Monographs on Physics) Oxford University Press, Oxford (2014).

[38] A. A. Zyuzin and A. A. Burkov, *Phys. Rev. B* **86**, 115133 (2012).

Chapter 3

Two-Dimensional, Layered Materials as Catalysts for Oxygen Reduction Reaction

Debdyuti Mukherjee and S. Sampath*

*Department of Inorganic and Physical Chemistry,
Indian Institute of Science,
Bangalore 560012, India*
**sampath@1pc.iisc.ernet.in*

Abstract. This chapter describes the recent developments in the area of two-dimensional, layered compounds for oxygen reduction reaction (ORR). After a brief introduction to the ORR, the thermodynamics involved and the mechanistic details of the reaction are given, followed by the use of layered materials as catalysts. The catalytic activity of graphene and its analogues such as silicene, chalcogenides, carbides, borides, borocarbonitrides and layered double hydroxides are discussed. Towards the end, a summary is given along with possible way forward.

3.1 Introduction

The increase in green house emissions, high demand for energy, rapid consumption of fossil fuels and growing environmental issues have led to the exploration of renewable and environmentally sustainable energy sources. In this context, development of high capacity energy devices such as fuel cells and metal–air batteries are in demand and are given top priority to fulfil the requirements of various applications, for example, in electric vehicles [1, 2]. These advanced energy conversion technologies involve reduction of oxygen as one of the electrode reactions [3–5]. Oxygen reduction reaction (ORR) is also a very important reaction in life processes such as biological respiration [6]. In aqueous solutions, ORR occurs mainly by two pathways, one is the direct 4-electron reduction from O_2 to H_2O and the second one involves 2 step 2-electron processes to produce H_2O through peroxide. It is also likely that the reaction stops at the peroxide stage (H_2O_2) on certain catalysts. In non-aqueous, aprotic media, one electron reduction from O_2 to superoxide (O_2^-) is preferred [7]. Very often, it is suggested that the ORR is the limiting reaction in the performance of electrochemical energy systems. The most efficient catalysts known till date for ORR are, Pt and palladium (Pd)-based materials [8]. The cost and abundance of these noble metals have led to an intense search for alternate, noble metal-free catalysts over the past several decades [9]. The materials that have been studied include noble metals and alloys, carbon-based materials, quinones and their derivatives, transition metal macrocyclic compounds, transition metal chalcogenides and transition metal carbides, nitrides and oxides.

In this focussed chapter, we initiate the discussion with the fundamentals of ORR including the requirement of a catalyst and reaction mechanisms, followed by recent studies on different types of catalysts, particularly emphasising on two-dimensional (2D), layered materials such as transitional metal chalcogenides, layered hydroxides, oxides, etc. Summary section is given at the end.

3.2   Fundamentals and Requirements of a Catalyst

Electrochemical ORR mechanism is quite complex and involves many intermediates depending upon the nature of the electrode material, electrocatalyst, electrolyte and the solvent used. As mentioned above, ORR takes place by either a 4-electron pathway or a 2-electron pathway in aqueous media [10–12]. The mechanism involves adsorption of molecular O_2 on to the electrocatalyst surface followed by the activation of O–O bond and subsequent cleavage to produce adsorbed atomic O on the electrode surface. This results in H_2O as the product either in a direct 4-electron pathway or in two consecutive steps of two electrons each (4-electron process). On the other hand, if the

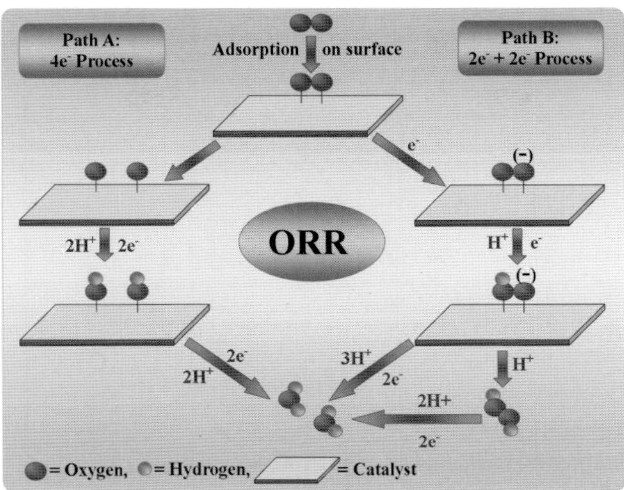

Figure 1. Schematic representation of the mechanism of ORR. (Adapted from Ref. [12]).

Table 1. Standard electrode potentials of reactions involved in electrochemical ORR at different pH values in aqueous media [11].

Electrolyte	Reaction	Standard potential/V
Aqueous acidic	$O_2 + 4H^+ + 4e^- \rightarrow 2H_2O$	1.229
	$O_2 + 2H^+ + 2e^- \rightarrow H_2O_2$	0.70
	$H_2O_2 + 2H^+ + 2e^- \rightarrow 2H_2O$	1.76
Aqueous alkaline	$O_2 + 2H_2O + 4e^- \rightarrow 4OH^-$	0.401
	$O_2 + H_2O + 2e^- \rightarrow HO_2^- + OH^-$	−0.065
	$HO_2^- + H_2O + +2e^- \rightarrow 3OH^-$	0.867

adsorbed molecular oxygen does not undergo O–O bond cleavage, the product stops at peroxide (2-electron transfer) stage. Figure 1 gives a schematic representation of ORR mechanisms in aqueous media. The standard electrode potential values for ORR in aqueous media at different pH values are given in Table 1.

An essential requirement for a heterogeneous catalyst-based on Sebatier principle is that the electrocatalyst should bind the key intermediate strong enough so that the reagents will likely bind and

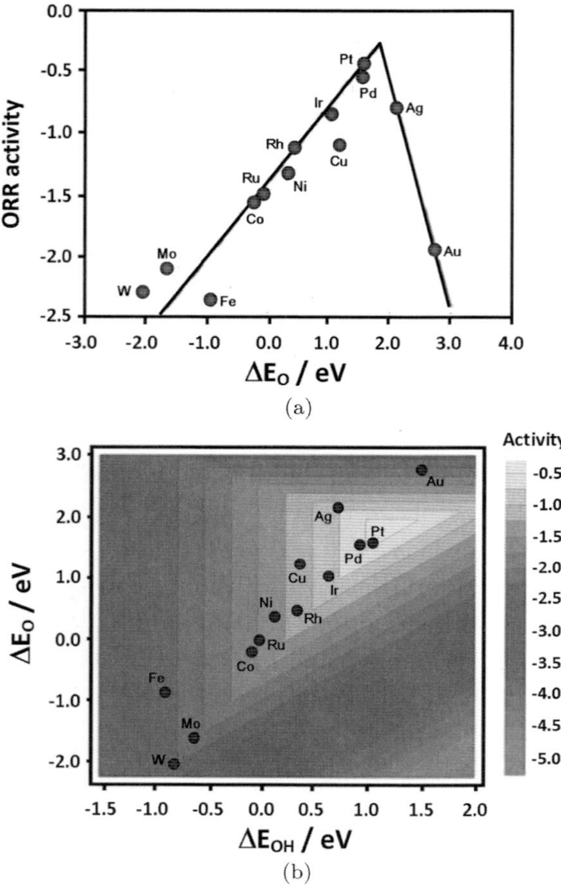

Figure 2. Trends in oxygen reduction activity plotted as a function of binding energies of (a) O and (b) both the O and the OH on various catalysts. (Adapted from Ref. [14]).

react, but weak enough so that the product will easily desorb. In other words, the catalyst should be thermoneutral with respect to the reaction concerned [13, 14]. Hence, it is important to find out the adsorption ability of the catalysts to various species involved in the reaction. Figure 2(a) shows the plot of activity for ORR vs. binding energy for various metallic catalysts towards oxygen. It should, however, to be pointed out that the adsorption of various intermediates is to be taken into account in predicting the activity.

Particularly, for a multi-electron transfer reaction such as ORR, it is quite important to understand the coverages of various species since the surface concentration (coverage, θ) would determine the adsorption parameters such as ΔH and ΔG values.

3.3   Electrocatalysts — Noble Metals

Figure 2(a) gives a correlation between the activity of a catalyst with oxygen binding energy, which reveals that metallic Pt and Pd are the best electrocatalysts for ORR [14]. In this context, it should be noted that it is not just the oxygen binding energy that determines the activity of a catalyst for reduction of oxygen, and the OH (product) binding energy is equally important. Figure 2(b) shows that the binding energies for O and OH are almost linearly correlated to each other for several elements [14]. The figures also reveal that though Pt and Pd are the most active towards ORR, they do not possess thermoneutrality, and on the contrary, some first-row transition metals such as Ni, Co show binding energies to be close to zero. Thus, the activity of Pt and Pd can be further improved by formation of bimetallic phases, for example, Ni, Co, Fe and Cr monolayer on the surface of Pt [15]. These alloys are shown to have smaller oxygen binding energies than that of pure Pt [16, 17]. Xu *et al.* [16] reported that a Pt layer on top of $Pt_3Co(111)$ possesses very low oxygen binding energy (0.38 eV), which is less than that on pure Pt(111). Another set of Pt-alloys with Y and Sc have been shown to exhibit good ORR activity (Figure 3) [18].

3.4   Non-noble Metal Catalysts

Though Pt-based materials are regarded as the best electrocatalysts for ORR-based on low over potential and high exchange current density, they are reported to show very sluggish kinetics [19], prone to deactivation by CO, susceptible to fuel cross-over from the anode, poor durability under electrochemical conditions, etc. [20]. Of course, excessive cost of Pt and limited resources are other major barriers that restricts commercialisation of Pt-based fuel cells and metal–air batteries for commercial applications [19].

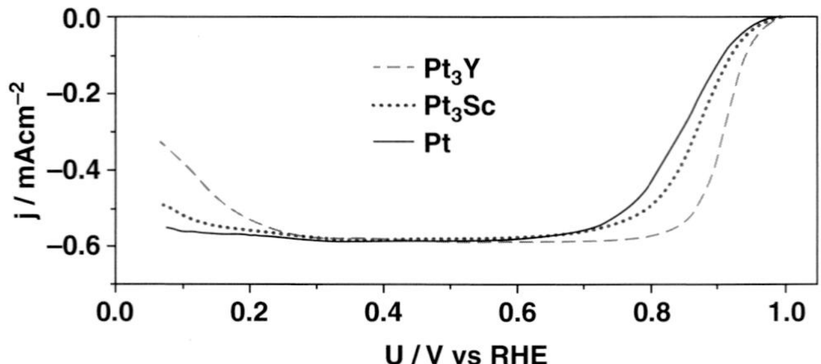

Figure 3. Electrocatalytic activity (current vs. potential) of Pt, Pt_3Sc and Pt_3Y in O_2 saturated 0.1 M $HClO_4$ solution at a scan rate of $20\,\mathrm{mVs^{-1}}$ and at 1600 rpm rotation speed. (Adapted from Ref. [18]).

Recently, considerable efforts have been devoted to develop ORR catalysts that can potentially replace Pt-based electrodes [21–23]. Various forms of carbon-based materials are reported as efficient [24–26], whose activity could be further improved by doping with non-metals such as B, N, P and S, that are known to activate π electrons of carbon [27–36]. Due to the difference in electronegativity, the heteroatoms induce localised positive or negative charges on adjacent carbon atoms and enhance the oxygen adsorption capacity as well as activity [37, 38]. It is reported that co-doping of two heteroatoms into carbon network show better activity as compared to the corresponding single-atom doped catalyst. Wang *et al.* [31] have reported that B/N co-doped vertically aligned carbon nanotubes (CNTs) (VA-BCN nanotubes) exhibit excellent ORR activity as compared to the corresponding N-doped and B-doped CNTs. This is attributed to the synergistic effects of the co-doped heteroatoms. Other than carbonaceous materials, transition metal-based carbides [39, 40], nitrides [41–43] and oxides [44–46] have attracted considerable attention as effective ORR electrocatalysts. It is found that the activity depends on the morphology of the catalyst in certain cases [40]. Recently, titaniumcarbonitride (TiCN, a transition metal carbonitride) [47] has been shown to be an excellent and highly stable electrocatalyst for ORR and is shown to be durable for zinc-air batteries. Transition metal containing macrocycles such as porphyrins and phthalocyanines are of great interest for ORR since 1960s [48]. It has been observed that they exhibit excellent activity in different media [49–53] and are also shown to be highly selective for ORR in presence of methanol and CO, which is an important factor for direct methanol and hydrogen fuel cells [54, 55]. The chemical stability of the macrocyclic catalysts are impressive, they are regularly used in biological fuel cells but in acidic conditions, demetalisation is occasionally observed. As already mentioned, carbon-based systems that show efficient ORR activity, contain functional groups such as quinones [10]. It is long known that quinone-based functional groups can catalyse ORR via 2-electron pathway, producing H_2O_2 [56–58]. This is efficiently used in various industrial applications, such as pulp and paper, textiles, chemical synthesis, as well as environmental remediation such as waste-water treatment. Matsuda *et al.* [59] have reported quinone-based systems as catalysts for ORR and further used it in $Li–O_2$ batteries.

3.5 2D Layered Materials

2D materials have recently been reported extensively for several applications in electronics, photonics and electrochemical energy storage devices due to their unique properties. Various 2D materials that include graphene, boron nitride, layered hydroxide, transition metal chalcogenides, oxides, etc. are being studied. Recent additions in this direction include silicene, silicon carbide (SiC) and transition metal diborides. The following schematic gives an overview of 2D layered materials that have been reported for ORR. The present section will describe the efforts made in this direction, based on recent literature.

3.5.1 Graphene

Graphene is a 2D layer network of sp^2-hybridised carbons that has attracted enormous attentions as a promising catalyst for electrochemical ORR. Discovery of graphene has opened up a new generation of 2D materials for fundamental science and potential technology [60, 61]. Large surface area as well as high electrical conductivity has been the main aspect implicated in the improvement

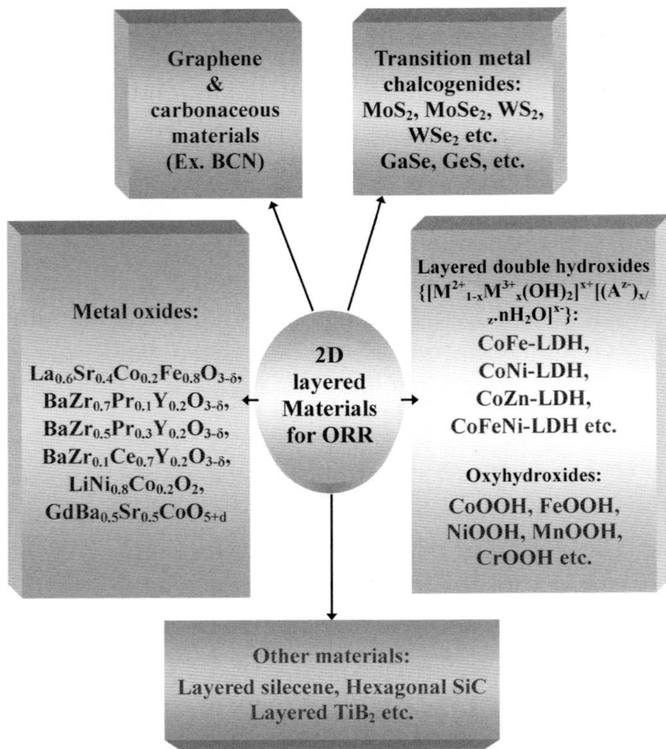

Scheme 1. Various 2D, layered materials studied for ORR.

of its catalytic activity. Having similarities with CNTs in terms of structure and properties, graphene is very attractive and an alternate candidate for several electrocatalytic reactions, sometimes reported to be even superior to CNTs. The 2D, planar geometry with one atomic thickness facilitates electron transport very effectively. Additionally, doping of heteroatoms such as nitrogen into graphene sheets further modulates the electronic properties [62] and enhances its activity towards various electrocatalytic reactions. It has been reported by different groups [63, 64] that N-doped graphene (NG) serves as excellent, highly stable ORR catalyst, which is tolerant to crossover and CO poison effects.

Figure 4 reveals the high electroactivity of graphene, particularly when doped with nitrogen and improvement in both onset potential

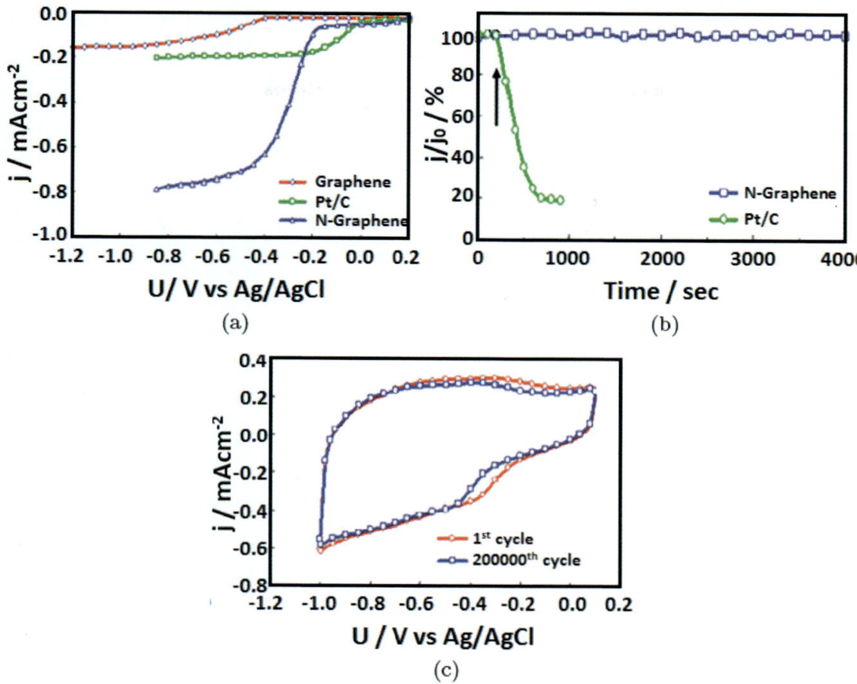

Figure 4. (a) Linear sweep voltammograms for ORR in air-saturated 0.1 M KOH on graphene (red), Pt/C (green) and N-graphene electrodes (blue) at a rotation speed of 1000 rpm and scan rate of $10\,\mathrm{mV/s}$. (b) Chronoamperometric response on Pt/C (circle line) and N-graphene (square line) electrodes to CO, upon addition of 10% (v/v) CO at $0.4\,\mathrm{V}$ (arrow indicates the time of addition of CO); j_o defines the initial current. (c) Cyclic voltammograms of N-graphene electrode before (red) and after (blue) 200,000 cycles at $25°\mathrm{C}$ at a scan rate of $100\,\mathrm{mV/s}$. (Adapted from Ref. [64]).

and current density are observed. It should be quickly noted that the steady-state catalytic current density for N-graphene catalyst is almost three times higher than that of Pt/C catalyst over a wide potential range. Upon incorporation of N atom in the graphene network, the ORR mechanism is strongly influenced, showing a one step 4 electron reduction pathway while pristine graphene catalyses 2-step 2 electron pathway. The inactivity of N-graphene [64] for CO in also pointed out by the authors (Figure 4(b)). The stability of the material for cycling is revealed in Figure 4(c). The results indicate that NG is highly efficient and extremely stable for ORR with no CO poisoning effect. It has also been observed that the NG possesses many active sites that coordinate with other elements like Fe, Co, Ni which further improve its catalytic activity towards ORR [65–67].

Parvez *et al.* [65] have reported Fe embedded NG as an efficient ORR catalyst. The composite is synthesised by heating carbon-nitride graphene (CN-G, obtained from cyanamide and graphene oxide precursor) along with Fe-salt ($FeCl_3$) at high temperatures (900°C) (Figure 5(a)). Figure 5(b) depicts the comparative ORR activity among pristine NG obtained by heating CN-G (without Fe salt) at different temperatures. It is observed that the NG-900 heated at 900°C exhibits better catalytic activity than those of ND-800 and NG-1000, and is attributed to the fact [65] that NG-900 contains more graphitic N over pyridinic N as compared to the other two [65]. Incorporation of Fe in NG-900 shows good catalytic activity and the amount of Fe in NG-900 largely affects the ORR activity. It is observed that Fe content up to 5 wt.% improves the activity and further increase is reported to be detrimental. It is likely that Fe coordinates with the available pyridinic N and when the availability of pyridinic N is exhausted (10–15 wt.% of Fe), excess non-coordinated Fe is present in the catalyst and it does not contribute towards ORR (Figure 5(c)). Hence, it is confirmed that the improvement of ORR activity of N-graphene is mainly due to the graphitic N, and the synergistic coupling between Fe and NG further improves the activity.

Ning *et al.* [67] have reported $CuCo_2O_4$ nanoparticles supported on N-doped reduced graphene oxide ($CuCo_2O_4$/N-rGO) as a highly efficient catalyst for ORR. The composite is prepared by a simple hydrothermal reaction from the precursors (GO, $NH_3 \cdot H_2O$ as N source, Cu^{2+} and Co^{2+} salt as Cu and Co source) (Figure 5(d)). The uniform distribution of the nanoparticles on N-rGO matrix is confirmed from transmission electron microscope (TEM) (Figure 5(e)). Figure 5(f) reveals the ORR activity of the $CuCo_2O_4$/N-rGO composite and it is almost comparable to the state of the art material, Pt/C.

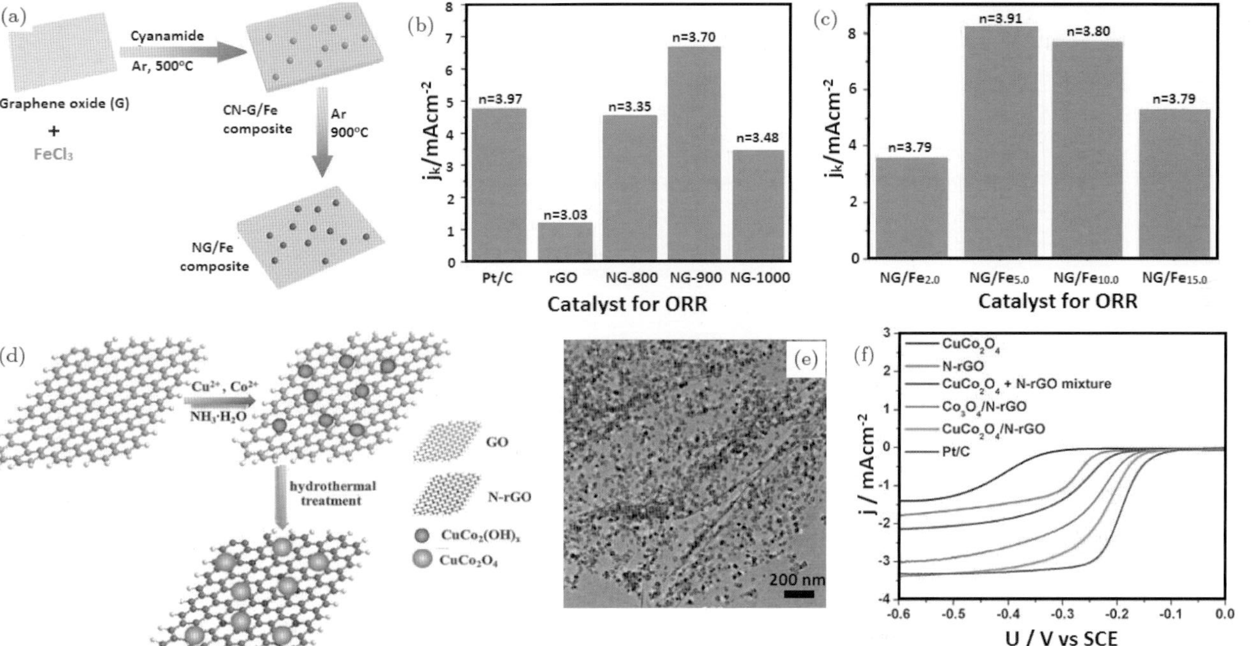

Figure 5. (a) Schematics of synthesis of Fe nanoparticle embedded NG (NG/Fe). (b) and (c) show the comparison of ORR activity of NG-T (NG is obtained by heating the precursors at different temperature T, $T = 800°C$, $900°C$ and $1000°C$) in oxygen saturated 0.1 M KOH, kinetic-limiting current density (j_k) value is determined at -0.4 V vs. Ag/AgCl. (Adapted from Ref. [65]). (d) Schematics of synthesis of $CuCo_2O_4$ embedded N-doped reduced graphene oxide, (e) corresponding TEM image, (f) electrocatalytic activity of $CuCo_2O_4$/N-rGO towards ORR in comparison with their pristine counter-parts and Pt/C. The experiments are carried out in oxygen saturated 0.1 M KOH solution at a scan rate of $5\,mVs^{-1}$ and rotation speed of 1600 rpm. (Adapted from Ref. [67]).

3.5.2 Borocarbonitrides

As mentioned in the earlier section, co-doping of heteroatoms, with different electronegativities, in carbon matrix improves the catalytic activity quite efficiently as compared to singly-doped carbon. Doping results in localised charges on the carbon network. The catalytic activity of the materials strongly depends upon the amount of carbon content, present in it. Wang *et al.* [29] reported $B_xC_yN_z$-graphene (with different x, y and z values) as efficient ORR catalysts. It is observed that with increasing carbon content in BCN-graphene, the catalytic activity increases (Figure 6(a)). Density functional theory (DFT) studies have been carried out to corroborate the experimental data and it is observed that with co-doping of heteroatoms (B and N), the highest occupied molecular orbital–lowest unoccupied molecular orbital (HOMO–LUMO) band gap decreases initially (for $B_{12}C_{77}N_{11}$), but with over-doping of B and N, the band gap increases and becomes very large for heavily doped graphene ($B_{38}C_{28}N_{34}$). This results in an increase in conductivity, as well as ORR catalytic activity upon light-doping of B and N atoms ($B_{12}C_{77}N_{11}$) and the activity decreases subsequently. Figure 6(b) exhibits the relationship between the electrocatalytic activity towards ORR and the HOMO–LUMO energy gap for BCN-graphene with different carbon contents. Also, upon initial doping of heteroatoms (B and N) in the graphene network, kinetic stability of the material decreases due to a decrease in the band gap, which implies that the state of the doped graphene is energetically favourable to accept electrons by LUMO as well as to extract electrons from HOMO, resulting in high chemical reactivity [68].

The atomic spin densities and charge densities have also been calculated for BCN-graphene model structures [29] and are listed in Table 2. The data reveals that pure (C_{100}) and heavily doped ($B_{38}C_{28}N_{34}$) graphene do not have any spin density. Although $B_{38}C_{28}N_{34}$-graphene contains most of the atoms (37%) with high charge density (>0.15) due to heavy doping, yet it shows poor ORR activity as all the atoms are distributed in the boron nitride (BN) phase and the BN cluster is electrically insulating, resulting in slow electron transfer. In contrast, $B_7C_{87}N_6$ and $B_{12}C_{77}N_{11}$ have several

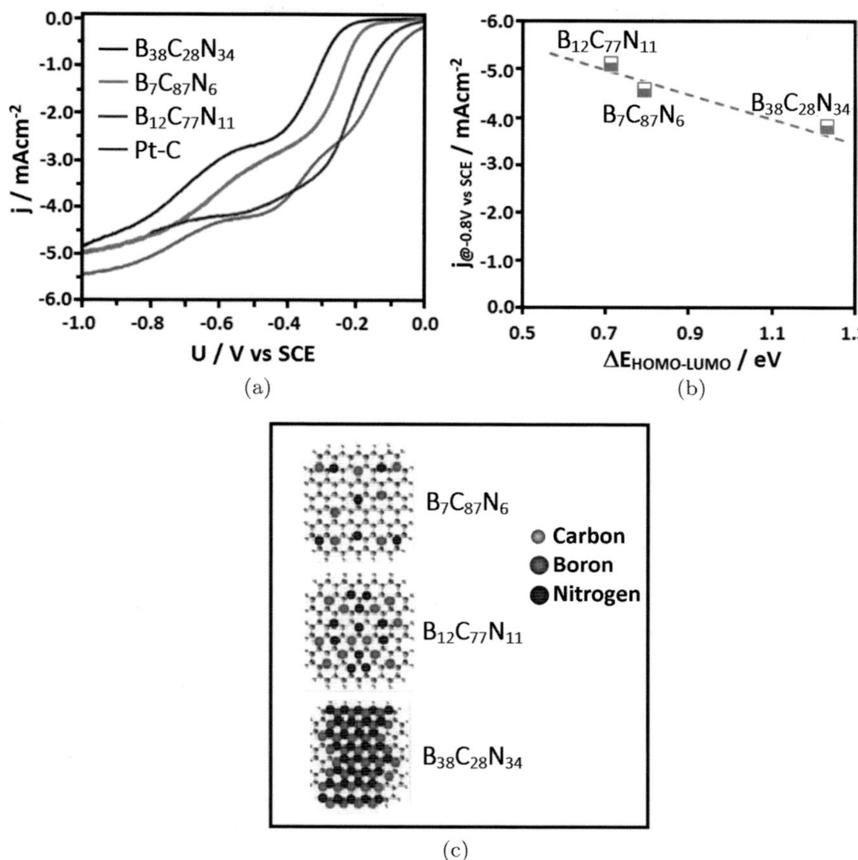

Figure 6. (a) Voltammograms on BCN-graphene towards ORR with different compositions in oxygen-saturated 0.1M KOH solution at $10\,\mathrm{mVs^{-1}}$. (b) ORR activities at a potential of $-0.8\,\mathrm{V}$ vs. SCE vs. HOMO-LUMO energy gap, calculated by DFT for BCN-graphene sheets with different carbon content. (c) BCN-model with different compositions. (Adapted from Ref. [29]).

carbon atoms with relatively high spin density as well as high charge density as compared to pure graphene due to doping, and they provide active sites for catalytic reduction of oxygen [37, 69]. If both spin and charge densities are considered, the number of possible catalytic sites in $B_{12}C_{77}N_{11}$ is larger than that of the number of sites available in $B_7C_{87}N_6$. Hence, $B_{12}C_{77}N_{11}$-graphene shows very good ORR catalytic performance among all the samples.

Table 2. HOMO-LOMO energy gap, spin density and charge density of pure and B and N-doped-graphene structures ([29]).

Structure	$\Delta E_{\mathrm{HOMO-LUMO}}$/eV	Maximum spin density	Maximum charge density
C_{100}	0.80	0	0.19 (4.0%)
$B_7C_{87}N_6$	0.79	0.33 (3.6%)	0.38 (10.8%)
$B_{12}C_{77}N_{11}$	0.71	0.36 (2.8%)	0.56 (21.6%)
$B_{38}C_{28}N_{34}$	1.23	0	0.34 (37.0%)

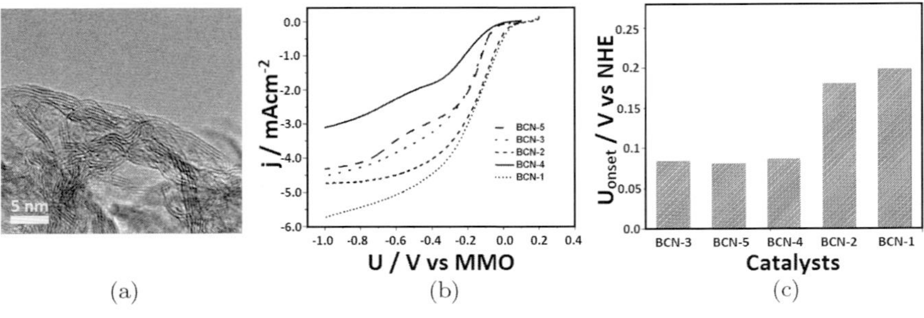

Figure 7. (a) HR TEM image of few layer BCN-1 nanosheets, (b) activity of different $B_xC_yN_z$ towards ORR recorded at a rotation speed of 1500 rpm and at a scan rate of $5\,\mathrm{mVs}^{-1}$ and (c) variation of onset potential for ORR as a function of composition of BCN catalyst. (Adapted from Ref. [70]).

Moses *et al.* [70] have reported composition dependent studies on ORR activity using few layers $B_xC_yN_z$ nanosheets. Layer-type morphology with an interlayer distance of 0.35 nm is observed from high resolution TEM (HRTEM) studies (Figure 7(a)). It is clear from their studies that the composition with high carbon content ($B_{0.15}C_{0.72}N_{0.12}$ or BCN-1) exhibits relatively better activity in terms of both onset potential as well as current density, as compared to other compositions ($B_{0.32}C_{0.42}N_{0.25}$, $B_{0.43}C_{0.31}N_{0.26}$, $B_{0.37}C_{0.39}N_{0.23}$, $B_{0.39}C_{0.43}N_{0.18}$, $B_{0.38}C_{0.24}N_{0.36}$ or BCN-2, BCN-3, BCN-4, BCN-5 and BCN-6, respectively) (Figures 7(b) and 7(c)). This is reported to be due to high carbon content in BCN-1 imparting good electronic conductivity as well as diffusion path ways as compared to the other compositions. Further, it is observed that BCN with low carbon content catalyses ORR by two step process (Figure 7(b)) to reduce oxygen while BCN-1, BCN-2 with high carbon content proceeds through one step process, to directly produce water.

3.5.3   Layered transition metal oxides

Graphene is the first 2D material experimentally explored and has given rise to a host of other materials based on chalcogenides, oxides, etc. Transition metal oxides are an important class of wide band-gap semiconducting materials composed of oxygen atoms bound to transition metals. These are generally known for their different catalytic activities and semiconducting properties. Their diverse structural variation affects the surface energy and influence chemical properties. Properties based on relative acidity-basicity and electronic structure are highly influenced by the nature of coordination of the metal centre and oxygen ions. The resulting structural defects greatly influence the catalytic properties.

Metal oxide-based materials have been considered as efficient cathode materials in fuel cells particularly, from the stability point of view [71–79]. Although these are highly stable in strong acidic and oxidative atmosphere, their large band gaps restrict their use in electrochemical catalysis. Hence, it is necessary to modify the materials to improve their catalytic activity. Zakaria *et al.* [71] have reported layer-by-layer formation of nanoporous nickel oxide flakes — graphene composites and used them as efficient ORR catalyst. Figure 8(a) gives a schematic representation of NiO-GO composites, where NiO flakes are enclosed by GO sheets in layer-by-layer manner as observed in the TEM pictures of the composite (Figure 8(b)). Figure 8(c) shows that the ORR-catalytic activity is enhanced upon composite formation and it is found to depend on the relative concentration of NiO and GO. Further, it is observed that when the ratio of NiO to GO is 3:1, best ORR activity is exhibited as compared to other compositions (NiO:GO ratio = 1:1 and 1:3). The large proportion of metal oxide provides comparatively large surface area, resulting in decreased of mass transport resistance that facilitates access of the electrolyte to active sites for ORR [71]. Transition metal oxides not only behave as catalysts for ORR but also act as supports for ORR catalysts. Chaisubanan *et al.* [73] have

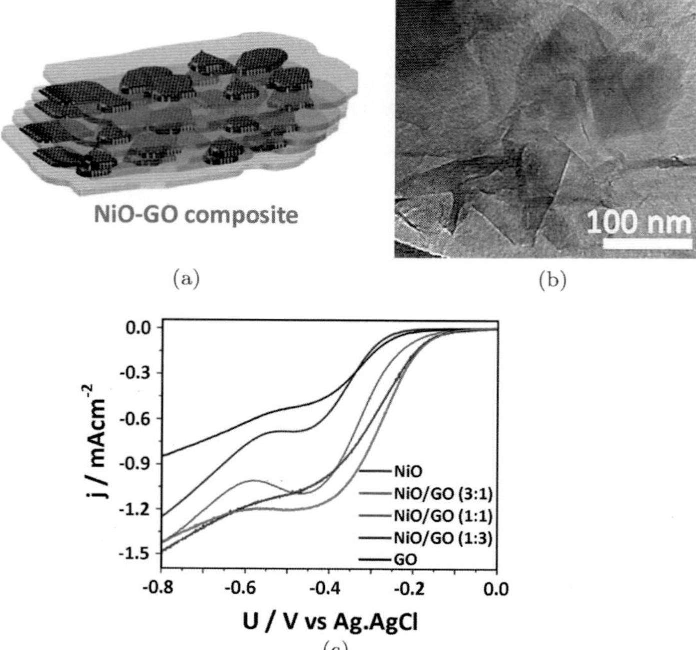

Figure 8. (a) Schematic representation of layer-by-layer NiO-GO hybrid material, (b) TEM of NiO-GO composite, (c) comparative study of electrocatalytic activity of different compositions of NiO-GO composite along with the pristine counterpart in oxygen saturated 0.1 M KOH solution. (Adapted from Ref. [71]).

reported observations where in layered transition metal oxides, MO_2 (M = Ti, Ce, Mo) affect the catalytic activity of PtCo/C towards ORR. The electrochemically active surface area, internal contact resistance are reported to get affected favourably to enhance the performance.

Layered mixed metal oxides have been reported as efficient ORR catalysts that could be used in fuel cells [74–77]. Mixed oxides are attractive since the presence of different metals introduces defects in the system, that occasionally results is good catalytic activities. Layer-type $LiNi_{0.8}Co_{0.2}O_2$ (LNCO) [78] is an efficient ORR catalyst that has been used in proton-conducting solid oxide fuel cells. Figure 9(a) shows the fuel cell performance of layered LNCO in combination with $BaZr_{0.1}Ce_{0.7}Y_{0.2}O_{3-\delta}$ (BZCY) electrolyte. The

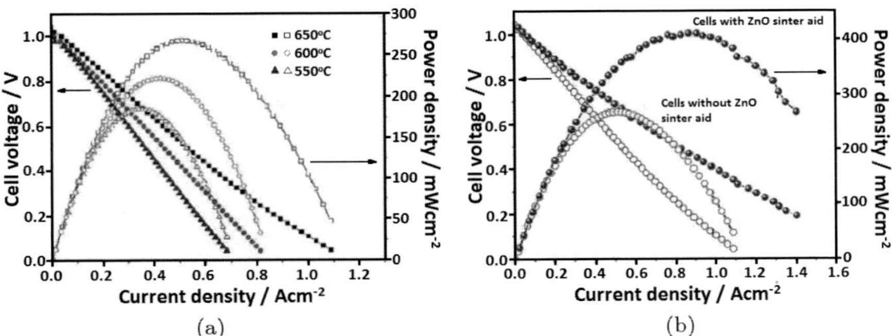

(a) (b)

Figure 9. Fuel cell characteristics (voltage–current density and power density-current density) of layered LNCO cathode with (a) BZCY electrolyte and (b) BZCY electrolyte along with 1% ZnO. (Adapted from Ref. [78]).

electrolyte film is spray deposited, followed by sintering at 1500°C for 3 h. This procedure has been shown to produce peak power density ($P_{\max}$) value of 268 mWcm^{-2} at 650°C under H$_2$/air flow. The cell performance is further improved by incorporating 1% of ZnO as the sinter-aid for BZCY electrolyte and the sintering temperature could be lowered to 1400°C. The high porosity of the anode microstructure and reduced ohmic resistance are reported to be responsible for the improvement in performance [78]. Peak power density of 410 mWcm^{-2} at the same temperature is obtained (Figure 9(b)).

Kim *et al.* [79] have reported a layered perovskite oxide GdBa$_{0.5}$Sr$_{0.5}$Co$_2$O$_{5+d}$ (GBSCO) as ORR catalyst (cathode) for intermediate temperature solid oxide fuel cells (Figure 10) and studied the effect of Fe doping on Co site on the fuel cell performance. It is observed that with increasing Fe content in GdBa$_{0.5}$Sr$_{0.5}$Co$_{2-x}$Fe$_x$O$_{5+d}$ ($x = 0, 0.5, 1$), fuel cell performance is improved. Peak power density value obtained using GdBa$_{0.5}$Sr$_{0.5}$Co$_{2-x}$Fe$_x$O$_{5+d}$ (GBSCF5) with $x = 1$ GdBa$_{0.5}$Sr$_{0.5}$Co$_{1.0}$O$_{5+d}$ (GBSCF10) cathode is $\sim$2 Wcm^{-2} at 650°C, whereas the $P_{\max}$ values for $x = 0$ and $x = 0.5$ (GBSCO and GBSCF5, respectively) are around 1.7 Wcm^{-2} and 1.8 Wcm^{-2}, respectively, under identical conditions. The difference in cell performance is inversely related to the area-specific resistance (ASR) values and it is found that the ASR value is lowest in the case of GBSCF10 and increases with a decrease in Fe content [79].

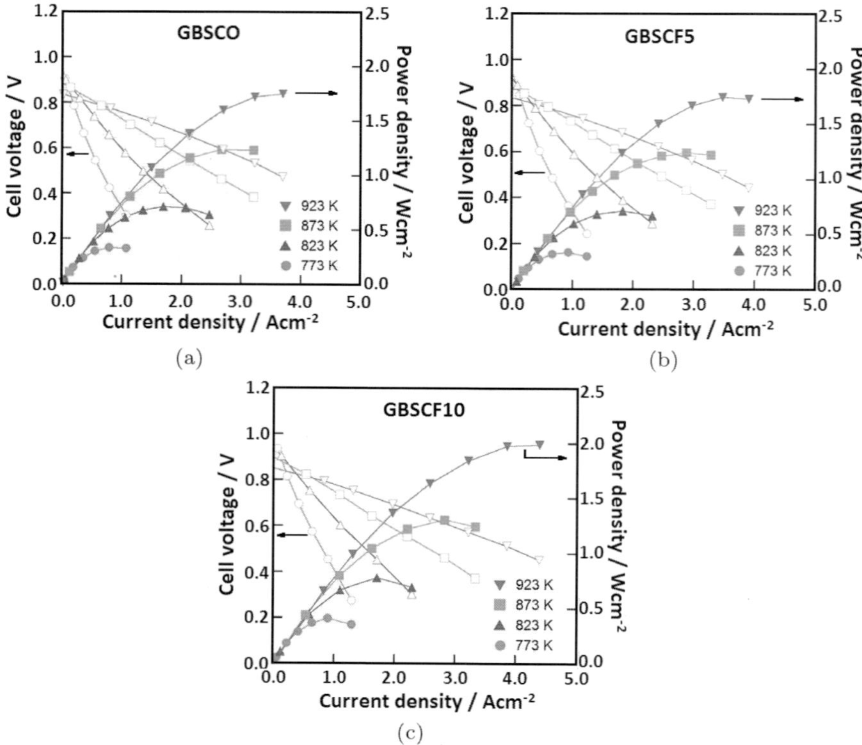

Figure 10. Fuel cell performance (I–V curves and corresponding power density curves) of single cells (GBSCF$_5$ (cathode)/GDC/NiO–GDC) at various temperatures: (a) $x = 0$, (b) $x = 0.5$, and (c) $x = 1.0$. (Adapted from Ref. [79]).

3.5.4   2D chalcogenides

2D layered transition-metal dichalcogenides (TMD) possess general formula MX_2, where M is a transition metal, Mo, W, Re, etc., and X is a chalcogen (i.e., S, Se, or Te). Due to their unique structural, chemical and favourable redox properties, they are used for various electrochemical applications. Extensive studies have been carried out, mainly on hydrogen evolution reaction (HER) [80–86], batteries [80, 87] and supercapacitors [88]. In the present section, we focus on the electrocatalytic activity of 2D TMDs towards ORR. Rao *et al.* [89] have reviewed the use of certain layered materials for various electrochemical studies including ORR.

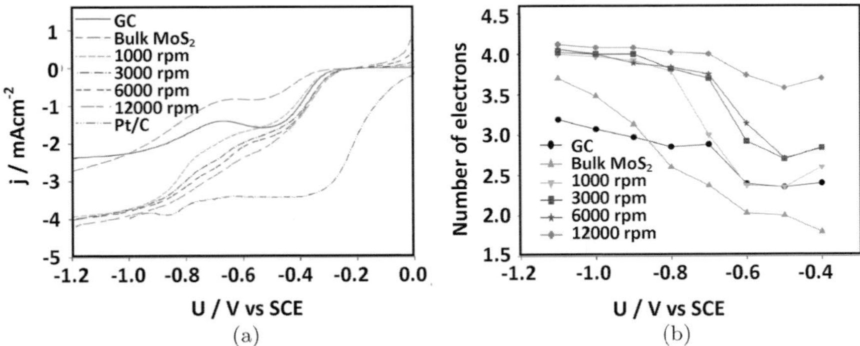

Figure 11. (a) Dependence of electrocatalytic activity on the particle size of MoS$_2$ (as mentioned in the text of the paper, different sized MoS$_2$ has been collected after 1000, 3000, 6000 and 12000 rpm centrifugation speed) in oxygen saturated 0.1 M KOH at a scan rate of $10\,\mathrm{mVs}^{-1}$ and with a rotation speed of 900 rpm. (b) A number of electrons transferred during ORR for the different sized MoS$_2$ particle at different applied potentials. The catalysts are (Adapted from Ref. [90]).

Very few reports are published on the use of pristine 2D layered TMDs as ORR catalysts [90, 91]. Wang *et al.* [90] reported ORR activity on pure MoS$_2$ that is enhanced upon decreasing the size of the particles (Figure 11(a)). More importantly, it is observed that for bulk and large sized MoS$_2$, the number of electrons transferred during ORR highly depends on the applied potential. The potential dependence is not prominent for small sized particles (Figure 11(b)) suggesting that the electronic structure based on particle size affects the ORR mechanism. ORR activity has also been followed on chemically exfoliated MoSe$_2$, WS$_2$ and WSe$_2$ [91], but there is no real significant enhancement of the electroactivity is observed. The ORR activity of MoSe$_2$ is comparatively better than that of WS$_2$ and WSe$_2$ (Figure 12) though the reason is unclear at present.

Improvement in activity [92] has been made possible upon formation of composites with conducting NG (Figure 13(a)). The activity for ORR has improved dramatically on incorporating oxygen into MoS$_2$ sheets [93]. Oxygen incorporation is carried out by treating aqueous colloids of MoS$_2$ with different amounts ('x' μL) of hydrogen peroxide for $\sim$2 h, followed by washing with pure water to remove

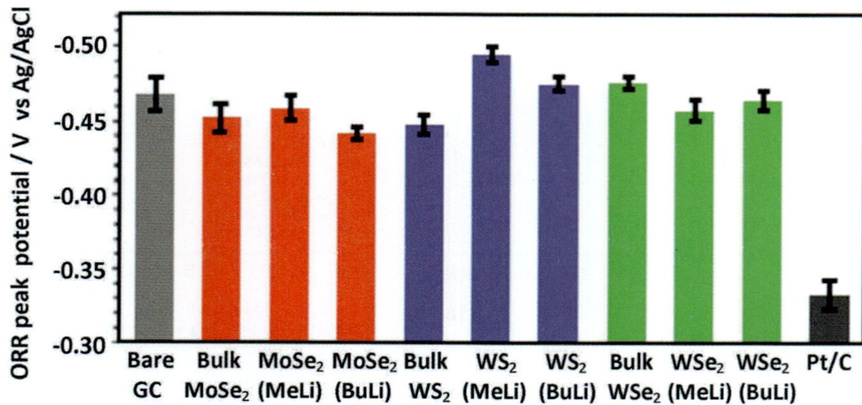

Figure 12. The plot of peak potential for ORR (vs. Ag/AgCl reference electrode) on different 2D chalcogenides (red: MoSe$_2$, blue: WS$_2$ and green: WSe$_2$), Bare glassy carbon and Pt–C are also given. (Adapted from Ref. [91]).

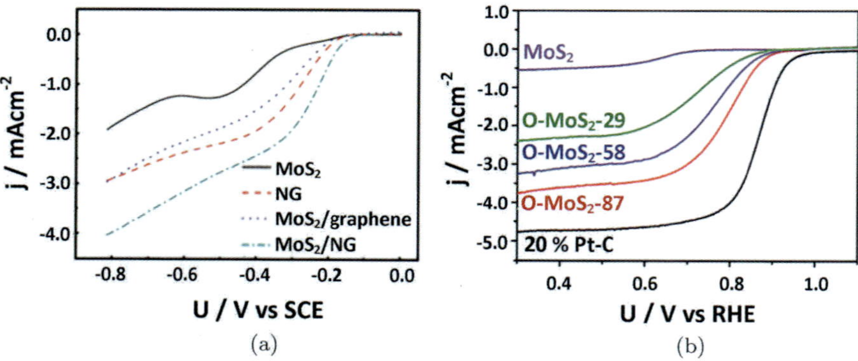

Figure 13. (a) Linear sweep voltammograms in oxygen saturated 0.1 M KOH solution on various modifications on MoS$_2$. Scan rate used is $10\,\mathrm{mVs^{-1}}$, rotation speed = 1600 rpm. (Adapted from Ref. [92]) and (b) Voltammograms on oxygen incorporated MoS$_2$ sheets (scan rate: $10\,\mathrm{mVs^{-1}}$, rotation speed: 2400 rpm). (Adapted from Ref. [93]). O-MoS$_{2-x}$ denotes the extent of oxygen. Please see the text for details.

excess hydrogen peroxide and by-products, and the resulting product is denoted as 'O-MoS$_2$-x'. The value of x is varied from $29\,\mu\mathrm{L}$, $58\,\mu\mathrm{L}$ to $87\,\mu\mathrm{L}$ and it is observed that the ORR activity depends on the x value (Figure 13(b)). Incorporation of oxygen not only improves the

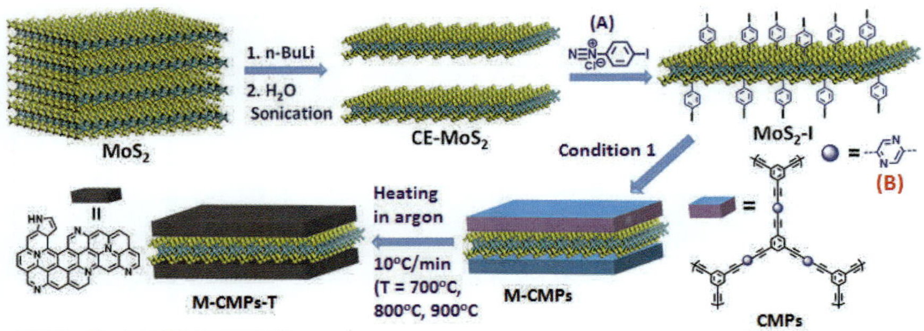

Figure 14. Schematic representation of the synthesis of MoS_2 coupled conjugated microporous polymer, followed by MoS_2 embedded N-doped porous carbon nanosheets. (Adapted from Ref. [94].)

activity, but also triggers the transformation process from 2-electron to 4-electron pathway.

Recently, Yuan *et al.* [94] have reported MoS_2 embedded highly N-doped porous carbon nanosheets, obtained from MoS_2-conjugated microporous polymer hybrid (M-CMPs) precursor, as an excellent catalyst for ORR. The hybrid of MoS_2 and N-doped porous carbon nanosheets are obtained by annealing M-CMPs at high temperatures. The sample is denoted as M-CMPs-T where T represents the annealing temperature, 700°C, 800°C and 900°C. Synthetic procedure of MCMPs-T is schematically shown in Figure 14. Figure 15(a) represents the TEM image of M-CMP-800 which shows the MoS_2 nanosheets spread on carbon matrix (pointed in the figure). Although pristine MoS_2 is not a highly efficient catalyst for ORR, upon forming hybrid with N-doped porous carbon sheets, a large increase in ORR catalytic activity is observed. Figures 15(b) and 15(c) reveal the catalytic activity of M-CMPs-T towards ORR. It is also observed that the precursor M-CMPs produces the most efficient catalyst upon heat treatment at 800°C (Figure 15(b)) and the catalyst (M-CMPs-800) exhibits almost comparable activity with that of Pt/C, showing a difference in half-wave potential (HWP) of only 11 mV (Figure 15(c)).

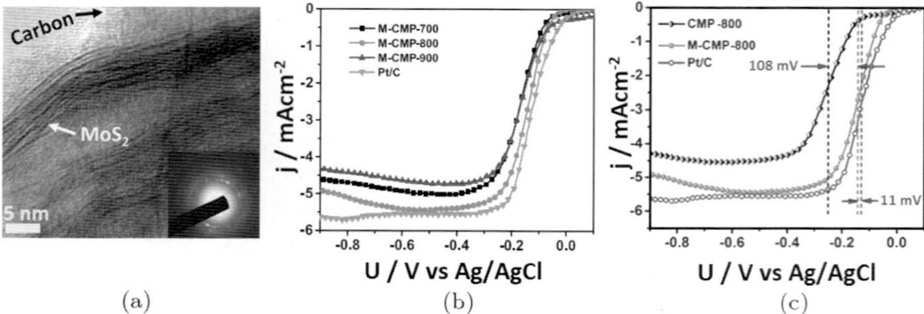

Figure 15. (a) TEM image of M-CMPs, (b) comparison of ORR activity among M-CMPs-T obtained by heating and pyrolising M-CMPs at different temperatures, (c) comparative ORR activity of M-CMP-800 with pristine CMP-800 and Pt/C. Linear sweep voltammograms are carried out in oxygen saturated 0.1 M KOH solution at a scan rate of $5\,\mathrm{mV\,s^{-1}}$ and at 1600 rpm of rotation speed. (Adapted from Ref. [94].)

Other than the conventional 2D-layered TMD materials (General formula MX_2), other 2D-chalcogenide materials [95] such as layered GaSe and GeS are shown to exhibit multifunctional catalytic activity towards ORR, OER and HER. The morphology shows the layer-like structure for both GaSe and GeS (Figures 16(a) and 16(b)). Figure 16(c) depicts the electrocatalytic activity of GaSe and GeS towards ORR. Though the activity is not very good, this opens up certain novel layered materials for electrocatalysis.

3.5.5 Layered double hydroxides and oxyhydroxides

Layered double hydroxides (LDHs), known as hydrotalcites are a class of 2D anionic clays made up of positively charged brucite-like $(Mg(OH_2))$ host layers and an exchangeable charge-balancing interlayer anions [96]. They are extensively studied in the areas of catalysis. LDHs are expressed by the general formula $\{[M^{2+}_{1-x}M^{3+}_x(OH)_2]^{x+}[(A^z)_{x/z} \cdot nH_2O]^{x-}\}$ [97], where the divalent metal ions, M^{2+} (such as Mg^{2+}, Fe^{2+}, Co^{2+}, Cu^{2+}, Ni^{2+}, or Zn^{2+}) and trivalent metal ions, M^{3+} (such as Al^{3+}, Cr^{3+}, Ga^{3+}, Mn^{3+} or Fe^{3+}) are octahedrally coordinated with the hydroxyl groups in the brucite — like layers. The molar ratio of the trivalent and divalent cations (value of x) normally varies between 0.2 and 0.4 [98].

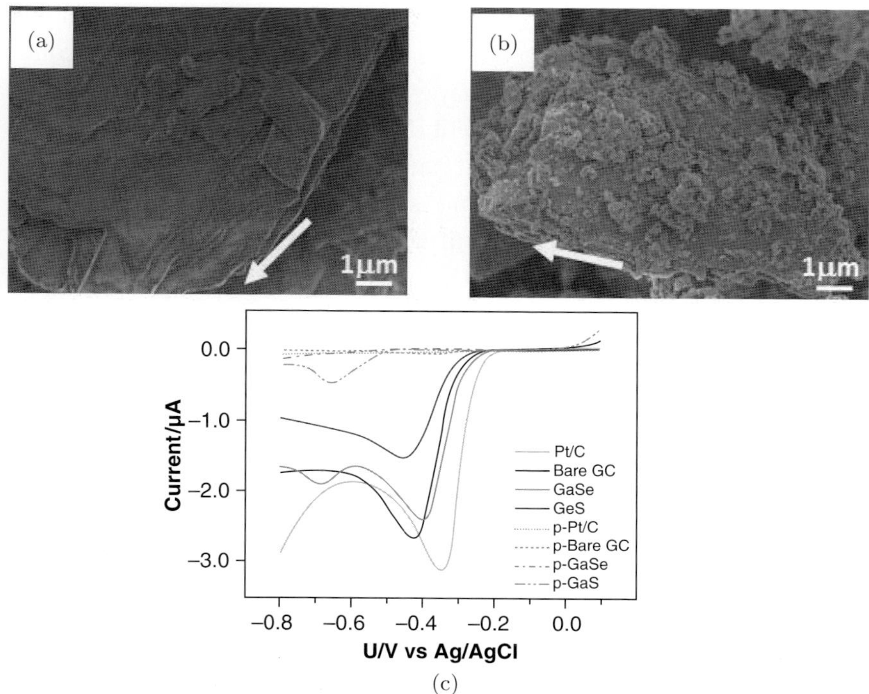

Figure 16. SEM of (a) GaSe and (b) GeS (c) linear sweep voltammograms of ORR on GaSe, GeS, glassy carbon and Pt–C. Solid lines denote oxygen saturated and dotted lines denote nitrogen saturated 0.1 M KOH. Scan rate used is $5\,\mathrm{mVs^{-1}}$. (Adapted from Ref. [95].)

Normally, in any LDH, CO_3^{2-} acts as the interlayer charge compensating anion (i.e., A^{z-}) because of its high affinity to the hydroxide layers [99]. The mixed-valent properties with the different molar ratio (different × values) and the layer-type structure of LDHs open up enormous areas of applications in the areas of adsorption, intercalation, devices, etc. Another attractive property of LDH materials is that they can be easily converted to the corresponding mixed metal oxides with a same molar ratio of trivalent to divalent metal ions by calcination treatment [100]. High surface area is an important attribute to this class of compounds.

Though there are reports on the use of calcined LDHs for ORR [100, 101], herein, we consider only the pristine LDH since the calcined materials do not retain the layered structure of LDH. Cobalt and iron centres are found to be unique with the favourable electronic structure for ORR catalysis. Zhang *et al.* [102] have reported that Co–Fe based LDH (Co is divalent and Fe is trivalent) along with polydopamine spheres (PDAS) act as a highly efficient ORR catalyst in alkaline medium. Polydopamine is added to improve the catalytic activity since it is known that nitrogen containing species enhance the ORR rate. It is observed that this hybrid material catalyses ORR by a 4-electron pathway to produce H_2O. Comparative studies carried out with pristine LDH and with PDAS reveal that the hybrid material performs better than the pristine counterpart and is attributed to the synergistic effect between LDH and PDAS (Figure 17). Ternary NiCoFe LDH with pre-oxidation treatment to improve the valence state of Co from $+2$ to $+3$ has been found to perform well towards ORR [103]. The pre-oxidation treatment changes only the oxidation state of Co and the phase remains intact as LDH. The electroactivity, however improves significantly (Figure 18). This opens up a way of tuning the ORR activity on materials containing earth abundant elements.

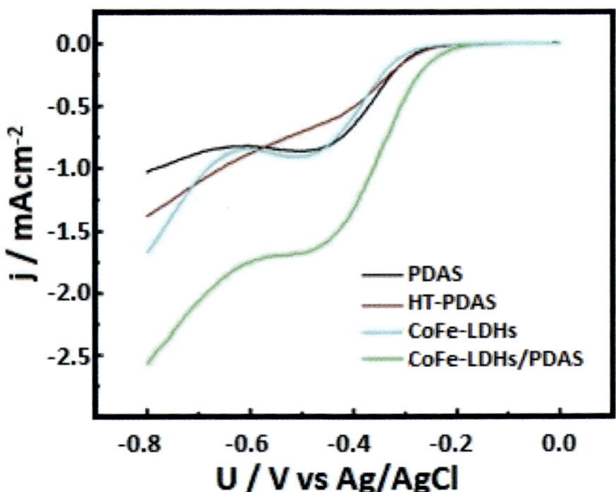

Figure 17. Electrocatalytic activity of CoFe-LDH/PDAS and related compounds towards ORR in an oxygen saturated 0.1 M of KOH solution at a scan rate of $10\,mVs^{-1}$ and with the rotation speed of 1600 rpm. (Adapted from Ref. [102].)

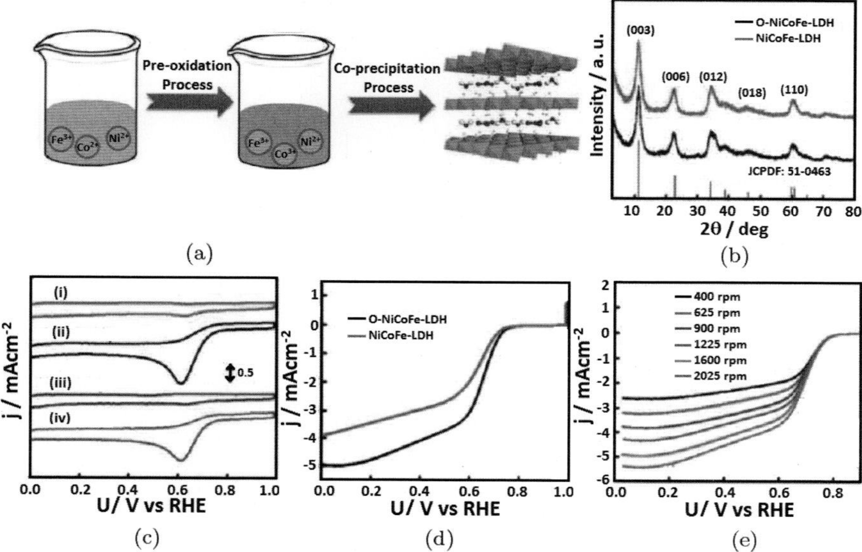

Figure 18. (a) Schematics of synthetic procedure of O–NiCoFeLDH, (b) X-Ray diffraction pattern of pre-oxidised ternary LDH (black) and corresponding LDH without pre-oxidation, (c) voltammograms of O–NiCoFeLDH ((i) in presence of N_2 and (ii) O_2) and NiCoFeLDH ((iii) in presence of N_2 and (iv) O_2), (d) activity of LDH catalysts towards ORR, (e) linear sweep voltammogram of O–NiCOFe–LDH at different rotation speed. (Adapted from Ref. [103].)

As for layered oxyhydroxides, there has been a recent report on ORR though this class of compounds are well known for catalysing oxygen evolution reaction (OER) [104]. The electrocatalytic activity of five different oxyhydroxides namely, CoOOH, CrOOH, FeOOH, NiOOH and MnOOH have been studied and it is found that CoOOH possesses the lowest over potential in alkaline medium. However, the activity is far inferior to Pt-based catalysts.

3.5.6 Layered transition metal boride (TiB_2)

Several transition metal diborides possess layer structure and show remarkable properties like good chemical stability, high thermal conductivity excellent refractory property, etc., that have led to their use in several areas. However, their use in electrochemistry remains very limited. Recently [105], TiB_2 has been proposed as

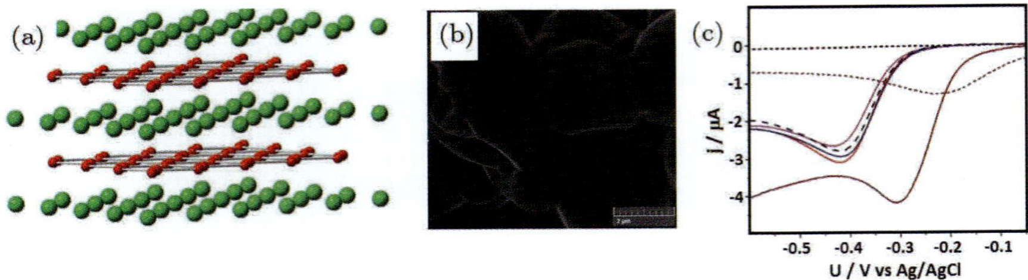

Figure 19. (a) The structure of TiB_2 with hexagonally aligned B atoms (red) and planes of Ti atom (green) in alternate layers, (b) SEM of TiB_2 sheets (c) electrocatalytic activity towards ORR in oxygen saturated (solid lines) and in the presence of dissolved oxygen (dotted lines) 0.1 M KOH. TiB_2 (orange–red), TiB_2/NaNAFT (blue), TiB_2/BuLi (pink), Pt–C (wine red) and bare glassy carbon (black dashed line) electrodes. (Adapted from Ref. [105].)

an electrocatalyst for ORR and HER. The authors have used TiB_2 modified with sodium naphthalenide or butyllithium to force exfoliation of the layers. Subsequently, electrocatalytic activities have been studied towards ORR and HER. Unmodified TiB_2 has also been used for comparative purposes. Figure 19(a) shows the structure of layered TiB_2, where B atoms (red) are hexagonally arranged in between two planes of Ti atoms (green). The layer-type morphology is observed in the SEM pictures (Figure 19(b)). Figure 19(c) shows the voltammograms for ORR in aqueous alkaline medium. The solid lines in this figure represent polarisation curve in an oxygen-saturated 0.1 M KOH solution, whereas the dotted lines represent the same in presence of dissolved oxygen. Peaks around $-0.4\,V$ vs. Ag/AgCl reference electrodes are due to ORR catalysed by TiB_2. It is also found that modification of TiB_2 does not change its electrochemical properties. This is the first report on TiB_2 for electrocatalysis and though the activity is nowhere close to Pt, it opens up an area where there are possibilities of improvement.

3.5.7   Silicene and SiC

Silicene is Si analogue of graphene and is predicted to be a gapless semiconductor [106]. The sp^2 hybridisation along with weak overlap of $3p_z$ orbitals leads to very reactive surface and amenable for adsorption of various species. The presence of sp^2 and sp^3 mixed

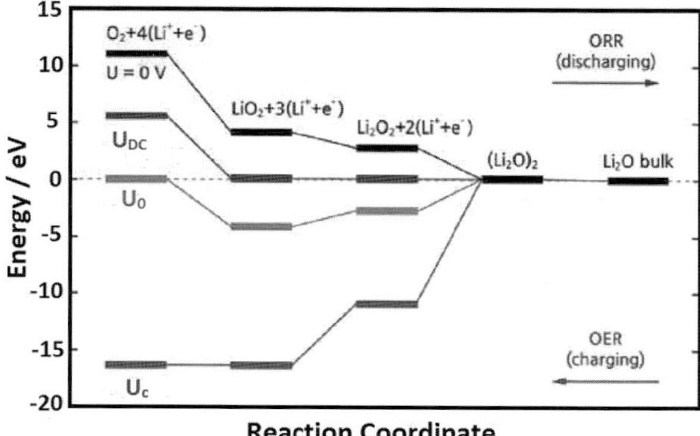

Figure 20. Free energy diagram for ORR and OER on silicene at zero potential ($U = 0\,V$), at maximum discharge potential (U_{DC}) of 1.36 V, minimum charge potential (U_C) of 6.84 V and at a theoretical equilibrium potential (U_0) of 2.76 V. (Adapted from Ref. [109]).

hybridisation [107] imparts the ability to be reactive and hence the variety of functionalisation has been proposed. Padova *et al.* [108] have reported the growth of aligned, high aspect ratio silicene nanoribbons. Based on first principle calculations on the adsorption of ORR intermediates, it is shown that [109] this material can be an excellent ORR catalyst. In contrast to graphene, it is expected to exhibit both ORR as well as OER activity. This might lead to an important catalyst for rechargeable metal–air battery systems [109]. The adsorption energies of various intermediates involved in Li-air battery system is given in Figure 20. The data reveals that both ORR and OER take place on silicene without any additional catalyst. Further, it is observed that the intermediate, as well as the product upon ORR in Li–O_2 battery possess lower energy than that of the initial reactants, which indicates spontaneity for the reduction of oxygen by silicene. Similar behaviour is observed for OER as well. Unlike graphene which promotes ORR only on defect sites, pristine silicene is predicted to be a good material for promoting adsorption and dissociation of oxygen and hence a four electron pathway may be expected [110].

Another Si-based layered material, SiC has been explored to understand its possible electroactivity towards ORR, by model calculations. High thermal conductivity, critical electrical field and carrier mobility make SiC an attractive candidate for several electronic applications. It has been suggested that equiatomic SiC possesses active sites that are larger than several doped carbon materials [111]. The layered SiC sheets are found to be energetically favourable and are stable as compared to cubic SiC, particularly when the number of layers is less than four. They also reveal weak van der waal's interactions between the layers as in graphite. 2D SiC sheets of thickness 1.5–2.0 nm have been prepared using solvent exfoliation method and shown to reveal photoluminescence properties [112]. The DFT studies on layered SiC [111, 113] expose the efficiency of 2D SiC sheets towards ORR activity in both acidic and alkaline media [111]. Figures 21(a) and 21(b) give the energy

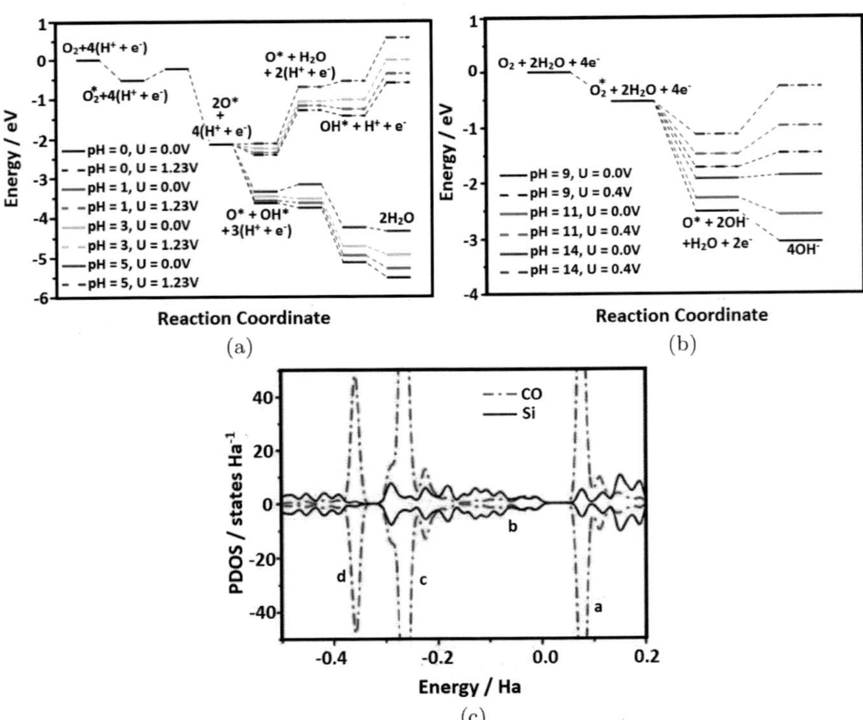

Figure 21. Schematic energy profile for ORR pathway (a) in acidic media, (b) in alkaline media (relative to the energy of $O_2 + 2H_2$ in acidic media and $O_2 + 2H_2O$ in alkaline media) on a single-layer SiC and (c) partial density of states for CO, adsorbed on single-layer SiC. (Adapted from Ref. [111].)

profile diagram for ORR on SiC monolayer in both acidic and alkaline media. It is revealed that the ORR is exothermic and depends on the pH of the media. Adsorption and dissociation of oxygen are the most important steps in ORR and the interaction between SiC and oxygen is to be conducive for the catalyst to show good activity. The electronic structure and the associated changes are investigated using spin polarised partial density of states (PDOS) and it is found that the O–O bond elongates and weakens thus leading to the assumption that it would be a good ORR catalyst. Further, the PDOS (Figure 21(c)) shows that the hybridisation between CO molecule and Si atom of SiC to be very small, indicating the negligible interaction between CO and SiC. This suggests that single layer SiC could catalyse ORR without CO poisoning, as an essential requirement for fuel cell applications [111] where CO poisoning is a major issue on the state of the art ORR catalyst, Pt.

3.6 Summary and Way Forward

There is a large number of publications in the area of ORR that consists of layered as well as non-layered materials including noble metals and their alloys. Though the thermodynamics seems to favour certain layered compounds for their excellent catalytic behaviour, still there are challenges to be overcome in terms of kinetics. There is hardly any catalyst that shows kinetic parameters (electrochemical rate constant, transfer coefficient, etc.) as good as Pt, if not better. Hence, there is enormous scope to search for new and novel materials, particularly based on carbides such as SiC, graphene analogue such as silicene that have been predicted to possess very favourable characteristics towards ORR. No doubt, there will be several such catalysts reported in the years to come.

References

[1] W. Vielstich, A. Lamm and H. Gasteiger, *Handbook of Fuel Cells: Fundamental, Technology and Applications*, Wiley, New York (2003).

[2] P. G. Bruce, S. A. Freunberger, L. J. Hardwick and J. M. Tarascon, *Nat. Mater.* **11**, 19 (2012).

[3] F. T. Wagner, B. Lakshmanan and M. F. Mathias, *J. Phys. Chem. Lett.* **1**, 2204 (2010).

[4] H. A. Gasteiger, S. S. Kocha, B. Sompalli and F. T. Wagner, *Appl. Catal. B* **56**, 9 (2005).

[5] W. Yu, M. D. Porosoff, J. G. Chen, *Chem. Rev.* **112**, 5780 (2012).

[6] D. Keilin, *Proc. R. Soc. London Ser. B* **100**, 129 (1925).

[7] C. Song and J. Zhang, *PEM Fuel Cell Electrocatalysts and Catalyst Layers*, Springer, London (2008).

[8] M. H. Seo, S. M. Choi, H. J. Kim and W. B. Kim, *Electrochem. Commun.* **13**, 182 (2011).

[9] L. Zhang, J. Zhang, D. P. Wilkinson and H. Wang, *J Power Sources* **156**(2), 171 (2006).

[10] E. Yeager, *J Mol Catal* **38**, 5 (1986).

[11] A. J. Bard and L. R. Faulkner, *Electrochemical Methods: Fundamentals and Applications.* Wiley, New York (1980).

[12] Y. Holade, N. E. Sahin, K. Servat, T. W. Napporn and K. B. Kokoh, *Catalysts* **5**, 310 (2015).

[13] A. B. Laursen, A. S. Varela, F. Dionigi, H. Fanchiu, C. Miller, O. I. Trinhammer, J. Rossmeisl and S. Dahi, *J. Chem. Educ.* **89**, 1595 (2012).

[14] J. K. Norskov, J. Rossmeisl, A. Logadottir, L. Lindqvist, J. R. Kitchin, T. Bligaard and H. Jónsson, *J. Phys. Chem. B*, **108**(46), 17886 (2004).

[15] A. V. Ruban, H. L. Skriver and J. K. Nørskov, *Phys. Rev. B* **59**, 15900 (1999).

[16] Y. Xu, A. V. Ruban and M. Mavrikakis, *J. Am. Chem. Soc.* **126**, 4717 (2004).

[17] J. R. Kitchin, J. K. Nørskov, M. A. Barteau and J. G. Chen, *J. Chem. Phys.* **120**, 10240 (2004).

[18] J. Greeley, I. E. L. Stephens, A. S. Bondarenko, T. P. Johansson, H. A. Hansen, T. F. Jaramillo, J. Rossmeisl, I. Chorkendorff and J. K. Norskov, *Nat. Chem.* **1**, 552 (2009).

[19] N. M. Marković, T. J. Schmidt, V. Stamenković and P. N. Ross, *Fuel Cells* **1**, 105 (2001).

[20] M. Winter and R. J. Brodd, *Chem. Rev.* **104**, 4245 (2004).

[21] Z. Peng and H. Yang, *J. Am. Chem. Soc.* **131**, 7542 (2009).

[22] M.-H. Shao, K. Sasaki and R. R. Adzic, *J. Am. Chem. Soc.* **128**, 3526 (2006).

[23] C. Wang, H. Daimon, T. Onodera, T. Koda and S. H. Sun, *Angew. Chem. Int. Ed.* **47**, 3588 (2008).

[24] L. Yang, S. Jiang, Y. Zhao, L. Zhu, S. Chen, X. Wang, Q. Wu, J. Ma, Y. Ma and Z. Hu, *Angew. Chem. Int. Ed.* **50**, 7132 (2011).

[25] G. Ma, R. Jia, J. Zhao, Z. Wang, C. Song, S. Jia and Z. Zhu, *J. Phys. Chem. C* **115**, 25148 (2011).

[26] S. Wang, D. Yu, L. Dai, D. W. Chang and J. B. Baek, *ACS Nano.* **5**, 6202 (2011).

[27] Z. Yang, Z. Yao, G. Li, G. Fang, H. Nie, Z. Liu, X. Zhou, X. Chen and S. Huang, *ACS Nano.* **6**, 205 (2012).

[28] R. Li, Z. Wei, X. Gou and W. Xu, *RSC Adv.* **3**, 9978 (2013).

[29] S. Wang, L. Zhang, Z. Xia, A. Roy, D. W. Chang, J. B. Baek and L. Dai, *Angew. Chem. Int. Ed.* **51**, 4209 (2012).

[30] Y. Xue, D. Yu, L. Dai, R. Wang, D. Li, A. Roy, F. Lu, H. Chen, Y. Liu and J. Qu, *Phys. Chem. Chem. Phys.* **15**, 12220 (2013).

[31] S. Wang, E. Iyyamperumal, A. Roy, Y. Xue, D. Yu and L. Dai, *Angew. Chem. Int. Ed.* **50**, 11756 (2011).

[32] L. Lai, J. R. Potts, D. Zhan, L. Wang, C. K. Poh, C. Tang, H. Gong, Z. Shen, J. Lin and R. S. Ruoff, *Energy Environ. Sci.* **5**, 7936 (2012).

[33] Y. Zheng, Y. Jiao, L. Ge, M. Jaroniec and S. Z. Qiao, *Angew. Chem. Int. Ed.* **52**, 3110 (2013).

[34] J. Liang, Y. Jiao, M. Jaroniec and S. Z. Qiao, *Angew. Chem. Int. Ed.* **51**, 11496 (2012).

[35] K. Raidongia, A. Nag, K. P. S. S. Hembram, U. Waghmare, R. Datta and C. N. R. Rao. *Chem. Eur. J.* **16**, 149 (2010).

[36] R. Kumar, K. Gopalakrishnan, I. Ahmad and C. N. R. Rao, *Adv. Func. Mater.* **25**(37), 5910 (2015).

[37] K. Gong, F. Du, Z. Xia, M. Durstock and L. Dai, *Science* **323**, 760 (2009).

[38] X. Fan, W. T. Zheng and J. L. Kuo, *RSC Adv.* **3**, 5498 (2013).

[39] H. Meng and P. K. Shen, *Electrochem. Commun.* **8**, 588 (2006).

[40] V. Kiran, K. Srinivasu and S. Sampath, *Phys. Chem. Chem. Phys.* **15**, 8744 (2013).

[41] J. Giner and L. Swette, *Nature* **211**, 1291 (1966).

[42] S. Isogai, R. Ohnishi, M. Katayama, J. Kubota, D. Y. Kim, S. Noda, D. Cha, K. Takanabe and K. Domen, *Chem. Asian J.* **7**, 286 (2012).

[43] J. Chen, K. Takanabe, R. Ohnishi, D. Lu, S. Okada, H. Hatasawa, H. Morioka, M. Antonietti, J. Kubota and K. Domen, *Chem. Commun.* **46**, 7492 (2010).

[44] Y. Gorlin, C. J. Chung, D. Nordlund, B. M. Clemens and T. F. Jaramillo, *ACS Catal.* **2**, 2687 (2012).

[45] F. Cheng, J. Shen, B. Peng, Y. Pan and Z. Tao, J. Chen, *Nat. Chem.* **3**, 79 (2011).

[46] J. Xu, P. Gao and T. S. Zhao, *Energy Environ. Sci.* **5**, 5333 (2012).

[47] V. G. Anju, R. Manjunatha, P. M. Austeria and S. Sampath, *J. Mater. Chem. A* **4**, 5258 (2016).

[48] R. Jasinski, *Nature* **201**, 1212 (1964).

[49] G. Lv, L. Cui, Y. Wu, Y. Liu, T. Pu and X. He, *Phys. Chem. Chem. Phys.* **15**, 13093 (2013).

[50] A. Morozan, S. Campedelli, A. Filoramo, B. Jousselme and S. Palacin, *Carbon* **49**, 4839 (2011).

[51] E. H. Yu, S. Cheng, B. E. Logan and K. Scott, *J. Appl. Electrochem* **39**, 705 (2009).

[52] Y. Jiang, Y. Lu, X. Lv, D. Han, Q. Zhang, L. Niu and W. Chen, *ACS Catal.* **3**, 1263 (2013).

[53] F. Demir, A. Erdogmus and A. Koca, *Phys. Chem. Chem. Phys.* **15**, 15926 (2013).

[54] D. Chu and R. Z. Jiang, *Solid State Ionics* **148**(3–4), 591 (2002).

[55] S. Baranton, C. Coutanceau, C. Roux, F. Hahn and J. M. Leger, *J. Electroanal. Chem.* **577**(2), 223 (2005).

[56] A. Huissoud and P. Tissot, *J. Appl. Electrochem.* **29**, 11 (1999).

[57] K. Tammeveski, K. Kontturi, R. J. Nichols, R. J. Potter and D. J. Schiffrin, *J. Electroanal. Chem.* **515**, 101 (2001).

[58] A. Sarapuu, K. Helstein, K. Vail and K. Tammeveski, *Electrochim. Acta* **55**(22), 6376 (2010).

[59] S. Matsuda, K. Hashimoto and S. Nakanishi, *J. Phys. Chem. C.* **118**, 18397 (2014).

[60] V. C. Tung, M. J. Allen, Y. Yang and R. B. Kaner, *Nat. Nanotechnol.* **4**, 25 (2009).

[61] J. T. Robinson, M. Zalalutdinov, J. W. Baldwin, E. S. Snow, Z. Wei, P. Sheehan and B. H. Houston, *Nano Lett.* **8**, 3441 (2008).

[62] D. Wei, Y. Liu, Y. Wang, H. Zhang, L. Huang and G. Yu, *Nano Lett.* **9**, 1752 (2009).

[63] K. Gopalakrishnan and C. N. R. Rao, *Mater. Res. Express,* **2**, 095503 (2015).

[64] L. Qu, Y. Liu, J.-B. Baek and L. Dai, *ACS Nano.* **4**(3), 1321 (2010).

[65] K. Parvez, S. Yang, Y. Hernandez, A. Winter, A. Turchanin, X. Feng and K. Müllen, *ACS Nano.* **6**(11), 9541 (2012).

[66] Z.-S. Wu, S. Yang, Y. Sun, K. Parvez, X. Feng and K. Müllen, *J. Am. Chem. Soc.* **134**, 9082 (2012).

[67] R. Ning, J. Tian, A. M. Asiri, A. H. Qusti, A. O. Al-Youbi and X. Sun, *Langmuir* **29**, 13146 (2013).

[68] J.-I. Aihara, *J. Phys. Chem. A* **103**(37), 7487 (1999).

[69] L. Zhang and Z. Xia, *J. Phys. Chem. C* **115**(22), 11170 (2011).

[70] K. Moses, V. Kiran, S. Sampath and C. N. R. Rao, *Chem. Asian J.* **9**, 838 (2014).

[71] M. B. Zakaria, V. Malgras, T. Takei, C. Li and Y. Yamauchi, *Chem. Commun.* **51**, 16409 (2015).

[72] J.-H. Kim, A. Ishihara, S. Mitsushima, N. Kamiya and K.-I. Ota, *Electrochim. Acta* **52**, 2492 (2007).

[73] N. Chaisubanan, M. Hunsom, H. Vergnes and K. Pruksathorn, *Energ. Convers. Manag.* **114**, 348 (2016).

[74] J. Dailly, S. Fourcade, A. Largeteau, F. Mauvy, J. C. Grenier and M. Marrony, *Electrochim. Acta* **55**, 5847 (2010).

[75] E. Fabbri, L. Bi, D. Pergolesi and E. Traversa, *En. Env. Sci.* **4**, 4984 (2011).

[76] L. Yang, C. Zuo, S. Wang, Z. Cheng and M. Liu, *Adv. Mater.* **20**, 3280 (2008).

[77] Z. Shao and S. M. Haile, *Nature* **431**, 170 (2004).

[78] L. Fan and P.-C. Su, *J. Power Sourc.* **306**, 369 (2016).

[79] J. Kim, A. Jun, J. Shin and G. Kim, *J. Am. Ceram. Soc.* **97**(2), 651 (2014).

[80] T. F. Jaramillo, K. P. Jørgensen, J. Bonde, J. H. Nielsen, S. Horch and I. Chorkendorff, *Science* **317**, 100 (2007).

[81] B. Hinnemann, P. G. Moses, J. Bonde, K. P. Jørgensen, J. H. Nielsen, S. Horch, I. Chorkendorff and J. K. Nørskov, *J. Am. Chem. Soc.* **127**, 5308 (2005).

[82] Y. Li, H. Wang, L. Xie, Y. Liang, G. Hong and H. Dai, *J. Am. Chem. Soc.*, **133**, 7296 (2011).

[83] Z. Yufei, Z. Yuxia, Y. Zhiyu, Y. Yiming and S. Kening, *Sci. Technol. Adv. Mater.* **14**, 043501 (2013).

[84] V. Kiran, D. Mukherjee, R. Jenjeti and S. Sampath, *Nanoscale* **6**, 12856 (2014).

[85] S. Sarkar S and S. Sampath, *Chem. Comm.* **50**, 7359 (2014).

[86] D. Kong, H. Wang, J. J. Cha, M. Pasta, K. J. Koski, J. Yao and Y. Cui, *Nano Lett.* **13**, 1341 (2013).

[87] H. Wang, H. Feng and J. Li, *Small* **10**, 2165 (2014).

[88] M. Pumera, Z. Sofer and A. Ambrosi, *J. Mater. Chem. A.* **2**, 8981 (2014).

[89] C. N. R. Rao, K. Gopalakrishnan and U. Maitra, *ACS App. Mater. Inter.* **7**, 7809 (2015).

[90] T. Y. Wang, D. L. Gao, J. Q. Zhuo, Z. W. Zhu, P. Papakonstantinou, Y. Li and M. X. Li, *Chem. Eur. J.* **19**, 11939 (2013).

[91] A. Y. S. Eng, A. Ambrosi, Z. Sofer, P. Simek and M. Pumera, *ACS Nano.* **8**, 12185 (2014).

[92] K. Zhao, W. Gu, L. Zhao, C. Zhang, W. Peng and Y. Xian, *Electrochim. Acta* **169**, 142 (2015).

[93] H. Huang, X. Feng, C. Du, S. Wu and W. Song, *J. Mater. Chem. A* **3**, 16050 (2015).

[94] K. Yuan, X. Zhuang, H. Fu, G. Brunklaus, M. Forster, Y. Chen, X. Feng and U. Scherf, *Angew. Chem. Int. Ed.* **55**, 6858 (2016).

[95] S. M. Tan, C. K. Chua, D. Sedmidubsky, Z. Sofer and M. Pumera, *Phys. Chem. Chem. Phys.* **18**, 1699 (2016).

[96] X. Long, Z. Wang, S. Xiao, Y. An and S. Yang, *Mat. Today*, **19**(4), 213 (2016).

[97] P. J. Sideris, U. G. Nielsen, Z. Gan and C. P. Grey, *Science* **321**, 113 (2008).

[98] Q. Wang and D. O'Hare, *Chem. Rev.* **112**, 4124 (2012).

[99] Z. Liu, R. Ma, M. Osada, N. Iyi, Y. Ebina, K. Takada and T. Sasaki, *J. Am. Chem. Soc.* **128**, 4872 (2006).

[100] R. Huo, W.-J. Jiang, S. Xu, F. Zhang and J.-S. Hu, *Nanoscale*, **6**, 203 (2014).

[101] A. Indra, P. W. Menezes, N. R. Sahraie, A. Bergmann, C. Das, M. Tallarida, D. Schmeiβer, P. Strasser and M. Driess, *J. Am. Chem. Soc.* **136**(50), 17530 (2014).

[102] X. Zhang, Y. Wang, S. Dong and M. Li, *Electrochim. Acta* **170**, 248 (2015).

[103] L. Qian, Z. Lu, T. Xu, X. Wu, Y. Tian, Y. Li, Z. Huo, X. Sun and X. Duan, *Adv. Ener. Mater.* 1500245 (2015).

[104] C. S. Lim, C. K. Chua, Z. S.ofer, K. Klımova, C. Boothroyd and M. Pumera, *J. Mater. Chem. A* **3**, 11920 (2015).

[105] C. S. Lim, Z. Sofer, V. Mazanek and M. Pumera, *Nanoscale* **7**, 12527 (2015).

[106] M. Houssa, G. Pourtois, M. M. Heyns, V. V. Afanas'ev and A. Stesmans, *ECS Trans.* **33**(3), 185 (2010).

[107] R. Wang, M. Xu and X.-D. Pi, *Chinese Phys. B* **24**(8), 086807 (2015).

[108] P. D. Padova, O. Kubo, B. Olivieri, C. Quaresima, T. Nakayama, M. Aono and G. L. Lay, *Nano Lett.*, **12**(11), 5500 (2012).

[109] Y. Hwang, K.-H. Yun and Y.-C. Chung, *J. Power Source* **275**, 32 (2015).

[110] V. O. Özçelik and S. Ciraci, *J. Phys. Chem. C* 2013, **117**(49), 26305.

[111] P. Zhang, B. B. Xiao, X. L. Hou, Y. F. Zhu and Q. Jiang, *Scientific Reports* **4**, 3821 (2014).

[112] S. S. Lin, *J. Phys. Chem. C* **116**(6), 3951 (2012).

[113] M. D. Ganji, R. Agheb, H. D. Ganji and S. Ashrafian, *J. Phys. Chem. Solid* **88**, 47 (2016).

Chapter 4

Phosphorene

Arpita Paul[*] and Umesh V. Waghmare[†]

*Theoretical Sciences Unit, Jawaharlal Nehru Centre for Advanced
Scientific Research, Jakkur, Bangalore 560064, India*
[*]*arpitapaul@jncasr.ac.in*
[†]*waghmare@jncasr.ac.in*

Abstract. Among the two-dimensional (2D) materials, phosphorene
is very attractive as an alternative to graphene and transition metal
dichalcogenides for use in future electronic devices, due to its finite band
gap and high carrier mobility. A remarkable anisotropy in the crystal
and electronic structure of phosphorene is associated with charge redistribution in the intra- and inter-layer regions. A suitable in-plane strain
can make phosphorene a potential photocatalyst, and also tune the
anisotropy in electrical conductance. The mutually orthogonal directions
of prominent electrical and thermal transport make phosphorene a
promising material for application in thermoelectric energy conversion.
The normal insulating state transforms to topological insulating state
through band-inversion upon application of external electric field.

4.1 Introduction

Two-dimensional (2D) layered materials, such as graphene, transition
metal dichalcogenides (with formula of MX_2; M = Mo, W, Nb and
Ta; X = S, Se and Te), silicene and germanane have attracted
tremendous attention for their potential for use in future generations
of nanoelectronic devices [1–7]. A high-performance device like field
effect transistor (FET) requires a material with moderate band gap
(a high current ON/OFF ratio) and high carrier mobility [6, 7].

137

Graphene has ultrahigh carrier mobility (250,000 cm^2 V^{-1} S^{-1}) due to symmetry avoided back scattering and a small value of carrier effective mass, and has been considered for FET devices [1, 6, 8]. But, the semimetallic nature of graphene results in relatively lower current ON/OFF ratio due to large OFF current in graphene-based FET devices [6]. This is one of the limitation of graphene in the field of field effect transistors [6]. In contrast, transition metal dichalcogenides (TMD) have very high ON/OFF ratio ($\sim 10^8$) due to its finite and tunable band gap (from 0.4 eV to 2.3 eV) [7, 9]. However, TMDs have relatively lower carrier mobility ($\sim$200 cm^2 V^{-1} S^{-1}) partly because of the localised nature of d electrons in valence and conduction bands, limiting their applications in electronic devices [10].

Recently discovered phosphorene, (a few layer black phosphorus) is quite attractive as a 2D material for the application in high-speed FET devices. Phosphorene has been studied enormously due to its highly anisotropic optoelectronic properties compared to other 2D materials. It has a band gap (direct) varying from 0.3 eV in bulk to 1.45 eV in monolayer, which covers a large range of electromagnetic spectrum and leads to its potential applications in optoelectronic devices [11]. It exhibits a high carrier mobility of 1000 cm^2 V^{-1} S^{-1} (at room temperature) and on–off ratio of $\sim 10^5$, and it is the only material other than graphite that can be mechanically exfoliated into an ultrathin nanosheet [9, 12].

Here, we review recent scientific progress on phosphorene including its structural, electronic, transport, adsorption and vibrational properties and their tunability with external fields (electric field and strain).

4.2 Crystal Structure

Phosphorene consists of a few layers of black phosphorus (as shown in Figure 1) in which two adjacent layers are connected by van der Waals (vdW) interaction [13]. It has a puckered honeycomb crystal structure with three phosphorus atoms covalently bonded together with sp^3 bonding [10, 14]. There are two orthogonal directions in phosphorene: arm-chair (a) and zigzag (b) edges. Monolayer

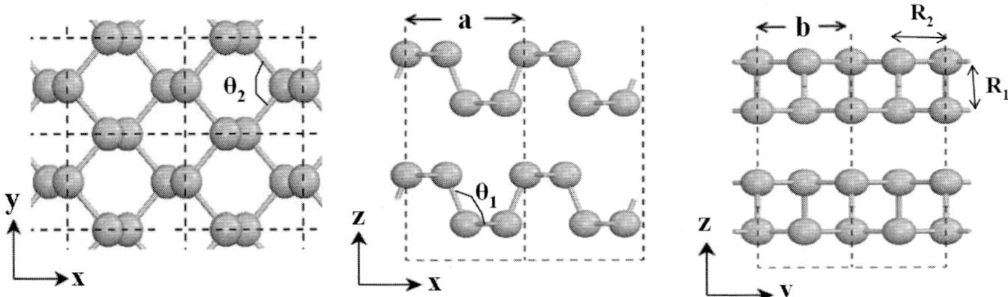

Figure 1. Crystal structure of phosphorene: R_1 and R_2 denote interplaner and intraplaner P$-$P bond lengths, θ_1 and θ_2 are the interplanar and intraplaner bond angles. Reprinted with permission from Ref. [10]. Copyright 2014 Creative Commons (visit http://creativecommons.org/licenses/by-nc-nd/4.0/).

Table 1. Effect of number of layers of black phosphorus on structural parameters.

Number of layers	a(Å)	b(Å)	R_1 (Å)	R_2(Å)	θ_1/θ_1' (°)	θ_2/θ_2' (°)
1	4.58	3.32	2.28	2.24	103.51	96.00
2	4.52	3.33	2.28	2.24	102.96	96.21/95.92
3	4.51	3.33	2.28	2.24	102.81/102.74	96.30/95.99
4	4.50	3.34	2.28	2.24	102.76/102.67	96.34/96.01
5	4.49	3.34	2.28	2.24	102.71/102.63	96.37/96.05
Bulk	4.47	3.34	2.28	2.25	102.42	96.16

phosphorene consists of two sublayers with unequal inter-sublayer and intra-sublayer bond lengths (R_1, R_2) and bond angles (θ_1, θ_2) between the nearest neighbour phosphorus atoms (Figure 1) [10]. The unit cell of bulk black phosphorus consists of the periodic structure of two monolayer phosphorene with AB stacking (relative displacement between two layers is ($0.0a$, $0.5b$)) [15].

Using first-principles based density functional theoretical (DFT) calculations with HSE06 functional, it has been shown in Table 1 that the value of a of phosphorene depends strongly on the number of layers whereas b remains largely independent [10]. The increment in the value of a by 2.4% from bulk to monolayer is primarily associated with the increase in bond angle θ_1, not with the bond lengths. R_1, R_2 and θ_1 remain independent of the number of layers. Also, it has been observed (from Monte Carlo simulations) that the change in energy of monolayer phosphorene due to the deformation in *a* parameter by 0.3 Å is about 5 meV/atom (thermal energy of 60 K) [16]. So, the *a* axis is relatively softer than *b* axis. The drastic change in the value of *a* from bilayer to monolayer (see Table 1) is a consequence of inter-layer interaction (0.46 eV) in phosphorene [10]. During the formation of the bulk structure by assembling monolayers, the covalent interaction between adjacent layers gets notably modified [16]. Such different trend in the in-plane lattice parameters (*a* and *b*) of a few-layer phosphorene as a function of a number of layers highlights the structural anisotropy of phosphorene.

4.3   Electronic Properties

4.3.1   Electronic structure

Monolayer phosphorene is a direct band-gap semiconductor with a band gap of 1.45 eV (experimentally measured) at Γ point like that in its bulk form (E_g=0.31 eV) [9]. The energy band gap of bulk falls in the infrared region, whereas monolayer phosphorene exhibits a peak in the PL spectrum at 1.45 eV with a narrow spread of energy (100 meV) [9]. Using DFT calculations (with HSE06 functional), it has been shown in the electronic structure of phosphorene that the valence band is contributed by p_z orbital whereas the conduction band is contributed by mixed p_z and s orbitals of phosphorus atoms (Figures 2(a) and 2(b)) [12, 17]. Band gap of few-layer phosphorene is a strong function of number of layers, but the topology of the band remains preserved unlike in TMDs [18]. The thickness independent band topology of phosphorene makes it a potential candidate for applications in optoelectronic devices. In the bilayer phosphorene (Figures 2(a) and 2(b)), band gap reduces to 1.02 eV (1.56 eV for monolayer, obtained from DFT calculations) and two extra valence (VB1) and conduction bands (CB1) begin to appear near Γ point (VB2 and CB2 were present for monolayer) [10]. VB1 has a bonding character (wave functions overlap in the inter-layer region) and is

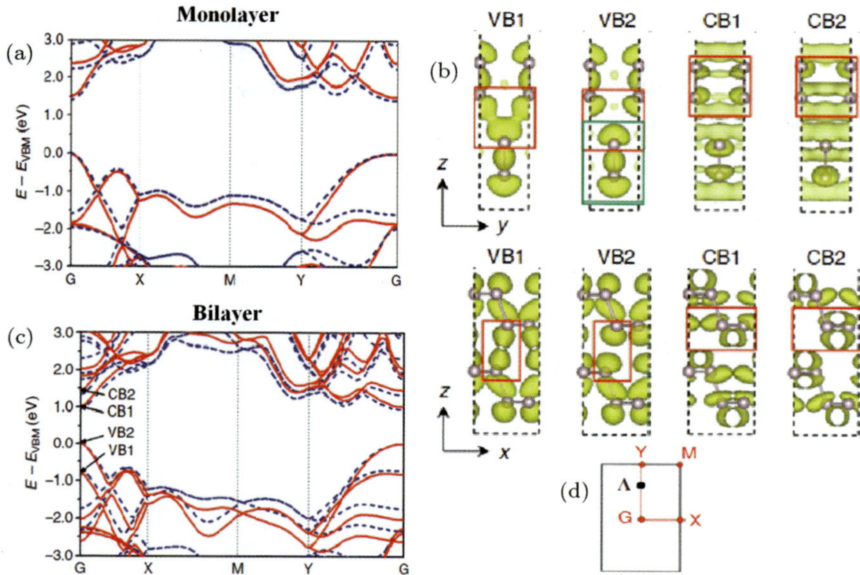

Figure 2. Electronic structure of monolayer (a) and bilayer (b) phosphorene calculated with HSE06 functional (red solid lines) and mBJ potential (blue dashed line), respectively, (c) spatial variation of the wave functions of CB1, CB2, VB1 and VB2 states are shown using isosurfaces of their density at 0.025 e/Å^3 (green and red rectangles denote antibonding and bonding features) and (d) Brillouin zone of monolayer phosphorene. Reprinted with permission from Ref. [10]. Copyright 2014 Creative Commons (visit http://creativecommons. org/licenses/by-nc-nd/4.0/).

energetically lower compared to VB2 which has an anti-bonding character (as shown in Figure 2(c)). Similar feature has been noticed for the two lowest lying conduction bands (CB1 and CB2), at Γ point (as shown in Figure 2(c)). The overlap of wave functions in the inter-layer region becomes the mode of interaction between adjacent layers and it explains the reduction of a parameter from the bulk to the monolayer (as mentioned in previous section). The increment in number of layers strengthens the inter-layer interaction and gives dispersive band structure which results in decreasing the band gap [10].

As the topmost valence and lowest lying conduction bands are highly dispersive along Γ–X direction (as shown in Figures 2(a) and 2(b)), the effective masses of electron and hole for monolayer are small (0.15 m_0 and 0.17 m_0, respectively), which remain almost unchanged with the number of layers (see Table 2). However, near Γ point in the Γ–Y direction of monolayer phosphorene, the valence band becomes flat and gives the hole effective mass of about 6.35 m_0 (0.71 m_0 for bulk), and it changes dramatically with the number of layers. The electron effective mass in the Γ–Y direction for monolayer is 1.12 m_0, and it remains independent of the number of layers [10]. This strong directional dependence of carrier effective mass also influences carrier mobility making it anisotropic.

4.3.2 Carrier mobility

Electronic transport properties of phosphorene depend primarily on the carrier mobility, i.e., on the curvature of band dispersion quantified in carrier effective masses. However, the deformation potential and elastic constant (in the direction of propagation of longitudinal acoustic wave) are also crucial in determining carrier mobility [10]. It is well known within a phonon limited scattering model, the 2D carrier mobility is as [10, 19]:

$$\mu_{2D} = \frac{e\hbar^3 C_{2D}}{k_B T m_c^* m_d^* E_l^2}, \tag{4.1}$$

where m_c^* is the effective mass of hole or electron in the transport direction and m_d^* is the effective mass of hole or electron in the xy plane $(\sqrt{m_x^* m_y^*})$. E_l (E_{1x} and E_{1y}) represents the deformation potential constant, which measures the change in energy of conduction band minimum (CBM) or valence band maximum (VBM) with an application of compressive or tensile strain in the transport direction. C_{2D} is the elastic constant of longitudinal acoustic wave in the direction of propagation.

Carrier mobilities using the phonon limited scattering model (estimated using DFT calculations with HSE06 functional) are given in Table 2 [10]. It is clear that holes are more mobile than electrons. The mobilities of electrons and holes in few-layer phosphorene along the arm-chair ($\boldsymbol{a}$) direction are 2–4 times larger than the mobilities

Table 2. Carrier effective mass (m_x^* and m_y^*), carrier mobility (μ_x and μ_y in 10^3 cm^2 V^{-1} s^{-1}) and deformation potential (E_{1x} and E_{1y}) of phosphorene with varying number of layers.

Carrier Type	Number of Layers	$m_x^*(m_0)$ (Γ–X)	$m_y^*(m_0)$ (Γ–Y)	E_{1x}(eV)	E_{1y} (eV)	μ_x	μ_y
e	1	0.17	1.12	2.72 ± 0.02	7.11 ± 0.02	1.10–1.14	~0.08
	2	0.18	1.13	5.02 ± 0.02	7.35 ± 0.16	~0.60	0.14–0.16
	3	0.16	1.15	5.85 ± 0.09	7.63 ± 0.18	0.76–0.80	0.20–0.22
	4	0.16	1.16	5.92 ± 0.18	7.58 ± 0.13	0.96–1.08	0.26–0.30
h	1	0.15	6.35	2.50 ± 0.06	0.15 ± 0.03	0.64–0.70	10–26
	2	0.15	1.81	2.45 ± 0.05	1.63 ± 0.16	2.6–2.8	1.3–2.2
	3	0.15	1.12	2.49 ± 0.12	2.24 ± 0.18	4.4–5.2	2.2–3.2
	4	0.14	0.97	3.16 ± 0.12	2.79 ± 0.13	4.4–5.2	2.6–3.2

along zigzag (b) direction (except for monolayer). For monolayer phosphorene, the effective mass of electron and deformation potential of the CBM are ten and three times larger in the b direction compared to a direction, respectively. The value of elastic constant (in the propagation direction of longitudinal acoustic wave) is independent of the type of carriers and it increases with increasing the number of layers. Elastic constant in a direction (28–146 Jm^{-2}) is four times smaller than the value along b direction (101–479 Jm^{-2}) which highlights the softness of a axis [10]. As the carrier mobility is inversely proportional to effective mass and square of deformation potential, electronic mobility along a direction (~ 1000 cm^2 V^{-1} S^{-1}) is one order of magnitude larger compared to that along b direction (~ 80 cm^2 V^{-1} S^{-1}). With increasing number of layers, the electronic mobility in the ab plane decreases due to the increase in the effective mass and deformation potential.

In spite of a large effective mass of holes along b direction (6.35 m$_0$) for monolayer phosphorene, the extremely small value of deformation potential (E_{1y}) constant of VBM (0.15 ± 0.03 eV) gives a rather high hole mobility ($10{,}000$–$26{,}000$ cm^2 V^{-1} S^{-1}) compared to a direction (640–700 cm^2 V^{-1}S^{-1}). The sudden drop in the deformation potential constant (E_{1y}) of valence band in the b direction from monolayer to bilayer phosphorene has been explained from the nature of wave functions associated with the topmost valence band in the ab plane of phosphorene [10]. The wave function of VB2 state (see Figure 2(c)) is isolated in the b direction. The effect of deforming the crystal structure along b direction has little effect on its energy resulting in a small value of deformation potential constant. However, the wave function of VB2 state along a direction gets influenced easily by strain applied along same direction, the energy of VBM changes largely due to the change of intra-layer overlap [10]. Due to strong overlap of wave functions of conduction band states along a and b directions (as shown in Figure 2(c)), the deformation of crystal causes significant perturbation of the energy level. The deformation potential constant is higher for conduction band compared to valence band due to this strong overlap of the wave functions [10].

4.4   Effect of Strain

4.4.1   Transport properties

The direction of higher electrical conductivity is switchable by applying in-plane strain [14]. Electrons have large mobility along arm-chair direction as the conduction band is highly dispersive along Γ–X direction compared to zigzag direction (as shown in Figure 3(a)). It has been shown using DFT calculations with generalised gradient approximated PBE functional that the anisotropy in electronic

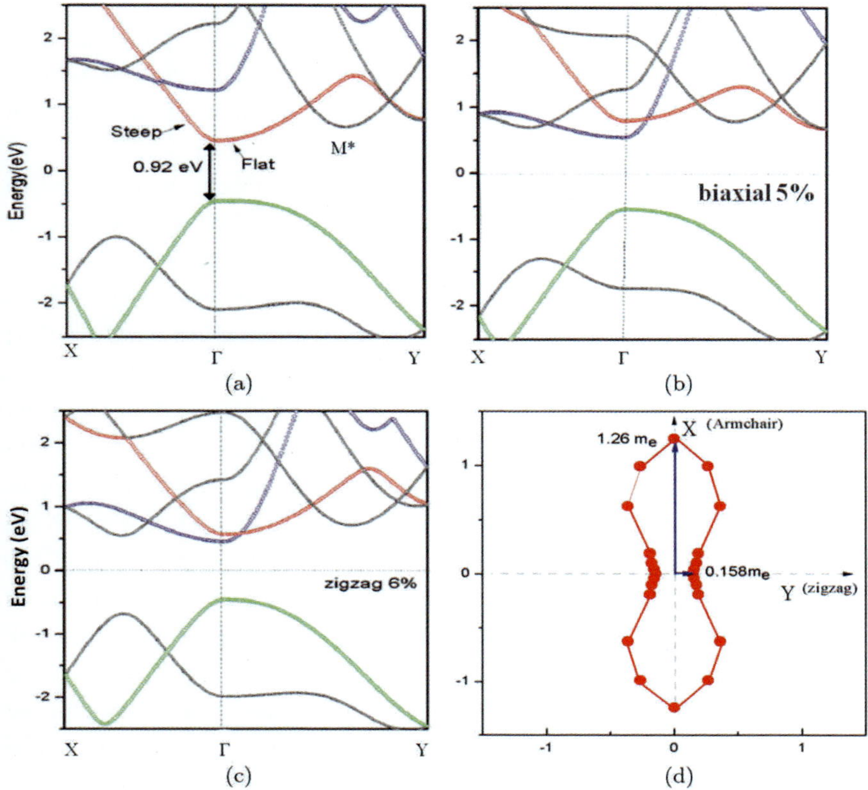

Figure 3. Band structure of (a) monolayer phosphorene with (b) 5% biaxial strain and (c) 6% zigzag uniaxial strain, (d) variation of electronic effective mass in the *ab* plane at 5% biaxial strain (the length of the blue arrow represents absolute value of effective mass). Reprinted with permission from Ref. [14]. Copyright 2014 American Chemical Society.

mobility can be tuned by changing the ordering of conduction bands with the application of in-plane strain on monolayer phosphorene [14]. The second conduction band (as shown in blue colour in Figure 3(a)) is energetically higher than the first conduction band, but it has an opposite feature of band dispersion compared to the first. The second conduction band can be lowered in energy by applying biaxial and uniaxial strains (as shown in Figures 3(b) and 3(c)). Under 5% biaxial and uniaxial strain along zigzag direction, it is noticed that electrons become heavier along arm-chair direction compared to zigzag direction (see Figure 3(d)). In the zigzag direction of monolayer phosphorene under 5% biaxial and uniaxial strains, the electronic mobility at $300\,\mathrm{K}$ becomes 10 times larger than that along arm-chair direction [14]. The same trend is found in bilayer phosphorene. However, the uniaxial strain along arm-chair direction has a negligible effect on switching the order of bands.

Application of strain in the direction perpendicular to surface can also engineer the electronic structure of bilayer phosphorene, and it can be easily validated through experiments compared to in-plane strain [15]. It has been found from DFT calculations (with HSE06 functional) that out of plane compressive strain converts AB stacked bilayer phosphorene from a direct band-gap semiconductor ($E_g = 1.03\,\mathrm{eV}$) to an indirect band-gap semiconductor ($E_g = 0.59\,\mathrm{eV}$) [15]. The process of recombination between electron and hole thus gets slower, as the radiative recombination is not allowed in indirect band-gap semiconductor. The reduction in inter-layer distance from d_0 (uncompressed) to $0.92\,d_0$ shifts the position of CBM from Γ to M^* at $k = (0.0, 1/3, 0.0)2\pi/b$ (Figure 3(a)) while the position of VBM remains unaltered. The stabilisation of CBM at M^* occurs due to strong inter-layer overlap as is seen from the conduction band charge density plot at M^* (Figure 4(a)). The intra-layer overlap of wave functions of the conduction band (CB1) (in the lower panel of Figure 2(c)) is quite weak in unstrained bilayer phosphorene.

In addition, it is noted that the dispersion of conduction band at M^* is isotropic in the $k_x - k_y$ plane (see Figure 4(b)) [15]. The isotropic band dispersion removes the anisotropy of electronic mobility in the ab plane ($m_{e,x}^* = 0.25\ m_0$, $m_{e,y}^* = 0.32\ m_0$). The

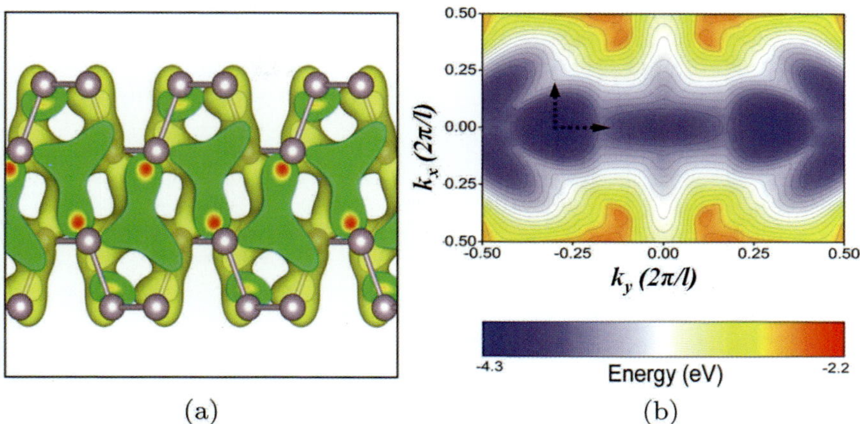

Figure 4. (a) Conduction band charge density at M* (isosurface was set to 0.01 e/Å^3) and (b) contour plot of energy of the first conduction band as a function of k_x and k_y at $d = 0.96\, d_0$ (conduction band minimum is denoted by a set of axis). Reprinted with permission from Ref. [15]. Copyright 2015 American Chemical Society.

value of deformation potential constant of CBM at M* (0.92 d_0) along arm-chair direction is 0.28 eV (3.66 eV at Γ). Along the zigzag direction, E_{1x} (deformation potential constant) of CBM at M* reduced to half of the value at Γ (unstrained). This highlights the weak coupling between electron and the deformation associated with the longitudinal acoustic phonon mode along the arm-chair direction. As the inter-layer distance decreases ($0.98d_0$), the occupancy of the conduction band state at M* also increases (CBM shifts from Γ to M*) and electronic mobilities along arm-chair and zigzag directions reach upto 60,000 cm^2 V^{-1} S^{-1} and 4000 cm^2 V^{-1}S^{-1} (as shown in Figure 5), respectively. It has been suggested that highly conducting electronic states at M* can be experimentally achieved by introducing small amount of electron doping in the bilayer phosphorene [15].

4.4.2 Photocatalytic properties

We now discuss the effect of in-plane strain on the band gap of monolayer phosphorene and its application as a photocatalyst. The CBM of a photocatalyst should be located more negative than the

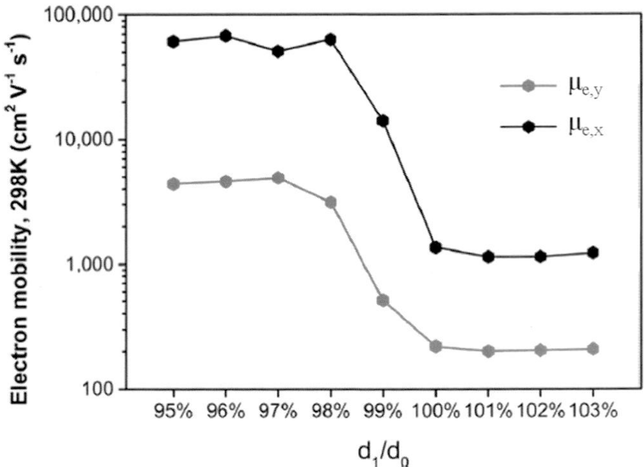

Figure 5. Dependence of electron mobility ($\mu_{e,x}$ and $\mu_{e,y}$) on the distance between two layers of phosphorene (CBMs are at Γ and M^* for d_1/d_0 above 0.98 and below 0.98, respectively). Reprinted with permission from Ref. [15]. Copyright 2015 American Chemical Society.

redox potential of H^+/H_2 reaction (HER) and VBM should lie more positive than the redox potential of O_2/H_2O reaction (OER). Moreover, the band gap of a photocatalyst must be larger than 1.23 eV with suitable band alignment [17]. For unstrained monolayer phosphorene, the band edge alignment of CBM and VBM with respect to NHE shows that VBM and CBM lies more negative than O_2/H_2O and H^+/H_2 redox potentials, respectively (see Figure 6(b)) [17]. This limits use of phosphorene as a photocatalyst for OER reactions. However, the redox potential of water also depends on the pH value.

The standard oxidation potential O_2/H_2O in a solution is given as [20]:

$$E^{ox}_{O_2/H_2O} = -5.67 + p\text{H} \times 0.059 \, \text{eV}. \tag{4.2}$$

In pH $= 8.0$ solution, phosphorene shows favourable band position (as shown in Figure 6(c)). However, unstrained phosphorene (in pH $= 8.0$ solution) absorbs both visible and UV light as can be seen from the calculated optical spectra (Figure 6(d)) [17]. Optical

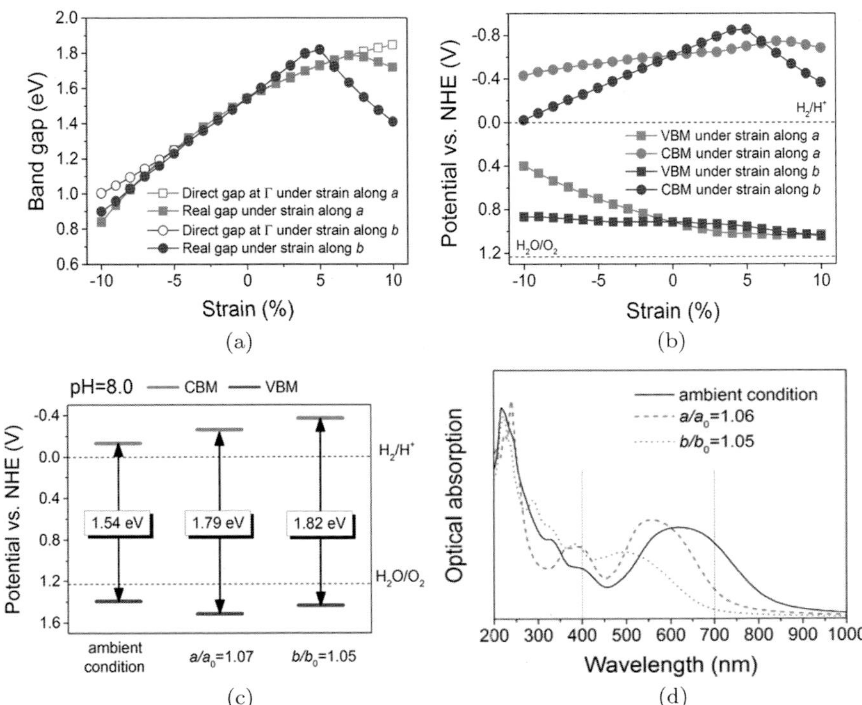

Figure 6. Theoretically (using HSE06 functional) calculated (a) band gap, (b) VBM and CBM of phosphorene as a function of uniaxial strain along a and b directions (vacuum level is the reference potential and dashed lines in (b) are the standard water redox potential), (c) energy alignment of VBM and CBM of phosphorene in $pH = 8.0$ solution and (d) calculated optical spectra of phosphorene under ambient condition, 7% and 5% uniaxial strains along a and b directions, respectively (dashed lines in (c) are the standard water redox potential in $pH = 8.0$ solution). Reprinted with permission from Ref. [17]. Copyright 2014 American Chemical Society.

absorption peak in the visible light wavelength range becomes more pronounced under the application of uniaxial strain along a and b directions (Figure 6(d)).

The band gap of phosphorene first increases for tensile strain along a as well as in b direction but after certain percentage of strain band gap starts decreasing (as shown in Figure 6(a)). Compressive strains along both directions reduces the band gap and cannot be used for photocatalysis [17]. The band gap is maximum for

7% (1.79 eV) and 5% (1.82 eV) tensile strain along a and b axes, respectively. The tensile strain along a axis shifts the position of VBM from Γ to Y point and makes phosphorene an indirect band-gap semiconductor.

As the charge transfer from Γ to Y point across indirect band gap is assisted by phonon, the energy conversion by solar energy becomes less efficient. On the other hand, phosphorene remains a direct band gap semiconductor at all values of tensile strain along b direction. The band edge alignment of phosphorene with respect to the NHE shows similar problem like unstrained phosphorene (see Figure 6(b)). In $pH = 8.0$ solution, strained phosphorene (7% and 5% tensile strain along a and b directions) also gives the favourable band edge alignment for water splitting (Figure 6(c)) [17]. It has been noticed that strained phosphorene absorbs visible light more efficiently (less absorbing UV light) than unstrained phosphorene (as shown in Figure 6(d)) [17].

4.5   Topological Phase Transition

Using DFT-PBE calculation, it was shown that the four layer phosphorene ($E_g = 0.07$ eV) undergoes topological phase transition through band inversion between conduction (C1 with Γ_{1c} symmetry) and valence band (V1 with Γ_{8v} symmetry) states under the application of external electric field (F) in a direction perpendicular to the layer (as shown in Figure 7(c)) [21]. At zero external electric field, four layers of phosphorene is a normal insulator with Z_2 topological invariant equal to zero (see Figure 7(e)). From the charge density isosurfaces (Figure 7(c)), it can be seen that the atomic orbitals of Γ_{8v} and Γ_{1c} are distributed along the out of plane and in-plane bonds of each P sublayers, respectively (Figure 7(c)).

Under the application of electric field, the energy of band gets shifted due to the stark effect and band gap starts reducing with increasing electric field (Figure 7(b)). At 0.3 V/Å electric field (F_C), the conduction and valence bands start overlapping with each other at Γ point and band inversion energy ($\Delta_{inv} = E_{\Gamma 1c} - E_{\Gamma 8v}$) becomes zero, and it is the critical field for topological phase transition

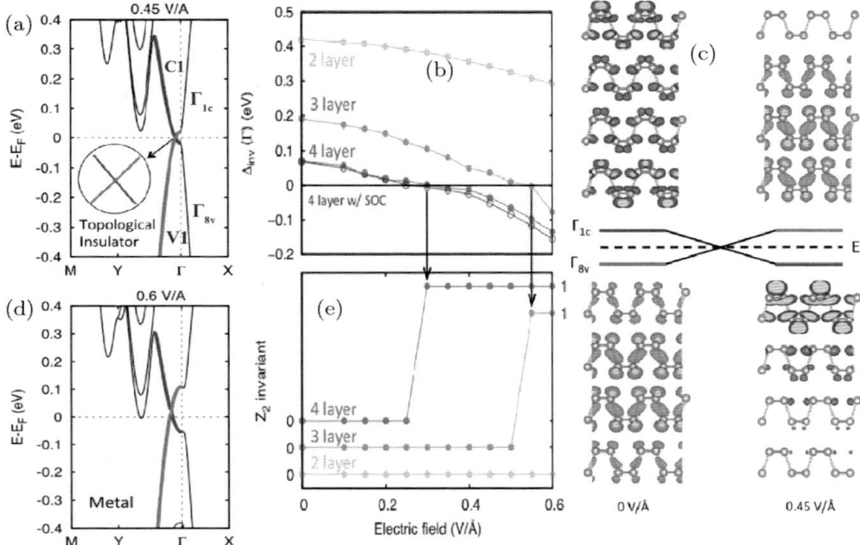

Figure 7. Electronic structure of four-layer phosphorene under the application of external electric field of (a) 0.45 V/Å, (d) 0.6 V/Å (using PBE functional), (c) evolution of valence (Γ_{8v}) and conduction (Γ_{1c}) band wave functions (isosurface at 0.03 au) with the application electric field from 0 V/Å to 0.45 V/Å, (b) band inversion energy Δ_{inv} (PBE level) of 2-, 3- and 4-layer phosphorene at Γ point for different values of electric field (negative value of Δ_{inv} represents non-trivial topological phase) and (e) Z_2 invariant of 2-, 3-, 4-layer phosphorene. Reprinted with permission from Ref. [21]. Copyright 2015 American Chemical Society.

(Figure 7(b)). With further increase in the electric field (0.45 V/Å), the Γ_{1c} state becomes occupied and Γ_{8v} state becomes unoccupied (as shown in Figures 7(a) and 7(c)). Charges get accumulated at one edge of few-layer phosphorene, as expected from broken spatial inversion symmetry. The band inversion energy becomes negative and Z_2 topological invariant (calculated using method introduced by Fu and Kane [22]) becomes 1 (Figures 7(b) and 7(e)) [21]. In addition, the Dirac cone associated with band crossing along Γ–Y direction makes few-layer phosphorene a semimetal with two nodes at $\Lambda(0, \pm y_D)$ at the Fermi level (as shown in Figures 7(a) and 2(d)). This band crossing along Λ point is preserved by fractional translational symmetry ($\tau_y = b/2$ for AB stacking) as conduction and

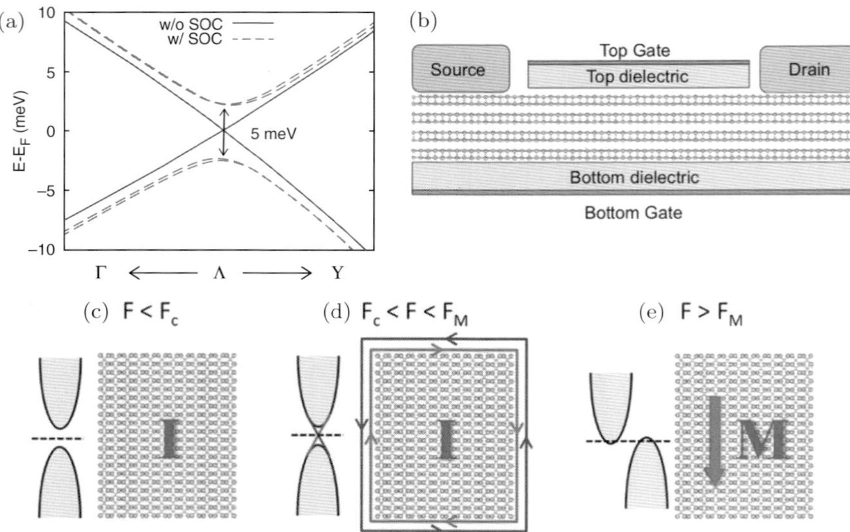

Figure 8. (a) Band structure of four-layer phosphorene at Λ point with the introduction of SOC, (b) schematic diagram of a duel gated topological field effect transistor and the charge and spin currents flowing through the phosphorene channel for (c) $F < F_c$, (d) $F_c < F < F_M$ (e) $F > F_M$ (I and M represent metallic and insulating states, respectively). Reprinted with permission from Ref. [21]. Copyright 2015 American Chemical Society.

valence bands have different symmetry [21]. Due to the absence of fractional translational symmetry along Γ–X direction, there is no band crossing, marking another manifestation of anisotropy in the ***ab*** plane. With the inclusion of spin-orbit coupling (SOC), a small gap (5 meV) at the Fermi level appears, lifting the degeneracy at Λ point and it makes few-layer phosphorene a topological insulator (as shown in Figure 8(a)). However, the effect of SOC is small as the strength of SOC for P atom is weak and it does not influence the band inversion energy and critical field. As the spatial inversion symmetry is not present after the application electric field, the bands will be spin-split at Λ point due to the effect of SOC. This feature gives rise to a quantum spin Hall effect in four-layer phosphorene [21].

As the strength of applied field increases, the VBM and CBM shift continuously in the upward and downward directions, respectively, and it shifts the Dirac cone along Y direction (see Figures 7(b) and 7(d)). At 0.6 V/Å electric field (F_M), another valley of C1 band touches Fermi level at a point along Γ–Y direction (Figure 7(d)). Hence, four-layer phosphorene becomes a topological metal as topological feature at Λ point (Z_2 invariant) remains unaltered (see Figure 8(e)). For bilayer phosphorene, topological phase transition is not observed (see Figures 7(b) and 7(e)) within this range of fields, as the band gap is large at zero field. For trilayer phosphorene, the critical field for topological phase transition is 0.55 V/Å (see Figures 7(b) and 7(e)). A field effect topological transistor (FETT) can be constructed by using the topological properties of four-layer phosphorene (Figure 8(b)). Liu *et al.* [21] proposed a model of dual gated FETT formed of few-layer phosphorene, arguing that FETT remains in "OFF" state when the external electric field is less than the critical field (F_C) as the system persists in normal insulating state below F_C (Figure 8(c)). When electric field is greater than F_M, the system becomes metallic with current flowing throughout the sheet and it switches the FETT in "ON" state (Figure 8(e)). However, quantum spin hall state with the net spin current at the edge occurs at a field between F_C and F_M (Figure 8(d)). However, electrons can thermally excite to higher energy states easily before reaching the critical field, as the thermal energy needed to make such transition is 5 meV (58 K). To verify the quantum spin hall state, it is necessary to keep the experimental setup at low temperature ($<$58 K) (otherwise the system can jump from insulating phase to metallic phase directly) [21].

4.6 Gas Sensing

Adsorption properties of 2D materials (graphene, TMDs and phosphorene) have been studied extensively due to their high chemical activity and high surface to volume ratio [23–25]. The adsorptions of CO, CO_2, NO, NO_2 and NH_3 molecules on phosphorene have been studied recently using first-principles based DFT calculations [25]. In the ground state of gas molecule adsorbed phosphorene (obtained using DFT calculations with an LDA functional), CO,

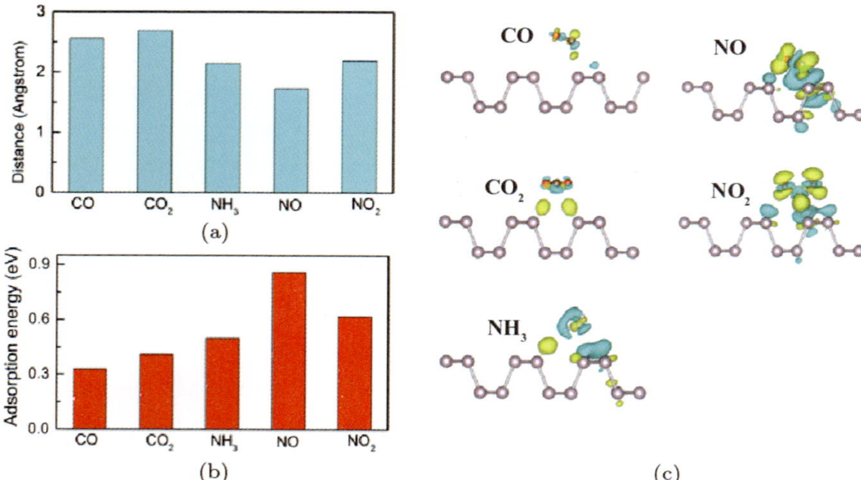

Figure 9. (a) Distance between gas molecules (CO, CO$_2$, NH$_3$, NO and NO$_2$) and phosphorene layer (b) adsorption energy and (c) charge transfer (isosurface value is set to 0.001 e/Å^3) for each gas adsorbed phosphorene (yellow part and blue part of isosurface denote electron gain and loss, respectively). Reprinted with permission from Ref. [25]. Copyright 2014 American Chemical Society.

CO$_2$, NO$_2$ and NH$_3$ molecules adsorb above the phosphorene layer at a distance of 2.56 Å, 2.69 Å, 2.2 Å and 2.14 Å, respectively (as shown in Figure 9(a)) without making any direct bond to phosphorus atom [25]. On the other hand, nitrogen atom of NO molecule makes a direct bond with phosphorus atom (1.73 Å) in the ground state. The adsorption energy ($E_a = E_{\text{gas}} + E_{\text{phosphorene}} - E_{\text{gas+phosphorene}}$) of NO molecule (0.863 eV/unit cell) is larger compared to other molecules, i.e., NO$_2$ (0.62 eV/unit cell), CO (0.325 eV/unit cell) and CO$_2$ (0.41 eV/unit cell) and NH$_3$ (0.5 eV/unit cell) (as shown in Figure 9(b)). A higher adsorption energy of NO in phosphorene compared to other 2D materials [23, 26] highlights phosphorene is particularly sensitive to NO gas.

Adsorption property of phosphorene depends on the charge transfer between phosphorene layer and a gas molecule. The induced charge carrier in phosphorene due to adsorption of gas molecules can also influence its electronic and transport properties. The charge transfer is defined as $\Delta\rho = \rho_{tot}(r) - \rho_p(r) - \rho_{gas}(r)$, where $\rho_{tot}(r)$,

$\rho_p(r)$ and $\rho_{gas}(r)$ are the charge densities of the phosphorene with adsorbed molecule, pristine phosphorene and gas molecule, respectively. It has been noticed that NO molecule transfers a charge (0.2 e) to phosphorene layer (as shown in Figure 9(c)), greater than the other gas molecules ($\Delta\rho(CO) = 0.03$ e, $\Delta\rho(CO_2) = 0.04$ e) [25]. Smaller amount of charge transfer from gas molecule (CO, CO_2, NH_3) to phosphorene sheet causes weaker binding with phosphorene. The gas molecules with low adsorption energy will have the least effect on the electronic properties. Adsorption of paramagnetic NO and NO_2 gas molecules introduce a higher level of carrier doping in phosphorene, which is expected from their large values of adsorption energy. From the spin density distribution, it is seen that spin polarised p electrons of N atom in NO and NO_2 molecules are primarily located on gas molecules (as shown in Figures 10(b) and 10(c)) and the value of total magnetic moment is $1\mu_B$ [25].

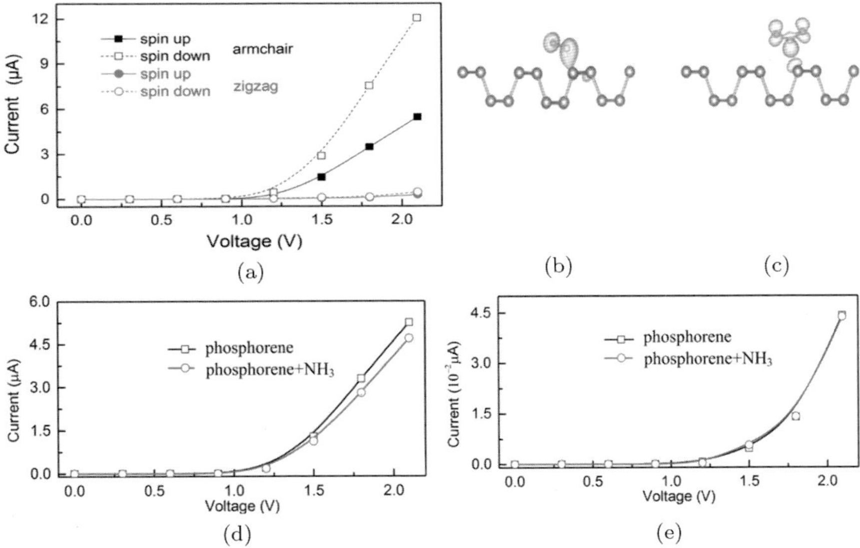

Figure 10. (a) *I–V* characteristics (spin-polarised) of NO adsorbed phosphorene, (b)–(c) spin density distributions on NO and NO_2 molecules, *I–V* characteristics of NH_3 adsorbed phosphorene in comparison with pristine phosphorene (d) along the arm-chair and (e) zigzag directions. Reprinted with permission from Ref. [25]. Copyright 2014 American Chemical Society.

To visualise the effect of adsorption of NH_3 and NO_2 gas molecules on transport properties of phosphorene, I–V characteristics of gas adsorbed phosphorene have been studied using DFT calculations (see Figures 10(a)–10(e)) [25]. Current flowing through pristine phosphorene at a bias voltage of 1.5 V along arm-chair and zigzag directions are (1.28 μA and 0.0047 μA): they differ by orders of magnitude due to the marked anisotropy in electronic mobility in the ab plane. As the electronic mobility along zigzag direction is very small compared to arm-chair direction, the current along zigzag direction is two order of magnitude smaller than that along arm-chair direction. For NH_3 adsorbed phosphorene (Figures 10(a)–10(e)), current decreases by 11% compared to pristine phosphorene along arm-chair direction due to the reduction in the available conduction channels of various energy bands (presence of back scattering centre after NH_3 adsorption) [25]. Current along zigzag direction remains unaffected after the adsorption of NH_3 molecule. In contrast, spin polarised current (up spin and down spin) for NO adsorbed on phosphorene (in the arm-chair direction, bias voltage 1.5 V, Figure 10(a)) increases by 9% in the spin up direction and 88% in the spin down direction compared to that in pristine phosphorene [25]. This increment of spin polarised current is attributed to the presence of N $2p$ states near the Fermi level (which was not present in pristine case). In the zigzag direction, current ($\sim 10^{-2}$ μA at 1.5 V) for both spin up and spin down directions increase by an order of magnitude compared to pristine phosphorene ($\sim 10^{-3}$ μA at 1.5 V) [25]. This high level of sensitivity of transport properties to adsorbed gas molecules makes phosphorene a good candidate for superior gas sensing.

It has been noticed that metal (Al, Fe, Co, Ni, Pd, Pt, Ag and Au) doped phosphorene exhibits much stronger NO affinity compared to pristine phosphorene [27]. Fe-doped phosphorene shows a half-metallic behaviour [27]. N atom of NO molecule orients towards the dopant atom in phosphorene layer in the minimum energy structure. For NO molecule adsorbed metal (Al, Fe, Co, Ni, Pd, Pt, Ag and Au) doped phosphorene, adsorption energy increases by a factor of 1.5–8 compared to NO adsorbed phosphorene which indicates promoted interaction upon metal substitution [27]. These cationic dopants (metal atom) allow multiple NO molecules per adsorption site, facilitating the increase in gas sensitivity compared to pristine phosphorene [27]. Also, it has been predicted using first-principles calculations combined with statistical thermodynamics that phosphorene is unusually sensitive to NO_2 gas molecule compared to graphene with an adsorption density around $\sim 10^{15}\,cm^2$ (for the 4.8 nm thick phosphorene nanosheet when exposed to 20 p.p.b. NO_2 at 300 K) [25, 28].

4.7 Electron–Phonon Coupling and Thermoelectricity

The unit cell of monolayer phosphorene contains four phosphorus atoms. Out of its 12 vibrational normal modes, nine modes are optical, six of which are Raman active. Three optical phonon modes having A_g^1, A_g^2 and B_{2g} symmetry are the most prominent modes observed in Raman scattering experiments [12]. A_g^1 mode involves atomic displacements in a direction perpendicular to the phosphorene sheet whereas A_g^2 and B_{2g} involve in-plane atomic displacements along arm-chair and zigzag directions, respectively (as shown in Figure 11).

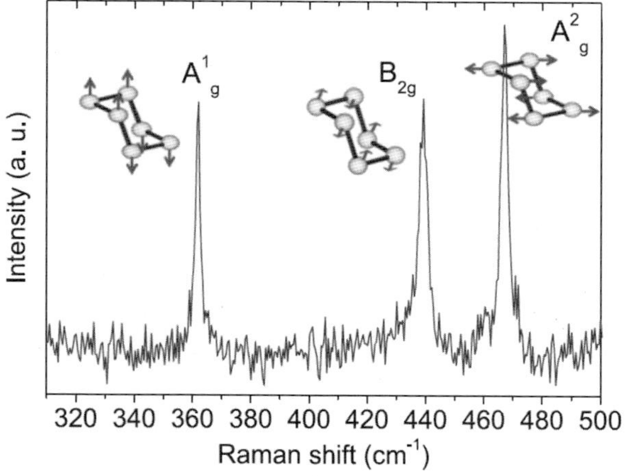

Figure 11. Raman spectra of monolayer phosphorene [12].

Effects of electron and hole doping in phosphorene on these three phonon modes have been studied experimentally (Raman spectroscopy) and theoretically (DFT calculations) [12]. It was noticed that A_g^1 and A_g^2 modes soften significantly with electron doping, whereas the effect of hole doping, on them is vanishingly small (as shown in Figures 12(a), 12(b), 12(d) and 12(e)). The

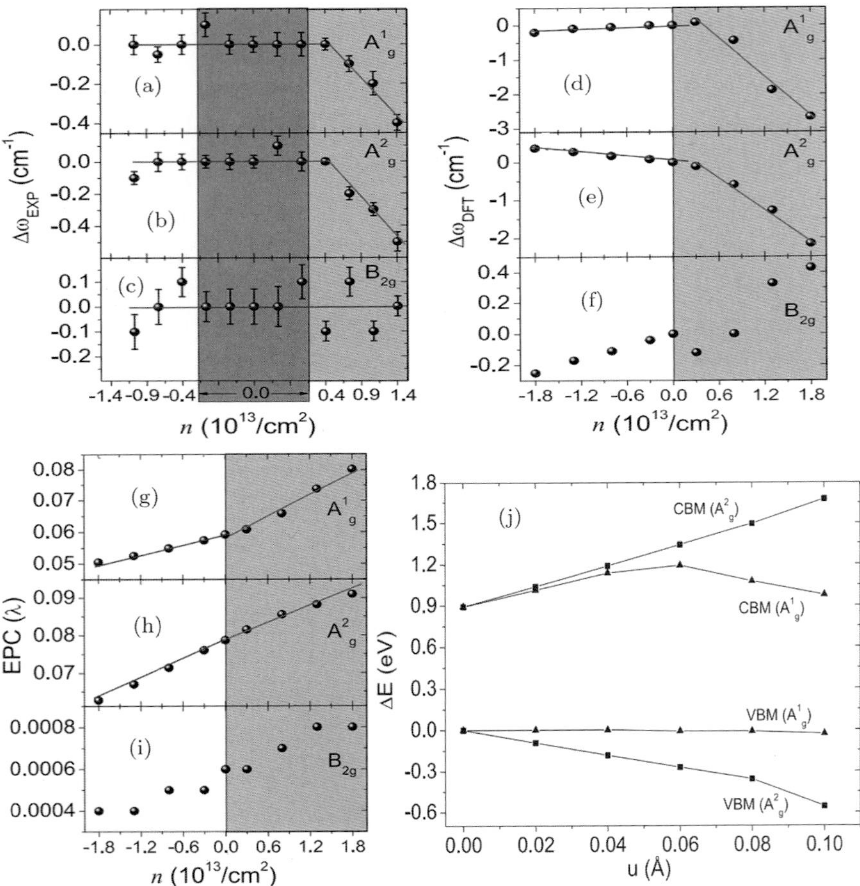

Figure 12. Changes in phonon frequencies (a)–(c) measured experimentally, (d)–(f) obtained from theoretical calculations as a function of carrier concentration, (g)–(i) electron–phonon coupling constant for different values of carrier concentration and (j) evolutions of VBM and CBM with structural distortion (obtained by freezing the atomic displacements of A_g^1 and A_g^2 modes) [12].

frequency of B_{2g} mode remains unaffected by either of the electron or hole doping (as shown in Figures 12(c) and 12(f)). To get insight into the dependence of phonon frequency on carrier doping, electron–phonon coupling (EPC) constants of these three phonon modes have been estimated [12]. EPC of a phonon mode ν with frequency ω at wave vector q is [29]:

$$\lambda_{q\nu} = \frac{2}{\hbar\omega_{q\nu}N(\epsilon)}\sum_{k}\sum_{ij}|g_{k+q,k}^{q\nu,ij}|^2$$

$$\times\,\delta(\epsilon_{k+q,i}-\epsilon_F)\delta(\epsilon_{k,j}-\epsilon_F), \tag{4.3}$$

where $N(\epsilon_F)$ is the density of states at Fermi level, and the EPC matrix element is:

$$|g_{k+q,k}^{q\nu,ij}| = \left(\frac{\hbar}{2M^*\omega_{q}\nu}\right)^{1/2}\langle\psi_{k+q,i}|\Delta V_{q\nu}|\psi_{k,j}\rangle, \tag{4.4}$$

where $\psi_{k,i}$ is the electronic wave function of ith band at wave vector k, and M^* is the effective mass associated with the phonon mode. $\Delta V_{q\nu}$ is change in the self-consistent potential caused by atomic displacements associated with the phonon mode.

The values of λ of phonon modes with A_g symmetry are two orders of magnitude greater than the value for B_{2g} mode. As A_g modes preserve the symmetry of crystal structure, it couples strongly with all electronic states (non-zero matrix element according to Eq. 4.4) compared to B_{2g} mode which breaks the lattice symmetry (B_{2g} mode is orthogonal to A_g modes). Calculation reveal that value of λ of A_g^1 mode is more affected by electron doping rather than hole doping, while A_g^2 mode gets affected by both electron and hole doping (Figures 12(g)–12(i)) [12]. This asymmetry in the dependence of λ on electron and hole doping has been explained using two facts. Firstly, as the wave functions at VBM and CBM have different symmetry (as shown in Figures 2(c) and 7(a)), electron–phonon coupling matrix element gives different values for electron and hole. Secondly, the variation of CBM and VBM as a function of structural distortions associated with A_g^1 and A_g^2 modes (as shown in Figure 12(j)) shows that CBM gets affected by structural distortions according to both

A_g modes whereas VBM is mostly gets affected by freezing of A_g^2 mode.

In the polarised Raman spectra, it has been noticed that fully symmetric A_g modes exhibit unusual dependence on angle and polarisation of light, which has been described by assuming complex values of Raman tensor elements [30]. This experimental finding has been explained [12] by using the asymmetry in electronic structure and dielectric constant (associated with the symmetric modes) as discussed below. The states near VBM and CBM along Γ–Y direction (zigzag direction) are more affected by low level of doping as the states are closer in energy compared to other direction (as shown in Figure 2(a)). At Y point, VBM and CBM are contributed by p_x and p_y orbitals. The relative phases between wave functions of the states (VBM and CBM) at Y and Γ point can be expressed as a Hermitian matrix (3D rotation of space made of two p orbitals), which is responsible for complex value of Raman tensor element. Also, the anisotropy in the Raman tensor gets reflected in optical dielectric tensor elements, which were obtained by freezing $A_g^1(\Delta\varepsilon_{xx} = 0.05, \Delta\varepsilon_{zz} = 1.3)$ and $A_g^2(\Delta\varepsilon_{xx} = 0.2, \Delta\varepsilon_{zz} = 1.5)$ modes [12].

From the dispersion of acoustic mode of monolayer phosphorene (Figure 13(a)), it is evident that acoustic phonon modes (TA,

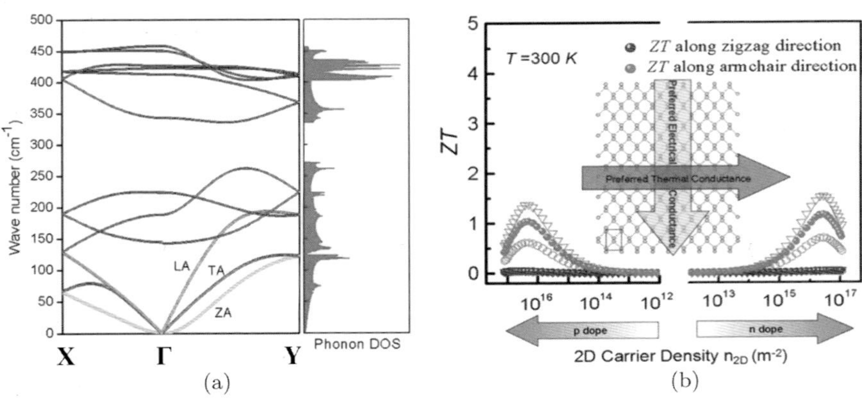

Figure 13. (a) Phonon dispersion and density of states of monolayer phosphorene and (b) thermoelectric figure of merit as a function of carrier doping density at 300 K. Reprinted with permission from Ref. [31]. Copyright 2014 American Chemical Society.

LA and ZA) are strongly dispersed along zigzag direction than along arm-chair direction [31]. TA and LA modes are transverse and longitudinal acoustic modes and they involve in-plane atomic displacements. ZA mode involves out of plane atomic displacements and have parabolic dispersion compared to other two acoustic modes (linear dispersion). The group velocity of the LA acoustic mode is highly anisotropic in the *ab* plane. Along zigzag direction, the group velocity (8397 m/s) is two times greater than the value along arm-chair direction (4246 m/s) [31] (anisotropy has been also found in the value of elastic constant discussed in earlier section). Thermal conductivity mostly depends on the group velocity of acoustic modes and phonon mean free path (which depend on phonon dispersion and scattering mechanism) [32, 33]. For isotropic phonon scattering, the mean free path along arm-chair direction becomes smaller compared to zigzag direction due to its smaller group velocity. The calculated values of thermal conductivities of monolayer phosphorene are $36 \, \mathrm{W \, m^{-1} \, K^{-1}}$ in the arm-chair direction and $110 \, \mathrm{W \, m^{-1} \, K^{-1}}$ along zigzag direction [31, 33].

The prominent electronic transport and poor thermal conductivity along arm-chair direction make phosphorene a good thermoelectric material [34]. The value of thermoelectric figure of merit (ZT) becomes large in the combination of high electrical conductance and low thermal conductance. From first-principles calculations [31], it has been shown that the value of ZT is 2.5 in the arm-chair direction (Figure 13(b)) at 500 K with carrier doping density of $2 \times 10^{16} \, \mathrm{m^{-2}}$ and estimated energy conversion efficiency is 15–20% [31, 34]. At room temperature, the ZT value still remains greater than 1.0. ZT value of phosphorene is much better compared to graphene which has a low ZT, due to its very high thermal conductivity ($\sim$2000 W/(mK)) [35].

4.8   Defects

Point defects and stacking fault alter the local crystal structure, and can have strong influence on electronic properties of phosphorene. We first discuss about different types of possible point defects in phosphorene (see Figure 14).

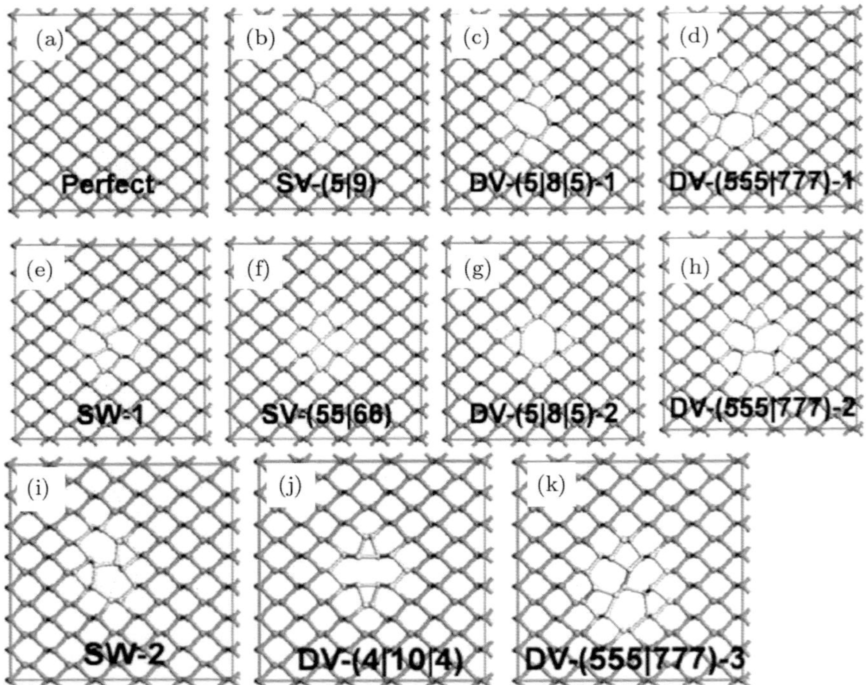

Figure 14. Structures of (a) perfect and (b)–(k) defective phosphorene in the 5×7 supercell. Defect in phosphorene includes Stone–Wales defects: (e) SW-1 and (i) SW-2; single vacancy defects: (b) SV-(5|9) and (f) SV-(55|66); double vacancy defects: (c) DV-(5|8|5)-1, (g) DV-(5|8|5)-2, (d) DV-(555|777)-1, (h) DV-(555|777)-2, (k) DV-(555|777)-3 and (j) DV-(4|10|4)). Reprinted with permission from Ref. [36]. Copyright 2015 American Chemical Society.

The possible 10 kinds of defects (Figure 14) in monolayer phosphorene (due to its low symmetry structure) are Stone–Wales (SW-1 and SW-2), single vacancy (SV-(5|9) and SV-(55|66)), double vacancy (DV-(5|8|5)-1, DV-(5|8|5)-2, DV-(555|777)-1, DV-(555|777)-2, DV-(555|777)-3 and DV-(4|10|4)) defects (Figure 14) [36]. However, double vacancy defect DV-(4|10|4) (see Figure 14(j)) is newly found in phosphorene, in comparison with other 2D materials [36]. The two types of Stone–Wales defects are generated by rotating two phosphorus bonds (one bond between the two sublayers and other bond within one sublayer) by 90°. In double vacancy defects, two nearest neighbour phosphorus atoms (from one sublayer or

Table 3. Formation energy (eV) and total magnetic moment (μ_B) of phosphorene with Stone–Wales and point defects.

Defect	E_f	μ
SW-1	1.012	0.003
SW-2	1.322	0.003
SV-(5|9)	1.626	0.980
SV-(55|66)	2.025	0.000
DV-(5|8|5)-1	1.906	0.002
DV-(5|8|5)-2	3.041	0.002
DV-(4|10|4)	2.137	0.000
DV-(555|777)-1	2.081	0.002
DV-(555|777)-2	2.350	0.002
DV-(555|777)-3	2.613	0.003
perfect	0.000	0.000

two sublayers) are removed from phosphorene. From estimates of adsorption energies (as given in Table 3) of several defects in phosphorene (using DFT calculations under GGA approximation), it has been found that SW-1 has the lowest formation energy (1.012 eV) and DV-(5\|8\|5)-2 has highest formation energy (3.041 eV) [36]. The newly found DV-(4\|10\|4) defect (which was not present in graphene and silicene) is stable and has a low formation energy about 2.137 eV. High density of unoccupied localised states (compared to pristine phosphorene) in the electronic structure is signature of defective phosphorene and they can be easily recognised by using STM images at positive bias [36].

The dangling bond of a phosphorus atom (two covalent bond) near SV-(5\|9) defect (see Figure 14(b)) gives rise to localised magnetic moment (0.98 μB) in phosphorene (Table 3). The DV-(5\|8\|5)-1 defect has shown to be effective in reducing the energy loss via non-radiative electron hole recombination [37]. It has been discussed in previous section that electrons strongly couple to optical phonon modes with A_g symmetry. In the presence of double vacancy defect, the band gap (1.86 eV) increases by 19% from the value of pristine phosphorene band gap (1.56 eV). Due to the lower symmetry induced by defects, electrons couple to more number of phonon modes than in pristine phosphorene. However, the strength of electron–phonon coupling of defective phosphorene (1.5 times smaller than pristine) decreases as phosphorene relaxes the strain induced by defect [37] (similar to graphene nanoribbons [38]). Thus, both increase in band gap and reduction in electron–phonon coupling decrease the probability of losing energy via non-radiative electron-hole recombination [37].

Similar to in-plane strain, stacking fault can also drive a transition from state direct to with the one with indirect band gap. Three types of stacking of two monolayers are possible. In AB and AC stacking, relative displacements between two monolayers are (0.5, 0.0) and (0.5, 0.5) in the **ab** plane, respectively (as shown in Figures 15(a) and 15(c)). AA stacking is formed without displacing two layers relatively within the **ab** plane. AB stacked bilayer (mostly seen in bulk black phosphorus) is more energetically favourable compared to AA and AC stacking (Figure 15(c)) [39]. It has been shown using diffusion Monte Carlo techniques that the binding energy (inter-layer distance) of AB stacked bilayer is 3 (1.1) times larger (smaller) than AA stacked bilayer [16]. From the calculated charge density plot ($\Delta\rho$ is the charge density difference induced by assembling the bulk system from bilayers), it has been shown (Figures 16(a) and 16(b)) that the different kinds of redistribution of charges in intra-layer and inter-layer portion for the two different stacking give distinctive binding energies [16]. This suggests that inter-layer interaction in black phosphorus is much complicated than a simple vdW interaction.

Recently, one metastable Aδ stacking has been discovered theoretically (Figure 15(b)). Aδ stacking is composed of two monolayers, which are displaced in the **ab** plane by (δ,0.0) ($\delta = 0.281$ a) [39]. As a result, structural symmetry is reduced from *Pbcm* (AB stacking) to *P2/m* space group [39]. Aδ stacked bilayer is energetically higher by 31 meV/unit cell than the AB stacking (Figure 15(c)). Optimised lattice parameter in the arm-chair direction (a) for Aδ stacking is 0.03 Å larger than the value for AB stacking while the change in

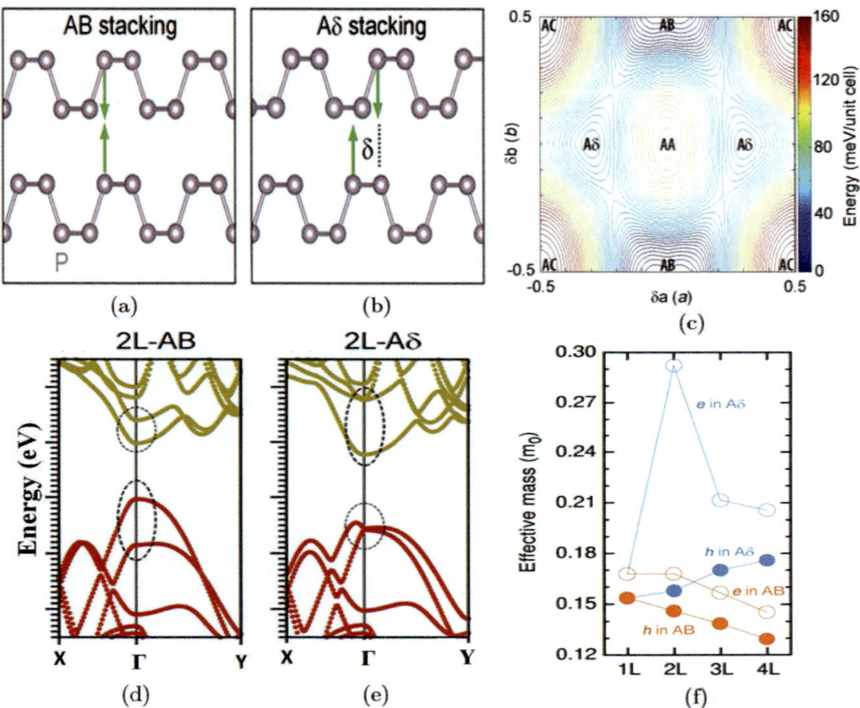

Figure 15. Atomic structure of bilayer phosphorene with (a) AB and (b) Aδ stacking, (c) contour plot of potential energy surface of bilayer phosphorene as a function of relative displacements (δa and δb) along a and b directions, (d)–(e) band structures calculated with HSE06 functional for bilayer phosphorene with AB and Aδ stacking respectively (dashed circles in (d) and (e) represents the opposite trends of splitting of VBM and CBM at Γ point) and (f) carrier effective mass of bilayer phosphorene with AB and Aδ stacking. Reprinted with permission from Ref. [39]. Copyright 2016 American Chemical Society.

b parameter is insignificant. This structural change influences the electronic properties (electronic mobility, band gap). For Aδ stacking, VBM shifts from Γ point to $(0.05k_x, 0.0k_y)$ point and bilayer phosphorene becomes an indirect band gap semiconductor (see Figure 15(e)) [39]. Effective mass of holes increases (see Figure 15(f)) with increasing number of layers whereas the effective mass of electrons increases largely from 1L to 2L and then decreases for the Aδ stacking. This change in the effective mass also modifies carrier mobilities along the arm-chair direction. In addition, level splitting at

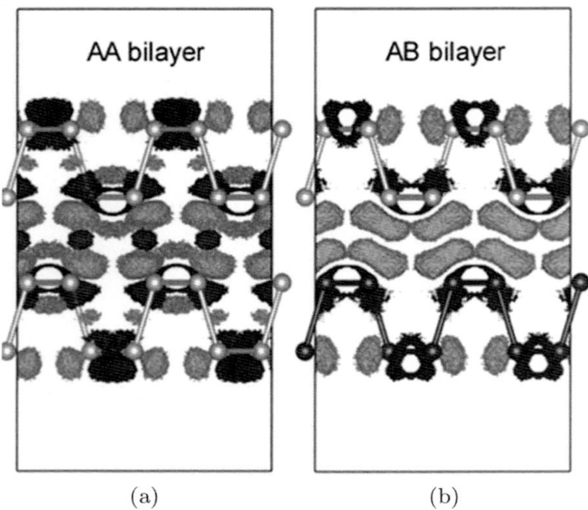

Figure 16. Electron density difference (induced by assembling bulk system from bilayers) for (a) AA and (b) AB stacked bilayer phosphorene using isosurfaces. Reprinted with permission from Ref. [16]. Copyright 2015 American Chemical Society.

CBM and VBM (Figure 15(d) and 15(e)) shows opposite trends for different stacking of monolayers. When bilayer is formed, the splitting of energy corresponding to the eigenstate ψ (of VBM or CBM) of monolayer is equal to the overlap between the states ψ in the lower (ψ_{lower}) and upper (ψ_{upper}) layer [39]. This again highlights that the overlap of wave functions of two neighbouring layers (in addition with vdW interaction) can affect the electronic structure drastically.

4.9 Summary

High charge carrier mobility ($\sim 26,000$ cm^2 V^{-1} S^{-1} for holes and ~ 1000 cm^2 V^{-1} S^{-1} for electrons) and a desirable band gap (1.45 eV) make monolayer phosphorene a potential candidate for making nanoelectronic devices. Phosphorene has highly anisotropic crystal structure, electronic and transport, thermoelectric properties and these behaviours make it unique among the 2D materials (graphene, TMDs). The anisotropy in electronic and transport properties is switchable by applying in-plane strain. Phosphorene is potentially good photocatalyst and gas sensor due to its finite band gap and high surface to volume ratio. External electric field perpendicular to layers converts four-layer phosphorene to a topologically insulating state. Point defects introduce magnetism in phosphorene and help reduce the energy loss via non-radiative electron-hole recombination. The inter-layer interaction in few-layer phosphorene is quite non-trivial and associated with the charge redistribution within the layer and in the inter-layer region.

Acknowledgements

Arpita Paul is thankful for research fellowship from the Department of Science and Technology, India. Umesh V. Waghmare acknowledges support from a J C Bose National Fellowship of the Department of Science and Technology, Govt. of India and US Air Force AOARD grant no. FA 238615-10002.

References

[1] K. S. Novoselov *et al.*, *Nature* **438**, 197 (2005).
[2] Q. H. Wang *et al.*, *Nat. Nanotechnol.* **7**, 699 (2012).
[3] P. Vogt *et al.*, *Phys. Rev. Lett.* **108**, 155501 (2012).
[4] E. Bianco *et al.*, *ACS Nano.* **7**, 4414 (2013).
[5] C. Berger *et al.*, *J. Phys. Chem. B* **108**, 19912 (2004).
[6] L. Liao *et al.*, *Nature* **467**, 305 (2010).
[7] B. Radisavljevic *et al.*, *Nat. Nanotechnol.* **6**, 147 (2011).
[8] K. Bolotin *et al.*, *Solid State Commun.* **146**, 351 (2008).
[9] H. Liu *et al.*, *ACS Nano.* **8**, 4033 (2014).
[10] J. Qiao *et al.*, *Nat. Commun.* **5**, 4475 (2014).
[11] S. Das, M. Demarteau and A. Roelofs, *ACS Nano.* **8**, 11730 (2014).
[12] B. Chakraborty *et al.*, *2D Materials* **3**, 015008 (2016).
[13] S. Appalakondaiah *et al.*, *Phys. Rev. B* **86**, 035105 (2012).
[14] R. Fei and L. Yang, *Nano Lett.*, **14**, 2884 (2014).
[15] H. M. Stewart, S. A. Shevlin, C. R. A. Catlow and Z. X. Guo, *Nano Lett.* **15**, 2006 (2015).
[16] L. Shulenburger *et al.*, *Nano Lett.* **15**, 8170 (2015).
[17] B. Sa *et al.*, *The J. Phys. Chem. C* **118**, 26560 (2014).
[18] Y. Zhang *et al.*, *Nat. Nanotechnol.* **9**, 111 (2014).
[19] S. Bruzzone and G. Fiori, *Appl. Phys. Lett.* **99**, 222108 (2011).
[20] V. Chakrapani *et al.*, *Science* **318**, 1424 (2007).

[21] Q. Liu *et al.*, *Nano Lett.* **15**, 1222 (2015).

[22] L. Fu and C. L. Kane, *Phys. Rev. B* **74**, 195312 (2006).

[23] O. Leenaerts, B. Partoens and F. M. Peeters, *Phys. Rev. B* **77**, 125416 (2008).

[24] F. K. Perkins *et al.*, *Nano Lett.* **13**, 668 (2013).

[25] L. Kou, T. Frauenheim and C. Chen, *J. Phys. Chem. Lett.* **5**, 2675 (2014).

[26] S. Zhao, J. Xue and W. Kang, *Chem. Phys. Lett.* **595596**, 35 (2014).

[27] N. Suvansinpan *et al.*, *Nanotechnology* **27**, 065708 (2016).

[28] S. Cui *et al.*, *Nat. Commun.* **6**, 8632 (2014).

[29] B. Chakraborty *et al.*, *Phys. Rev. B* **85**, 161403 (2012).

[30] H. B. Ribeiro *et al.*, *ACS Nano.* **9**, 4270 (2015).

[31] R. Fei *et al.*, *Nano Lett.* **14**, 6393 (2014).

[32] C. Jeong, S. Datta and M. Lundstrom, *J. Appl. Phys.* **111**, 093708 (2012).

[33] Z. Luo *et al.*, *Nat. Commun.* **6**, 8572 (2015).

[34] L. Kou, C. Chen and S. C. Smith, *J. Phys. Chem. Lett.* **6**, 2794 (2015).

[35] D. Dragoman and M. Dragoman, *Appl. Phys. Lett.* **91**, 203116 (2007).

[36] W. Hu and J. Yang, *J. Phys. Chem. C* **119**, 20474 (2015).

[37] R. Long, W. Fang and A. V. Akimov, *J. Phys. Chem. Lett.* **7**, 653 (2016).

[38] B. F. Habenicht, O. N. Kalugin and O. V. Prezhdo, *Nano Lett.* **8**, 2510 (2008).

[39] S. Lei *et al.*, *Nano Lett.* **16**, 1317 (2016).

Chapter 5

2D van der Waals Hybrid: Structures, Properties and Devices

Md. Ali Aamir, Tanweer Ahmed, Kimberly Hsieh, Saurav Islam,
Paritosh Karnatak, Ranjit Kashid, Phanibhusan Singha Mahapatra,
Jayanta Mishra, Tathagata Paul, Avradip Pradhan, Kallol Roy,
Anindita Sahoo and Arindam Ghosh*

*Department of Physics, Indian Institute of Science,
Bangalore 560012, India*
** arindam@physics.iisc.ernet.in*

Abstract. The isolation of graphene and subsequently other two-dimensional (2D) materials like transition metal dichalcogenide's (TMDCs) and topological insulators have led to the discovery of a plethora of new fundamental phenomena and concepts in atomically thin materials. While single layer of graphene has no bandgap and can host high mobility carriers, TMDCs show a thickness dependent bandgap, but respond sensitively to optical illumination. It is however, often desirable, to combine these contrasting individual material properties for enhanced functionality. An emerging method to achieve this is called the van der Waals epitaxy, where atomic and/or molecular layers of different 2D materials are physically stacked on top of each other in appropriate sequences. In this chapter, we have reviewed various aspects pertinent to 2D van der Waals hybrids. In Sec. 5.2, we introduce some of the methods of fabricating such heterostructures, and in Sec. 5.3 we discuss about characterising them. In Secs. 5.4–5.6, we discuss several electrical, optical and thermal phenomena related to these hybrids. In Sec. 5.7, we review some of the emerging phenomena in these systems. The van der Waals hybrids seem to hold great promise in the context of both unanticipated fundamental discoveries and device applications, and we are still only at the incipient stages.

5.1 Introduction

After the first fabrication of a field effect transistor from graphene in 2004 [3], both material science and device engineering entered a new era. Graphene is the thinnest material which can be conceived and yet it has a unique combination of physical properties. This single layer of carbon atoms arranged in a hexagonal lattice is the strongest known material, best electrical and thermal conductor and mechanically flexible and transparent. Graphene, however, is not alone anymore. The same approach of creating ultra-thin field effect devices was soon extended to the molecular layers of other two-dimensional (2D) materials, where individual layers are weakly attached to the neighbouring ones by a small van der Waals force. Starting in 2010 [6] with MoS_2 (molybdenum disulphide), a member of the transition metal dichalcogenide (TMDC) family, a new generation of electronic devices consisting of other TMDCs, hexagonal boron nitride (h-BN), layered bismuth chalcogenides, or even layered oxides, have become the subject of intense fundamental and applied research.

The key driving force to the emergent architecture of what is now known as "2D electronics", is the availability of diverse physical properties from a common genre of material. Many of the TMDCs are semiconductors and have properties that are complementary to graphene. MoS_2, for example, has a thickness-dependent band gap of 1.2 eV–1.9 eV [6], which is necessary for designing a digital logic gate. So far, more than ten different 2D semiconductors (with band gap values from a few meV's to several eV's) have been experimentally isolated, and there are potentially many more that could be isolated in the near future. Because of this broad catalogue of materials, it is always possible to find a 2D semiconductor that is suitable for a certain application. This material advantage is not confined just to electronics or optoelectronics, but percolates to other applications as well. Ultra-thin films of Bi_2Se_3 or Bi_2Te_3, which belong to the bismuth chalcogenide family, are materials that cannot only be used as the active channel in 2D electronics, but also as efficient thermoelectrics, which can hence have applications beyond that of a simple field effect transistor.

While individual 2D materials continue to exhibit a plethora of interesting physical phenomena in electronics, light-matter interaction, magnetism, catalysis and bio-compatible designs to new topological states of matter, researchers are now trying to identify ways in which properties of two or more 2D materials may be combined. A protocol with rapidly increasing popularity is based on vertical stacking of atomic/molecular layers of dissimilar 2D crystals. The resulting composite structures are called 'van der Waals heterostructures' because van der Waals interaction is the primary force that holds the hybrid together. When the order of the constituent 2D-crystal layers are swapped or changed, it is hoped that new materials, with properties not found in their component parts could be made. The weak van der Waals interactions are usually not strong enough to perturb the unique electronic or structural properties of the individual layers. Instead, as different layers of materials are stacked on one another their physical and quantum properties can combine, interfere or cancel one another, leading to the new properties.

The first realisation of van der Waals hybrids dates back to 2010 when Dean *et al.* [7] fabricated the first graphene transistor on a boron nitride (BN) substrate. Since then van der Waals hybrids have not only been implemented to improve the performance of electronic or optoelectronic devices, but also discover new fundamental phenomena. Subsequently, the 2D van der Waals hybrids were used as the platform to demonstrate electrical tunability of metal insulator transitions [8], strong Coulomb drag with broken electron-hole symmetry in double-layer graphene heterostructures [9], tunnelling and resonant tunnelling effect in graphene–BN–graphene devices [10–12], single Coulomb impurity driven Fermi-edge transmission resonance in graphene–BN heterostructures [4], realisation of robust topological insulators with graphene and chalcogenides BiTeX (X = Cl, Br, and I) [13], strong light matter interactions in heterostructures made of TMDCs [14], existence of long-lived excitons [15] and valley polarised inter-layer excitons [16] in monolayer $MoSe_2$–WSe_2 heterostructures, tunable Schottky barriers in van der Waals metal semiconductor junctions [17], ultra-fast charge transfer (CT) mechanism in MoS_2–WS_2 heterostructures [18], and many more.

Observation of many of these phenomena is due to the unique devices designed from van der Waals hybrids, where new functionality is realised by addressing individual component of the hybrid through electrical, mechanical or optical means. Vertical field effect transistors with graphene, MoS_2, $Bi_2Sr_2Co_2O_8$ [19], graphene–WS_2 [20] and graphene–MoS_2 [21] heterostructures for flexible device applications, tunable photoresponse with ultrahigh gain in graphene–MoS_2 heterostructures [22–24], atomically thin p–n junction in MoS_2–WSe_2 heterostructures [25], picosecond photodetectors with graphene–WSe_2–graphene heterostructures, and ultra-low noise graphene-BN heterostructures for electronic applications [26], illustrate the capability of device architecture with the van der Waals hybrids [22].

This chapter provides an overview of the current development in the synthesis, device design, characterisation and multi-functionality of 2D van der Waals hybrids. This is not meant to be a review of the entire field, which would be an impractically complex task given the diverse nature of its impact on numerous areas of science and engineering. Our goal here is to identify few key hybrids to demonstrate the design concepts, flexibility and the functional details for specific applications. In Section 5.2, we discuss some of the popular methods of preparing the hybrids, which include vertical stacking of layers, as well as the multicomponent lateral network of 2D materials. Section 5.3 explains some of the characterisation protocols. Sections 5.4–5.6 are dedicated to electrical, optical/optoelectronic and thermal properties and phenomena, respectively. Section 7 concludes the chapter with an account of emerging phenomenology and outlook.

5.2   Fabrication of Hybrids from 2D Atomic and Molecular Membranes

At present, the most popular methods to realise multicomponent van der Waals hybrids are based on the vertical assembly of mechanically exfoliated 2D atomic membranes (the micromechanical transfer method) and chemical vapour deposition. Although new methods with self-assembly [27] or organic/inorganic heteroepitaxy created by physical vapour deposition [28] of organic molecules have been demonstrated, we will discuss the micromechanical transfer and chemical vapour deposition techniques in greater detail here.

5.2.1   Micromechanical transfer method

One of the key factors behind the initial rise of this research field is the simplicity of achieving high material quality by mechanical exfoliation. Indeed, the "peeling" of atomically thin membranes from bulk three-dimensional (3D) crystals has democratised materials science because high-quality samples can be prepared without expensive equipment. The mechanical exfoliation-based technique has been adopted for realising the first generation (and still the best quality) of van der Waals heterostructure devices.

The first van der Waals heterostructure was demonstrated with graphene and h-BN heterostructures. Dean *et al.* used mechanical transfer process (Figure 1(a)) to place a graphene on h-BN substrate and showed that the carrier mobility and charge inhomogeneity in graphene are improved by nearly one order of magnitude compared to the devices, where graphene is placed on the SiO_2 substrate. The

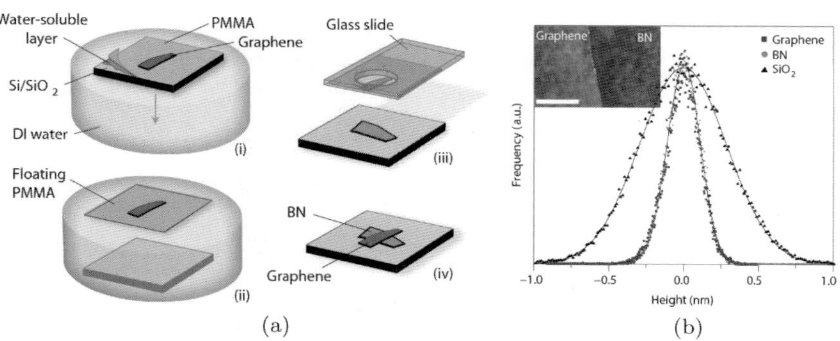

Figure 1. Demonstration of van der Waals heterostructure: (a) A graphene on h-BN heterostructure fabrication process. (b) Histogram of height distribution obtained by atomic force microscopy at different places of fabricated structure compare the smoothness of the graphene, h-BN substrate and SiO_2. (Adapted from Ref. [29].)

improved characteristics is attributed to the atomic scale smoothness of the BN surface (Figure 1(b)), which is also free of dangling bonds and charge traps. This study established the idea of van der Waals heterostructures as a possible mean to obtain improved or new device functionality.

Subsequently, the control, quality and versatility in the fabrication of van der Waals hybrids have increased dramatically, with several variations are now available for damage-free lifting and accurate translational and rotational alignment of the atomic layers. Utilising the meniscus of an epoxy or polydimethylsiloxane (PDMS) droplet for lifting and alignment is shown and explained in Figure 2 [5].

The process detail of such controlled micromechanical fabrication method is presented in Figure 3 [30]. Atomically, thin layers ($0.3\,\mathrm{nm} < t < \mathrm{few\,nm}$) are stable on substrate and remain attached due to van der Waals force, but cannot stand free and prone to mechanical damages such as tearing, folding because of stress, etc. Using the support of a flexible substrates (coated with a polymer sacrificial layer (SL)), these flakes can be transferred from one place to the other while avoiding damages. A specially designed substrate

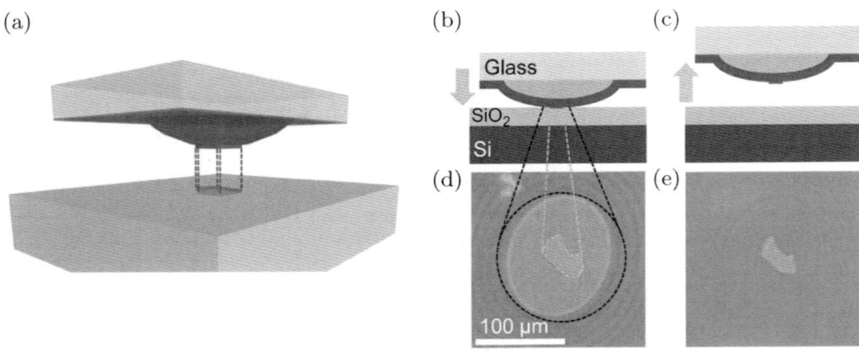

Figure 2. Controlled heterostructure fabrication technique: (a) A transparent mask with global minima (inverted spherical cap shape) to control well defined contact area. (b) and (d) Mask and bottom substrate in contact with each other. (c) and (e) When mask is separated. Flake gets attached to the mask. This process is repeated for multiple flakes to stack them one after the other (see Figure 3). (Adapted from Ref. [5].)

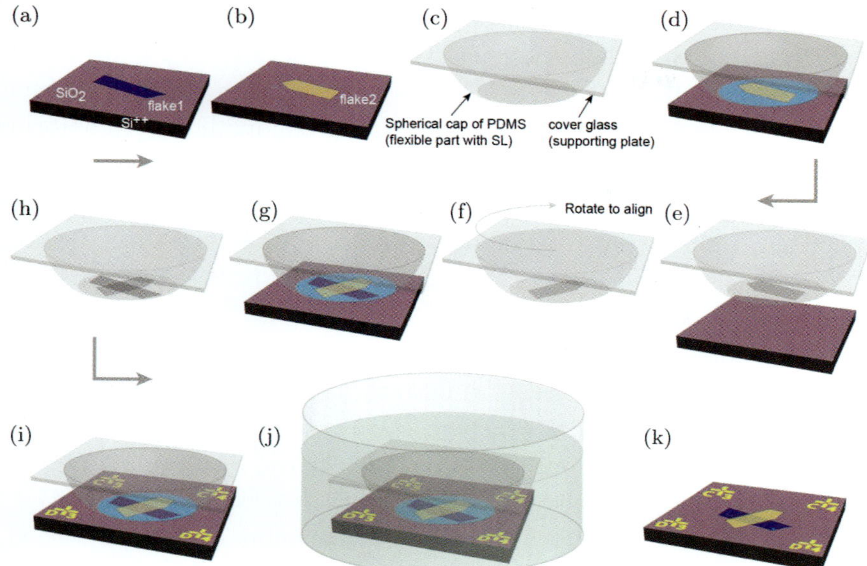

Figure 3. **Detail process flow of controlled heterostructure fabrication:** (a) and (b) Two flakes to be stacked to make a bilayer heterostructure. (c) Schematic of the mask used for stacking. (d) Mask and lower substrate in contact with each other. (e) After detaching, the flake gets attached to the SL coated on mask. (f)–(h) Stacking of 2nd layer following rotation and translation alignment. (i) Stacked layers in contact with pre-pattern substrate. (j) Mask and pre-patterned substrate combination in respective solution to dissolve SL. (k) Stack on pre-patterned substrate after dissolving SL. It remains attached because of van der Waals attractions. (Adapted from Ref. [30].)

which is coated with a SL allows to attach several mono- or multi-layer materials in any sequence. Strong interlayer van der Waals attractions between different layers of materials also help in the attachment process. While stacking the layers the relative position and orientation are controlled using a high precision mechanical micromanipulator.

Every attachment of a flake to the mask follow through a three step process and termed as align, contact, and detach. These are illustrated in schematics shown in Figures 3(c)–3(e). During the contacting process, the lower substrate temperature is maintained using a temperature controller. Before contacting, relative alignment is corrected with the precision stages on which the bottom substrate and mask remain mounted. When the mask comes in contact to SiO_2 a distinct colour appears (generally green, here cyan Figure 3(d)) along with circular interference fringes outside this region (not shown in scheme). The flexibility of the contacting region allows to conform at all area of contact including the sides of the flakes showing uniform colour. It is observed that failure in the attachment process likely to occur if the conformation has not been allowed, especially for thicker flakes. Same process steps are repeated to attach next flake (Figures 3(f)–3(h)).

After completion of stacking the assembly is brought in contact to a pre-patterned Si/SiO_2 substrate kept at relatively higher temperature (Figure 3(i)). The substrate is then cooled down to the room temperature without detaching from mask. The mask and pre-pattern assembly is moved into a solution to dissolve SL (Figure 3(j)). The mask gets separated and the stack remains on pre-patterned substrate because of van der Waals attraction between the stack and the SiO_2 (Figure 3(k)).

The optical images of heterostructures made from various layered materials such as graphene, BN, molybdenum disulphide (MoS_2), tungsten diselenide (WSe_2), and topological insulators such as BSTS ($Bi_{1.6}Sb_{0.4}Te_2Se$) are shown the Figure 4. All heterostructures

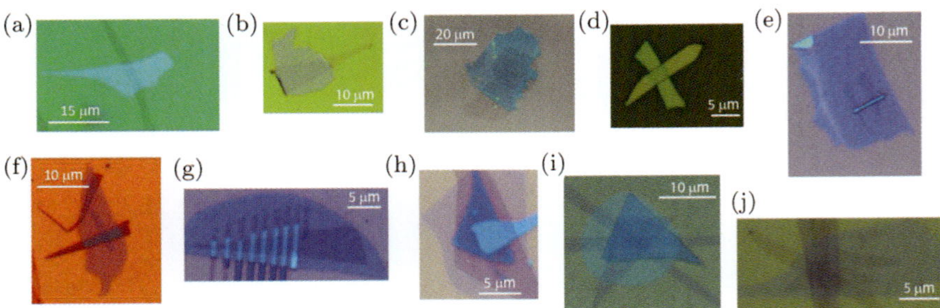

Figure 4. Various types of van der Waals heterostructures: (a)–(j) Double to multiple layers of heterostructures. Materials used are graphene, MoS_2, WSe_2, BN, topological insulators such as BSTS ($Bi_{1.6}Sb_{0.4}Te_2Se$). All stacks are on 285 nm SiO_2 substrate. The different colours arise because of spin coating of PMMA of different thickness figure ((a), (b), (i) and (j)), different settings in digital filters in the camera while taking images (f). A colour enhancement scheme is applied in certain flakes of figures shown in figures (f)–(j) to enhance the visibility of different layers. (Adapted from Ref. [30].)

presented here are laid on 285 nm SiO_2 grown on Si. The scale bar associated with each structure indicate the dimension of the layers. Use of a flexible substrate coated with a proper polymer SL shows close to 100% yield in stacking process. High precision mechanical micromanipulator with high-resolution microscope ease the alignment process while attaching the layers.

5.2.2   Chemical vapour deposition and lateral hybrids

Chemical vapour deposition (CVD) has been one of the most important and reliable techniques for the synthesis of 2D materials, with the potential to grow high quality material with controlled layer number over large area. Following initial success in synthesising graphene through decomposition of methane in the presence of copper catalyst [31], and later directly on insulating substrates [32], CVD growth of other layered crystals, such as the TMDCs [33], can now be achieved on a variety of substrates [34].

Recently, the CVD technique has been extended to synthesise both vertical and lateral hybrids of 2D materials. Figure 5 illustrates a protocol to grow an in-plane heterostructure of graphene and BN [35]. BN is first grown on Cu/Ni foils using ammonia–borane (NH_3–BH_3) as a precursor then patterns are etched using reactive ion etching (RIE) with Ar ions. Graphene is then grown on the etched regions by passing methane (CH_4) gas. Raman spectroscopy of the heterojunction shows that all intrinsic vibration modes from BN and graphene are found at the interface.

The synthesis steps of a vertical heterostructure of MoS_2 and WS_2 is illustrated in Figure 6 [36]. Molybdenum trioxide (MoO_3) powder is used as a precursor for the CVD synthesis of MoS_2 while a mixture of tungsten and tellurium is required for the growth of WS_2. Tellurium is added to accelerate the melting of tungsten. Higher temperatures of $\sim$850°C favour the growth of vertically stacked bilayers while in-plane lateral hybrids dominate when the synthesis is carried out at $\sim$650°C.

The key factors for larger single crystal of 2D materials by the CVD method are the suppression of nucleation and well-maintained growth. Similar challenges remain for controllable CVD

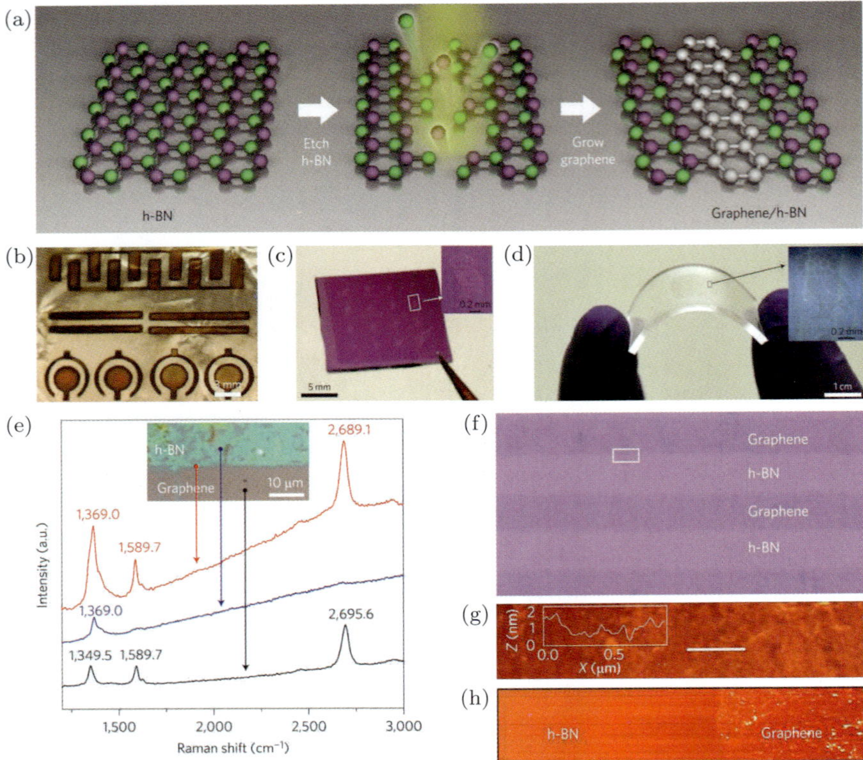

Figure 5. A CVD-grown graphene/BN in-plane heterostructure: (a) Schematic showing the fabrication of lateral heterostructures of graphene–BN by chemical vapour deposition (CVD). (b) Optical image of graphene–BN hybrids patterned as combs, bars and rings. Light areas indicate BN and dark areas indicate graphene. (c) and (d) Graphene–BN patterns transferred on Si^{++}/SiO_2 and flexible PDMS substrates, respectively. (e) Raman spectra of the individual regions (black: graphene and blue: BN) and their interface (red). (f) Optical image of alternating stripes of graphene and BN. (g) and (h) AFM height topography and corresponding current sensing image from the corresponding regions in (f). The height difference is negligible but the two regions can be differentiated in the current sensing image. (Adapted from Ref. [35].)

growth and integration of heterostructures of 2D materials. With continuous advancement in optimisation of surface topology, growth temperatures and concentration of the precursors, and other parameters, CVD is among the most promising routes to large-area devices of van der Waals heterostructures.

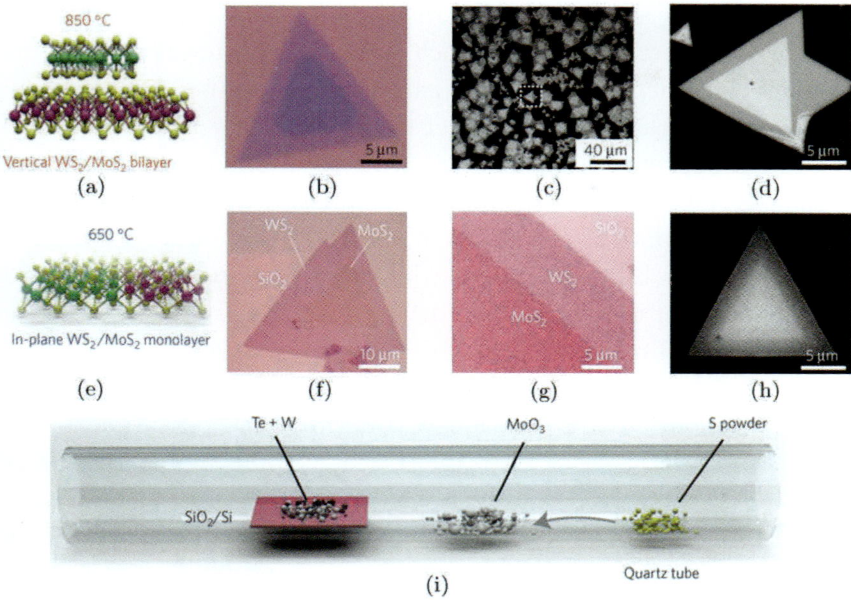

Figure 6. CVD-grown WS$_2$/MoS$_2$ vertical and in-plane heterostructures: (a)–(d) Schematic, optical and SEM images of vertical WS$_2$/MoS$_2$ heterostructures grown at ∼850°C. (e)–(h) Schematic, optical and SEM images of in-plane WS$_2$/MoS$_2$ heterojunctions synthesised at ∼650°C. The green, purple and yellow spheres in (a) and (e) represent W, Mo and S atoms, respectively. (g) is an optical image with enhanced optical contrast demonstrating the abrupt interface between WS$_2$ and MoS$_2$. (i) Schematic of the growth process of both heterostructures. (Adapted from Ref. [36].)

5.3   Interface Characterisation: Microscopy and Electrical Properties

5.3.1   Raman spectroscopy

Raman spectroscopy is one of the most important tools to characterise individual 2D materials, which not only reveals the crystallinity and Raman-active phonon modes, but also used to estimate the number of layers accurately. Raman spectroscopy of 2D heterostructures has been an active topic of research, where both hybridisation and layer alignment can play a significant role.

To illustrate the role Raman spectroscopy in the context of van der Waals heterostructure we will consider a hybrid of two graphene layers attached at different orientations [37]. Mechanically exfoliated bilayer graphene, which is generally obtained in Bernal (AB) stacking, exhibits characteristic band structural properties and Raman signatures that are distinct from single layer graphene (SLG). In twisted bilayer graphene (tBLG), the interaction and a rotational mismatch between the two individual constituent layers, makes the electronic and optical properties of tBLG differ from both SLG and Bernal stacked bilayer graphene [37–39]. This includes rotational angle dependent development of a Van Hove singularity and electron localisation that causes the Fermi velocity to go to zero as the relative rotation angle between the layers goes to zero.

Raman spectroscopy provides a powerful characterisation tool for understanding the structure-property correlation in this kind of 2D hybrid structures [40]. Figure 7 shows rotation-dependent Raman spectra of tBLG hybrids, where the position, spectral width and intensity of the 2D and G peaks show non-monotonic variation with the mismatch angle. Theoretical analysis reveals that the observed behaviour can be explained by changes in electronic band structure due to the inter-layer interaction, such as rotational-angle dependent Van Hove singularities. By combining the information of the 2D peak width, 2D peak location, and the 2D-to-G intensity ratio, one can deduce the general rotational angles in tBLG hybrids.

5.3.2   Transmission electron microscopy

Transmission electron microscopy (TEM) has been another powerful tool that serves in multiple characterisation schemes for van der Waals heterostructure. TEM has been successfully used to evaluate the misorientation angle in tBLG [37]. Moreover, the cross-sectional TEM images of vertical hybrids reveal the quality of the atomic interface between heterogeneous materials [41]. The interfaces between assembled layered 2D materials are usually pristine, as shown in Figure 8(a) which consists of alternate stacking of three bilayers of graphene and three bilayers of BN. The pristine assembly is believed to be a result of the lipophilic properties of 2D atomic layers. During assembly the hydrocarbons adsorbed on these surfaces

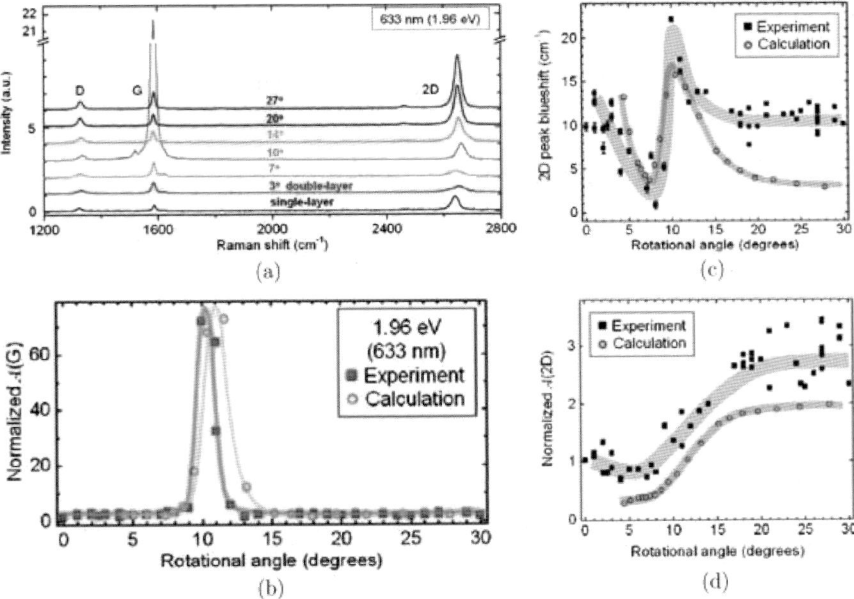

Figure 7. Raman spectroscopy of tBLG: (a) Raman spectra for tBLG for different rotational mismatch angle. (b) Raman G peak intensity integral as a function of rotational angle. Resonance in G peak occurs when the laser energy is close to the momentum mismatch between the two layers. (c) Non-monotonic behaviour of 2D peak blue-shift as a function of rotational angle. (d) Dependence of normalised 2D peak intensity on rotational angle. (Adapted from Ref. [37].)

are repelled and accumulate into small pockets [42], Figures 8(b) and 8(c).

5.3.3   The Graphene–TMDC interface

The structural and electrical properties of the interface between semiconducting 2D materials and graphene have been of particular interest. Scanning tunnelling microscopy has been one of the important tools to characterise the interface, but can be corrupted by the presence of water vapour or surface adsorbates. Nevertheless, an annealed hybrid of chemical vapour deposited graphene on MoS_2 shows weak Moire pattern lattice mismatch and random alignment of graphene with respect to the MoS_2 substrate [44].

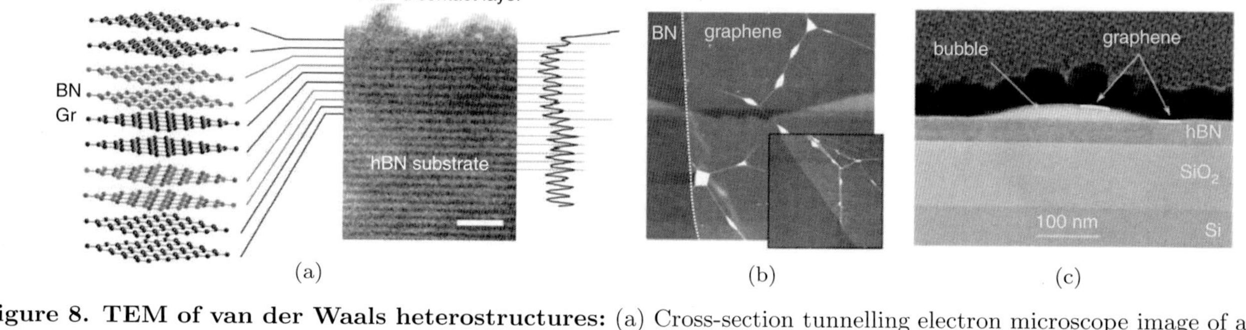

Figure 8. TEM of van der Waals heterostructures: (a) Cross-section tunnelling electron microscope image of a stack consisting of three bilayers of graphene and three of BN. (Adapted from Ref. [43].) (b) AFM topographical image of graphene on BN, shows polymer bubble formation due to the self-cleansing mechanism. (c) TEM cross-sectional image shows a polymer bubble trapped between graphene and BN. (Adapted from Ref. [42].)

Another important aspect of the graphene and TMDC (MoS_2) interface is the formation of the Schottky barrier, which can impact the applications where charge transport across the interface is important, for example, in photodetectors [22]. In graphene barristor with graphene–MoS_2 (GM) Schottky junction [2], the barrier height can be changed by 0.4 eV by tuning the work function of graphene. With increasing back gate potential (V_{BG}), work function of graphene decreases resulting in the decrease of ϕ_B with increasing reverse current from graphene to MoS_2 across the Schottky junction. The mechanism behind this tunability of ϕ_B within a vast range of 50 meV–440 meV (Figure 9(b)) is demonstrated schematically with band diagrams of graphene and MoS_2 in Figure 9(a). On the other hand, for graphene–MoS_2–graphite (GMG) device, ϕ_B can be tuned from 200 meV to −20 meV by modulating the work function of graphene (Figure 9(b)). A metal-semiconductor junction having negative Schottky barrier [45, 46] means that the work function of the metal is less than the electron affinity of the semiconductor and the junction behaves as an ohmic contact with very low contact resistance. The negative barrier height observed in GMG device indicates that low resistive ohmic contact can be formed by proper selection and geometry of 2D material heterostructure which will increase the device efficiency for application purpose.

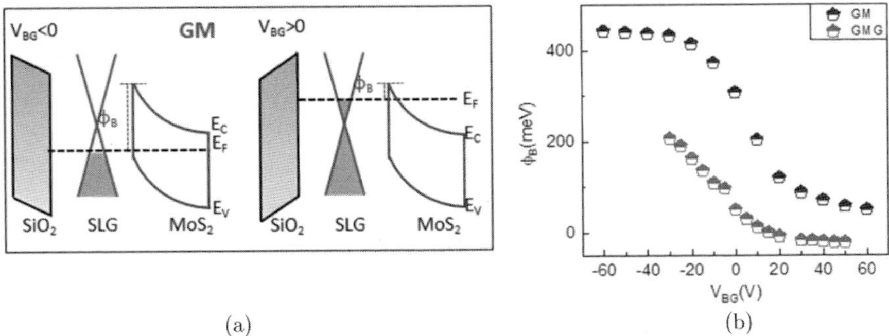

(a)   (b)

Figure 9. Schottky barrier at the interface of graphene and MoS_2: (a) Band diagrams of graphene and MoS_2 to explain the mechanism of changing the Schottky barrier height with varying back gate potential for GM device at reverse bias voltage ($V_{SD} < 0$). (b) The Schottky barrier height (ϕ_B) with varying back gate potential, V_{BG} for GM and GMG devices. (Adapted from Ref. [2].)

5.4   Electrical Devices and Phenomena from van der Waals Heterostructures

5.4.1   Ultra-high mobility and ballistic transport in graphene field-effect transistors (FET)

One of the most common van der Waals heterostructure consists of vertically stacked layers of graphene and BN. Atomically thin graphene is highly susceptible to external sources of disorder from the underlying substrate, the adsorbates or the polymer residues from device processing. A remedy to this is to encapsulate graphene on both sides with BN, which is a large band gap insulator, and hence does not participate in charge transport. BN has a small lattice mismatch (1.7%) with graphene, has far fewer traps or defects than the Si/SiO_2 substrate and provides an atomically flat surface [7]. Therefore, such an assembly can significantly lower the susceptibility of the conduction channel to the external fluctuations by physically separating them from the scattering species and modifying the local dielectric environment.

The contact to graphene is made by etching away all layers and exposing a one-dimensional (1D) edge of graphene on which the contact metal can be deposited. This technique results in graphene devices that display extremely high mobility values up to $140,000\,cm^2/Vs$ at room temperature [47], mainly limited by the scattering from acoustic-phonon, shown in Figures 10(b) and 10(c). Such high-quality devices have enabled the investigation of various novel phenomena [48–50], and even led to the emergence of new physics like the viscous flow of carriers in graphene [51].

5.4.2   Electronic properties of bilayer Graphene–BN heterostructures

Bilayer graphene (BLG), made of two layers of graphene sheets in Bernal stacking, is fundamentally different from an individual graphene layer. The most striking property is that a tunable band gap can be induced in BLG by applying an electric field perpendicular to its plane [53]. The induced band gap is nearly linear to applied electric field and can be remarkably large, up to $250\,meV$ on

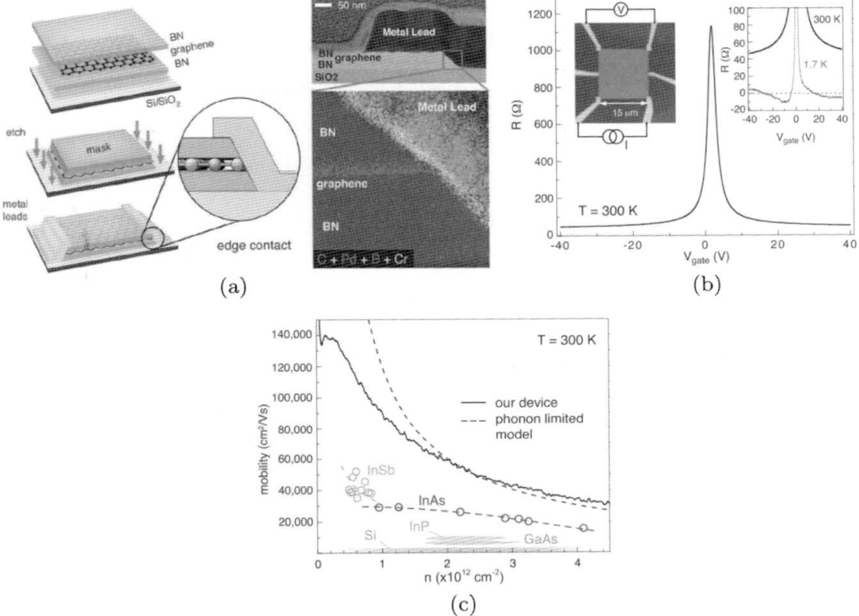

Figure 10. Ultra-high mobility encapsulated graphene FET: (a) Edge contact with graphene encapsulated with BN on both sides. Schematic of the procedure (left). High-resolution bright-field cross section scanning transmission electron microscope image and the false-colour electron energy-loss spectroscopy map (right) reveal the details of the contact region. (b) Resistance vs. gate voltage characteristics of an edge-contacted graphene device. The negative resistance at low temperature is a result of ballistic transport (right inset). (c) The solid curve shows the room temperature mobility as a function of number density, which is close to the phonon limited mobility. A comparison with other high-quality transistors is also shown. (Adapted from Ref. [47].)

application of 3 V/nm of electric field [54]. As a result, BLG can be electrically driven from a metallic state, with high carrier mobility, to a strongly insulating state, thus being suitable for a diverse range of novel device applications [55–57].

In order to apply large electric fields of the order 1 V/nm, a dual-gated FET geometry is required. Vital to such a FET device is the choice of gate dielectric to induce high gate electric fields by providing large capacitance and minimal leakage current, for which, oxide based dielectrics have been very successful [54, 58, 59]. Although it enables excellent electrostatics, the oxides however also degrade the surface quality of the BLG channel. Specifically, the oxide based dielectrics deteriorate the characteristics of graphene by presenting a chemically active buffer layer [60–62], charge traps [63–65] and surface roughness [66–68]. A BN-on-BLG binary heterostructure can be used for this purpose [29, 47], because dielectric quality of BN is comparable to that of SiO_2 [69, 70], and allows very high electric fields with low gate voltages.

In Figure 11, the construction and characteristics of a BN-gated BLG FET is described in detail. Figures 11(a) and 11(b) show the device image and its cross-sectional schematic, where 10 nanometre BN layer is the top gate dielectric for the BLG FET on a SiO_2/Si^{++} substrate [52]. The electrical leads to the BLG channel is created via edge contacting method described in the previous section. Each gate contributes independently to number density and transverse electric displacement field to give a combined number density n and electric displacement field D as:

$$D\,(\mathrm{V/nm}) = [C_{\mathrm{BG}}(V_{\mathrm{BG}} - V_{B0}) - C_{\mathrm{TG}}(V_{\mathrm{TG}} - V_{T0})]/2\epsilon_0,$$

$$n\,(\mathrm{m^{-2}}) = [C_{\mathrm{BG}}(V_{\mathrm{BG}} - V_{B0}) + C_{\mathrm{TG}}(V_{\mathrm{TG}} - V_{T0})]/e,$$

where C_{XG}, V_{XG}, V_{X0} are capacitance of the dielectric from the gate, the gate voltage applied and the gate voltage required to offset extrinsic doping, respectively, and X is B or T for back gate and top gate, respectively.

Figure 11(c) shows the resistance of the device as a function of V_{TG} for several fixed values of V_{BG} at an electron temperature of $\sim 100\,\mathrm{mK}$. When V_{TG} is tuned so that the number density is zero, the resistance takes a maximum value, and this state is called a charge neutrality point (CNP). The position of CNP depends on the doping induced by the environment and back gate and therefore shifts with V_{BG}. Furthermore, the maximum value of resistance R_{CNP} changes with V_{BG} such that it increases when the transverse electric field increases, reflecting the widening of band gap. In the metallic state, the resistance is as low as 300 Ω, whereas at CNP with high electric field, it can become very large, nearly 30 MΩ at $D = 0.67\,\mathrm{V/nm}$.

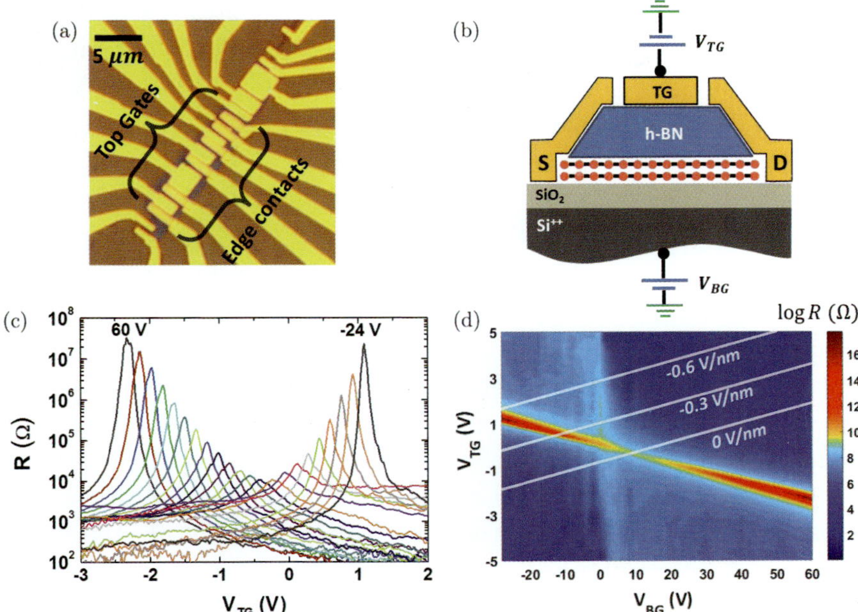

Figure 11. Characteristics of a dual-gated bilayer graphene–BN heterostructure: (a) Optical micrograph and (b) Cross-sectional schematic of BN and bilayer graphene hybrid FET device with edge contacts. The BN plays the role of a top gate dielectric. (c) Resistance as a function of top gate V_{TG} for fixed values of back gate V_{BG} from 60 V to -24 V in steps of 4 V at an electron temperature of ~ 100 mK. The resistance is a maximum R_{CNP} when number density is tuned to zero, at points called charge neutrality point (CNP) and R_{CNP} increases with increasing electric field. (d) 2D colour map of natural logarithm of resistance as a function of V_{BG} and V_{TG}. The white lines mark points with fixed electric displacement field (as labelled) and varying number density. (Adapted from Ref. [52].)

This can be compared with oxide based device in Ref. [59], where a similar magnitude of resistivity is reached at a much higher electric field of 1.9 V/nm.

Owing to the dual-gated geometry, number density can be independently controlled while the electric field is fixed at a large value. This is done by varying V_{BG} and V_{TG} according to the following equation obtained by fixing D in the above expression for D:

$$V_{\mathrm{TG}} = \frac{C_{\mathrm{BG}}}{C_{\mathrm{TG}}} V_{\mathrm{BG}} - \frac{C_{\mathrm{BG}}}{C_{\mathrm{TG}}} V_{B0} + V_{T0} + 2D\epsilon_0/C_{\mathrm{TG}}.$$

The 2D colour plot of resistance as a function of both V_{BG} and V_{TG} is shown in Figure 11(d). It also shows the (V_{BG}, V_{TG}) configuration for fixed electric field displacement $D = 0\,V/nm$, $D = -0.3\,V/nm$, and $D = -0.6\,V/nm$, superimposed as white lines.

5.4.3   Contact resistance

Van der Waals hybrids have been implemented in reducing the metal–channel contact resistance, which is essential for high-speed field effect transistors. The large contact resistance of MoS_2FETs due to high Schottky barrier at the metal–MoS_2 contacts compromises the device performance. Recently, Liu *et al.* [71] have shown that graphene can be used as a contact to MoS_2 devices which is significantly more electrically transparent, and allow low-resistance ohmic contacts. By tuning the gate voltage the Fermi energy of graphene can be modified to reduce the Schottky barrier height, leading to further reduction in the contact resistance. As illustrated in Figure 12, transparent electrical contacts, along with BN encapsulation, have been instrumental in the observation of metal–insulator transition, extremely high carrier mobility (up to $30{,}000\,cm^2/Vs$) and Shubnikov-de Haas oscillations in MoS_2 FETs [71, 74].

Figure 13 illustrates a design of MoS_2–graphene vertical FET by Shih *et al.* [75]. This architecture uses graphene channel with tunable the Schottky barrier height at graphene–MoS_2 junction, thereby achieving high carrier mobility ($\sim 600\,cm^2/Vs$) and current on–off ratio (~ 30) at room temperature.

5.4.4   Electrical noise in 2D heterostructures

Slow time-dependent fluctuations in the electrical resistance, or the low-frequency noise, of electronic devices is of fundamental interest and reveals the transport dynamics and the nature of scattering of the charge carriers [64, 76–87]. The low frequency $1/f$ noise, where the power spectral density of noise varies with frequency as $1/f$, has also been extensively used as a diagnostic tool to determine the quality of electronic devices [88, 89].

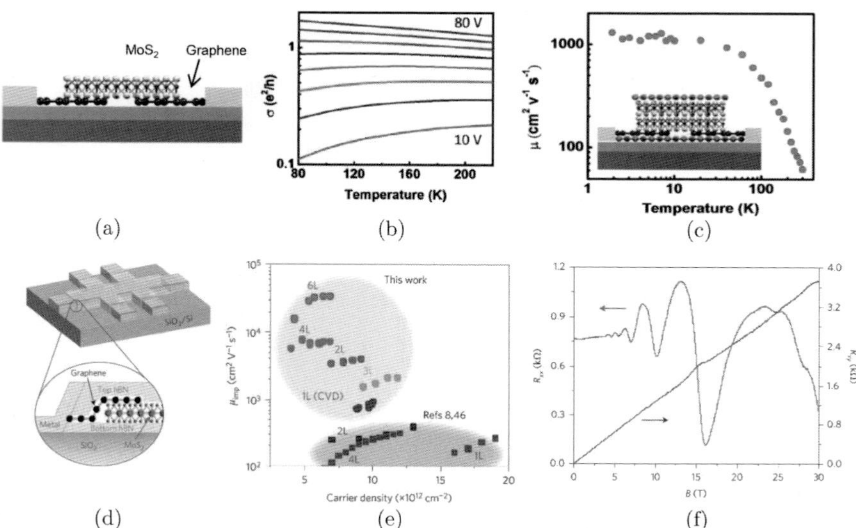

Figure 12. High mobility MoS$_2$ FET from van der Waals hybrids:
(a) Schematic of graphene contacted two terminal monolayer MoS$_2$ device on
Si/SiO$_2$ substrate. (b) Temperature dependence of sheet conductivity of the
device shown in (a) at different back gate voltages (V_{BG}). For $V_{BG} < 50$ V, conductivity increases with increasing temperature indicating insulating behaviour,
whereas for $V_{BG} > 50$ V, it shows metallic behaviour. (c) Temperature dependence
of mobility with quite high value ($\sim 1 \times 10^3$ cm^2/Vs) at low temperature in the
BN encapsulated graphene contacted MoS$_2$ device whose schematic is shown
here. (d) Schematic of BN encapsulated multi-terminal graphene contacted MoS$_2$
device. (e) Very high intrinsic impurity limited mobilities ($\sim 10^4$ cm^2/Vs) for
the device of figure (d) with different thicknesses of MoS$_2$ at low temperature
comparing with bare MoS$_2$ devices on SiO$_2$ substrate as reported in Refs. [72, 73].
(f) R_{xx} and R_{xy} of BN encapsulated 4-layer MoS$_2$ device with varying magnetic
field obtained from Hall measurement at 0.3 K displaying the Shubnikov–de
Haas oscillations. ((a)–(c) adapted from Ref. [71] and (d)–(f) adapted from
Ref. [74].)

5.4.4.1 *1/f noise in graphene and graphene-hybrids*

Charge transport in graphene devices is excessively susceptible to
any external fluctuations due to their proximity to local charged
scattering centres and poor screening. It is now mostly understood
that the 1/f noise in graphene transistors is also due to random
fluctuations in the local electrostatic environment of the channel or
the contacts, particularly in the supporting substrate [4, 82, 83, 85,

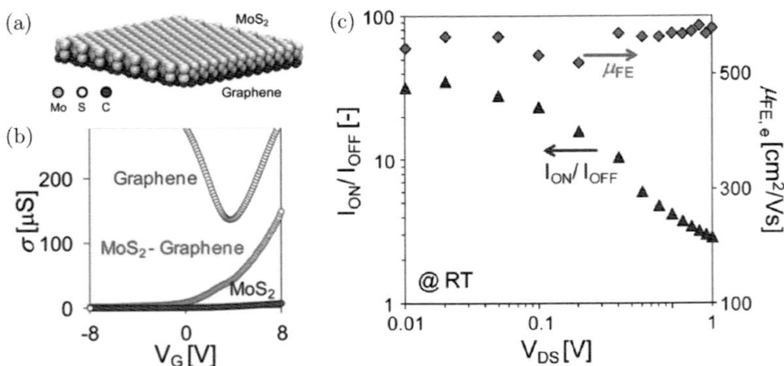

Figure 13. Electrical transport in vertical graphene–MoS2 heterostructure: (a) Schematic of MoS$_2$–graphene vertical heterostructure. (b) The comparison between the transfer characteristics of graphene, MoS$_2$ and MoS$_2$–graphene field effect transistors (FET). The MoS$_2$–graphene hybrid shows higher on–off ratio. (c) Current on–off ratio and electron field effect mobility with varying source-drain bias for MoS$_2$–graphene FET at room temperature. (Adapted from Ref. [75].)

90, 91]. Intermittent exchange of charge between the channel and the trap states at the substrate-channel interface leads to correlated carrier number and mobility fluctuations and slow fluctuations in the source-drain current. Detailed analysis of gate voltage-dependence of noise reveals that inter-trap migration of charge may also contribute to noise which is sensitive to the electronic properties, such as the dispersion relation, carrier mobility, etc.

Van der Waals heterostructures can potentially modify the susceptibility of the conduction channel to external fluctuations by physically separating them from the scattering species or by modifying the screening properties by changing the local dielectric environment. Heterostructures also allow designing novel contacting methods [47, 71, 92, 93]. The study of electrical noise in 2D hybrids thus offers new possibilities and has been reported widely.

This can be illustrated with the vertical hybrid structures is graphene BN. BN has fewer traps or defects than the Si/SiO$_2$ substrate and provides an atomically flat surface. This not only leads to improved carrier mobility [7, 94], but also results in a 10–100 times drop in the noise magnitude in graphene on hBN devices compared to graphene on Si/SiO_2 [91, 95, 96].

5.4.4.2 *Noise TMDC and MoS_2-BN hybrids*

The differences in band structure, carrier mobility, etc. between graphene and TMDC, leads to a very different noise characteristics in TMDC FETs. In recent studies [87, 97–101], with MoS_2 and WSe_2 FETs, noise magnitude N in TMDC FETs was found to decrease with increasing gate voltage (V_{BG}) as, $N \propto 1/(V_{BG} - V_T)^2$, where V_T is the threshold voltage of the onset of conduction (Figure 14(a)). Such a behaviour is a signature of McWhorter model of carrier number fluctuation [87, 98, 99, 101]. Since most of the early noise measurement in TMDC FETs were carried out with the channel on SiO_2 substrate, it was readily recognised that the random exchange of charge with the traps either on the surface or at SiO_2 interface drives the fluctuation in carrier number.

The unresolved issues in these studies were two-fold: (1) Role of surface traps as opposed to those at the SiO_2 interface, and (2) the role of the TMDC-metal contacts. It has been shown conclusively that the electrical transparency of the metal-TMDC contacts is crucial to high carrier mobility in TMDC FETs [87, 98, 102, 103], and that mobility increases dramatically when the contacts are annealed at elevated temperatures prior to measurement [87, 98, 104]. The possibility of contact origin of noise in TMDC FETs is strengthened by the lack of 'scaling' of the noise magnitude when it is normalised by the device/channel area (Figure 14(b)). To resolve these issues, a series of van der Waals heterostructures were studied, where a part of TMDC (MoS_2) channel was covered with BN, and hence remained uncontaminated during processing of the devices (inset of Figure 14(c)). Figure 14(c) shows the gate voltage dependence of noise (after annealing) from both protected and unprotected regions of the MoS_2 channel. The agreement in the noise magnitude establishes that the surface traps have very little effect on the observed noise magnitude. The contacts, however, were observed to have a great effect on the noise magnitude in TMDC FETs. This

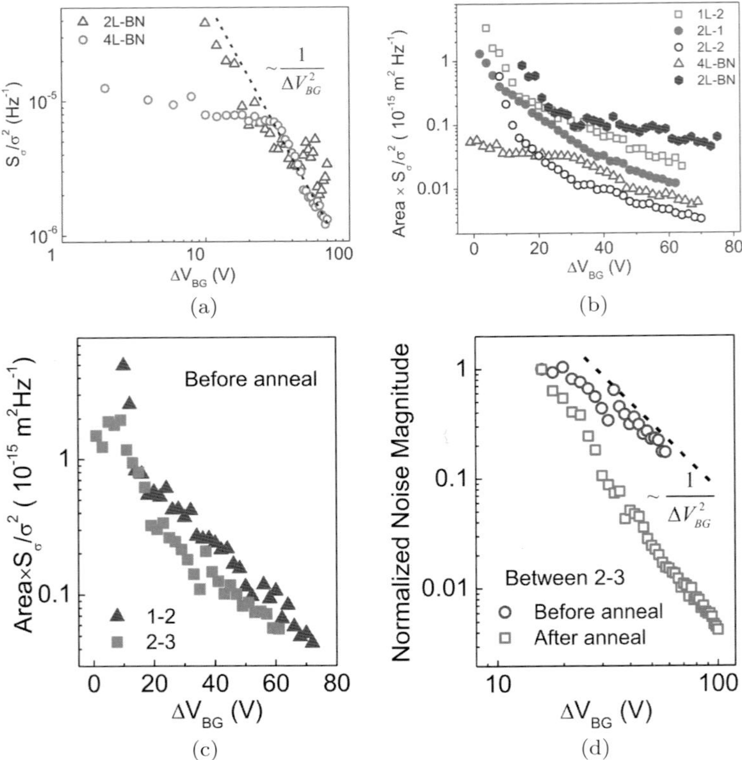

Figure 14. Noise in BN-MoS$_2$ hybrid FETs: (a) Gate voltage (V_{BG}) dependence normalised noise magnitude showing in MoS$_2$ FET on a BN substrates. The $(1/\Delta V_{BG})^2$ scaling of noise suggests McWhorter-type number fluctuation as the main source of noise. (b) Channel area scaled noise magnitude in several bare (unannealed) MoS2 and MoS2/BN FETs. The wide device-to-device variation suggests channel region is possibly not the primary source of noise. (c) Plot of area scaled noise magnitude vs. gate voltage for the unprotected (1–2) and protected (2–3) channel of a MoS$_2$ channel that is partially covered by BN. The inset shows a schematic of the device, where the channel region between contacts 2 and 3 is protected from surface contaminants by using a BN overlayer. The noise magnitude from both protected and unprotected regions of the channel are almost identical, indicating that metal–MoS$_2$ contacts are the dominant source of noise in the unannealed devices. (d) The decrease of normalised noise magnitude with annealing in BN protected channel (2–3). A similar effect is seen in the unprotected MoS$_2$ channel (1–2) as well. Annealing reduces both contact resistance and contact noise in MoS$_2$ FETs. (Adapted from Ref. [87].)

is illustrated in Figure 14(d), where the magnitude of noise in the MoS_2 channel before and after annealing is compared.

5.4.4.3 *Noise in bilayer graphene–hBN heterostructures*

The tunability of band gap in BLG impacts the electrical noise in these devices in a characteristic way. This has been explored with dual gated BLG heterostructures, where the band gap and Fermi energy of the channel can be tuned independently [26, 105]. A BN layer acts as the top gate dielectric allowing access to large transverse electric fields across the BLG (also see Figure 11).

Figure 15(a) shows a dual gated BLG device whose resistance fluctuations are studied. Figure 15(b) shows the number density variation of average resistance alongside normalised variance of resistance noise for a few transverse electric fields at 77 K. At low electric fields, the normalised noise amplitude has a weak modulation even when the Fermi energy is tuned to CNP, because there would still exist many current paths through well-connected puddles. The number of

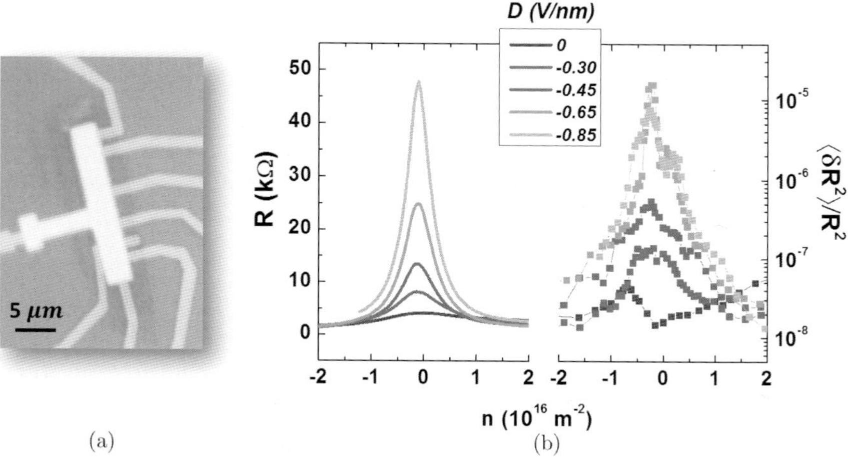

Figure 15. Low-frequency noise in dual-gated bilayer graphene–BN heterostructure: (a) Optical micrograph of dual-gated BLG on SiO_2 substrate and BN as top gate dielectric (b) Left panel — Resistance as a function of number density for fixed electric displacement field. Right panel — normalised variance of resistance as a function of number density for the corresponding electric fields. (Adapted from Ref. [105].)

well-connected puddles decrease rapidly at high electric fields, leaving very few conducting channels. The normalised resistance noise in this regime has a very strong variation of up to three orders of magnitude with number density because the conductance is strongly modified when the transmission through these channels fluctuates with time. When compared with the change in resistance, which is only a factor of 25 for the largest electric field shown in Figure 15(b), it is clear that resistance noise is significantly more sensitive to band gap than the average resistance [105].

5.4.4.4   *Contact noise*

The question that whether noise arises from the channel or the contacts, in atomically thin heterostructures, is both of fundamental and practical significance. This is of further importance in van der Waals heterostructure devices, where carrier mobility is often very high, and scattering processes are less frequent within the channel itself.

As we discussed in Section 5.4.3, the metal–graphene junction is complex and has been thoroughly researched [106–115], thus the relative contributions of the channel and the contacts to the device resistance are well understood. In spite of much work on the contact resistance, the magnitude or the microscopic origin of the contact noise in graphene FETs is not well understood. This has led to conflicting claims, where some attributed the $1/f$ noise in graphene transistors primarily to noise generated within the channel region [64, 116], whereas other investigations claim a strong contribution from the contact [117, 118].

Recently, a variety of graphene devices with different mobilities, substrates and configurations of metal contacts have been studied to demonstrate that exceptionally large contact noise arises due to the fluctuating charge trap potential at the oxide substrate underneath the metal contacts [91]. The contact region is more susceptible to the external potential fluctuations because metal contacts to atomically thin materials can strongly modify the chemical and the band structure of the material. This can lead to significant noise generated at the contacts.

The origin of noise at the contacts can be quantified by studying the scaling of contact noise, $\langle(\Delta R_C)^2\rangle$, with the contact resistance R_C, i.e., $\langle(\Delta R_C)^2\rangle \propto R_C^{\gamma}$. γ depends on the type of contacts and $\gamma \approx 1$ for interface type contacts or $\gamma \approx 3$ for constriction-type contacts to metals or 3D semiconductors [88]. Whereas $\gamma \approx 2$ for metal contacts to purely 2D materials that are limited by current crowding, i.e., $L_C \gg L_T$, where L_C is the length of the contact and L_T is the length within which the majority of charge injection occurs from the metal contact to the channel region. Such a dependence is shown for a graphene Hall bar device in Figure 16, which is followed over four decades of noise magnitude. While the exact dependence may vary with the details of the contacts, it is clear that noise has a power law dependence on the contact resistance and highlights the importance of minimising the contact resistance.

The most striking outcome of these experiments is the conclusion that the intrinsic channel noise in high-mobility graphene–BN heterostructures is probably immeasurably small. This is confirmed from the dependence of the normalised magnitude of specific contact noise $\langle(\Delta R_C)^2\rangle \cdot W^2$, shown in Figure 16(c) as a function of specific contact resistance $R_C \cdot W$. The specific contact noise depends only on the substrate type or on the contact metal used and has no

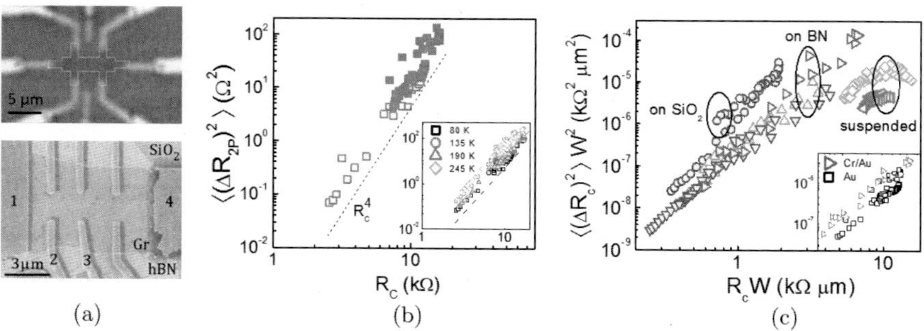

Figure 16. Contact noise in high mobility graphene — BN FETs: (a) Hall bar device with graphene on SiO_2 (top) and electron microscope image of graphene on hBN with invasive linear contacting geometry. (b) $\langle(\Delta R_{2p})^2\rangle$ as a function of contact resistance R_C, shows R_C^4 dependence. This dependence is valid for temperatures down to $\sim$80 K. (inset). (c) Comparison of contact noise in various devices, noise depends strongly on the substrate. (Adapted from Ref. [91].)

dependence on the carrier mobility itself. This indicates that the role of the channel in noise generation is minimal and the contact noise is affected by the degree of separation of the contacts from the fluctuating charge trap potential or by the metal species used.

5.4.5   Emerging device architectures

5.4.5.1   *Memory devices*

Vertical van der Waals heterostructures can also be configured as logic or memory devices with appropriate stacking sequences. In a recently published work [19], Yu *et al.* have demonstrated the working principle of such a vertical field effect transistor (VFET) which consists of a vertical stack of MoS_2 and graphene. A schematic with a cross-sectional view of the device geometry is presented in Figure 17(a). In Figure 17(b), a cross-sectional TEM image of the device shows different layers of graphene and MoS_2, confirming homogeneity and cleanliness of the interface. In this geometry, MoS_2 serves as an intrinsically n-doped channel. The VFET shows a very high ON current density $\sim 5000 \, A \, cm^{-2}$. The on–off ratio is found to be $\sim 10^3$.

The concept of vertical transport can be generalised further to form perpendicular logic devices by vertically stacking materials with different kinds of doping [120]. It is possible to create a logic inverter by sequentially stacking graphene, $Bi_2Sr_2Co_2O_8$ (BSCO), graphene and MoS_2. A typical schematic and a cross-sectional view of such a device is presented in Figures 17(e) and 17(f). In this architecture, BSCO and MoS_2 serves as p-type and n-type channels, respectively. The inverter characteristics of the device is shown in Figure 17(g), with a constant supply voltage $V_{DD} = 2 \, V$ applied on bottom graphene layer and with top gate grounded. The Si back gate and the middle graphene layer are used as Input and output terminals, respectively. The maximum gain of the logic inverter is found to be ~ 1.7.

In a more recent work, Choi *et al.* [119] have used BN as a tunnel barrier, in a perpendicular heterostructure of MoS_2, BN and graphene, where MoS_2 and graphene can be used as both channel and charge trapping layer. As explained in Figure 18, the integrated

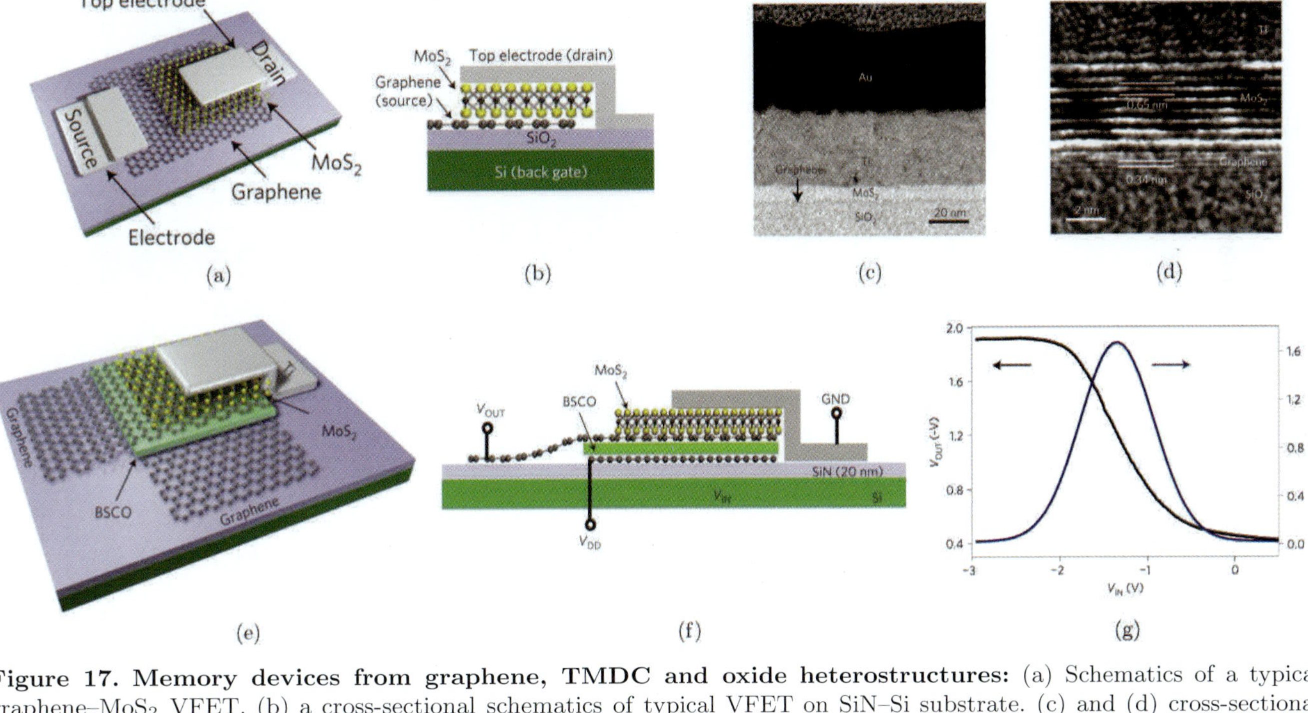

Figure 17. Memory devices from graphene, TMDC and oxide heterostructures: (a) Schematics of a typical graphene–MoS$_2$ VFET. (b) a cross-sectional schematics of typical VFET on SiN–Si substrate. (c) and (d) cross-sectional TEM image showing individual layers of graphene and MoS$_2$. (e) a schematics of a typical graphene–BSC–graphene–MoS$_2$ complimentary inverter. (f) cross-sectional view of the same of graphene–BSCO–graphene–MoS$_2$ vertical heterostructure for complementary inverter. (g) V_{out} vs. V_{in} of graphene–BSCO–graphene–MoS$_2$ logic inverter showed in the black trace, blue trace shows the gain of the logic inverter. (Adapted from Ref. [19].)

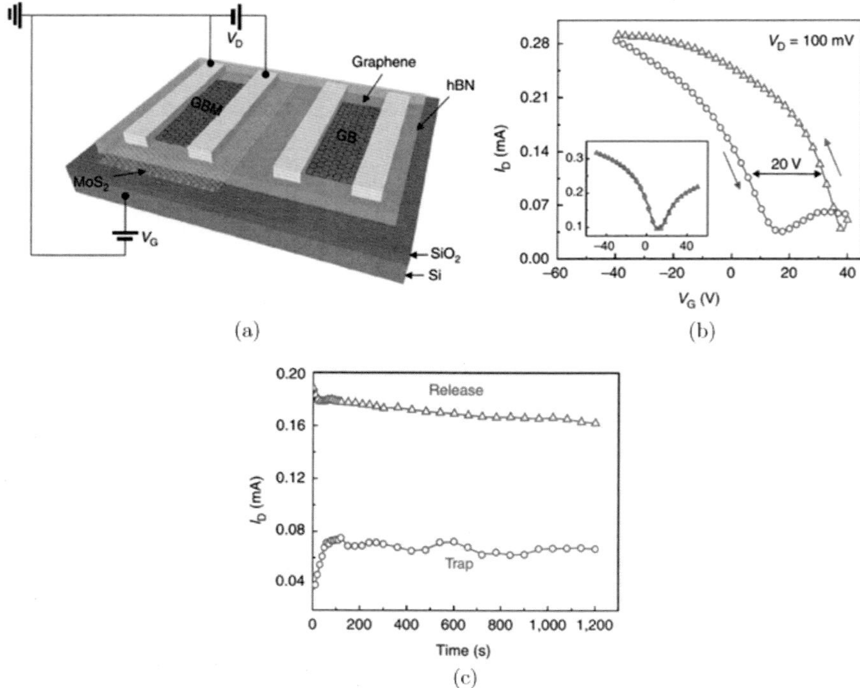

Figure 18. Charge trap-based non-volatile memory from graphene, BN and MoS$_2$: (a) Typical device schematics and circuit diagram for measurements of graphene–BN–MoS$_2$ and graphene–BN vertical heterostructure. (b) Hysteretic Ids–Vg transfer characteristics of a GBM device, the inset shows non-hysteretic transfer characteristics from GB device without charge trapping layer. Evidently, existence MoS$_2$ bottom layer is responsible for the hysteretic behaviour. (c) Time series of Ids at Vg = 15 V during charge trapping and de-trapping. At this gate voltage a 25 V gate voltage pulse of the 15 ms duration is applied to take the device to release mode (on state) and a −55 V gate voltage pulse of same duration is applied to take it back to trapping mode. (Adapted from Ref. [119].)

heterostructure works as a FET and shows a large hysteresis in its transfer characteristics and therefore can be used as a memory.

In a similar device architecture, as schematically described in Figure 19, a large hysteresis in electrical transport through graphene (topmost layer in the graphene/BN/MoS$_2$ trilayer on Si/SiO$_2$ substrate), can be obtained by controlled CT between the graphene and the MoS$_2$ layers. A switch can thus be designed by taking Dirac point resistance of graphene (during forward sweep) as ON state,

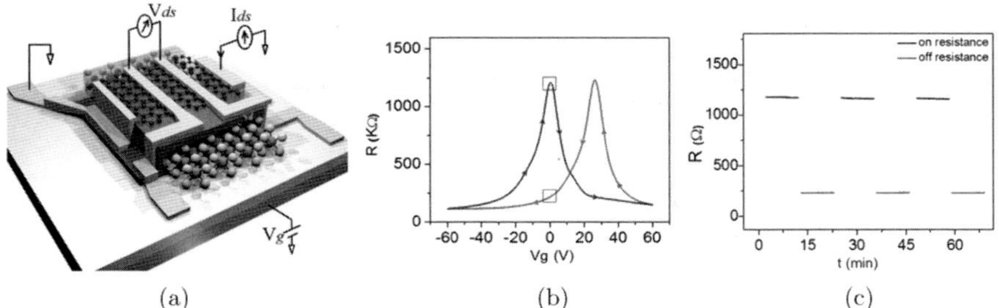

Figure 19. ON–OFF switch from van der Waals heterostructure of graphene, BN and MoS$_2$: (a) A schematic representation of a graphene–BN–MoS$_2$ vertical heterostructure with added circuit diagram. (b) Hysteretic R–Vg characteristics of the device measured in 4 probe technique. (c) time series of on and off state taken at $Vg = 0$. $+60$ V gate pulse was applied to go from on to off state and a -60 V gate pulse was applied to reach on state from off state.

while the resistance of graphene at the same point during backward gate voltage sweep defines the OFF state. Figure 19(c) shows long time stability of ON–OFF cycles at room temperature. The room temperature on–off ratio of the memory is found to be ∼6.

5.4.5.2 *Tunnelling transistors*

New designs of tunnelling transistors are among the most important implementation of van der Waals heterostructure. Recent construction of graphene–BN based vertical tunnelling field-effect transistor [10], vertically stacked graphene–MoS$_2$ heterostructure as logic transistors [119], and graphene barristor with graphene–silicon interface [121] have opened up new possibilities for logic gate with reduced size in flexible electronics.

In graphene–BN–graphene vertical field effect tunnelling transistor [10], the current on–off ratio (∼50) and on-current density (Figures 20(a) and 20(b)) are however still rather low. In graphene barrister (Figures 20(c) and 20(d)) with graphene–silicon interface [121], high current ON–OFF ratio (∼10^5) has been obtained at room temperature by modifying the work function of graphene with varying gate potential which tunes the Schottky barrier height from ∼0.25 eV to ∼0.45 eV at graphene–silicon junction (Figure 20(e)). The tunability of the barrier potential leads to the

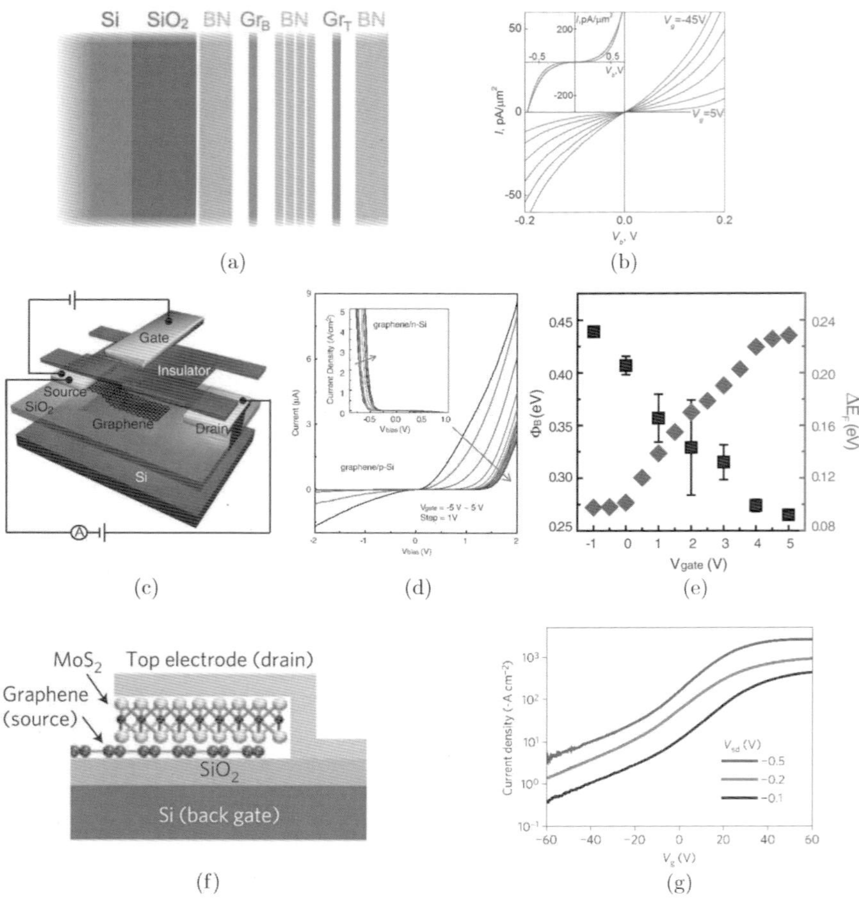

Figure 20. van der Waals heterostructures for tunnelling transistors:
(a) Schematic of graphene–BN–graphene vertical field effect tunnelling transistor.
(b) Transport characteristics at different back gate potential of device in figure
(a) displaying quite low on-current density. (c) Schematic structure of graphene
barristor with graphene–Si Schottky junction. (d) Transport characteristic of p-
type and n-type (inset) graphene baristor at different gate voltages with the
red arrow indicating the direction of increasing gate voltages. The plot shows
higher on-current density than graphene–BN–graphene transistor. (e) Tuning
of Schottky barrier height (ϕ_B) with varying gate potential. (f) Schematic of
vertically stacked graphene–MoS$_2$ heterostructure on Si/SiO$_2$ substrate. (g) The
transfer characteristic of the device shown in figure (f) at different source-drain
bias voltages showing a very high on–off ratio ($\sim$1500) and huge on-current
density ($\sim$2600 A/cm^2 at $V_{\rm sd} = -0.5$ V and $V_g = 60$ V) at room temperature.
((a) and (b) adapted from Ref. [10]; (c)–(e) adapted from Ref. [121] and (f) and
(g) adapted from Ref. [19].)

barristor action. However, although the mobility of this device is very large $\sim 10^3 \, \text{cm}^2/\text{Vs}$, the ON-current density is still rather low (Figure 20(d)).

New device structures [2] have been proposed by replacing graphene–silicon interface with graphene–MoS_2 (GM) in the barristor device, or BN with MoS_2 (GMG) in vertical tunnel transistors (Figures 21(a) and 21(b)). The on-current density increases giving rise to a very high on–off current ratio. Therefore, high mobility, high on–off ratio and low power flexible logic gate is possible to achieve with graphene–MoS_2 heterostructure, since the charge transport across the device can be modulated by tuning the Schottky barrier height with varying gate potential. The current density achieved in GM device at room temperature is $\sim 3500 \, \text{A/cm}^2$ at $V_{\text{SD}} = 1 \, \text{V}$

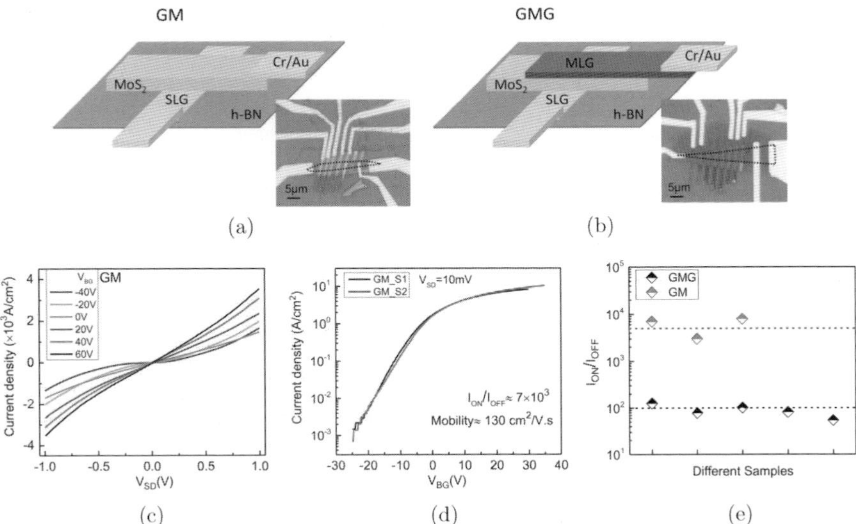

Figure 21. Vertical transistor with graphene and MoS_2 hybrids: Schematic of two types of graphene barristors, (a) with SLG–MoS_2 heterostructure (GM) and (b) with SLG–MoS_2–MLG vertical heterostructure (GMG) together with their respective optical microscope images. (c) Room temperature transport characteristics of GM at different back gate voltages. (d) The transfer characteristics of two different GM devices showing high mobility as well as very high current on–off ratio at room temperature. (e) Current on–off ratio ($I_{\text{ON}}/I_{\text{OFF}}$) for GMG and GM which shows almost two orders of magnitude increase of $I_{\text{ON}}/I_{\text{OFF}}$ in case of GM device. (Adapted from Ref. [2].)

and $V_{BG} = 60\,V$ with very high on–off current ratio $\approx 5 \times 10^3$ (Figures 21(c) and 21(d)) which is comparable with that of vertically stacked graphene–MoS_2 heterostructure reported by Choi *et al.* [119]. Due to high contact resistance of MoS_2–metal electrode and the absence of tunnelling current through graphene–MoS_2 contact, the off-current of GM device is quite low. As a result, the current on–off ratios (I_{ON}/I_{OFF}) of GM devices are almost two orders of magnitude higher than that of GMG devices (Figure 21(e)). Very high mobilities have been achieved for both GMG and GM devices at room temperature which are comparable for both types of devices and range between 100 and $800\,cm^2/Vs$.

Recently, a band-to-band tunnelling field effect device from double-layer MoS_2 hybrid have also been demonstrated [122]. Using highly doped germanium as the source and atomically thin MoS_2 as the channel, such vertical heterostructure exhibiting subthreshold swing as low as $3.9\,mV/decade$ at the switching of the transistor has been achieved.

5.4.5.3 *Graphene–STO heterostructures*

Oxides are emerging as new partners for 2D materials in van der Waals heterostructure for electronic devices. The performance of graphene-based electronic devices is strongly influenced by the quality of graphene as well as the supporting substrate which determines the carrier mobility, intrinsic doping, etc. In graphene transistors, to achieve very high doping densities, substrate with high dielectric constant is required. The high dielectric constant of the substrate is also beneficial in screening the Coulomb impurities leading to increasing the carrier mobility. In this context, strontium titanate ($SrTiO_3$), is an ideal substrate for graphene transistor. $SrTiO_3$ (STO) is a band insulator with very high dielectric constant. At room temperature, the dielectric constant of STO is 300 which increases to $\sim$30,000 at cryogenic temperature.

Couto *et al.* [123] have shown that in graphene on STO, the conductivity does not depend on temperature (Figure 22(a)), which indicates that the long range Coulomb potential is effectively screened by the high dielectric constant of STO. The carrier mobility

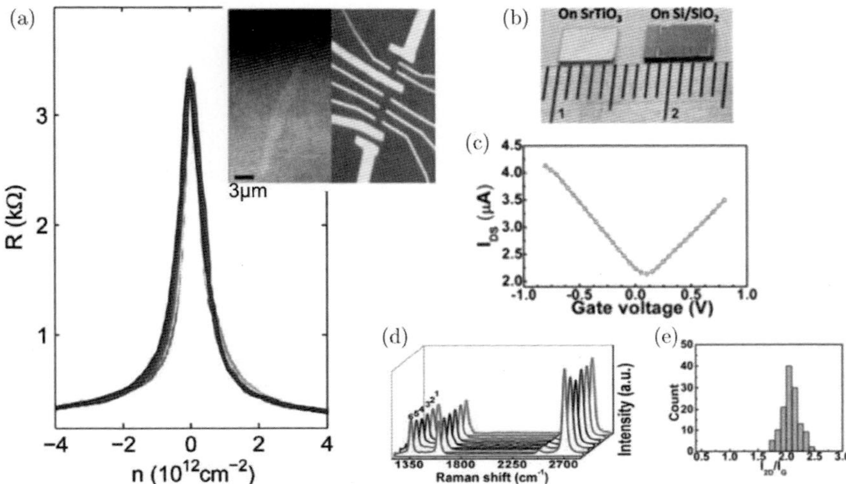

Figure 22. Graphene–SrTiO3 hybrid transistors: (a) Square resistance of graphene on STO with varying carrier density which is estimated by Hall measurement at different temperatures from $250\,\mathrm{mK}$ to $50\,\mathrm{K}$ showing the temperature independence of graphene conductivity on STO. The inset displays the optical microscope image of graphene on STO and the final device structure. (b) CVD grown graphene film on STO and transferred CVD graphene on SiO_2. (c) The transfer characteristic of graphene on STO in air. (d) Raman spectroscopy at different places on graphene. (e) Histogram of I_{2D}/I_G ratio showing an average value of 2 indicating good quality single layer graphene. ((a) adapted from Ref. [123] and (b)–(e) adapted from Ref. [124].)

in graphene on STO substrate is however limited by the resonant scattering mechanism. The short range potential responsible for resonant scattering cannot be screened by the high dielectric constant of STO, which results in similar conductivity and mobility in graphene on STO as on SiO_2.

High quality single layer graphene has been grown on STO by CVD process [124] with carrier mobility $\sim 1000\,\mathrm{cm^2/Vs}$ at room temperature in air with operating voltage lower than graphene FET on SiO_2 substrate (Figures 22(b) and 22(c)). Raman spectroscopy performed on CVD grown graphene on STO shows a uniform high quality graphene (Figures 22(d) and 22(e)). The direct growth of graphene on transparent STO (Figure 22(b)) substrate indicates the possible application of transfer-free transparent electronics. Complex oxide heterostructure such as $LaAlO_3$–$SrTiO_3$ has also been used as a substrate for graphene for the purpose of controlling the heterostructure properties [125].

5.5   Light-matter Interaction in van der Waals Hybrids

Since many of the 2D materials, especially the TMDCs, are semi-conductors, the light-electric interconversion with van der Waals heterostructures is a rapidly growing field of research. Several fundamental questions, for example, the extent of energy band renormalisation due to screening at the heterojunction between two dissimilar atomically thin semiconductor membranes, are being addressed with multiple combinations of 2D materials [1, 18, 24, 126, 127]. Moreover, the nature of excitons across the interface, also known as the CT excitons, as well as mechanism that separates the photo-generated carriers remain to be fully understood.

From the perspective of device application, the van der Waals hybrids are emerging as a new class of high-sensitivity photode-tectors, where a small intensity of electromagnetic radiation, from near infrared to the ultraviolet regime, is converted to measurable electric current [22, 24, 128–131] Most 2D materials interact strongly with electromagnetic radiation, with absorption coefficient as large as $10^7\,\mathrm{m^{-1}}$ in graphene [132], but quality of photodetectors from individual atomic/molecular layers are often limited due to sub-optimal combination of thickness, carrier mobility and band gap. Layered hybrids have shown significant promise to overcome some of these difficulties.

5.5.1   Band renormalisation in van der Waals heterostructures

Electronic band gap and excitonic binding energy in TMDC such as in $MoSe_2$ is strongly dependent on the screening by the substrate. Ugeda *et al.* [1] have shown theoretically that the electronic band gap, as well as exciton binding energy in $MoSe_2$, are reduced when a layer BLG is used as substrate. Figure 23 outlines the theoretical and experimental results on the substrate-mediated modification in

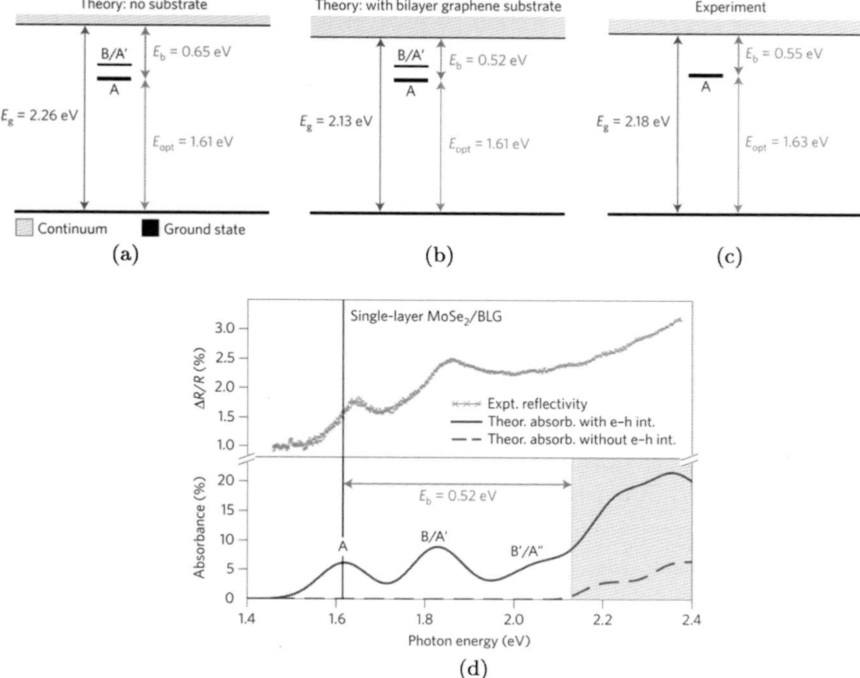

Figure 23. Band renormalisation in graphene–TMDC heterostructures: Energy levels of MoSe$_2$ theoretically calculated by using GW plus the-Salpeter (GW-BSE) methods. (a) Without any substrate. (b) on BLG. (c) experimentally observed. (d) reflectivity measurement on MoSe$_2$ on BLG and calculated optical absorption with and without electron-hole interaction including substrate screening from BLG. (Adapted from Ref. [1].)

optical response in MoSe$_2$. Calculations based on GW-BSE method reveal that the binding energy of excitons in MoSe$_2$ is reduced from 0.65 eV to 0.52 eV in BLG substrate. This reduction has been confirmed experimentally by reflectivity measurement (Figure 23(d)), as well as by photoluminescence (PL) measurement [1].

5.5.2   Charge transfer excitons

When subjected to optical illumination hybrids of dissimilar semiconducting layers from, for example MoS$_2$ and WS$_2$ [18], give rise to a new class of excitonic quasiparticles. A salient feature of such van der Waals interfaces is the poorly screened Coulomb potential that can give rise to bound electron–hole pairs across the interface, i.e., CT or inter-layer excitons. The origin of CT excitons lies in the existence of type II heterojunction of the TMDC hybrid in which the conduction band minimum and valance band maximum lie in two different materials. Hong *et al.* [18] have probed the CT processes in photoexcited MoS_2–WS_2 heterostucture using PL and femtosecond pump-probe spectroscopy. The CT is inferred from the quenching of PL at the A-exciton resonance for MoS_2–WS_2 heterojunction, compared to that of the individual layers (see Figure 24).

The CT excitons at the interface in van der Waals hybrids are similar to that at the molecular donor/acceptor interface in organic photovoltaics, although electronic delocalisation in the latterin real space can dictate charge carrier separation. At the 2D semiconductor heterojunction, however, delocalisation in momentum space due to strong exciton binding has been suggested to assist in parallel momentum conservation in CT exciton formation [133].

Recently, similar quenching of PL has also been observed in graphene–MoS_2 [24] and graphene–WS_2 [127] heterostructure at room temperature. Figure 25 shows that the extent of quenching increases with increasing thickness of the TMDC layer. Li *et al.* [134]

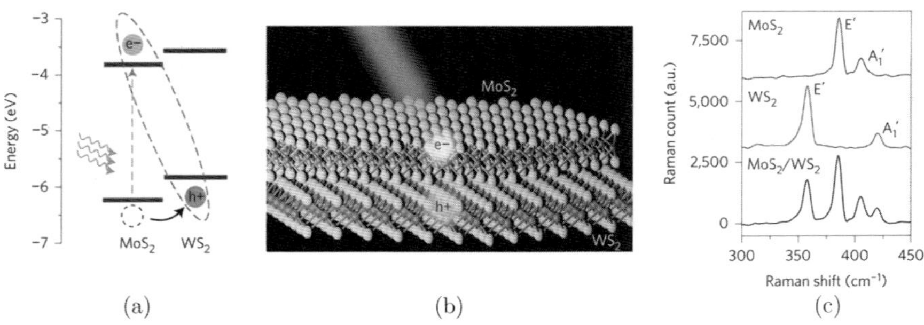

(a)   (b)   (c)

Figure 24. Charge transfer excitons in TMDC-based binary van der Waals hybrid: (a) Type II band alignment of MoS_2–WS_2 heterojunction. Upon photoexcitation in MoS_2, 'A' exciton leads to the formation of electrons and holes in different layers. (b) A schematic of electrons and holes in two separate layers as a result of photoexcitation. (c) PL spectra of individual MoS_2, WS_2 and MoS_2–WS_2 heterostructure. The strong PL at A-excitonic resonance in individual monolayers of MoS_2 and WS_2 is significantly quenched in MoS_2–WS_2 heterostructure due to CT between MoS_2 and WS_2 layers. (Adapted from Ref. [18].)

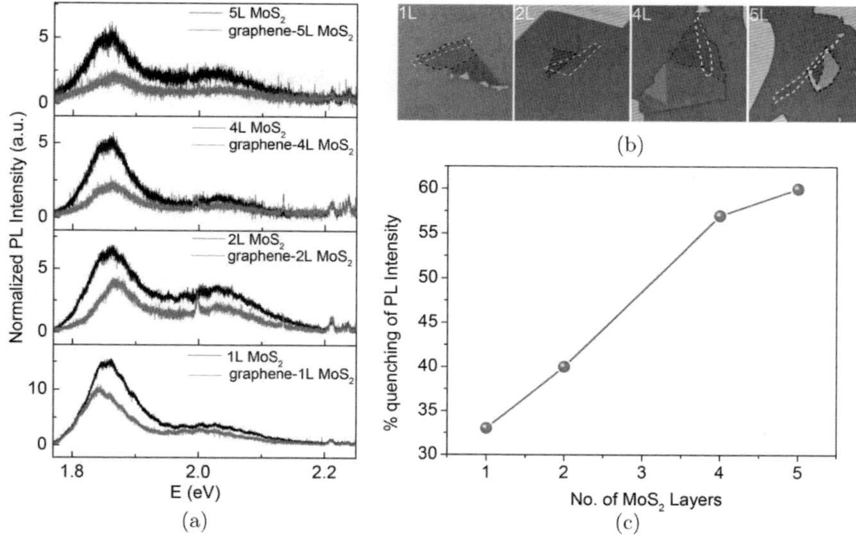

Figure 25. **Thickness-dependent quenching of PL in graphene–MoS$_2$ binary hybrids:** (a) Room temperature PL spectra of monolayer graphene–(variable thickness) MoS$_2$. (b) Optical microscope image of monolayer graphene–MoS$_2$ binary hybrid. Graphene and MoS$_2$ are indicated by white and black coloured outline, respectively. (c) Percentage quenching of PL intensity as a function of the thickness of MoS$_2$. (Adapted from Ref. [136].)

and Shih *et al.* [135] argue that the PL intensity quenches as a consequence of exciton dissociation due to built-in potential originating from Schottky barrier at graphene–TMDC interface. Following exciton dissociation, the photo-generated hole moves to valance band of MoS$_2$ whereas Schottky barrier prevents the transfer of photoelectron into graphene, which ultimately reduces the quantum yield of PL. These results indicate that CT between graphene–MoS$_2$ hybrids is the primary reason for PL quenching.

5.5.3   Photodetection

5.5.3.1   *Vertical optically-active photodetector and ultra-fast photoresponse*

While photodetection with individual 2D materials, such as graphene and TMDCs, is a matured subject [137–142], van der Waals heterostructures have triggered a paradigm shift in the photodetection technology.

By combining graphene with a strongly light absorbing material one can increase the performance of these photo detectors. The light absorbing material also provides the photodetector with the spectral selectivity. Sandwiching a tunnel barrier between two graphene layers allows photo-exited hot carriers of one graphene layer to tunnel into the other graphene layer, introducing a strong photo-gating effect to the channel conductance. Recently, reported graphene/Ta_2O_5/graphene heterostructure devices demonstrate hot carrier assisted room temperature broadband photoresponce from visible to mid infrared range, but sacrifices the high responsivity of selective spectral photodetectors [129].

Recently, Britnell et $al.$ [14] have reported graphene–WS_2–graphene vertical heterostructure devices to operate at a high external quantum efficiency (EQE) of 30%. The schematics of their device geometry and optical image of such a heterostructure are shown in Figure 26(a). I–V characteristics of the sample is shown in Figure 26(b) with and without 1.95 eV Laser illumination. Sample bias is applied across the top and bottom graphene layer. The I–V characteristics is strongly nonlinear without optical illumination, but becomes linear in the presence of light, and resistance decreases by more than three order of magnitude. The gate voltage dependence of the photo current can be readily understood from the band diagram of the heterostructure presented in Figure 26(c). In presence of a built in electric field, created by gate voltage or source-drain bias or both, photo-generated electrons and holes can reach the top and bottom graphene layer, giving rise to the finite photocurrent. EQE is a figure of merit for photodetectors and solar cells, which is defined by the ratio of no. of detected (measurable) electron-hole pairs to the number of incident photons. Figure 26(d) presents a plot of EQE as a function of the incident laser power.

In a recent work Massicotte et $al.$ [143] showed, using time resolved photocurrent measurement, that in vertically integrated graphene/WSe_2/graphene photodetectors, photoresponse time can be as short as 5.5 picosecond, for extremely thin layers of WSe_2.

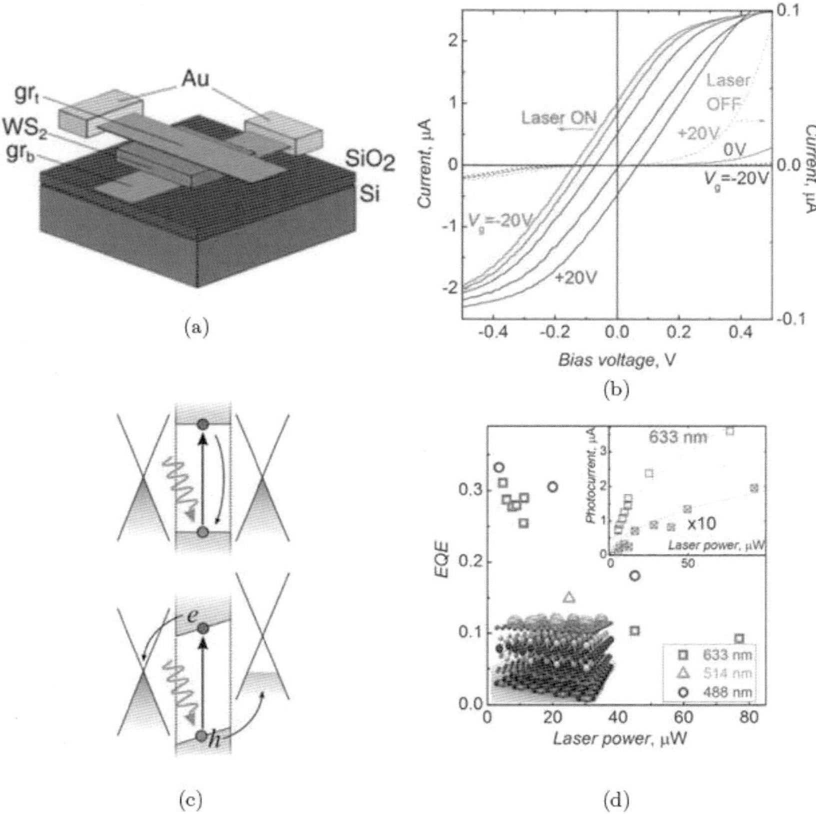

Figure 26. A graphene–TMDC vertical heterostructure for photodetection: (a) a typical device schematics of graphene–WS$_2$–graphene vertical photodetector. (b) vertical current vs. source-drain voltage characteristics of the device with laser on and off at different Vg. (c) a schematic presentation of the energy band diagram of the photodetector explaining the origin of the light response. It also shows how the direction of drift of the photo-generated carriers can be controlled by tuning source-drain bias and Vg. (d) dependence of EQE on laser power, inset shows power dependence of photocurrent, for different wavelength. (Adapted from Ref. [14].)

Figure 27(a) shows a typical device schematic. In a photocurrent autocorrelation measurement with $300\,\mu$W power and 1.2 V source-drain bias, the normalised photocurrent as a function of delay interval between two light pulses is plotted in Figure 27(b). Photocurrent is normalised by dividing the absolute value of the photocurrent with

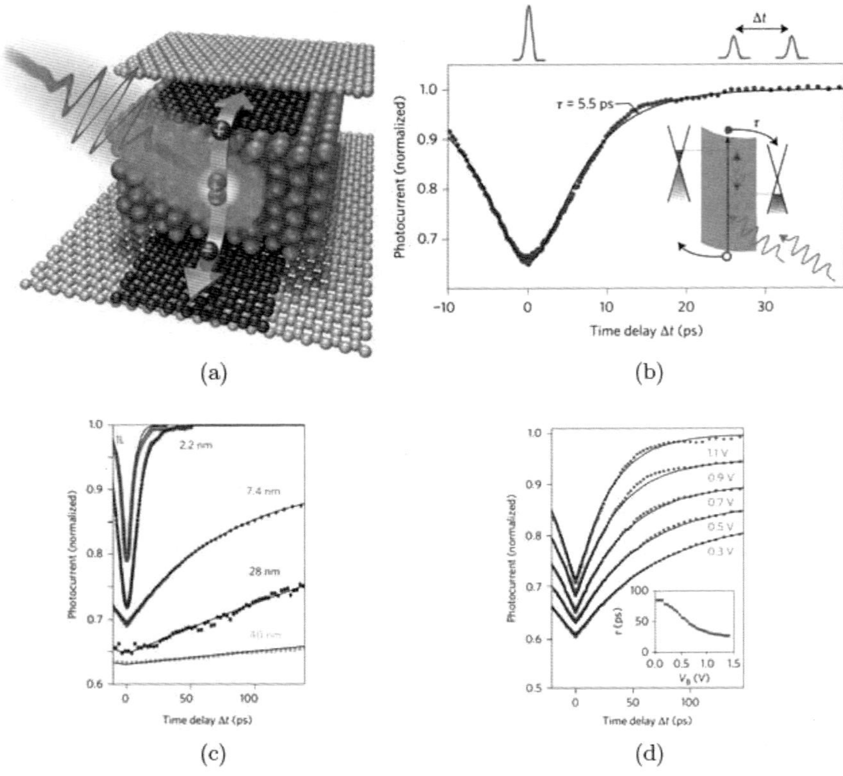

Figure 27. A graphene–WSe$_2$–graphene vertical photodetector: (a) Schematics presentation of the BN encapsulated graphene–WSe$_2$–graphene photodetector. (b) Photocurrent autocorrelation measurement data showing normalised photocurrent as a function of time delay between two pulses. (c) Time resolved photocurrent measurement data for different thicknesses of WSe$_2$. (d) Time resolved photocurrent measurement data for different source drain bias voltage. Response time can be tuned by controlling thickness of WSe$_2$ and bias voltage. (Adapted from Ref. [143].)

its value at large Δt, when the photocurrent saturates. The response time can be controlled by tuning the vertical channel length and drift velocity of the photo-generated charge carriers and hence by changing the WSe$_2$ thickness and the source-drain bias. Figures 27(c) and 27(d) show that photoresponse time increases for thicker WSe$_2$ and smaller bias, respectively.

5.5.3.2 *Planar photodetectors*

Following the cross-layer design, it was soon realised that physically separating the region of photocarrier generation and that of charge transport, can lead to a very different class of detectors that can exhibit extremely large photoresponsivity. The planar photosensitive hybrids have been achieved in multiple ways. Konstantatos *et al.* [144] showed that a 2D array of lead sulphide (PbS) quantum dots self-assembled on graphene (Figure 28(a)) can induce extremely high in-plane photoresponse (responsivity $\sim 10^7$ AW^{-1}, Figure 28(b)) in graphene. In these detectors, photogenerated holes from PbS quantum dots are transferred to graphene by the graphene–PbS interfacial electric field, but photogenerated electrons remain trapped. Since the effective carrier transit time across the graphene channel is low ($\sim 0.1\,\mu$s) because of high carrier mobility in graphene, whereas the lifetime of the electron-hole pair exceed $\sim 1\,$s, a massive photogain mechanism leads to high responsivity and high specific detectivity ($\sim 10^{13}$ Jones). The response band of the detector can also be tuned by tailoring the excitonic gap of PdS QDs by choosing different size.

Similar mechanism has allowed large photoresponsivity in graphene–MoS$_2$ van der Waals heterostructure (Figure 29), however with more direct control over the CT process with a global back gate [22]. The photoresponsivity as high as $\sim 10^{10}$ AW^{-1} at 130 K and $\sim 10^8$ AW^{-1} at 300 K has been attained with these devices (Figure 29(b)). The gate electric field assists in trapping the holes at the MoS$_2$/SiO$_2$ interfaces, while the photogenerated electrons are transferred to graphene. The trapping of holes also leads to a persistent photoconductivity, which can be removed only by applying an appropriate gate pulse (Figure 29(c)). The controlled trapping of photogenerated carrier allow these heterostructures to be used as optoelectronic memory devices [22].

Inclusion of a BN dielectric layer between graphene and MoS$_2$ (Figure 30(a)), prevents the transferred carriers to relax back into MoS$_2$ when the illumination is turned off. This leads to a non-volatile and non-relaxing memory device and demonstrated in Figure 30(b). Interestingly, even though the optoelectronic responsivity of these heterostructures is very large, the intrinsic noise, which is mostly 1/f

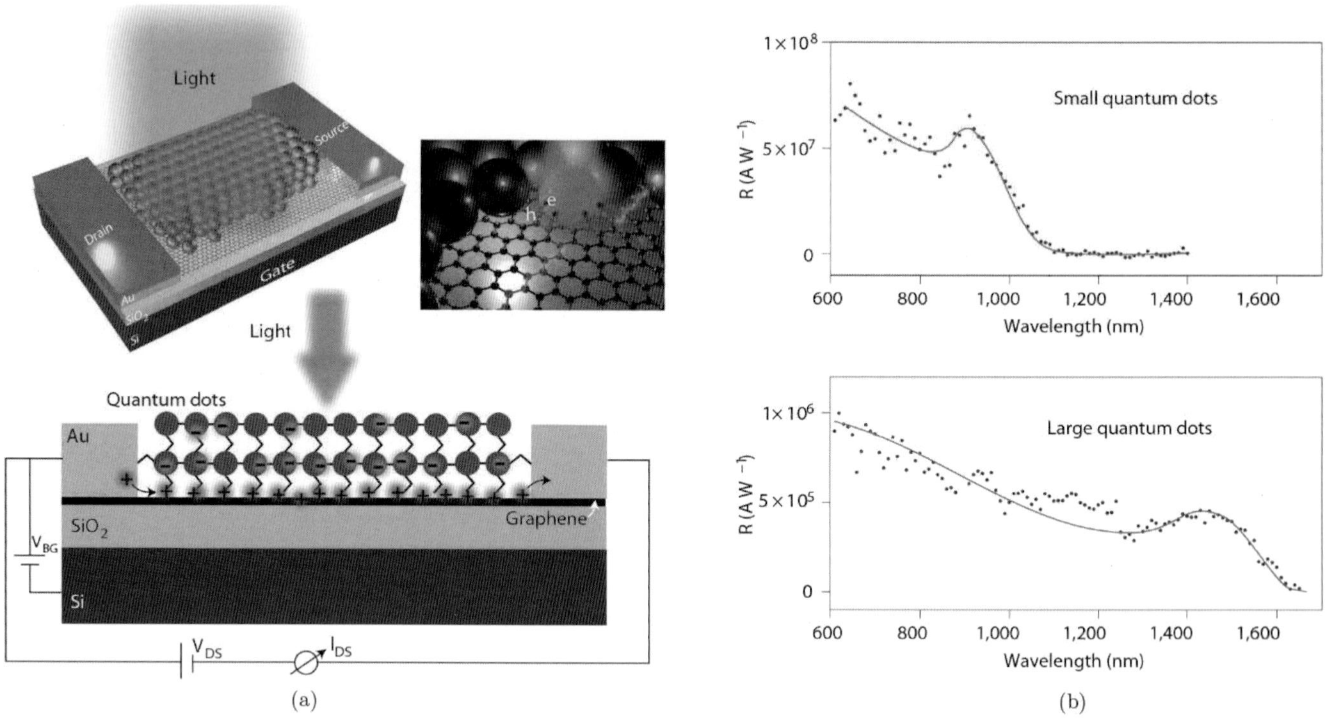

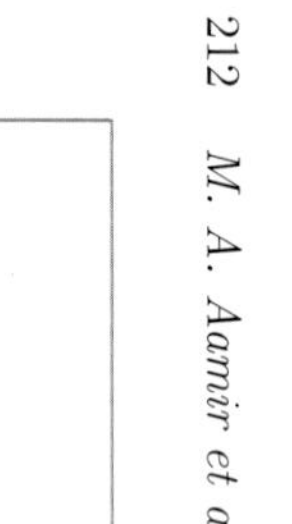

Figure 28. Graphene–PbS photodetector with ultrahigh gain: (a) PbS quantum dots (QDs) on graphene devices made on a Si/SiO₂ substrate. The scheme also indicate the CT process in presence of light. (b) Photoresponsivity as at different wavelength using Small QDs (top panel) and large QDs (bottom panel). (Adapted from Ref. [144].)

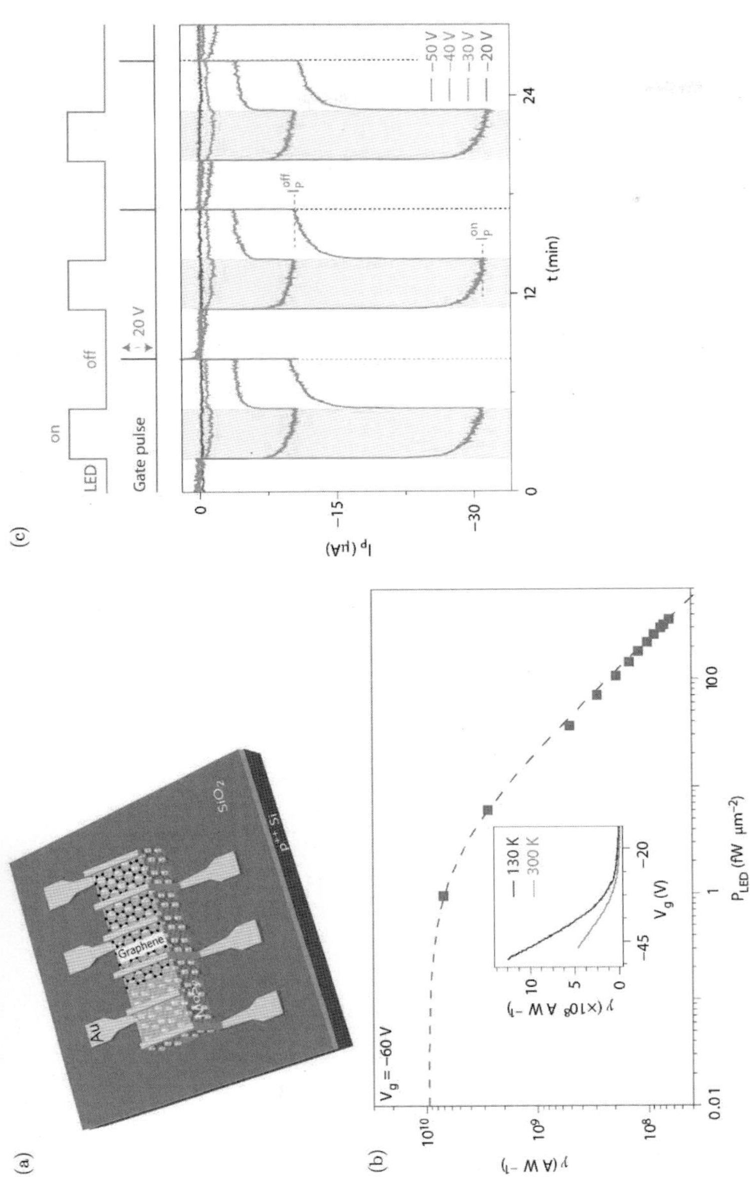

Figure 29. Photoresponse with graphene MoS$_2$ heterostructures: (a) A schematic of graphene on few-layer MoS$_2$ device made on Si/SiO$_2$ substrate. (b) Photoresponsivity (γ) with the variation of optical power at 130 K temperature. Inset shows the tenability of photoresponse as a function of back gate voltages (V_g) at low temperature and room temperature. (c) Tunable persistent photoresponse at various gate voltages. Multiple cycles demonstrate stable operation of the device. (Adapted from Ref. [22].)

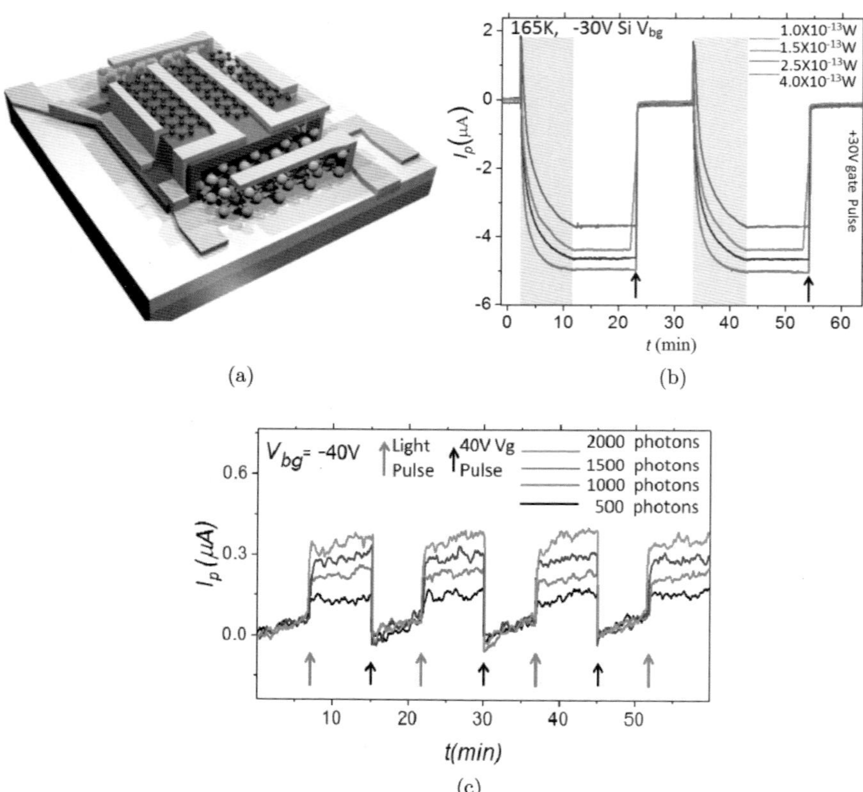

(a)   (b)   (c)

Figure 30. A non-volatile and non-relaxing optoelectronic memory with graphene, BN and MoS$_2$ trilayer van der Waals hybrid: (a) A schematic of graphene–BN–MoS$_2$ trilayer van der Waals heterostructure. (b) Time series of light ON–OFF at a fixed back gate voltage, but varying the excitation power. (c) Time series in the persistent mode after illuminating with light pulses. The intrinsic noise limits the sensitivity of these devices to around 500 photons.

in nature, limits the photon sensitivity to around 500 (in the visible range) (Figure 30(c)).

5.5.4 Photodetection with 2D–0D composites

5.5.4.1 *Plasmonic photodetection*

Novel routes to photodetection has been achieved with cross-dimensional hybrids of 2D materials, for example, graphene and metallic nanoparticles. The plasmonic resonances in the metallic nanostructures allow intrinsically fast and colour sensitive photodetection. The large near-field enhancement of the electromagnetic intensity at the plasmon resonance is exploited to create excess electron-hole pair in graphene as well as other 2D semiconductors.

Echtermeyer *et al.* [145] use plasmonic structures decorated graphene photo transistors (Figures 31(a)–31(d)) to enhance photovoltage generation by concentrating electric filed in the metal-graphene p–n junctions. The enhanced concentration of electric field is also demonstrated computationally for different shapes and structure of the nanoscale metallic elements (Figures 31(e)–31(g)). The sensitivity with respect to the angle of polarisation of the incident light is achieved with finger like structures (Figures 31(e) and 31(f)), which show high photovoltage generation in transverse polarisation (TR) compared to longitudinal polarisation (L) of light. A maximum photoresponse of $\sim$10 mA W^{-1} is attained in these devices.

A more efficient plasmonic coupling of light to graphene is demonstrated recently by Paria *et al.* [146] by using vertically arranged silver nanoparticle dimers separated by a single layer of graphene (Figure 32). The atomic scale spacing between the dimers allow field enhancement factor of around 100 for blue (470 nm) or green (535 nm) illumination (Figure 32(d)), but reduces to $\sim$10 for red light which is away from the plasmon resonance of silver nanoparticles. The field enhancement is also observed in Raman spectroscopy as shown in Figure 32(a). Photocurrent measurement shows maximum responsivity of $\sim$3 AW^{-1}, which is almost 300 times more than the response shown by Echtermeyer *et al.* [145].

5.5.5 Emerging optoelectronic device architectures with van der Waals heterostructure

5.5.5.1 *Atomically thin p–n junctions*

Atomically thin p–n junctions are natural consequence of van der Waals heterostructures. Lee *et al.* assembled a monolayer n-type semiconducting TMDC (MoS$_2$) and another monolayer p-type

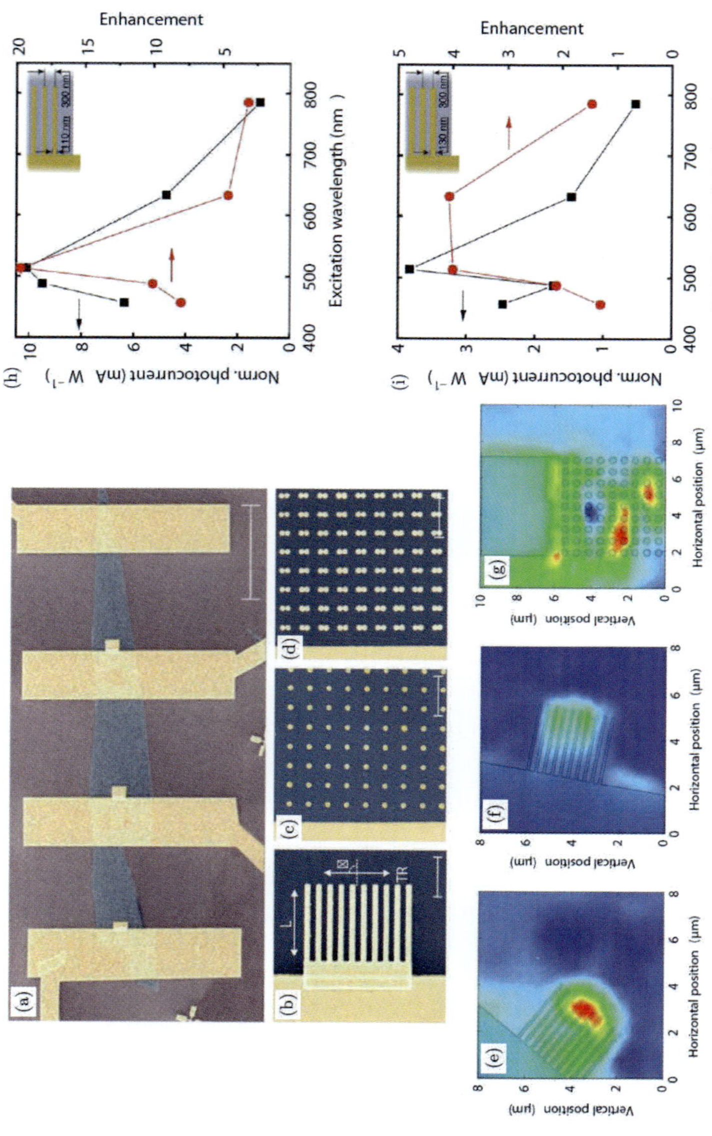

Figure 31. Coupling light to graphene using plasmonic structures: (a) Graphene (blue) on, SiO$_2$ (purple) with Ti/Au contacts (yellow), scale bar 20 mm. (b)–(d) Different types of plasmonic structures where L and TR indicate L and TR of incident light, respectively. Scale bar 1 mm. (e)–(g) Spatial map of photovoltage response. (e) and (f) 514 nm illumination with TR and L polarisation, respectively. Illumination wavelength 633 nm, TR polarisation. (g) Photovoltage response in nanodot array. Illumination wavelength 633 nm, TR polarisation. Colour scale indicate -4 mV = blue, and 12 mV = red. (h) and (i) Normalised photocurrent and enhancement coefficient for finger structure with 300 nm pitch having 110 nm and 130 nm width, respectively (inset). (Adapted from Ref. [145].)

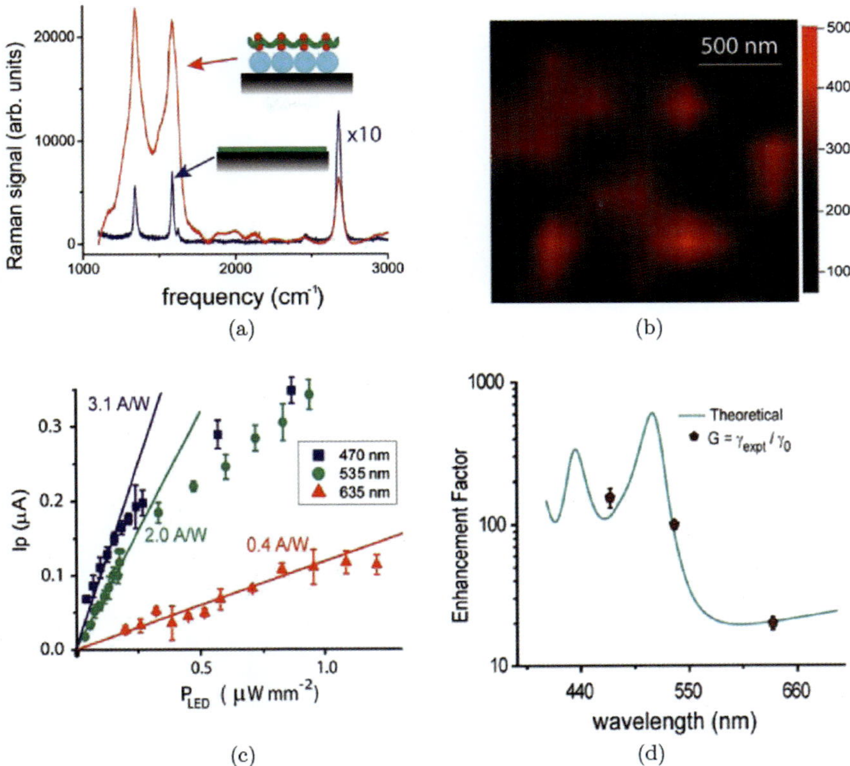

Figure 32. Ultrahigh field enhancement using plasmonic dimers:
(a) Raman spectra of graphene without (blue) and with (red) the presence silver
dimers. Inset shows the typical device scheme where red dots are the silver
dimers separated by a graphene layer (green line). The complete structure is
made on 500 nm silver bids (cyan colours) placed on Si/SiO_2 substrate (grey
shade at bottom). (b) Spatial map of Raman signal of the device with silver
dimers. (c) Photocurrent measurement at different illumination power. Results for
three different wavelengths are shown. (d) Experimentally obtained enhancement
factor (black solid circles) in comparison with theoretical analysis (cyan curve)
at different wavelengths. (Adapted from Ref. [146].)

TMDC (WSe_2) on Si/SiO_2 substrate (Figure 33(a)), and obtain
rectifying current-voltage characteristics (Figure 33(b)). The tunability in $I_{ds} - V_{ds}$ is attained by controlling the tunnel assisted
recombination of majority carriers in the layers. Inset of Figure 33(b)
shows the transfer characteristics of the individual layers establishing
the n and p-type nature of MoS_2 and WSe_2, respectively. Figure 33(c)

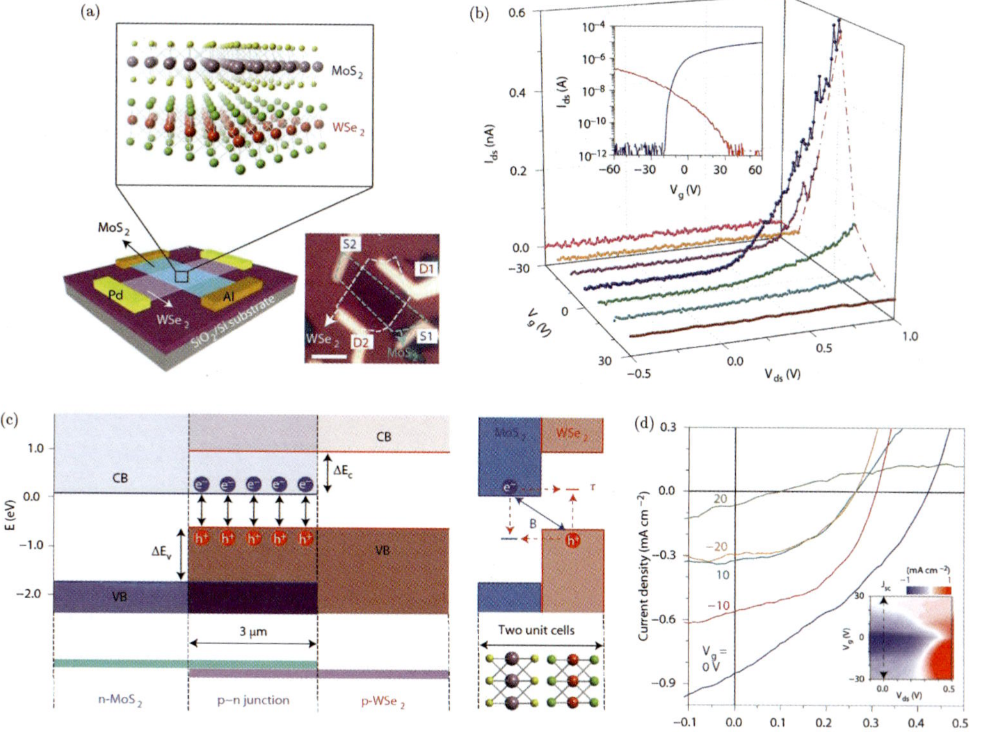

Figure 33. Atomically thin p–n junction: (a) Monolayer MoS_2–WSe_2 heterostructure. Left panel is a schematic, right panel indicates optical image. Inset shows the stack scheme in the overlap region. (b) Tunable rectification characteristics of the p–n junction. Inset indicates transfer characteristics of individual layers. (c) Energy band diagram in lateral (left panel) and vertical (right panel) directions. (d) photoresponse at different gate voltages. Inset shows colour representation of photo current (Ip) as a function of gate voltage (V_g) and source-drain bias (V_{ds}). (Adapted from Ref. [25]).

depicts the energy band alignment of the p–n junction in both lateral (left panel) and vertical directions (right panel). In forward bias condition electrons (e−) and holes (h+) recombine with each other following either Shockley–Read–Hall or Langevin processes as indicated by red and blue arrow, respectively, in the right panel. A realignment in the Fermi energy causes a band offsets for electrons (ΔE_C) and holes (ΔE_V) (left panel). Photovoltaic effect in these p–n junctions are demonstrated by Lee *et al.* [25] (Figure 33(d)), which can be tuned by an external back gate.

5.5.5.2 *Light emitting diodes*

Apart from photodetection, a new generation of van der Waals hybrids has now been fabricated for creating a tunable source of light. Earlier designs of LEDs with TMDC materials involved lateral p–n junctions on a single layer of TMDC (e.g., MoS_2 [147] or WSe_2 [148]) in a dual gated geometry. The vertical assemblies of graphene, BN and TMDC, where the TMDC acts as the central recombination layer, simplifies the gate assembly and also enhanced efficiency ($\sim$10%) in light emission [149]. A control on the frequency of the emitted light can also be obtained by selecting a TMDC layer of appropriate band gap.

5.6 Thermal Management with van der Waals Hybrids

5.6.1 Cross-plane thermal transport in crossed bilayer devices

The in-built anisotropy of van der Waals heterostructures have been utilised not only in tuning the directionality of charge transport, but also in transport of heat. Figure 34 shows a device architecture for cross plane thermoelectricity measurement in graphene–BN–graphene hybrid structure. The second harmonic voltage difference, which is a direct consequence of oscillating temperature difference, between the two graphene layers is measured while an alternating current is sent along the upper graphene layer. The temperature difference between the two layers is measured by Raman spectroscopy.

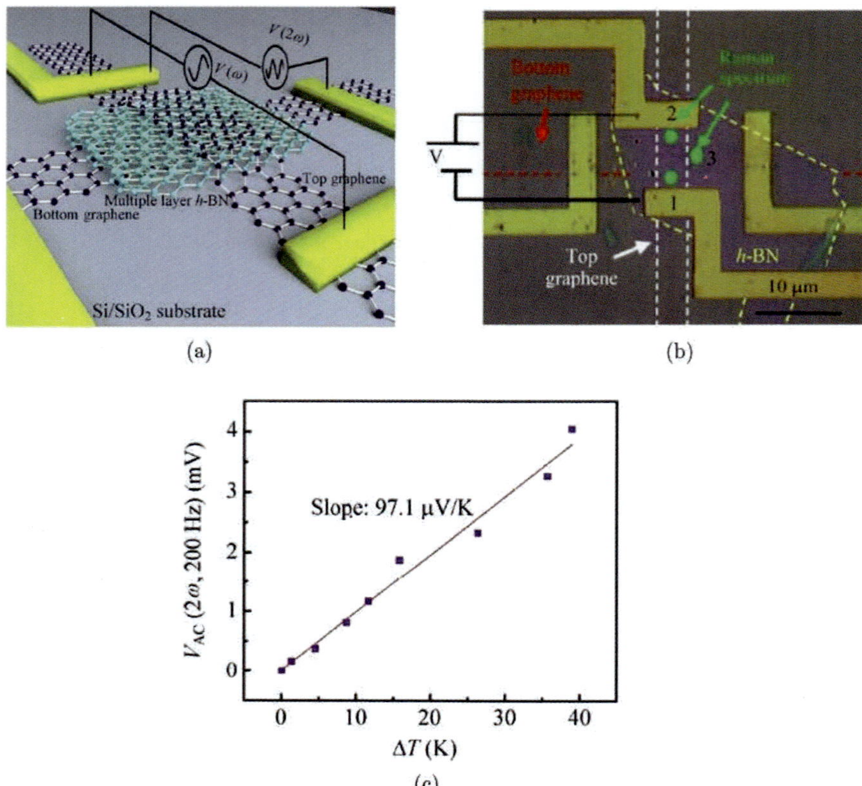

Figure 34. Thermal properties of graphene–BN heterostructure devices: (a) shows the device schematic of Graphene/hBN/Graphene heterostructure for thermoelectric measurements. (b) Optical image of the device and the locations of the Raman spectroscopy (green circles) for temperature calibration. (c) 2nd harmonic voltage drop between the two layers as a function of temperature difference between the two graphene layers. The slope gives the Seebeck Coefficient of the Graphene/BN/Graphene heterostructure. (Adapted from Ref. [150].)

The measured Seebeck coefficient for graphene–BN–graphene is approximately $98\,\mu$V/K at room temperature [150].

5.6.2   Tuning thermalisation of electrons

The graphene/BN/graphene van der Waals hybrid have also been employed in tuning thermalisation of optically excited carriers in graphene. In graphene, this electron thermalisation occurs in extremely fast due to strong electron–electron interaction. Figure 35 shows the schematic of graphene/BN/graphene device under optical excitation [151].

In the presence of optical excitation, the inter-layer photocurrent increases with laser power and goes from linear to super linear as the photon energy or the bias voltage are reduced which show two distinct processes of thermalisation: Thermionic emission and direct tunnelling of the photo excited carriers. For low bias voltage to photon energy ratio, the optically excited carriers remain in graphene and quickly reach thermal equilibrium due to scattering. The high energy carriers in the tail of the resultant thermal distribution overcome the high BN barrier and their population increases exponentially leading to the super linear dependence of photocurrent with laser power. In contrast, at high bias voltage to photon energy ratio, the effective potential barrier of the BN is largely reduced, and as a result the photo excited carriers tunnel through the BN barrier before they scatter with in plane carriers. The photocurrent increases linearly with laser power in this regime.

5.7   Emerging Phenomena and Outlook

The improved material quality and structural tunability in the van der Waals heterostructures have been pivotal in the demonstration of several fundamental phenomena in solid state physics with unprecedented control. Examples include the observation of Coulomb drag [9] and the fractional quantum Hall effect [152]. In addition, several new phenomena arising from many-body effects in electronic transport to 2D superconductivity, have triggered parallel channels of research. Some of these phenomena are described briefly below.

Fermi edge singularity

The observation of many-body phenomena in graphene that arises due to interactions, e.g., the discovery of the fractional quantum Hall effect [153, 154], broken symmetry states at zero filling factor [155], or spin-valley quantum Hall ferromagnetism [156], requires high-mobility suspended graphene or graphene on BN to minimise the effect of external disorder.

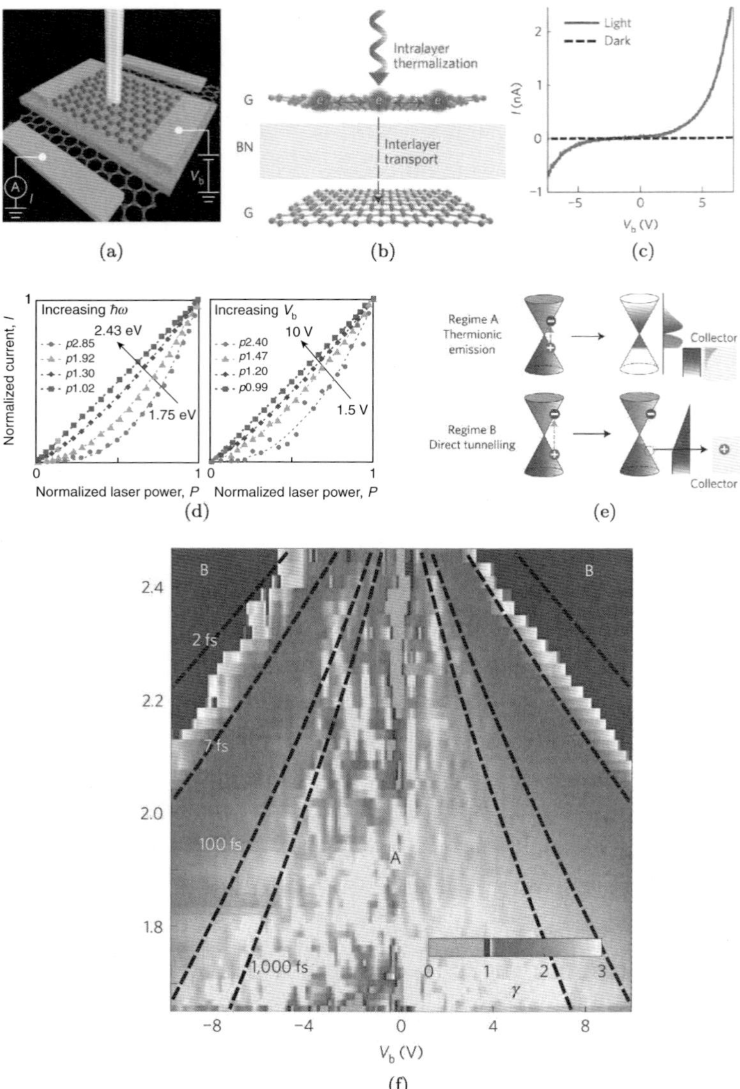

Figure 35. Thermalisation in graphene–BN–graphene vertical assembly: (a) The schematic of graphene/BN/graphene device to tune the thermalisation of carriers when subjected to optical excitation. (b) Schematic of Intra-layer thermalisation and Inter-layer transport of optically excited carriers in the graphene. (c) Inter-layer photocurrent with and without light. (d) Normalised photocurrent as a function of laser power for different photon energy and bias voltage. (e) The two distinct processes, thermionic emission and direct tunnelling of the photo excited carriers. (f) Dependence of photocurrent on both bias voltage and photon energy. (Adapted from Ref. [151]).

The ability of graphene to screen the external Coulomb potential can affect the strength of electron–electron interaction and even alter the physical properties like the dielectric constant [157] or the Fermi velocity [158]. The screening properties can also give rise to phenomena in which the disorder itself plays a central role such as the Anderson orthogonality catastrophe [159, 160], local moment formation [161, 162] and the Kondo effect [163, 164].

The Fermi edge singularity is one of the intriguing results of attractive Coulomb interaction between a charged impurity and Fermi sea of electrons (or holes). In the unitary limit, the impurity is totally screened, and forms a many-body state with one additional channel of transport. Fabrication of ultra-high mobility graphene–BN van der Waals heterostructure has recently resulted in the demonstration of Fermi edge singularity in graphene [4]. As shown in Figure 36, the effect is manifested as abrupt switchings in the conductance $\approx e^2/h$, when the defect level and graphene Fermi energy are in resonance (Figure 36(a)). These abrupt switchings occur when an electron is captured by the defect, which leads to an attractive interaction with the hole gas in graphene. The quasi-bound state between the impurity potential and graphene opens up a conduction

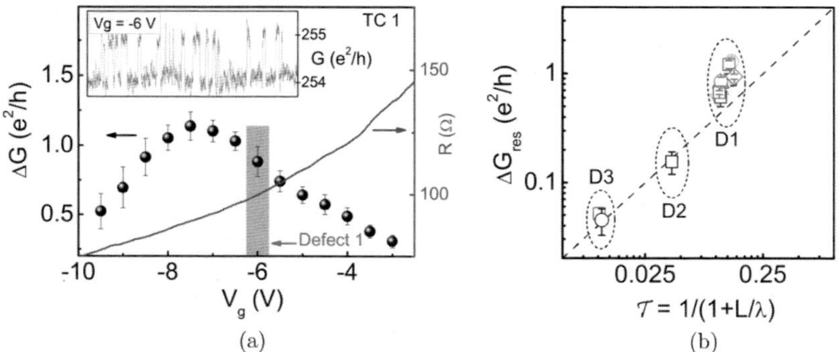

Figure 36. Fermi edge singularity in high mobility graphene–BN heterostructures: (a) Conductance jumps (ΔG) as a function of back gate voltage (V_g). $\Delta G \approx e^2/h$ when $V_g \approx -6\,V$ (shaded region) and decreases away from resonance. Solid line shows resistance as a function of V_g. Inset shows the jumps in conductance (ΔG) vs. V_g. (b) The change in conductance at resonance for multiple devices as a function of the corresponding transmission probability (T). The dashed line has a slope of 4. (Adapted from Ref. [4].)

channel and thus the conductance rises by $\approx e^2/h$. The conductance change decreases from $\sim e^2/h$ in lower mobility of graphene due to conventional impurity scattering in graphene (Figure 36(b)), emphasising the importance of high mobility of van der Waals heterostructures in discovering new fundamental phenomena.

Secondary Dirac points, Hoffstadter butterfly

h-BN has a small lattice mismatch (1.7%) with graphene which creates the possibility of creating a moiré pattern by aligning the crystallographic directions of the two crystals. This results in an emergent superlattice due to the substrate potential, the periodicity is of the order of tens of nanometres [29, 165]. When such a 2D superlattice is subjected to a magnetic field and the two length scales are comparable, it can result in the formation of a fractal energy spectrum known as Hofstadter's butterfly [166].

This behaviour was observed for both single [165] and bilayer graphene [29] that was aligned over BN (Figures 37(a) and 37(b)). The side Dirac peak resulting from the extra Dirac point in single layer graphene is shown in Figure 37(c).

2D Superconductivity in van der Waals heterostructures

The high electronic quality of graphene–BN heterostructures has allowed the creation of ballistic Josephson junctions [48, 167]. This was first achieved with superconducting MoRe contacts [48], shown in Figure 38(a). MoRe contacts show low contact resistance, have a high critical temperature ($\approx 8\,\text{K}$) and critical field ($\approx 8\,\text{T}$). The critical supercurrent in graphene oscillates (Figure 38(b)) as a function of the number density due to the phase-coherent interference of charge carriers in the formed Fabry–Pérot cavity.

Other experiments on clean heterostructures have mapped out the spatial distribution of current in a graphene Josephson junction using Fourier mapping [168, 169]. When graphene is nearly charge neutral peaks in current are observed at the crystal edges, shown in Figure 38(c), demonstrating the spatial confinement of electron waves. Graphene heterostructures with MoRe contacts also allow

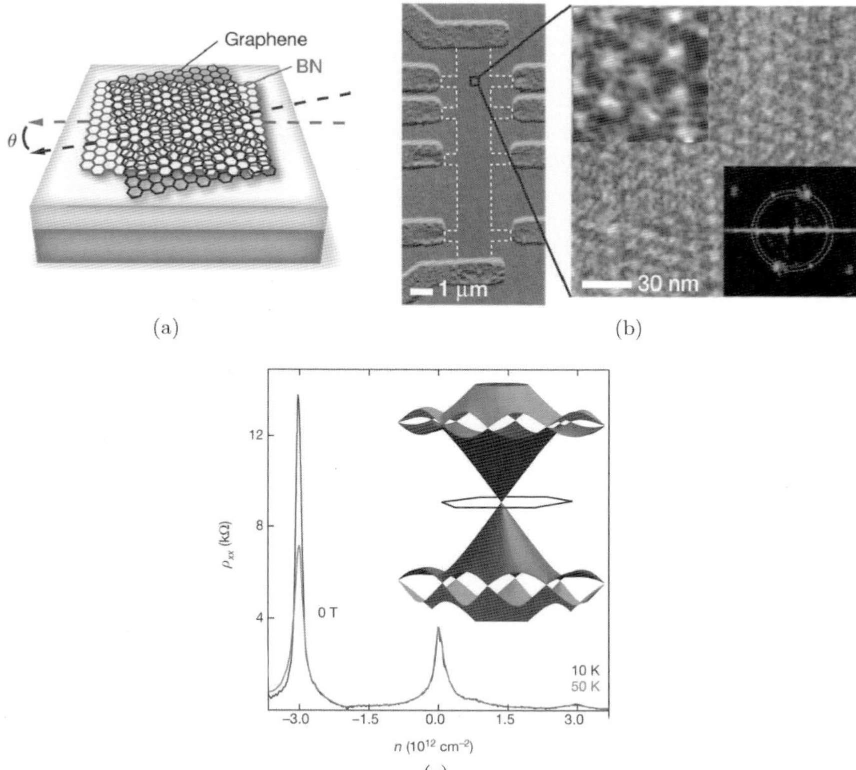

Figure 37. Secondary Dirac point in graphene–BN hybrids: (a) Moiré pattern in graphene on BN. (b) AFM topography of a BLG Hall bar on BN (left). Magnified image on the right shows a triangular lattice emerging from the moiré pattern. The fast Fourier transform on the bottom reveals periodicity $\approx$15.5 nm of the triangular lattice, indicating nearly zero angle mismatch between BLG and BN. (Adapted from Ref. [29]). (c) Resistance vs. gate voltage for a single layer graphene reveals the satellite maxima resulting from the secondary Dirac peaks. (Adapted from Ref. [165].)

the injection of a supercurrent in the quantum Hall regime [48, 170], creating the possibility for exploiting exotic topological properties.

The observation of superconductivity in heterostructures extends to beyond graphene–BN. NbSe$_2$ is a superconducting 2D crystal and it has been shown [171] that an artificially assembling two NbSe$_2$ crystals leads the formation of a short, vertical, transparent Josephson junction.

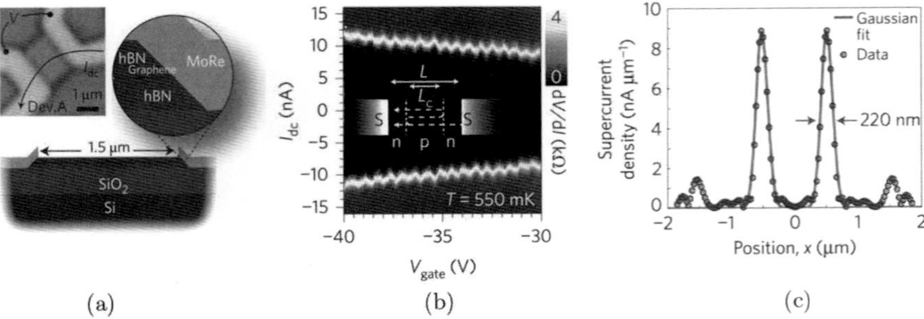

**Figure 38. High mobility graphene heterostructure with supercon-
ducting contacts:** (a) Graphene, encapsulated with BN, is edge-contacted by
molybdenum rhenium (MoRe) alloy. (b) The critical supercurrent oscillates as a
function of the number density within the formed Fabry–Pérot cavity. (Adapted
from Ref. [48].) (c) Superconducting interferometry in a graphene Josephson
junction. Suppercurrent peaks at the sample edges show the localisation of
electron wavefunction at the edge near the charge-neutrality point. (Adapted
from Ref. [168].)

*Hydrodynamic regime of charge and heat transport in van der Waals
heterostructures*

In very high quality electronic devices, it is possible to access a regime
in which the electron–electron interaction mean free path l_{ee} is the
shortest length scale, compared to the sample size or the mean free
path. In this regime, the charge carriers behave as a classical fluid
and are governed by the hydrodynamic equations [172]. The van der
Waals heterostructures of graphene and BN have allowed access to
such a regime of transport in graphene [51].

Graphene displays a weak electron–phonon scattering, and high
quality heterostructures are ideal candidate for probing the hydrodynamic regime. It has been seen that, under the above conditions,
carriers in graphene collectively behave as a highly viscous fluid. One
signature is that this leads to a charge backflow, see Figure 39(b) near
the injection contacts that can be measured (Figure 39(a)). The non-
local voltage in proximity to the injection contact changes sign with
both the carrier density and the temperature (Figure 39(a)) and
can be explained by the hydrodynamic equations [51, 173, 174]. The
hydrodynamic signatures are also visible near the charge neutrality
point and the decoupling of the thermal and the electronic flow result

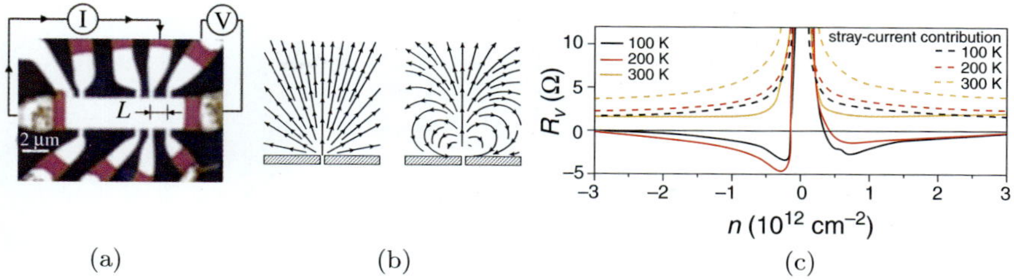

(a)   (b)   (c)

Figure 39. Non-local resistance and hydrodynamic transport in graphene heterostructure: (a) Non-local measurement geometry in an edge contacted graphene encapsulated with BN. (b) Current injection in a normal Fermi liquid (left). Current injection in a viscous Fermi liquid. (c) One signature of the viscous flow is the negative vicinity voltage in the appropriate temperature range. (Adapted from Ref. [51].)

in a large enhancement of the thermal conductivity and subsequent violation of the Wiedemann–Franz law [175, 176].

Inducing spin-orbit coupling (SOC) in graphene

Though graphene has very large spin relaxation lengths, the manipulation of spin currents is extremely difficult due to very weak intrinsic spin orbit coupling. Hydrogenation or addition of heavy atoms such as gold, indium, thallium have been shown to enhance SOC, but they introduce disorder into the system as well. Another method is to use proximity to WS_2 [177], which has been shown by Avsar *et al.* to increase SOC in Graphene to $\sim 17\,\mathrm{meV}$, allowing field controllable spin currents even at RT while also preserving high mobilities of graphene (Figure 40).

In conclusion, our goal in this chapter has been to outline the state of the art in the van der Waals heterostructure, with particular emphasis on the current structure of devices and their functionality. We have discussed the fabrication, characterisation and functional uniqueness of these devices, and also the domain of impact. The remarkable breadth of phenomenology that can be addressed with these devices has been indicated, although only a few could be discussed in some detail. We emphasise that the research with van der Waals hybrid structures is still in its infancy, as new class and quality of layered materials are isolated and processed. Clever combination of materials have already led to unexpected phenomena

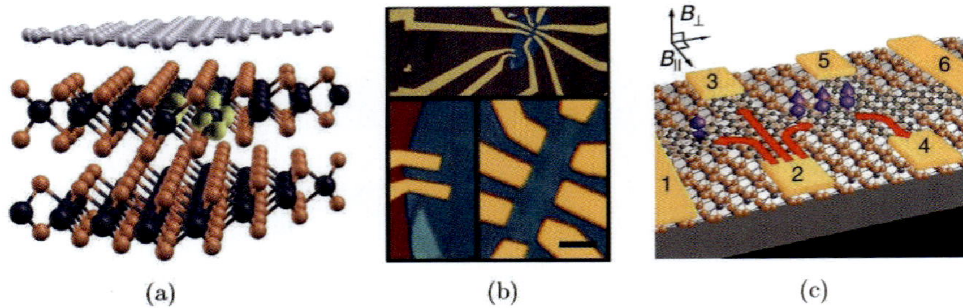

Figure 40. Graphene–WS₂ heterostructure for inducing spin-orbit interaction in graphene by proximity: (a) Schematic of multi-layer grapene-WS$_2$ heterostructure, where Carbon, Tungsten and Sulphur is represented by light grey, dark grey and orange, respectively, while yellow represents the highest unoccupied Sulphur vacancies. (b) Optical micrograph of a typical graphene–WS$_2$ device in hall bar geometry and a 2-probe WS$_2$ device. (c) Schematic of current flow that lead to non-local resistance. (Adapted from Ref. [177].)

and functionality, and it is very likely that there are many more waiting to be discovered.

References

[1] M. M. Ugeda, A. J. Bradley, S. F. Shi, F. H. da Jornada, Y. Zhang, D. Y. Qiu, W. Ruan, S. K. Mo, Z. Hussain, Z. X. Shen, F. Wang, S. G. Louie and M. F. Crommie, *Nat. Mater.* **13**, 1091 (2014).

[2] A. Sahoo, K. Roy and A. Ghosh, to be published (2017).

[3] K. S. Novoselov, *Science* **306**, 666 (2004).

[4] P. Karnatak, S. Goswami, V. Kochat, A. N. Pal and A. Ghosh, *Phys. Rev. Lett.* **113**, 026601 (2014).

[5] K. Kim, M. Yankowitz, B. Fallahazad, S. Kang, H. C. P. Movva, S. Huang, S. Larentis, C. M. Corbet, T. Taniguchi, K. Watanabe, S. K. Banerjee, B. J. LeRoy and E. Tutuc, *Nano Lett.* **16**, 1989 (2016).

[6] K. F. Mak, C. Lee, J. Hone, J. Shan and T. F. Heinz, *Phys. Rev. Lett.* **105**, 136805 (2010).

[7] C. R. Dean, A. F. Young, I. Meric, C. Lee, L. Wang, S. Sorgenfrei, K. Watanabe, T. Taniguchi, P. Kim, K. L. Shepard and J. Hone, *Nat. Nanotechnol.* **5**, 722 (2010).

[8] A. S. Mayorov, R. V. Gorbachev, S. V. Morozov, L. Britnell, R. Jalil, L. A. Ponomarenko, P. Blake, K. S. Novoselov, K. Watanabe, T. Taniguchi and A. K. Geim, *Nano lett.* **11**, 2396 (2011).

[9] R. V. Gorbachev, A. K. Geim, M. I. Katsnelson, K. S. Novoselov, T. Tudorovskiy, I. V. Grigorieva, A. H. MacDonald, S. V. Morozov, K. Watanabe, T. Taniguchi and L. A. Ponomarenko, *Nat. Phys.* **8**, 896 (2012).

[10] L. Britnell, R. V. Gorbachev, R. Jalil, B. D. Belle, F. Schedin, A. Mishchenko, T. Georgiou, M. I. Katsnelson, L. Eaves, S. V. Morozov, N. M. Peres, J. Leist, A. K. Geim, K. S. Novoselov and L. A. Ponomarenko, *Science* **335**, 947 (2012).

[11] L. Britnell, R. V. Gorbachev, A. K. Geim, L. A. Ponomarenko, A. Mishchenko, M. T. Greenaway, T. M. Fromhold, K. S. Novoselov and L. Eaves, *Nat. commun.* **4**, 1794 (2013).

[12] A. Mishchenko, J. S. Tu, Y. Cao, R. V. Gorbachev, J. R. Wallbank, M. T. Greenaway, V. E. Morozov, S. V. Morozov, M. J. Zhu, S. L. Wong, F. Withers, C. R. Woods, Y.-J. Kim, K. Watanabe, T. Taniguchi, E. E. Vdovin, O. Makarovsky, T. M. Fromhold, V. I. Fal'ko, A. K. Geim, L. Eaves and K. S. Novoselov, *Nat. Nanotechnol.* **9**, 808 (2014).

[13] L. Kou, S.-C. Wu, C. Felser, T. Frauenheim, C. Chen and B. Yan, *ACS Nano.* **8**, 10448 (2014).

[14] L. Britnell, R. M. Ribeiro, A. Eckmann, R. Jalil, B. D. Belle, A. Mishchenko, Y. J. Kim, R. V. Gorbachev, T. Georgiou, S. V. Morozov, A. N. Grigorenko, A. K. Geim, C. Casiraghi, A. H. Castro Neto and K. S. Novoselov, *Science* **340**, 1311 (2013).

[15] P. Rivera, J. R. Schaibley, A. M. Jones, J. S. Ross, S. Wu, G. Aivazian, P. Klement, K. Seyler, G. Clark, N. J. Ghimire, J. Yan, D. G. Mandrus, W. Yao and X. Xu, *Nat. Commun.* **6**, 6242 (2015).

[16] P. Rivera, K. L. Seyler, H. Yu, J. R. Schaibley, J. Yan, D. G. Mandrus, W. Yao and X. Xu, *Science* **351**, 688 (2016).

[17] Y. Liu, P. Stradins and S.-H. Wei, *Science Advances* **2**, e1600069 (2016).

[18] X. Hong, J. Kim, S. F. Shi, Y. Zhang, C. Jin, Y. Sun, S. Tongay, J. Wu, Y. Zhang and F. Wang, *Nat. Nanotechnol.* **9**, 682 (2014).

[19] W. J. Yu, Z. Li, H. Zhou, Y. Chen, Y. Wang, Y. Huang and X. Duan, *Nat. Mater.* **12**, 246 (2013).

[20] T. Georgiou, R. Jalil, B. D. Belle, L. Britnell, R. V. Gorbachev, S. V. Morozov, Y.-j. Kim, A. Gholinia, S. J. Haigh, O. Makarovsky, L. Eaves, L.a. Ponomarenko, A. K. Geim, K. S. Novoselov and A. Mishchenko, *Nat. Nanotechnol.* **8**, 100 (2012).

[21] G.-H. Lee, Y.-J. Yu, X. Cui, N. Petrone, C.-H. Lee, M. S. Choi, D.-Y. Lee, C. Lee, W. J. Yoo, K. Watanabe, T. Taniguchi, C. Nuckolls, P. Kim and J. Hone, *ACS Nano.* **7**, 7931 (2013).

[22] K. Roy, M. Padmanabhan, S. Goswami, T. P. Sai, G. Ramalingam, S. Raghavan and A. Ghosh, *Nat. Nanotechnol.* **8**, 826 (2013).

[23] W. J. Yu, Y. Liu, H. Zhou, A. Yin, Z. Li, Y. Huang and X. Duan, *Nat. Nanotechnol.* **8**, 952 (2013).

[24] W. Zhang, C. P. Chuu, J. K. Huang, C. H. Chen, M. L. Tsai, Y. H. Chang, C. T. Liang, Y. Z. Chen, Y. L. Chueh, J. H. He, M. Y. Chou and L. J. Li, *Sci. Rep.* **4**, 3826 (2014).

[25] C.-H. Lee, G.-H. Lee, A. M. van der Zande, W. Chen, Y. Li, M. Han, X. Cui, G. Arefe, C. Nuckolls, T. F. Heinz, J. Guo, J. Hone and P. Kim, *Nat. Nanotechnol.* **9**, 676 (2014).

[26] C. Kumar, M. Kuiri, J. Jung, T. Das and A. Das, *Nano Lett.* **16**, 1042 (2016).

[27] B. Choi, J. Yu, D. W. Paley, M. T. Trinh, M. V. Paley, J. M. Karch, A. C. Crowther, C. H. Lee, R. A. Lalancette, X. Zhu, P. Kim, M. L. Steigerwald, C. Nuckolls and X. Roy, *Nano Lett.* **16**, 1445 (2016).

[28] G.-H. Lee, C.-H. Lee, A. M. van der Zande, M. Han, X. Cui, G. Arefe, C. Nuckolls, T. F. Heinz, J. Hone and P. Kim, *APL Mater.* **2**, 092511 (2014).

[29] C. R. Dean, L. Wang, P. Maher, C. Forsythe, F. Ghahari, Y. Gao, J. Katoch, M. Ishigami, P. Moon, M. Koshino, T. Taniguchi, K. Watanabe, K. L. Shepard, J. Hone and P. Kim, *Nature* **497**, 598 (2013).

[30] K. Roy, T. Ahmed, H. Dubey, T. P. Sai, R. Kashid, M. Shruti, K. Hsieh, S. Shamim and A. Ghosh, to be published.

[31] X. S. Li, W. W. Cai, J. H. An, S. Kim, J. Nah, D. X. Yang, R. Piner, A. Velamakanni, I. Jung, E. Tutuc, S. K. Banerjee, L. Colombo and R. S. Ruoff, *Science* **324**, 1312 (2009).

[32] A. Ismach, C. Druzgalski, S. Penwell, A. Schwartzberg, M. Zheng, A. Javey, J. Bokor and Y. G. Zhang, *Nano Lett.* **10**, 1542 (2010).

[33] Y. H. Lee, X. Q. Zhang, W. J. Zhang, M. T. Chang, C. T. Lin, K. D. Chang, Y. C. Yu, J. T. W. Wang, C. S. Chang, L. J. Li and T. W. Lin, *Adv. Mater.* **24**, 2320 (2012).

[34] R. Kitaura, Y. Miyata, R. Xiang, J. Hone, J. Kong, R. S. Ruoff and S. Maruyama, *J. Phys. Soc. Jpn.* **84** (2015).

[35] Z. Liu, L. L. Ma, G. Shi, W. Zhou, Y. J. Gong, S. D. Lei, X. B. Yang, J. N. Zhang, J. J. Yu, K. P. Hackenberg, A. Babakhani, J. C. Idrobo, R. Vajtai, J. Lou and P. M. Ajayan, *Nat. Nanotechnol.* **8**, 119 (2013).

[36] Y. J. Gong, J. H. Lin, X. L. Wang, G. Shi, S. D. Lei, Z. Lin, X. L. Zou, G. L. Ye, R. Vajtai, B. I. Yakobson, H. Terrones, M. Terrones, B. K. Tay, J. Lou, S. T. Pantelides, Z. Liu, W. Zhou and P. M. Ajayan, *Nat. Mater.* **13**, 1135 (2014).

[37] K. Kim, S. Coh, L. Z. Tan, W. Regan, J. M. Yuk, E. Chatterjee, M. F. Crommie, M. L. Cohen, S. G. Louie and A. Zettl, *Phys. Rev. Lett.* **108**, 246103 (2012).

[38] J. Hicks, M. Sprinkle, K. Shepperd, F. Wang, A. Tejeda, A. Taleb-Ibrahimi, F. Bertran, P. Le Fèvre, W. A. de Heer, C. Berger and E. H. Conrad, *Phys. Rev. B* **83**, (2011).

[39] G. H. Li, A. Luican, J. M. B. L. dos Santos, A. H. Castro Neto, A. Reina, J. Kong and E. Y. Andrei, *Nat. Phys.* **6**, 109 (2010).

[40] R. W. Havener, H. Zhuang, L. Brown, R. G. Hennig and J. Park, *Nano Lett.* **12**, 3162 (2012).

[41] S. J. Haigh, A. Gholinia, R. Jalil, S. Romani, L. Britnell, D. C. Elias, K. S. Novoselov, L. A. Ponomarenko, A. K. Geim and R. Gorbachev, *Nat. Mater.* **11** 764 (2012).

[42] A. V. Kretinin, Y. Cao, J. S. Tu, G. L. Yu, R. Jalil, K. S. Novoselov, S. J. Haigh, A. Gholinia, A. Mishchenko, M. Lozada, T. Georgiou, C. R. Woods, F. Withers, P. Blake, G. Eda, A. Wirsig, C. Hucho, K. Watanabe, T. Taniguchi, A. K. Geim and R. V. Gorbachev, *Nano Lett.* **14**, 3270 (2014).

[43] A. K. Geim and I. V. Grigorieva, *Nature* **499**, 419 (2013).

[44] H. Coy Diaz, R. Addou and M. Batzill, *Nanoscale* **6**, 1071 (2014).

[45] M. Tao, D. Udeshi, S. Agarwal, E. Maldonado and W. P. Kirk, *Solid-State Electronics* **48**, 335 (2004).

[46] D. Qiu and E. K. Kim, *Sci. Rep.* **5**, 13743 (2015).

[47] L. Wang, I. Meric, P. Y. Huang, Q. Gao, Y. Gao, H. Tran, T. Taniguchi, K. Watanabe, L. M. Campos, D. A. Muller, J. Guo, P. Kim, J. Hone, K. L. Shepard and C. R. Dean, *Science* **342**, 614 (2013).

[48] V. E. Calado, S. Goswami, G. Nanda, M. Diez, A. R. Akhmerov, K. Watanabe, T. Taniguchi, T. M. Klapwijk and L. M. Vandersypen, *Nat. Nanotechnol.* **10**, 761 (2015).

[49] M. Sui, G. Chen, L. Ma, W.-Y. Shan, D. Tian, K. Watanabe, T. Taniguchi, X. Jin, W. Yao, D. Xiao and Y. Zhang, *Nat. Phys.* **11**, 1027 (2015).

[50] P. Maher, L. Wang, Y. Gao, C. Forsythe, T. Taniguchi, K. Watanabe, D. Abanin, Z. Papic, P. Cadden-Zimansky, J. Hone, P. Kim and C. R. Dean, *Science* **345**, 61 (2014).

[51] D. A. Bandurin, I. Torre, R. Krishna Kumar, M. Ben Shalom, A. Tomadin, A. Principi, G. H. Auton, E. Khestanova, K. S. Novoselov, I. V. Grigorieva, L. A. Ponomarenko, A. K. Geim and M. Polini, *Science* **351**, 1055 (2016).

[52] M. A. Aamir, P. Karnatak and T. P. Sai and A. Ghosh, to be published (2017).

[53] E. V. Castro, K. S. Novoselov, S. V. Morozov, N. M. R. Peres, J. M. B. L. Dos Santos, J. Nilsson, F. Guinea, A. K. Geim and A. H. C. Neto, *Phys. Rev. Lett.* **99**, 8 (2007).

[54] Y. Zhang, T. T. Tang, C. Girit, Z. Hao, M. C. Martin, A. Zettl, M. F. Crommie, Y. R. Shen and F. Wang, *Nature* **459**, 820 (2009).

[55] M. T. Allen, J. Martin and A. Yacoby, *Nat. Commun.* **3**, 934 (2012).

[56] B. N. Szafranek, G. Fiori, D. Schall, D. Neumaier and H. Kurz, *Nano. Lett.* **12**, 1324 (2012).

[57] A. S. M. Goossens, S. C. M. Driessen, T. A. Baart, K. Watanabe, T. Taniguchi and L. M. K. Vandersypen, *Nano. Lett.* **12**, 4656 (2012).

[58] J. B. Oostinga, H. B. Heersche, X. Liu, A. F. Morpurgo and L. M. K. Vandersypen, *Nat. Mater.* **7**, 151 (2008).

[59] T. Taychatanapat and P. Jarillo-Herrero, *Phys. Rev. Lett.* **105**, 166601 (2010).

[60] D. B. Farmer, H. Y. Chiu, Y. M. Lin, K. A. Jenkins, F. Xia and P. Avouris, *Nano Lett.* **9**, 4474 (2009).

[61] I. Meric, C. R. Dean, A. F. Young, N. Baklitskaya, N. J. Tremblay, C. Nuckolls, P. Kim and K. L. Shepard, *Nano Lett.* **11**, 1093 (2011).

[62] S. Kim, J. Nah, I. Jo, D. Shahrjerdi, L. Colombo, Z. Yao, E. Tutuc and S. K. Banerjee, *Appl. Phys. Lett.* **94**, 062107 (2009).

[63] Y. M. Lin, J. Appenzeller, J. Knoch, Z. Chen and P. Avouris, *Nano Lett.* **6**, 930 (2006).

[64] I. Heller, S. Chatoor, J. Mannik, M. A. G. Zevenbergen, J. B. Oostinga, A. F. Morpurgo, C. Dekker and S. G. Lemay, *Nano Lett.* **10**, 1563 (2010).

[65] A. N. Pal, S. Ghatak, V. Kochat, E. S. Sneha, A. Sampathkumar, S. Raghavan and A. Ghosh, *ACS Nano.* **5**, 2075 (2011).

[66] M. Ishigami, J. H. Chen, W. G. Cullen, M. S. Fuhrer and E. D. Williams, *Nano Lett.* **7**, 1643 (2007).

[67] S. V. Morozov, K. S. Novoselov, M. I. Katsnelson, F. Schedin, D. C. Elias, J. A. Jaszczak and A. K. Geim, *Phys. Rev. Lett.* **100**, 016602 (2008).

[68] M. I. Katsnelson and A. K. Geim, Philosophical transactions. *Series A: Mathematical, Physical and Engineering Sciences* **366**, 195–204 (2008).

[69] R. Geick, C. H. Perry and G. Rupprecht, *Phys. Rev.* **146**, 543 (1966).

[70] N. Ohba, K. Miwa, N. Nagasako and A. Fukumoto, *Phys. Rev. B* **63** (2001).

[71] Y. Liu, H. Wu, H. C. Cheng, S. Yang, E. Zhu, Q. He, M. Ding, D. Li, J. Guo, N. O. Weiss, Y. Huang and X. Duan, *Nano Lett.* **15**, 3030 (2015).

[72] B. W. Baugher, H. O. Churchill, Y. Yang and P. Jarillo-Herrero, *Nano Lett.* **13**, 4212 (2013).

[73] A. T. Neal, H. Liu, J. Gu and P. D. Ye, *ACS Nano.* **7**, 7077 (2013).

[74] X. Cui, G. H. Lee, Y. D. Kim, G. Arefe, P. Y. Huang, C. H. Lee, D. A. Chenet, X. Zhang, L. Wang, F. Ye, F. Pizzocchero, B. S. Jessen, K. Watanabe, T. Taniguchi, D. A. Muller, T. Low, P. Kim and J. Hone, *Nat. Nanotechnol.* **10**, 534 (2015).

[75] C. J. Shih, Q. H. Wang, Y. Son, Z. Jin, D. Blankschtein and M. S. Strano, *ACS Nano.* **8**, 5790 (2014).

[76] G. MacFarlane, *Proceedings of the Physical Society. Section B* **63**, 807 (1950).

[77] F. N. Hooge, *Phys. Lett.* **29A**, 139 (1969).

[78] P. Dutta and P. M. Horn, *Rev. Mod. Phys.* **53**, 497 (1981).

[79] M.-Y. Luo, G. Bosman, A. Van Der Ziel and L. L. Hench, *IEEE Trans. Electron. Dev.* **35**, 1351 (1988).

[80] R. Jayaraman and C. G. Sodini, *IEEE Trans. Electron. Dev.* **36**, 1773 (1989).

[81] Y.-M. Lin and P. Avouris, *Nano Lett.* **8**, 2119 (2008).

[82] A. N. Pal and A. Ghosh, *Appl. Phys. Lett.* **95**, 082105 (2009).

[83] A. N. Pal and A. Ghosh, *Phys. Rev. Lett.* **102**, 126805 (2009).

[84] S. A. Imam, S. Sabri and T. Szkopek, *Micro Nano Lett. IET* **5**, 37 (2010).

[85] A. N. Pal, A. A. Bol and A. Ghosh, *Appl. Phys. Lett.* **97**, 133504 (2010).

[86] G. Xu, C. M. Torres, Y. Zhang, F. Liu, E. B. Song, M. Wang, Y. Zhou, C. Zeng and K. L. Wang, *Nano Lett.* **10**, 3312 (2010).

[87] S. Ghatak, S. Mukherjee, M. Jain, D. D. Sarma and A. Ghosh, *APL Mater.* **2**, 092515 (2014).

[88] L. Vandamme, *IEEE Trans. Electron. Dev* **41**, 2176 (1994).

[89] S. Shamim, S. Mahapatra, C. Polley, M. Y. Simmons and A. Ghosh, *Phys. Rev. B* **83**, 233304 (2011).

[90] V. Kochat, C. S. Tiwary, T. Biswas, G. Ramalingam, K. Hsieh, K. Chattopadhyay, S. Raghavan, M. Jain and A. Ghosh, *Nano Lett.* **16**, 562 (2015).

[91] P. Karnatak, T. P. Sai, S. Goswami, S. Ghatak, S. Kaushal and A. Ghosh, *Nat. Commun.* **7** (2016) 13703.

[92] Y. Matsuda, W.-Q. Deng and W. A. Goddard III, *J. Phys. Chem. C* **114**, 17845 (2010).

[93] S. M. Song, T. Y. Kim, O. J. Sul, W. C. Shin and B. J. Cho, *Appl. Phys. Lett.* **104**, 183506 (2014).

[94] P. J. Zomer, S. P. Dash, N. Tombros and B. J. van Wees, *Appl. Phys. Lett.* **99**, 232104 (2011).

[95] M. A. Stolyarov, G. Liu, S. L. Rumyantsev, M. Shur and A. A. Balandin, *Appl. Phys. Lett.* **107**, 023106 (2015).

[96] M. Kayyalha and Y. P. Chen, *Appl. Phys. Lett.* **107**, 113101 (2015).

[97] I.-T. Cho, J. I. Kim, Y. Hong, J. Roh, H. Shin, G. W. Baek, C. Lee, B. H. Hong, S. H. Jin and J.-H. Lee, *Appl. Phys. Lett.* **106**, 023504 (2015).

[98] T. Paul, S. Ghatak and A. Ghosh, *Nanotechnol.* **27**, 125706 (2016).

[99] J. Renteria, R. Samnakay, S. L. Rumyantsev, C. Jiang, P. Goli, M. S. Shur and A. A. Balandin, *Appl. Phys. Lett.* **104**, 153104 (2014).

[100] V. K. Sangwan, H. N. Arnold, D. Jariwala, T. J. Marks, L. J. Lauhon and M. C. Hersam, *Nano Lett.* **13**, 4351 (2013).

[101] X. Xie, D. Sarkar, W. Liu, J. Kang, O. Marinov, M. J. Deen and K. Banerjee, *ACS Nano.* **8**, 5633 (2014).

[102] J. Kang, W. Liu, D. Sarkar, D. Jena and K. Banerjee, *Phys. Rev. X* **4**, 031005 (2014).

[103] W. Liu, J. Kang, D. Sarkar, Y. Khatami, D. Jena and K. Banerjee, *Nano Lett.* **13**, 1983 (2013).

[104] D. Ovchinnikov, A. Allain, Y.-S. Huang, D. Dumcenco and A. Kis, *ACS Nano.* **8**, 8174 (2014).

[105] M. A. Aamir and A. Ghosh, to be published (2017).

[106] B. Huard, N. Stander, J. A. Sulpizio and D. Goldhaber-Gordon, *Phys. Rev. B* **78**, 121402 (2008).

[107] E. J. Lee, K. Balasubramanian, R. T. Weitz, M. Burghard and K. Kern, *Nat. Nanotechnol.* **3**, 486 (2008).

[108] K. Nagashio, T. Nishimura, K. Kita and A. Toriumi, *Appl. Phys. Lett.* **97** (2010).

[109] S. Russo, M. Craciun, M. Yamamoto, A. Morpurgo and S. Tarucha, *Physica E: Low-dimensional Systems and Nanostructures* **42**, 677 (2010).

[110] A. Venugopal, L. Colombo and E. Vogel, *Appl. Phys. Lett.* **96**, 013512 (2010).

[111] F. Xia, V. Perebeinos, Y.-m. Lin, Y. Wu and P. Avouris, *Nat. Nano* **6**, 179 (2011).

[112] S. M. Song, J. K. Park, O. J. Sul and B. J. Cho, *Nano Lett.* **12**, 3887 (2012).

[113] S. M. Song and B. J. Cho, *Carbon Letters* **14**, 162 (2013).

[114] C. Gong, S. McDonnell, X. Qin, A. Azcatl, H. Dong, Y. J. Chabal, K. Cho and R. M. Wallace, *ACS Nano.* **8**, 642 (2014).

[115] M. Ghatge, M. Shrivastava, *IEEE Trans. Electron. Dev* **62**, 4139 (2015).

[116] S. Rumyantsev, G. Liu, W. Stillman, M. Shur and A. A. Balandin, *J. Phys.: Condens. Matter.* **22**, 395302 (2010).

[117] G. Liu, S. Rumyantsev, M. Shur and A. A. Balandin, *Appl. Phys. Lett.* **100**, 033103 (2012).

[118] G. Liu, W. Stillman, S. Rumyantsev, Q. Shao, M. Shur and A. Balandin, *Appl. Phys. Lett.* **95**, 033103 (2009).

[119] M. S. Choi, G. H. Lee, Y. J. Yu, D. Y. Lee, S. H. Lee, P. Kim, J. Hone and W. J. Yoo, *Nat. Commun.* **4**, 1624 (2013).

[120] A. G. Loeser, Z. X. Shen, D. S. Dessau, D. S. Marshall, C. H. Park, P. Fournier and A. Kapitulnik, *Science* **273**, 325 (1996).

[121] H. Yang, J. Heo, S. Park, H. J. Song, D. H. Seo, K. E. Byun, P. Kim, I. Yoo, H. J. Chung and K. Kim, *Science* **336**, 1140 (2012).

[122] D. Sarkar, X. Xie, W. Liu, W. Cao, J. Kang, Y. Gong, S. Kraemer, P. M. Ajayan and K. Banerjee, *Nature* **526**, 91 (2015).

[123] N. J. Couto, B. Sacepe and A. F. Morpurgo, *Phys. Rev. Lett.* **107**, 225501 (2011).

[124] J. Sun, T. Gao, X. Song, Y. Zhao, Y. Lin, H. Wang, D. Ma, Y. Chen, W. Xiang, J. Wang, Y. Zhang and Z. Liu, *J. Amer. Chem. Soc.*, **136** 6574 (2014).

[125] M. C. Huang, G. Jnawali, J. F. Hsu, S. Dhingra, H. Lee, S. Ryu, F. Bi, F. Ghahari, J. Ravichandran, L. Chen, P. Kim, C. B. Eom, B. D'Urso, P. Irvin and J. Levy, *Apl Mater.* **3**, 062502 (2015).

[126] S. Latini, T. Olsen and K. S. Thygesen, *Phys. Rev. B* **92** (2015).

[127] H. Tan, Y. Fan, Y. Rong, B. Porter, C. S. Lau, Y. Zhou, Z. He, S. Wang, H. Bhaskaran and J. H. Warner, *ACS Applied Materials & Interfaces* **8**, 1644 (2016).

[128] B. Y. Zhang, T. Liu, B. Meng, X. Li, G. Liang, X. Hu and Q. J. Wang, *Nat. Commun.* **4**, 1811 (2013).

[129] C. H. Liu, Y. C. Chang, T. B. Norris and Z. Zhong, *Nat. Nanotechnol.* **9**, 273 (2014).

[130] H. Qiao, J. Yuan, Z. Xu, C. Chen, S. Lin, Y. Wang, J. Song, Y. Liu, Q. Khan, H. Y. Hoh, C. X. Pan, S. Li and Q. Bao, *ACS Nano.* **9**, 1886 (2015).

[131] Z. Xu, S. Lin, X. Li, S. Zhang and Z. Wu, arXiv, 1512.06867 (2016).

[132] R. R. Nair, P. Blake, A. N. Grigorenko, K. S. Novoselov, T. J. Booth, T. Stauber, N. M. R. Peres and A. K. Geim, *Science* **320**, 1308 (2008).

[133] X. Zhu, N. R. Monahan, Z. Gong, H. Zhu, K. W. Williams and C. A. Nelson, *J. Amer. Chem. Soc.* **137**, 8313 (2015).

[134] Y. Li, C.-Y. Xu, J.-K. Qin, W. Feng, J.-Y. Wang, S. Zhang, L.-P. Ma, J. Cao, P. A. Hu, W. Ren and L. Zhen, *Adv. Funct. Mater.* **26**, 293 (2016).

[135] C. J. Shih, Q. H. Wang, Y. Son, Z. Jin, D. Blankschtein and M. S. Strano, *ACS Nano.* **8**, 5790 (2014).

[136] R. V. Kashid, P. Munda, P. Deshpande, J. Mishra, K. Roy, M. Jain, A. Ghosh, A. Ghosh, to be published (2017).

[137] T. Mueller, F. Xia and P. Avouris, *Nat. Photon.* **4**, 297 (2010).

[138] T. J. Echtermeyer, L. Britnell, P. K. Jasnos, A. Lombardo, R. V. Gorbachev, A. N. Grigorenko, A. K. Geim, A. C. Ferrari and K. S. Novoselov, *Nat. Commun.* **2**, 458 (2011).

[139] N. M. Gabor, J. C. Song, Q. Ma, N. L. Nair, T. Taychatanapat, K. Watanabe, T. Taniguchi, L. S. Levitov and P. Jarillo-Herrero, *Science* **334**, 648 (2011).

[140] G. Jnawali, Y. Rao, H. Yan and T. F. Heinz, *Nano Lett.* **13**, 524 (2013).

[141] O. Lopez-Sanchez, D. Lembke, M. Kayci, A. Radenovic and A. Kis, *Nat. Nanotechnol.* **8**, 497 (2013).

[142] B. Y. Zhang, T. Liu, B. Meng, X. Li, G. Liang, X. Hu and Q. J. Wang, *Nat. Commun.* **4**, 1811 (2013).

[143] M. Massicotte, P. Schmidt, F. Vialla, K. G. Schadler, A. Reserbat-Plantey, K. Watanabe, T. Taniguchi, K. J. Tielrooij and F. H. Koppens, *Nat. Nanotechnol.* **11**, 42 (2016).

[144] G. Konstantatos, M. Badioli, L. Gaudreau, J. Osmond, M. Bernechea, F. P. G. de Arquer, F. Gatti and F. H. L. Koppens, *Nat. Nanotechnol.* **7**, 363 (2012).

[145] T. J. Echtermeyer, L. Britnell, P. K. Jasnos, A. Lombardo, R. V. Gorbachev, A. N. Grigorenko, A. K. Geim, A. C. Ferrari and K. S. Novoselov, *Nat. Commun.* **2**, 458 (2011).

[146] D. Paria, K. Roy, H. J. Singh, S. Kumar, S. Raghavan, A. Ghosh and A. Ghosh, *Adv. Mater.* 1751 (2015).

[147] O. Lopez-Sanchez, E. Alarcon Llado, V. Koman, A. Fontcuberta i Morral, A. Radenovic and A. Kis, *ACS Nano.* **8**, 3042 (2014).

[148] B. W. H. Baugher, H. O. H. Churchill, Y. Yang and P. Jarillo-Herrero, *Nat. Nanotechnol.* **9**, 262 (2014).

[149] F. Withers, O. Del Pozo-Zamudio, A. Mishchenko, A. P. Rooney, A. Gholinia, K. Watanabe, T. Taniguchi, S. J. Haigh, A. K. Geim, A. I. Tartakovskii and K. S. Novoselov, *Nat. Mater.* **14**, 301 (2015).

[150] C. C. Chen, Z. Li, L. Shi and S. B. Cronin, *Nano Res.* **8**, 666 (2015).

[151] Q. Ma, T. I. Andersen, N. L. Nair, N. M. Gabor, M. Massicotte, Chun H. Lui, A. F. Young, W. Fang, K. Watanabe, T. Taniguchi, J. Kong, N. Gedik, F. H. L. Koppens and P. Jarillo-Herrero, *Nat. Phys.* **12**, 455 (2016).

[152] C. R. Dean, A. F. Young, P. Cadden-Zimansky, L. Wang, H. Ren, K. Watanabe, T. Taniguchi, P. Kim, J. Hone and K. L. Shepard, *Nat. Phys.* **7**, 693 (2011).

[153] X. Du, I. Skachko, F. Duerr, A. Luican and E. Y. Andrei, *Nature*, **462**, 192 (2009).

[154] K. I. Bolotin, F. Ghahari, M. D. Shulman, H. L. Stormer and P. Kim, *Nature*, **462** 196 (2009).

[155] R. T. Weitz, M. T. Allen, B. E. Feldman, J. Martin and A. Yacoby, *Science*, **330** (2010) 812–816.

[156] A. F. Young, C. R. Dean, L. Wang, H. Ren, P. Cadden-Zimansky, K. Watanabe, T. Taniguchi, J. Hone, K. L. Shepard and P. Kim, *Nat. Phys.* **8**, 550 (2012).

[157] Y. Wang, V. W. Brar, A. V. Shytov, Q. Wu, W. Regan, H.-Z. Tsai, A. Zettl, L. S. Levitov and M. F. Crommie, *Nat. Phys.* **8**, 653 (2012).

[158] D. C. Elias, R. V. Gorbachev, A. S. Mayorov, S. V. Morozov, A. A. Zhukov, P. Blake, L. A. Ponomarenko, I. V. Grigorieva, K. S. Novoselov, F. Guinea and A. K. Geim, *Nat. Phys.* **7**, 701 (2011).

[159] P. W. Anderson, *Phys. Rev. Lett.* **18**, 1049 (1967).

[160] M. Hentschel and F. Guinea, *Phys. Rev. B.* **76**, 115407 (2007).

[161] B. Uchoa, V. N. Kotov, N. M. R. Peres and A. H. Castro Neto, *Phys. Rev. Lett.* **101**, 026805 (2008).

[162] P. S. Cornaglia, G. Usaj and C. A. Balseiro, *Phys. Rev. Lett.* **102**, 046801 (2009).

[163] K. Sengupta and G. Baskaran, *Phys. Rev. B* **77**, 045417 (2008).

[164] J.-H. Chen, L. Li, W. G. Cullen, E. D. Williams and M. S. Fuhrer, *Nat. Phys.* **7** 535 (2011).

[165] L. A. Ponomarenko, R. V. Gorbachev, G. L. Yu, D. C. Elias, R. Jalil, A. A. Patel, A. Mishchenko, A. S. Mayorov, C. R. Woods, J. R. Wallbank, M. Mucha-Kruczynski, B. A. Piot, M. Potemski, I. V. Grigorieva, K. S. Novoselov, F. Guinea, V. I. Fal'ko and A. K. Geim, *Nature* **497**, 594 (2013).

[166] D. R. Hofstadter, *Phys. Rev. B* **14**, 2239 (1976).

[167] M. Ben Shalom, M. J. Zhu, V. I. Fal'ko, A. Mishchenko, A. V. Kretinin, K. S. Novoselov, C. R. Woods, K. Watanabe, T. Taniguchi, A. K. Geim and J. R. Prance, *Nat. Phys.* **12**, 318 (2015).

[168] M. T. Allen, O. Shtanko, I. C. Fulga, A. R. Akhmerov, K. Watanabe, T. Taniguchi, P. Jarillo-Herrero, L. S. Levitov and A. Yacoby, *Nat. Phys.* **12**, 128 (2015).

[169] M. Tinkham, *Introduction to Superconductivity*, McGraw-Hill, New York (1996).

[170] F. Amet, C. T. Ke, I. V. Borzenets, J. Wang, K. Watanabe, T. Taniguchi, R. S. Deacon, M. Yamamoto, Y. Bomze, S. Tarucha and G. Finkelstein, *Science* **352**, 966 (2016).

[171] N. Yabuki, R. Moriya, M. Arai, Y. Sata, S. Morikawa, S. Masubuchi and T. Machida, *Nat. Commun.* **7**, 10616 (2016).

[172] M. J. M. de Jong and L. W. Molenkamp, *Phys. Rev. B* **51**, 13389 (1995).

[173] A. Tomadin, G. Vignale and M. Polini, *Phys. Rev. Lett.* **113**, 235901 (2014).

[174] R. Gurzhi, *Sov. Phys. Usp.* **11**, 255 (1968).

[175] J. Crossno, J. K. Shi, K. Wang, X. Liu, A. Harzheim, A. Lucas, S. Sachdev, P. Kim, T. Taniguchi, K. Watanabe and T. A. Ohki, K. C. Fong, *Science*, **351**, 1058 (2016).

[176] N. W. Ashcroft and N. D. Mermin, *Solid State Physics*, Harcourt College Publishers, Fort Worth (1976).

[177] A. Avsar, J. Y. Tan, T. Taychatanapat, J. Balakrishnan, G. K. Koon, Y. Yeo, J. Lahiri, A. Carvalho, A. S. Rodin, E. C. O'Farrell, G. Eda, A. H. Castro Neto and B. Ozyilmaz, *Nat. Commun.* **5**, 4875 (2014).

Chapter 6

Thermoelectric Energy Conversion in Layered Metal Chalcogenides

Satya N. Guin, Ananya Banik and Kanishka Biswas*

*New Chemistry Unit, Jawaharlal Nehru Centre for
Advanced Scientific Research (JNCASR),
Jakkur, Bangalore 560064, India*
**kanishka@jncasr.ac.in*

Abstract. Nearly, $\sim$65% utilised energy being lost as waste heat. Thermoelectric materials can directly and reversibly convert heat to electrical energy, thereby thermoelectrics has an important role to play in the future energy management. Recently, layered metal chalcogenides has gained a momentous attention in the field of thermoelectrics as some of them exhibit record high values of thermoelectric figure of merit due to their intrinsically low thermal conductivity. For example, layered Bi_2Te_3 and its alloys are known for their high thermoelectric performance near room temperature. Mention must be made that remarkably high thermoelectric figure of merit of $\sim$2.6 has been recently achieved in layered SnSe due to its ultra-low thermal conductivity. Herein, we discuss about the synthesis, structure and thermoelectric properties of important layered chalcogenides.

6.1 Introduction

Synthesis of new inorganic materials, characterisations and studies of their novel physical and chemical properties are the important aspects for both fundamental solid state chemistry and next generation technological developments. Thus, the prediction and design of new material is one of the important concerns [1–7]. In this

239

context, the dimensionality of material plays an important role to govern the chemical and physical properties. In an alternative way, not only the size of the material but also dimensionality (whether the structural connectivity is in zero-dimensional (0D) or one-dimensional (1D) or two-dimensional (2D) or three-dimensional (3D)) is the defining parameters for determining material properties [3–7]. The dimensionality of a material defines the crystal structure, thus can also modulate the electronic structure [3, 5]. In the last two decades, an extensive research is going on in the area of 0D quantum dots, 1D nanotubes, and nanowires, 2D single-atom thick materials, and 3D crystal solids [5]. However, special mention is required for 2D material as the research and development in this area is enormously growing. The discovery of graphene and its potential application ranging from electronics to catalysis encouraged the scientists across the globe in search of new 2D inorganic layered materials and the investigations of their exotic properties [1–9]. In a 2D material, a dimension of the material is restricted in one direction, thus the structure of the material is layered. The strong covalent bonds in the layer provide in-plane stability of 2D structure while the van der Waals interaction between the layers holds the layers together [3, 5]. 2D layered materials have shown promise in a variety of applications such as for next generation electronics, gas storage or separation, catalysis, high-performance sensors, support membranes, inert coatings and many others [1–7]. However, in the present chapter, we will be only focusing on the thermoelectric properties of layered chalcogenides. In the following section, we will first provide a brief discussion on the importance of thermoelectrics for the future energy management, followed by various discovered strategies for improving thermoelectric performance and then a discussion on the importance and uniqueness of layered material for the thermoelectric energy conversion.

Development of sustainable energy supply is one of the major challenges of current century [10–15]. Thus, research on the renewable energy resources and their utilisation for mankind is one of the important challenges in today's world. Thermoelectrics, the conversion of waste heat to electricity, has an important role to play for the future energy management [10–15]. The conversion efficiency of a thermoelectric material is related to Carnot Efficiency and dimensionless thermoelectric figure of merit, ZT as follows,

$$\eta = \frac{\Delta T}{T_H} \frac{\sqrt{1 + ZT} - 1}{\sqrt{1 + ZT} + \frac{T_C}{T_H}},$$

where, T_H and T_C are the respective temperatures of the hot and cold junctions and ΔT the difference between them [9–12]. ZT of a material can be expressed as $ZT = \sigma S^2 T / \kappa_{\text{total}}$, where σ is the electrical conductivity, S the Seebeck coefficient, T the absolute temperature and κ_{total} the total thermal conductivity of the material under consideration [10–15]. The quantity σS^2 is called power factor. κ_{total} of a material is depend on electronic (κ_{el}) and lattice (κ_{lat}) thermal conductivity, $\kappa_{\text{total}} = \kappa_{\text{el}} + \kappa_{\text{lat}}$. To maximise the ZT value, it is essential to have high σ, high S and low κ_{total} [10–15], but these three properties are interconnected and depends on the carrier concentration of the materials. During the last decade, different new strategies have been introduced to decouple these interrelated physical parameters. Two parallel approaches have been taken to achieve this target. First one is to optimise the thermoelectric properties of bulk thermoelectric materials and the second one is for low-dimensional solids [10–16]. Till date, some of the well-known approaches to tune the electronic transport in bulk material are by the modification of their electronic structure via the introduction of resonance levels and electronic band valley convergence. Thermal conductivity (κ_{lat}) can be reduced through solid solution alloying, second-phase nanostructuring, and the all scale phonon scattering approach [11, 14, 17].

The investigations on the thermoelectric properties of the layered materials have been mainly focused on two different class of solids: (a) bulk-layered materials and (b) ultrathin 2D nanostructures. The anisotropic nature of crystal structure and chemical bonding results in significant variation of the thermoelectric properties in different crystallographic directions for layered solids. Synthesis of single/few-layer nanosheets material by exfoliation of layered bulk materials or by bottom up wet chemical method and investigation of their thermoelectric properties is also important for fundamental materials science and technology design.

Primarily, the work of Hicks and Dresselhaus in 1993, inspired the researchers to study the thermoelectric properties in nanomaterials, which shows reduction of dimensionality from 3D to 2D structure (quantum wells) to 1D structure (quantum wires) and finally to 0D structure (quantum dots), can create new opportunities to vary S, σ and κ quasi-independently [16, 17, 19]. The dimensionality reduction offers the accessibility of new variable of length scale to control the materials properties. When the size of the material reduced to nanodimension, comparable to the spatial extent of the electronic wave function confining the electrons in one or more dimension, the quantum confinement arises. According to Mott–Jones relation, this can results in increase of the slope of the density of states (DOS) at the Fermi level [11, 16, 17, 19].

$$S = \frac{\pi^2}{3} \frac{k^2 T}{e} \left(\frac{\partial \log \sigma(E)}{\partial E} \right)_{E=E_F},$$

where S is Seebeck coefficient, $\sigma(E)$ is the electronic conductivity determined as a function of the band filling or Fermi energy E_F, k is Boltzmann constant, e is an electronic charge. Low-dimensional quantum-confined systems generally show sharp and dispersionless bands, hence by tuning the size of solids it is possible to place these states near the Fermi level, which could enable a marked increase in Seebeck coefficient. The presence of numerous interfaces and surfaces in nanostructured material should enhance phonon scattering, thereby can reduce thermal conductivity. The advantages of using nanostructured materials for thermoelectrics can be summarised in following points (a) enhancement in the density of states near E_F, which leads to enhancement of the Seebeck coefficient, (b) benefit of the anisotropic Fermi surfaces in the case of multielectronic band valley material, (c) increase the boundary scattering of phonons at the interfaces, without much carrier scattering at the interface, and (d) increased carrier mobility at a given carrier concentration when quantum confinement conditions are satisfied, so that modulation doping can be utilised [16, 17, 19]. In general, all these advantages collectively or distinctly favour to optimise the S, σ and κ independently in a single material in nanodimension [16, 17, 19]. The thermoelectric properties in nanoscale have been investigated theoretically as well as experimentally for various thin film superlattices, nanowires and nanotubes, quantum dots, and nanocomposites. Though, in the present chapter, we will be only discussing present development, understanding and future direction of thermoelectrics based on layered materials.

6.2 Bismuth Telluride

Bismuth telluride (Bi_2Te_3), a narrow-gap semiconductor with an indirect band gap of $\sim$0.15 eV, is well-known for its distinct layered crystal structure and excellent thermoelectric properties [12, 20–22]. H. Julian Goldsmid in 1954, first inspected Bi_2Te_3 as an effective material for thermoelectric refrigeration [23]. Bi_2Te_3 crystallises in a rhombohedral layered structure (space group R-*3m*) with five atoms in one unit cell. The lattice parameters of Bi_2Te_3 are $a = 0.4384$ nm and $c = 3.045$ nm [24]. The structure is formed by quintuple layers of Te(1)–Bi–Te(2)–Bi–Te(1) stacked by van der Waals interactions along the crystallographic c-axis (Figure 1).

The superscript (1) and (2) denote two different chemical states of Te. The interaction between adjacent quintuple layers is due to the weak van der Waals forces. Within the quintuple layers, the stronger bonds are of the covalent or partially ionic in nature.

Since the discovery of its extraordinary thermoelectric properties, Bi_2Te_3 has become a vital component for thermoelectric industry [24–26]. Bi_2Te_3-based materials are known to have the highest thermoelectric figure of merit, ZT–1.14 at room temperature [24]. It was obvious from the ZT expression that a significant improvement in ZT can be achieved through simultaneous enhancement of the power factor, $S^2\sigma$ and decrease in the thermal conductivity. Various approaches have been tried to improve the thermoelectric properties of Bi_2Te_3 and its alloys. These approaches comprised the composition change from its stoichiometry, the use of polycrystalline materials with different grain sizes, introduction

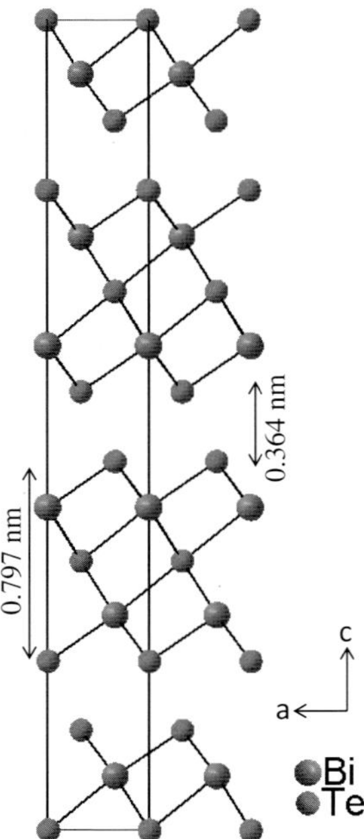

Figure 1. Crystal structure of Bi_2Te_3 viewed along the crystallographic *b* axis. Each QL constitutes five covalently bonded atomic planes.

of structural defects, and incorporation of different dopants, e.g., Sb or Se, into Bi_2Te_3 lattice. Alloying with Sb_2Te_3 and Bi_2Se_3 allowed the fine tuning of the carrier concentration along with a reduction in lattice thermal conductivity of Bi_2Te_3 [10]. The most commonly studied *p*-type composition is near $Bi_{0.5}Sb_{1.5}Te_3$, whereas *n*-type composition is close to $Bi_2(Te_{0.8}Se_{0.2})_3$ [10]. The Bi–Sb atomic disorder in $Bi_{0.5}Sb_{1.5}Te_3$ scatters the heat-carrying phonons, thereby decreasing κ_{lat} leading to high ZT values. Bi_2Te_3 nanostructures offer an exceptional model system for studying the preparation of nanostructured thermoelectric materials because bulk Bi_2Te_3-based materials are known to exhibit very high ZT values over the temperature range of 200–400 K. A remarkable increase in ZT–1.4 has been accounted in Bi_2Te_3-based nanostructured bulk materials prepared by the hot-pressing of ball-milled nanoparticles (Figure 2) [27]. The ZT enhancement for this system occurs from reduced the lattice thermal conductivity while maintaining analogues

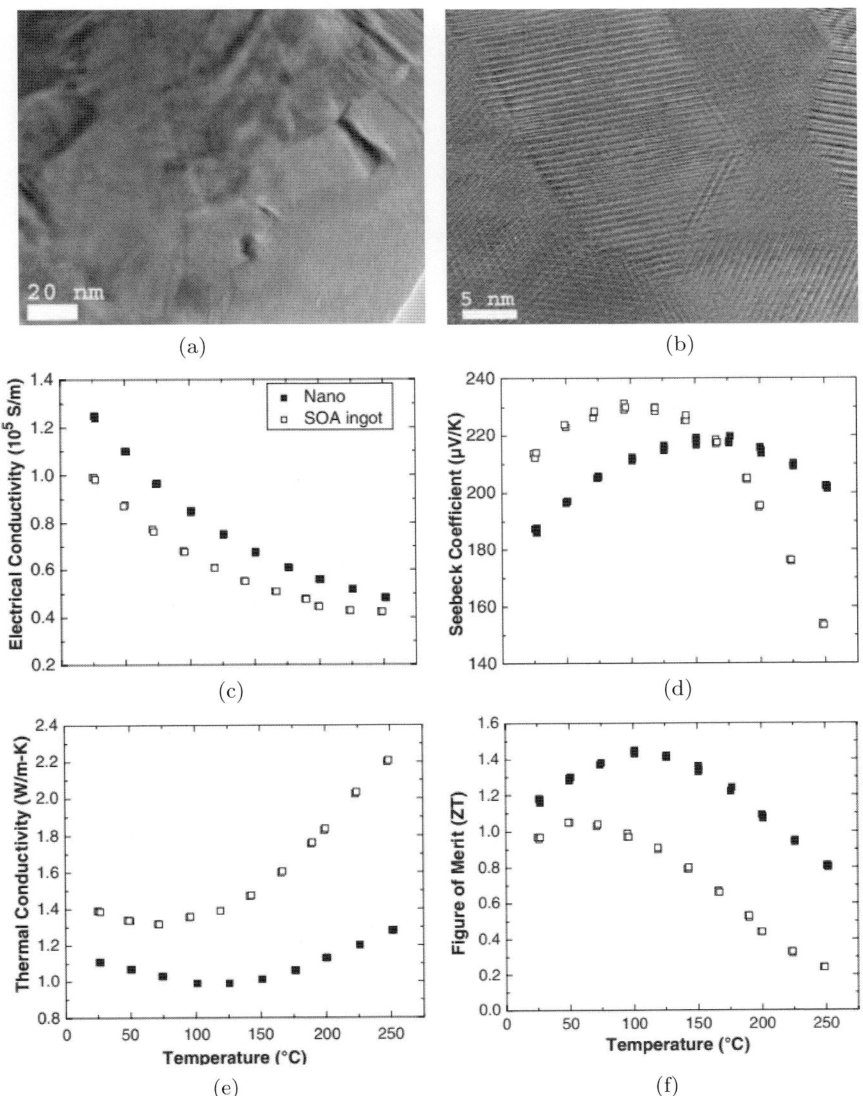

Figure 2. Thermoelectric properties of a hot-pressed nanocrystalline (NC) bulk bismuth antimony telluride sample. (a) Low-magnification TEM image showing the nanograins. (b) High-magnification TEM image showing the nanosize, high crystallinity, random orientation and clean grain boundaries. Temperature dependence of (c) σ, (d) S, (e) κ, and (f) ZT of a hot-pressed NC bulk sample (black squares) as compared with that of a state-of-the-art ingot (white squares). (Adapted with permission from Ref. [27]. Copyright 2012 AAAS.)

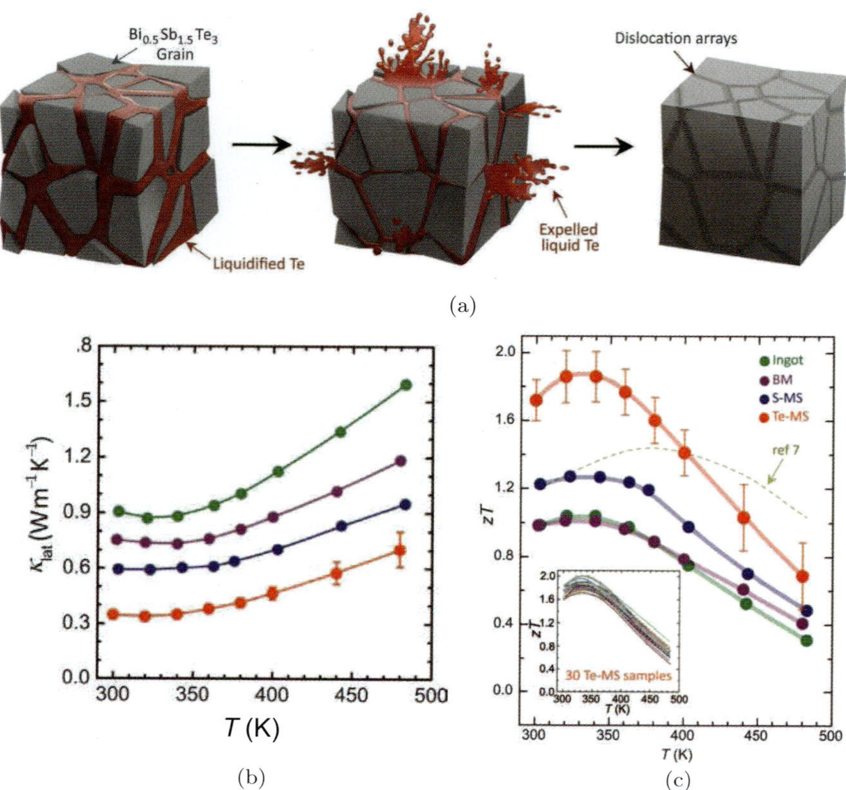

Figure 3. Generation of dislocation arrays at grain boundaries in Bi$_{0.5}$Sb$_{1.5}$Te$_3$. (a) Schematic illustration showing the generation of dislocation arrays during the liquid-phase compaction process. The Te liquid (red) between the Bi$_{0.5}$Sb$_{1.5}$Te$_3$ grains flows out during the compacting process and facilitates the formation of dislocation arrays embedded in low-energy grain boundaries. (b) thermal conductivity (κ_{lat}) for all samples. (c) ZT as a function of temperature for Bi$_{0.5}$Sb$_{1.5}$Te$_3$ alloys. (Adapted with permission from Ref. [28]. Copyright 2015 AAAS.)

power factor compared to that of the bulk p-type Bi$_{0.5}$Sb$_{1.5}$Te$_3$. Formation of dense dislocation arrays at low-energy grain boundaries by liquid-phase compaction in Bi$_{0.5}$Sb$_{1.5}$Te$_3$, effectively scatter mid-frequency phonons, resulting in a substantially lower lattice thermal conductivity (Figure 3) [28]. Full-spectrum phonon scattering with minimal charge carrier scattering significantly improved the ZT to 1.86 K at 320 K [28]. Furthermore, a thermoelectric cooler validated the performance with a maximum temperature difference of 81 K, which is much higher than current commercial Peltier cooling devices. Z. F. Ren's group has achieved ZT values of 1.04 for n-type $Bi_2Te_{3-y}Se_y$ nanostructured bulk alloy by reorienting the ab planes of the small crystals through repressing the as-pressed sample with composition $Bi_2Te_{2.7}Se_{0.3}$ [29].

The lattice spacing between layers has a direct relation with the atomic bond strength connecting the neighbouring layers [24]. Hence, the van der Waals gap between Te(1)–Te(1) corresponds to the largest d spacing of $\sim$0.37 nm. In addition to this, the strength and length of the Bi–Te(2) bond is not much different from the van der Waals gaps between Te (1)–Te(1). This weak binding between the Te layers is attributed to the cleavage along the plane perpendicular to the c-axis and the anisotropic thermal and electrical transport properties of Bi_2Te_3. According to theoretical study, ZT can be increased by a factor of $\sim$13 in a Bi_2Te_3 quantum well over the bulk value [15, 30]. Superlattice systems with low dimensionality have been proposed to achieve a record high ZT value as a result of the confinement effects on the electronic density of states. A ZT of 2.4 at room temperature was reported for a p-type Bi_2Te_3/Sb_2Te_3 superlattice grown by chemical-vapour deposition, a value which broke through the long-standing $ZT = 1$ record for Bi_2Te_3 alloys [31]. Because of the crystal structure, most chemically synthesised Bi_2Te_3 have 2D plate like shape [29]. Nanostructured Bi_2Te_3 has been prepared through top-down exfoliation technique. In order to isolate bismuth telluride quintuples and break them into atomic planes, researchers have employed a method, similar to the one used for exfoliation of single-layer graphene [24]. Through a mechanical cleavage, researchers have separated nanosheets from crystalline Bi_2Te_3. The process was repeated several times to obtain layers with just a few atomic planes. Owing to the specific structure of Bi_2Te_3 crystal along the c-direction, researchers have successfully verified the number of layers using atomic force microscopy (AFM) and micro-Raman spectroscopy. Top-down approaches for nanostructuring, such as ball-milling [27] and melt-spinning [32] can yield bulk quantities of high-ZT p-type alloys. These processes are energy consuming, and hence restrict them for the production of scalable quantity high-ZT Bi_2Te_3.

Devising low-energy-intensity scalable methods to obtain high-ZT materials of both n- and p-type is crucial for fabricating commercially viable thermoelectric devices and modules for transforming the fields of solid-state refrigeration and harvesting electricity from heat. Recently, bottom-up colloidal synthetic processes have been used for synthesising large-quantities of high-quality 2D materials. These chemical methods can produce uniform-sized materials with controlled sizes which provide the possibility of achieving further improvement in thermoelectric efficiency because in nanostructured materials both thermopower and thermal conductivity are dependent on the size and shape of the nanostructures [33]. Hence, chemically synthesised nanostructured Bi_2Te_3 has been investigated for thermoelectric applications. Ramanath's group has reported the realisation of both p- and n-type nanostructured bulk materials with ZT as high as 1.1, obtained by bottom-up assembly of rapidly synthesised single crystal nanoplates of sulphur-doped pnictogen chalcogenides (V_2VI_3) [33]. These synthetic strategies have exploited inexpensive organic solvents and metal salts as precursors ($BiCl_3$ and $SbCl_3$) for rapid preparation of large amount sulphur-doped pnictogen chalcogenide (V_2VI_3) nanocrystals to create single- and multicomponent nanostructured bulk thermoelectric materials (Figure 4). Microwave stimulation triggers the reaction between molecularly ligated chalcogen and pnictogen complexes with thioglycolic acid (TGA) in the presence of a high-boiling solvent. Single component pellets prepared from pnictogen chalcogenide nanoplates showed two-fold lower lattice thermal conductivity and high power factors, comparable to state-of-the-art bulk alloys. This remarkable charge-carrier-crystal and phonon-glass behaviour without alloying is made possible by combining nanostructuring and sub-atomic-percent sulphur doping. Nanostructured bulk Bi_2Te_3 obtained from single-component assemblies or nanoplate mixtures of different materials exhibit 250% higher ZT than their non-nanostructured bulk counterparts and state-of-the-art alloys (Figure 4).

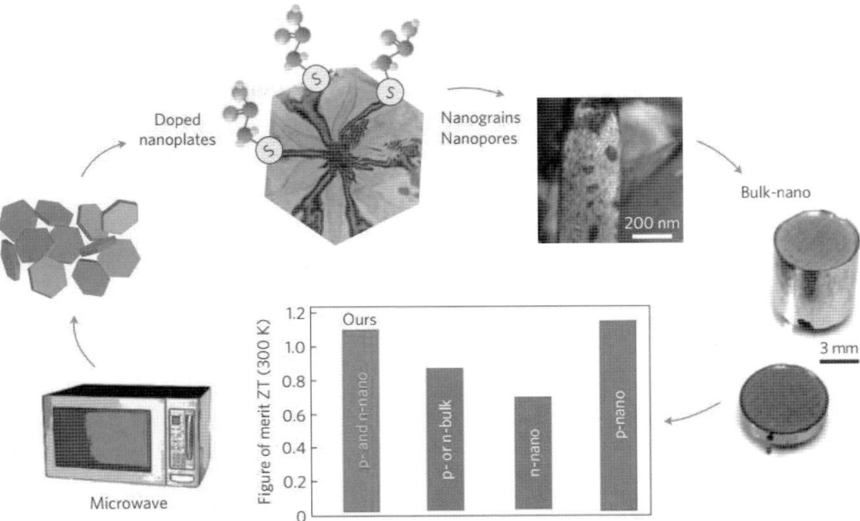

Figure 4. Schematic representation of the scalable synthesis used to obtain both n- and p-type bulk thermoelectric nanomaterials $Bi_2Te_{3-x}S_x$ with high ZT. Microwave synthesis of sulphur-doped nanoplates followed by cold-compaction and sintering yields up to $92 \pm 3\%$ dense bulk pellets with nanostructured grains. Sulphur doping controls the electrical conductivity, Seebeck coefficient and majority carrier type, while nanostructuring resulted in very low thermal conductivity. The combined effect is applicable to multiple pnictogen chalcogenide systems and their combinations. The graph compares the best ZT of p- and n-type nanomaterials with those of the best p- and n-bulk materials, denoted as p- or n-bulk nanoparticle-dispersed n-bulk, referred to as n-nano, and a p-type ball-milled alloy denoted as p-nano. (Adapted with permission from Ref. [33]. Copyright 2012 Nature Publishing Group.)

In another recent work, a facile solvothermal route was developed to prepare Bi_2Te_3 nanoplates and Bi_2Te_3 nanoflowers via a chemical self-assembly route using these nanoplates as building blocks [34a]. Furthermore, to improve thermoelectric performance, high-pressure route was used to sinter powder to form a pellet. The resultant product was mostly hexagonal plates with a uniform size of 100–200 nm diameter and thickness of $\sim$18 nm. The pellet prepared from flower-like Bi_2Te_3 has larger Seebeck coefficient and simultaneous smaller resistivity as compared to plate-like Bi_2Te_3. The room temperature resistivity of both the samples is almost double of that of n-type bulk Bi_2Te_3 ($1.4\,m\Omega\,cm$). Indeed such discrepancies in resistivity are often typical of nanocrystalline materials that possess different degrees of disorder and formation of grain boundary barriers [34a]. The large surface-to-volume ratio of nanograins effects in high trap-state densities at each grain boundary caused by defects and dangling bonds. These trap-states immobilise charge carriers and result in a reduced mobility. Then, the grain boundaries become electrically charged and result in a grain boundary potential barrier to electronic transport. Thus, $d\sigma/dT$ can become positive at certain temperatures for grain boundary potential barriers. This is an important difference from bulk Bi_2Te_3 materials, where the negative $d\sigma/dT$ ratio forces a maximum in ZT at 50°C, preventing its application at significantly higher temperatures. This behaviour of nanostructured materials has been recently predicted by Martin *et al.* [34b], where transport properties are dominated by grain boundary potential barrier scattering in combination with phonon scattering. The nanostructured Bi_2Te_3 shows *n*-type behaviour with a maximum S value of $\sim -150\,\mu V/K$. The observed power factor value of $8.6\,\mu Wcm^{-1}K^{-2}$ is much higher as compared to previously reported values for chemically synthesised Bi_2Te_3 nanoparticles [34c]. The room temperature thermal conductivity is $\sim 0.5\,Wm^{-1}\,K^{-1}$. As a result, a maximum ZT of 0.7 has been achieved at temperature of 180°C [34a].

Spark plasma sintering (SPS) is known to be a very useful technique for preparing nanostructured bulk materials due to its rapid heating and cooling rates, which allow fast sintering, preventing unnecessary grain growth arising from a long sintering process at high temperatures. Moreover, the grain growth and densification can be controlled by varying the SPS conditions, allowing the study of the effect of grain size and porosity on the thermoelectric properties of nanostructured bulk materials. Recently, Son *et al.* [35] reported a colloidal synthesis of ultrathin Bi_2Te_3 nanoplates with the thickness of 1–3 nm (Figure 5) and subsequent SPS processing in order to fabricate *n*-type nanostructured bulk thermoelectric materials.

Bi_2Te_3 nanoplates were synthesised by the reaction between bismuth dodecanethiolate and tri-*n*-octylphosphine telluride in the presence of oleylamine [35]. It was observed that grain size and

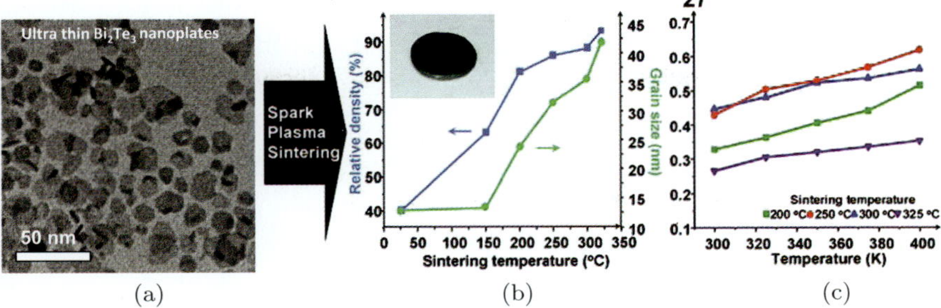

Figure 5. (a) TEM image shows ultrathin Bi_2Te_3 nanoplates. The density of Bi_2Te_3 nanostructured bulk materials was found to be strongly dependent on the sintering temperature. (b) Graph of relative densities (blue square) and grain sizes (red circle) has been calculated by the XRD patterns using the Scherrer equation vs. sintering temperature. (c) Temperature dependence of ZT of Bi_2Te_3 nanostructured bulk materials sintered at 200°C (green square), 250°C (red circle), 300°C (blue upward triangle), and 325°C (purple downward triangle). (Adapted with permission from Ref. [35]. Copyright 2012 American Chemical Society.)

density of Bi_2Te_3 nanostructured bulk materials were strongly dependent on the SPS temperature (Figure 5). Accordingly, the thermoelectric properties of Bi_2Te_3, nanostructured bulk materials sintered at various temperatures were characterised to get the most favourable sintering condition for highly efficient nanostructured thermoelectric materials. The highest ZT of 0.62 was achieved in sample sintered at 250°C, which is one of the highest values among those of reported n-type nanostructured thermoelectric materials prepared from chemically synthesised nanoparticles (Figure 5) [35].

6.3 Bismuth Selenide

Bismuth selenide (Bi_2Se_3) is a V–VI semiconductor with a narrow band gap of $\sim$0.3 eV, which has potential application in optical recording system, photoelectrochemical devices, and thermoelectrics [36–39]. Like Bi_2Te_3, it is also a 3D topological insulator with metallic surface states protected by the time reversal symmetry [40]. Bi_2Se_3 is an anisotropic layered material (R-3m space group and crystal structure is same as Bi_2Te_3) containing quintuple layers (QL) each of $\sim$1 nm thick and composed of five covalently bonded atomic planes [Se_2–Bi–Se_1–Bi–Se_2]. Atomically thin layers of Bi_2Se_3 are an important candidate for thermoelectrics. In addition, the attraction of Bi_2Se_3 also comes from its unique 2D electron gas like nature that covers the whole surface, thus contributing to the metallic surface states. This suppresses the electron backscattering, ensuring a high μ of $\sim$6000 $cm^2\,V^{-1}\,s^{-1}$ and a Fermi velocity of $\sim 5 \times 10^5$ m/s [40, 41]. The 2D electron gas and the effective phonon scattering at the grain boundaries and interfaces enable the Bi_2Se_3 single-layer-based composite to achieve improved σ and S as well as the reduction in κ, thereby leading to an enhancement of ZT as compared to bulk material (Figure 6).

Sun *et al.* [40] first reported the synthesis of isolated five-atom-thick Bi_2Se_3 single layers via a scalable intercalation/exfoliation strategy, which takes advantage of an intermediate precursor (Li-intercalated Bi_2Se_3 microplates). The Bi_2Se_3 single-layer-based (SLB) composite exhibited a significant enhancement in electric transport properties and a much lower thermal conductivity than that of the bulk material over the entire temperature range. Hence, the σ/κ ratio for the SLB composite is higher than that of the bulk material over the whole temperature range. $|S|$ for the SLB composite gradually increases from 90 μV/K at 300 K to 121 μV/K at 400 K, which is a factor of 1.2 greater than that of bulk material (98.5 μV/K). The power factor (σS^2) of the SLB composite is

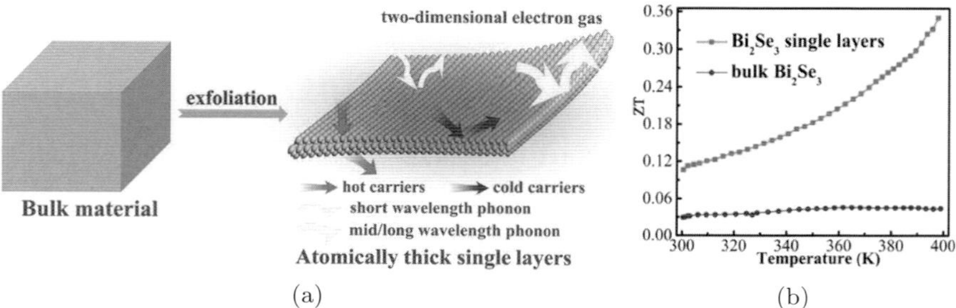

Figure 6. (a) Schematic representation of exfoliation technique and representation of transport properties of Bi_2Se_3 single layer. (b) Temperature dependent ZT of bulk and single layered Bi_2Se_3. (Adapted with permission from Ref. [40]. Copyright 2012 American Chemical Society.)

greater than that of the bulk material, which could be a benefit of the simultaneously increased σ and S values. Hence, ZT for the SLB composite is higher than that of the bulk material over the whole temperature range. In particular, the SLB composite exhibits $ZT = 0.35$ at $400\,\mathrm{K}$, which is 8 times larger than ZT of the bulk material and higher than the previously reported values for pure Bi_2Se_3 nanostructures, suggesting the superiority of the atomically thick SLB structure (Figure 6).

Jana *et al.* [41] have studied the thermoelectric properties of few-layer nanosheets of Bi_2Se_3, synthesised via a green ionothermal method in the water-soluble, room-temperature ionic liquid, 1-ethyl-3-methylimidazolium tetrafluoroborate ([EMIM][BF$_4$]). Ionothermal reaction of bismuth acetate and selenourea in [EMIM][BF$_4$] yields ultrathin (3–5 layers) Bi_2Se_3 nanosheets. The high electrical conductivity is attributed to the presence of surface states that offer high mobility and scattering-resistant carriers. In addition to this surface defects, nanoscale grain boundaries and interfaces effectively scatter the heat-carrying phonons, thereby decreasing κ_{total} to as low as $0.4\,\mathrm{Wm^{-1}\,K^{-1}}$ near room temperature. As a result, maximum ZT of 0.1 has been observed near room temperature.

6.4 Antimony Telluride

Antimony telluride (Sb_2Te_3) is a narrow band gap semiconductor with a band gap of $0.28\,\mathrm{eV}$. Sb_2Te_3 have a layered tetradymite structure similar to Bi_2Te_3 [42]. Because of high electrical conductivity ((3.11–5.3) $\times 10^5\,\mathrm{Sm^{-1}}$) and low thermal conductivity (1.6–$5.6\,\mathrm{Wm^{-1}\,K^{-1}}$) of Sb_2Te_3 bulk crystals at $300\,\mathrm{K}$, Sb_2Te_3 and its doped derivatives are considered as one of the best p-type thermoelectric materials for near room-temperature applications [43]. However, the S of the pristine Sb_2Te_3 single crystal is too low (83–$92\,\mu\mathrm{VK^{-1}}$) at $300\,\mathrm{K}$ [43]. S can be enhanced greatly by introducing nanostructures; for instance, $125\,\mu\mathrm{VK^{-1}}$ was attained for the thin film composed of Sb_2Te_3 nanoplates at room temperature. S of 147–$210\,\mu\mathrm{VK^{-1}}$ was achieved at 300–$420\,\mathrm{K}$ for the Sb_2Te_3 bulk samples composed of nanosheets prepared by spark plasma sintering [44]. Unfortunately, the measured σ values of both samples were in the order of $\sim\!10^2\,\mathrm{Sm^{-1}}$, which were far lower than that of the Sb_2Te_3 single crystal.

Hence, researchers are in search for the development of a simple method to synthesise Sb_2Te_3 nanostructures with well-defined morphologies and improved thermoelectric efficiency. Sb_2Te_3 single-crystalline nanosheets having edge lengths of 300–500 nm and thicknesses of 50–70 nm were rapidly synthesised by a microwave-assisted reaction of $SbCl_3$, Na_2TeO_3 and hydrazine hydrate in ethylene glycol at 200°C for 15 min [45]. For thermoelectric measurements, nanostructured bulk Sb_2Te_3 was prepared from Sb_2Te_3 nanosheets by spark plasma sintering method. Morphology of samples has been retained after SPS processing. High electrical conductivity $(2.49 \times 10^4\,\mathrm{Sm^{-1}})$, high Seebeck coefficient $(210\,\mu\mathrm{VK^{-1}})$ and low thermal conductivity $(0.76\,\mathrm{Wm^{-1}\,K^{-1}})$ at 420 K were achieved. A relatively high ZT of 0.58 at 420 K was obtained for the sintered Sb_2Te_3 sample prepared from nanosheets.

Dong *et al.* [46] have reported a rapid microwave-assisted solvothermal method for the preparation of Sb_2Te_3 nanosheets and the thermoelectric properties of nano-bulk Sb_2Te_3 prepared from this nanosheets. $SbCl_3$, TeO_2, PEG and $N_2H_4 \cdot 5H_2O$ were used to synthesise the nanostructured Sb_2Te_3 in presence of EG as a solvent. The thermoelectric properties of the Sb_2Te_3 tablets obtained by room-temperature pressing of powders were tested at the temperatures ranging from 300 K to 450 K. All Sb_2Te_3 nanostructured samples show the characteristics of p-type semiconductors. Sb_2Te_3 nanosheets exhibit Seebeck coefficient of $\sim\!194$–245 $\mu\mathrm{VK^{-1}}$ and power factor of 0.48–$1.14 \times 10^{-4}\,\mathrm{Wm^{-1}\,K^{-2}}$ in the temperature range of 300–450 K.

6.5 Solid Solutions and Nanocomposites of Bi_2Te_3, Bi_2Se_3 and Sb_2Te_3

Solid solutions and nanocomposites of Bi_2Te_3, Bi_2Se_3 and Sb_2Te_3 have been preferably considered for their high-performance thermoelectric properties [28, 47]. While the six electronic band valley degeneracy and the narrow energy gap of Bi_2Te_3 (and compounds in the same family) are attractive for good electrical conduction, its layered crystal structure leads to a poor thermal conductivity, making these layered materials most appropriate for thermoelectric applications [42]. Due to the isomorphic crystal structure of Bi_2Te_3 and Bi_2Se_3, the solubility of Se in Bi_2Te_3 results in a modification of the crystal lattice and electronic DOS, which is advantageous for reducing the onset of intrinsic conduction and is an important factor for the thermoelectric properties of Bi_2Te_3 alloys [48]. Additionally, the anisotropy of the layered crystal structure plays an important role, as reported for solid solutions of Bi_2Te_3 and Bi_2Se_3, and also for nanograins of $Bi_2Te_{3-x}Se_x$ aligned by the repressing of compacted pellets. Soni *et al.* [49] have reported thermoelectric performance of $Bi_2Te_{3-x}Se_x$ nanoplatelet (NP) composites synthesised by the polyol method. In a typical synthesis, a stoichiometric ratio of bismuth nitrate ($Bi(NO_3)_3 \cdot 5H_2O$), potassium tellurite ($K_2TeO_3 \cdot H_2O$), and sodium selenite (Na_2SeO_3) were used as precursors. The thermoelectric power, S ($-259\,\mu V/K$) and the figure of merit, ZT (0.54) were enhanced by nearly a factor of 4 for SPS pellets of $Bi_2Te_{2.7}Se_{0.3}$ in comparison to Bi_2Te_3 NP composites.

Min *et al.* [50] have reported thermoelectric properties of $Bi_2Te_{3-x}Se_x$ nanocomposite pellets. Figure 7 display the as prepared pellets and their thermoelectric properties according to the mixing ratio of the Bi_2Te_3 and Bi_2Se_3 nanoflakes. Because the anisotropic orientation of the nanoflakes can lead to different transport properties along the transverse directions, thick cylinder-shaped pellets (10 mm diameter and 15 mm thickness) were prepared. The electrical conductivities (σ) of the nanocomposites monotonically increased in the range of 200–440 S cm^{-1} at room temperature as the Bi_2Se_3 composition was raised (Figure 7(b)). The Seebeck coefficient was negative (Figure 7(c)), which indicates the major charge carriers were electrons. The absolute values of S increased as the measuring temperature was raised to 360 K, and then became almost temperature-independent at higher temperatures. S increased with increasing composition of the Bi_2Se_3 nanoflakes. The maximum value was found with the $(Bi_2Te_3)_{85}(Bi_2Se_3)_{15}$. The existence of the maximum S is

10 mm
15 mm
BT
BT90BS10
BT85BS15
BT80BS20
BS
Thermal conductivity
Electrical conductivity
Seebeck coefficient
(a)
Electrical conductivity (S/cm)
Temperature, T(K)
(b)
Seebeck coefficient (µV/K)
(c)
Power factor (mW/m·K²)
(d)
Thermal conductivity (W/m·K)
Klat
T, K
Temperature, T(K)
(e)
ZT
Temperature, T(K)
(f)

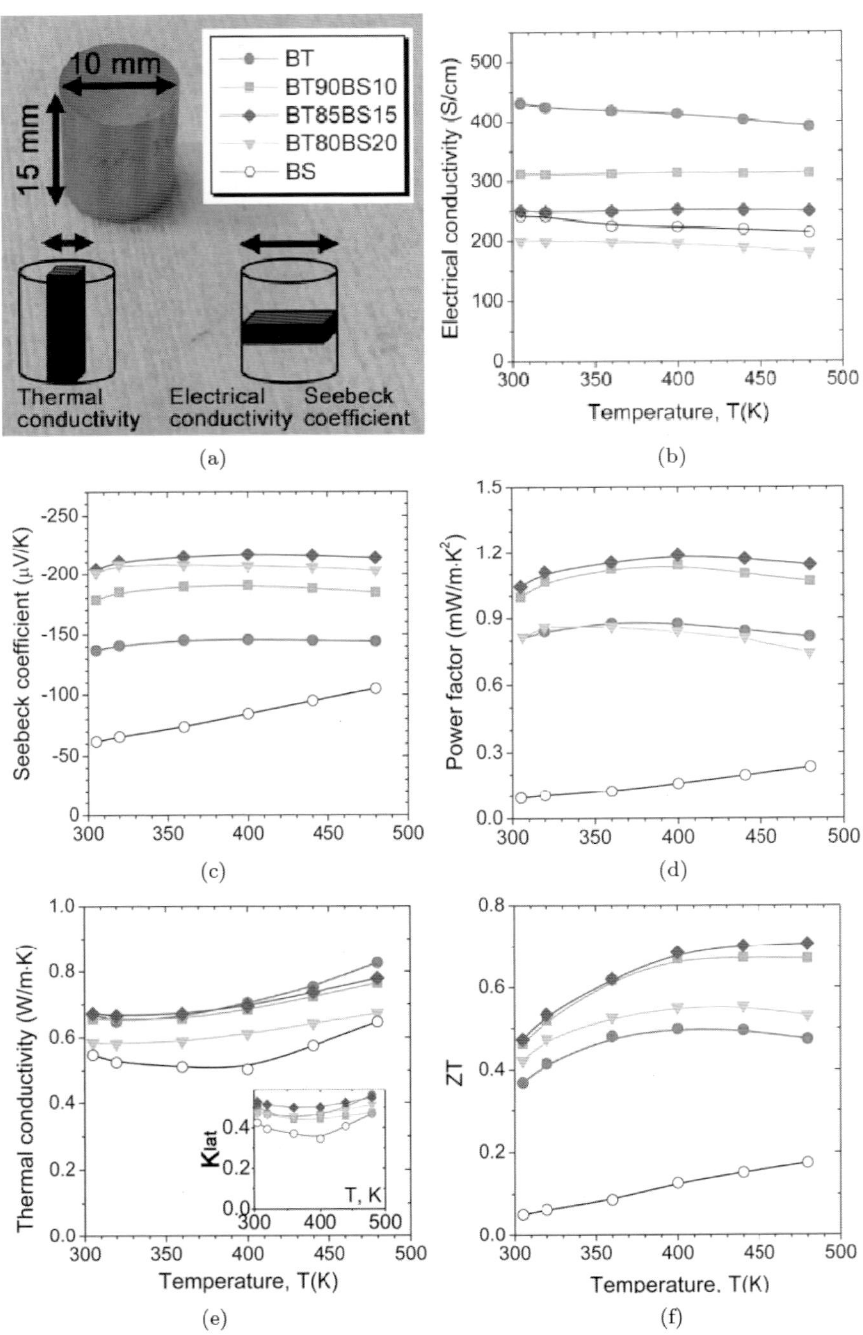

Figure 7.

attributed to the positively affecting carrier energy filtering effect and the negatively affecting density-of-state D(E) by the addition of the Bi_2Se_3 nanoflakes in the Bi_2Te_3 matrix. The maximum power factor ($1.2\,mW\,m^{-1}\,K^{-2}$ at 400 K) in this study was obtained with $(Bi_2Te_3)_{85}(Bi_2Se_3)_{15}$. The κ_{total} values of the nanocomposites ranged from $0.55\,W\,m^{-1}K^{-1}$ to $0.68\,W\,m^{-1}K^{-1}$ at room temperature, which is significantly lower than that of the bulk Bi_2Te_3 with microsized grains ($\sim 1.5\,Wm^{-1}\,K^{-1}$). This reduction of κ_{total} is mainly caused by the reduction of κ_{lat} due to the phonon scattering at the interfaces of the randomly alternating 5–20 nm sized nanograins. Room temperature κ_{lat} values of the nanocomposites decreased by about 40% (0.43–$0.52\,Wm^{-1}\,K^{-1}$). The reduction of the κ_{total} and the increase of the power factor resulted in maximum thermoelectric figure of merit (ZT) of ~ 0.71 at 480 K for $(Bi_2Te_3)_{90}(Bi_2Se_3)_{10}$ samples (Figure 7(f)).

Liu *et al.* [51] have reported high-thermoelectric performance in nanostructured $Bi_{2-x}Sb_xTe_3$, which is fabricated by hydrothermal methods followed by cold-pressing and sintering in an evacuated and encapsulated ampoule. The high value of σS^2 is attributed to enhanced electrical conductivity due to Sb doping. A significant decrease in κ_{lat} has also been observed for these samples. Hence, the sample with a nominal composition of $Bi_{2-x}Sb_xTe_3$, $x = 1.55$ exhibits a dimensionless figure of merit, ZT of 1.65 at 290 K and 1.75 at 270 K with significant improvement compared to that of the commercial state-of-the-art Bi_2Te_3 materials around room temperature (Figure 8).

Although conducting polymers usually have a power factor ranging from 10^{-6} to $10^{-10}\,Wm^{-1}\,K^{-2}$, which is three orders of magnitude lower than that of the state-of-the-art inorganic TE materials,

---

Figure 7. (a) Photograph of a sintered pellet of Bi_2Te_3 nanocomposite with thermoelectric measurement orientations. Thermoelectric properties of the nanocomposite pellets: (b) temperature dependence of electrical conductivities, (c) Seebeck coefficients, (d) power factors, (e) thermal conductivities (inset: lattice thermal conductivities), and (f) figure of merit (ZT) for nanocomposite of Bi_2Te_3 and Bi_2Se_3. (Adapted with permission from Ref. [50]. Copyright 2013 WILEY-VCH Verlag GmbH & Co).

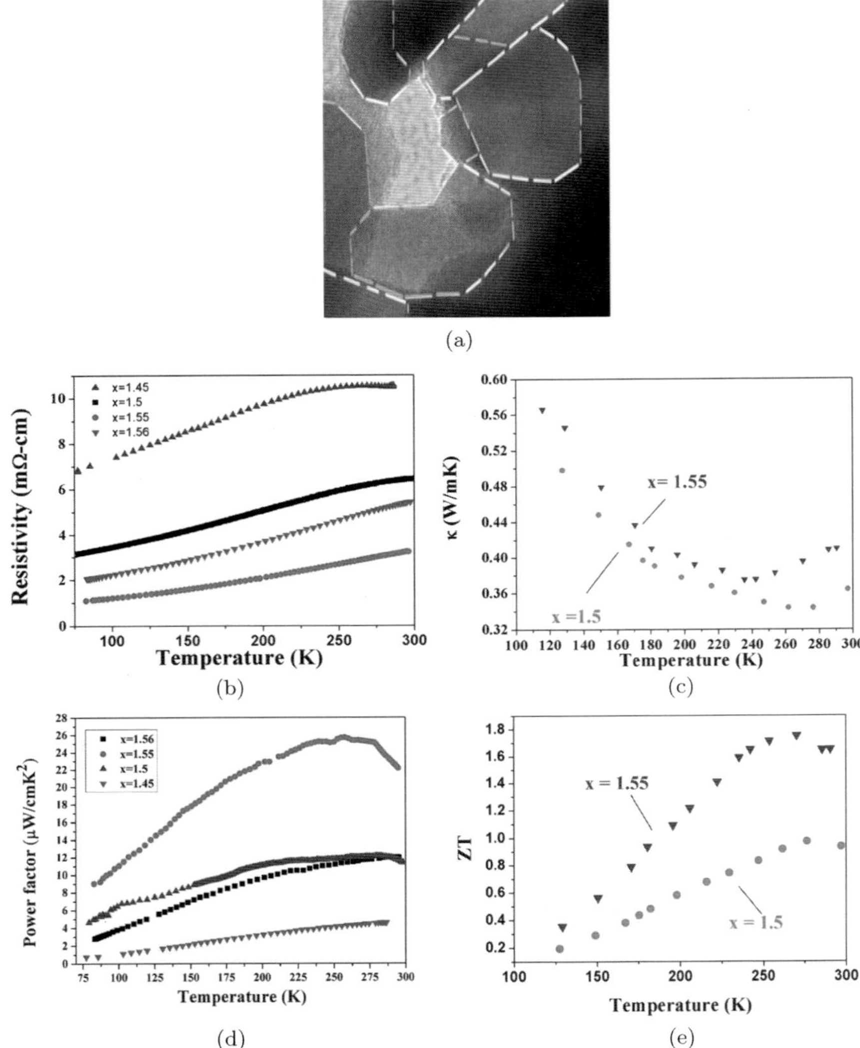

Figure 8. Temperature dependent thermoelectric properties of nanostructured $Bi_{2-x}Sb_xTe_3$ synthesised via hydrothermal method. (a) TEM images are showing the nanosized grains for $Bi_{2-x}Sb_xTe_3$, where $x = 1.55$. Temperature dependence of (b) electrical resistivity (σ), (c) thermal conductivity (κ), (d) power factor (σS^2) and (e) ZT are shown for compacted $Bi_{2-x}Sb_xTe_3$ with $x = 1.45$, 1.5, 1.55 and 1.56 sintered in an evacuated and encapsulated ampoule. (Adapted with permission from Ref. [51]. Copyright 2012 Royal Society of Chemistry).

they possess unique features such as low thermal conductivity, low density, low cost, easy synthesis, and relatively facile to be processed into flexible forms [52–54]. More attention has recently been paid toward the TE properties of conducting polymers and inorganic TE nanostructure/conducting polymer nanocomposites. Song *et al.* [55] prepared Bi_2Te_3/PEDOT:PSSTE composite films by a mixing method with different contents of the Bi_2Te_3 microparticle, and the highest power factor of $9.9\,\mu Wm^{-1}K^{-2}$ was accounted for the composite film with 10 wt.% Bi_2Te_3. Power factor is relatively low for this nanocomposite due to presence of micrometre sized Bi_2Te_3 particle. To overcome this problem, Du *et al.* [56] have studied thermoelectric properties of BST NS/PEDOT:PSS nanocomposite films, prepared from as synthesised Bi_2Te_3 based alloy, $Bi_{0.5}Sb_{1.5}Te_3$ nanosheets (NS) and PEDOT:PSS. The drop cast composite film containing 4.10 wt.% Bi_2Te_3 based alloy NSs showed electrical conductivity as high as 1295.21 S/cm, which is higher than that (753.8 S/cm) of a dimethyl sulphoxide doped PEDOT:PSS film prepared under the same condition and that of the Bi_2Te_3 based alloy bulk material (850–1250 S/cm).

6.6 Homologous Layered Intergrowth Chalcogenide

Layered intergrowth inorganic compounds in the homologous series are one of the interesting classes of material for thermoelectric application [2, 57–61]. Before we discuss the thermoelectric properties of this layered intergrowth material, first, we will briefly discuss about the structural aspects of compounds in homologous series.

In a homology, a series of structures can be built on the same structural principle with a certain module(s) expanding in various dimension(s) by regular increments. The modules are generally cluster blocks, infinite rods, or layers. These modules combine in different ways to form each member by coordination chemistry principles or structure building operators, reflection twinning, glide reflection twinning, cyclic twinning, unit cell intergrowth, etc. [2]. A homologous series is expressed in terms of a mathematical formula that is capable of producing each member [2, 57]. The use of phase homologies in solid state inorganic chemistry is a well-known approach to forecasting modular compounds with a predictable structure. The general representation of some of the series are $A_m(M'_{1+l}Se_{2+l})_{2m}(M''_{2l+n}Se_{2+3l+n})$ (A = alkali metal; M', M'' = main group element), $Cs_4(Bi_{2n+4}Te_{3n+6})$, $(BiQ_X)^{2-}$ $(Ag_xBi_{1-x}Q_{2-2x}X_{2x-1})_{n+1}$ (Q = S, Se; X = Cl, Br; $1/2 \times 1$), and $(Sb_2Te_3)_m(Sb_2)_n$ [2, 57, 62]. These materials are anisotropic in nature from structurally and as well as electronically, and this can have an optimistic impact on the thermoelectric property. The homologous series can be termed as "compound generating machine" and many of the predicted structure were successfully synthesised [57].

Apart from above mentioned homologous compounds, several layered intergrowth compounds in the homologous series of quasibinary $(A^{IV}B^{VI})_m(A^V_2B^{VI}_3)_n$ systems, where $A^{IV} =$ Ge, Sn, Pb; $A^V =$ Sb, Bi; and $B^{VI} =$ Se, Te, are predicted to show promising thermoelectric properties [2, 63]. These compounds usually crystalise in anisotropic layered tetradymite Bi_2Te_2S-type structures [63, 64]. They can also be viewed as intergrowths of PbTe-type rocksalt and Bi_2Te_3-type hexagonal phases, which are indeed natural heterostructures [63–65]. These compound are expected to show low thermal conductivity due to complex crystal structure and strong phonon scattering at the interfaces between the layers [63, 64]. In $Pb_mBi_{2n}Te_{3n+m}$ [that is, $(PbTe)_m(Bi_2Te_3)_n$] homologous series, many layered intergrowth compound exists in the PbTe–Bi_2Te_3-phase diagram (Figure 9(a)). Several members of this series were predicted or verified as topological insulator [65–70]. It is really challenging for solid-state chemists to synthesise these compounds in pure phases through high-temperature solid-state melting technique as the melting points of most of the compounds are incongruent (Figure 9(a)) [64]. Hence, peritectic reaction condition has been utilised previously to synthesise polycrystalline bulk compounds in order to avoid unwanted binary phases [67]. Recently, 2D ultrathin nanosheets of layered $PbBi_2Te_4$ ($m=1$, $n=1$), $Pb_2Bi_2Te_5$ ($m=2$, $n=1$), and $PbBi_6Te_{10}$ ($m=1$, $n=3$) compounds from homologous $Pb_mBi_{2n}Te_{3n+m}$ were synthesised by low temperature solution-based bottom up method and their thermoelectric properties

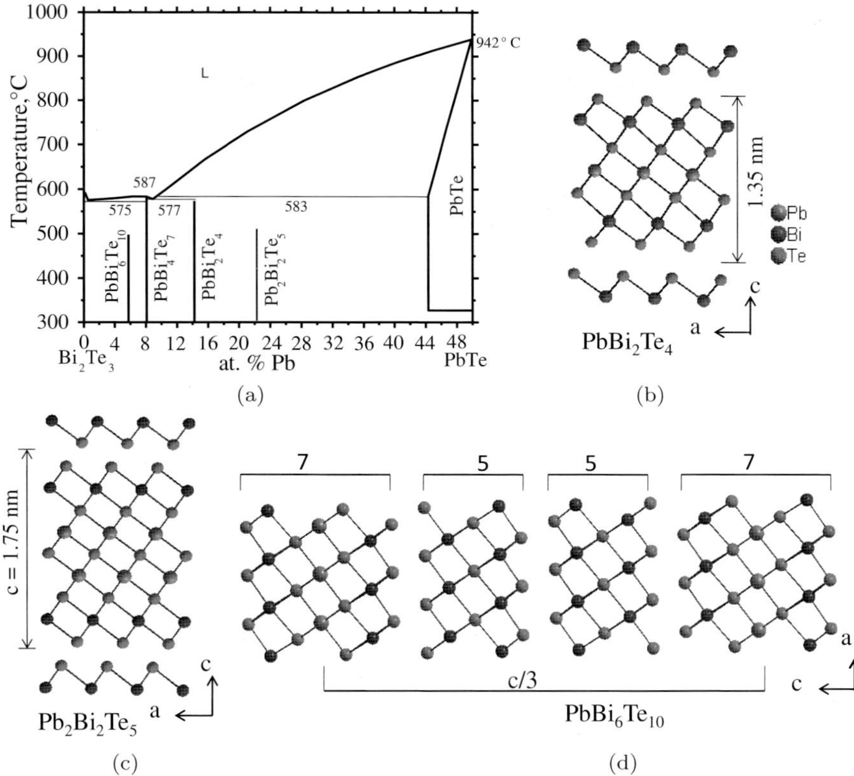

Figure 9. (a) Pseudobinary phase diagram of the PbTe–Bi$_2$Te$_3$ system showing the presence of several incongruently melting compounds. Crystal structures of PbBi$_2$Te$_4$, (b) and Pb$_2$Bi$_2$Te$_5$. (c) showing the 1.35 nm and 1.75 nm thick septuple and nonuple layers, respectively. (d) Crystal structure of PbBi$_6$Te$_{10}$ showing the stacking of quintuple layers of Bi$_2$Te$_3$ and septuple layers of PbBi$_2$Te$_4$. (Adapted with permission from Ref. [64]. Copyright 2015 WILEY-VCH Verlag GmbH & Co.)

were investigated [64]. All these aforementioned compounds possess anisotropic crystals structure and the layers are periodically stacked along the crystallographic c-axis by van der Waals interactions (Figures 9(b)–9(d)). A detail structural description can be found elsewhere [64, 65]. The nanosheets were characterised using various experimental techniques (Figure 10) [64]. AFM and transmission electron microscopy (TEM) measurement indicate the ultrathin nature of the nanosheets.

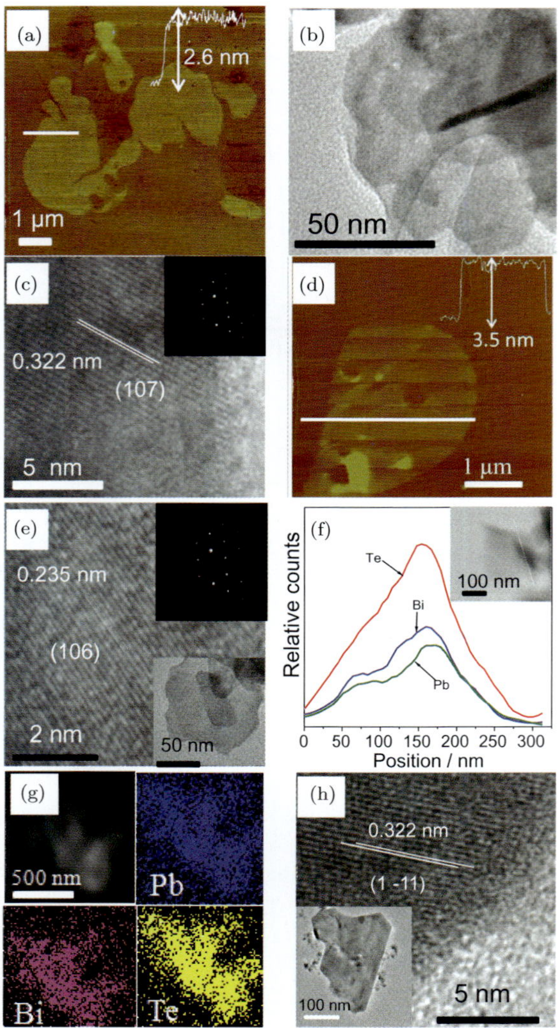

Figure 10. (a) AFM image of a PbBi$_2$Te$_4$ nanosheet. The inset in (a) shows an FESEM image of a PbBi$_2$Te$_4$ nanosheet. (b) TEM image of PbBi$_2$Te$_4$ nanosheet. (c) HRTEM image of a PbBi$_2$Te$_4$ nanosheet. The inset in (c) shows the SAED pattern of a single PbBi$_2$Te$_4$ nanosheet. (d) AFM image of a Pb$_2$Bi$_2$Te$_5$ nanosheet. (e) HRTEM image of a Pb$_2$Bi$_2$Te$_5$ nanosheet. The top inset shows the SAED pattern and the bottom inset shows a TEM image of a Pb$_2$Bi$_2$Te$_5$ nanosheet. (f) EDAX line scan for a single Pb$_2$Bi$_2$Te$_5$ nanosheet. The inset in (f) is the corresponding STEM image of the nanosheet. (g) EDAX colour mapping for Pb, Bi and Te of a Pb$_2$Bi$_2$Te$_5$ nanosheet during STEM imaging. (h) HRTEM image of a PbBi$_6$Te$_{10}$ nanosheet. The inset is the low magnification TEM image. (Adapted with permission from Ref. [64]. Copyright 2013 WILEY-VCH Verlag GmbH & Co.)

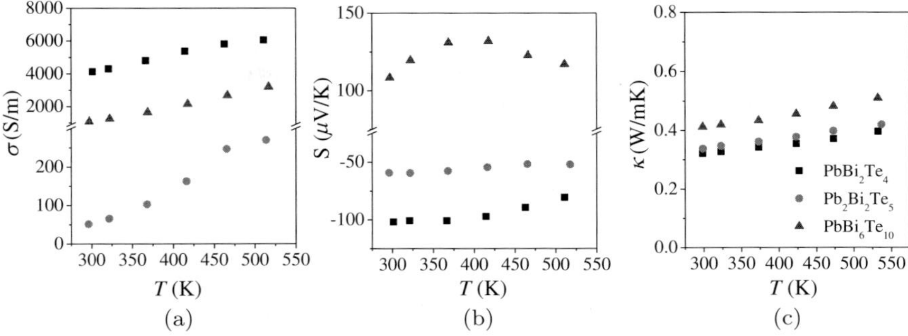

Figure 11. Temperature-dependent (a) electrical conductivity (σ), (b) Seebeck coefficient (S), and (c) thermal conductivity (κ) of $PbBi_2Te_4$ $Pb_2Bi_2Te_5$ and $PbBi_6Te_{10}$ nanosheets. (Adapted with permission from Ref. [64]. Copyright 2015 WILEY-VCH Verlag GmbH & Co.)

The electronic- and phonon-transport measurements on the nanosheets display superior electronic transport and low thermal conductivity compared to that of their bulk counterpart (Figure 11) [64]. Long periodic intergrowth structure and effective phonon scattering at the interface lead to low thermal conductivity for these 2D nanosheets (Figure 11).

$PbBi_2Se_4$ also belongs to quasi binary $(A^{IV}B^{VI})_m(A_2^V B_3^{VI})_n$ family. It is an intergrowth compound of PbSe (rocksalt) and Bi_2Se_3 (hexagonal) [71]. The few-layered 2D nanosheets of $PbBi_2Se_4$ has been synthesised by using a simple solution based reaction at low temperatures [71]. Electrical transport measurements on the ultrathin 3–5 septuple (SL) $PbBi_2Se_4$ shows a semiconducting behaviour (Figure 12). Optical band gap measurement indicates a band gap of 0.6 eV. The high electronic transport in ultrathin $PbBi_2Se_4$ is due to the presence of dominant surface states that offer high electrical mobility of charge carriers [71]. $PbBi_2Se_4$ nanosheets show superior electrical transport properties compared to that of the n-type bulk-layered compounds such as $Pb_5Bi_6Se_4$, $Pb_5Bi_{12}Se_{23}$ and $Pb_5Bi_{18}Se_{32}$ [67, 71].

6.7 Tin Selenide

SnSe is a layered material, adopt an orthorhombic crystal structure, which can be derived from a 3D distortion of the NaCl structure

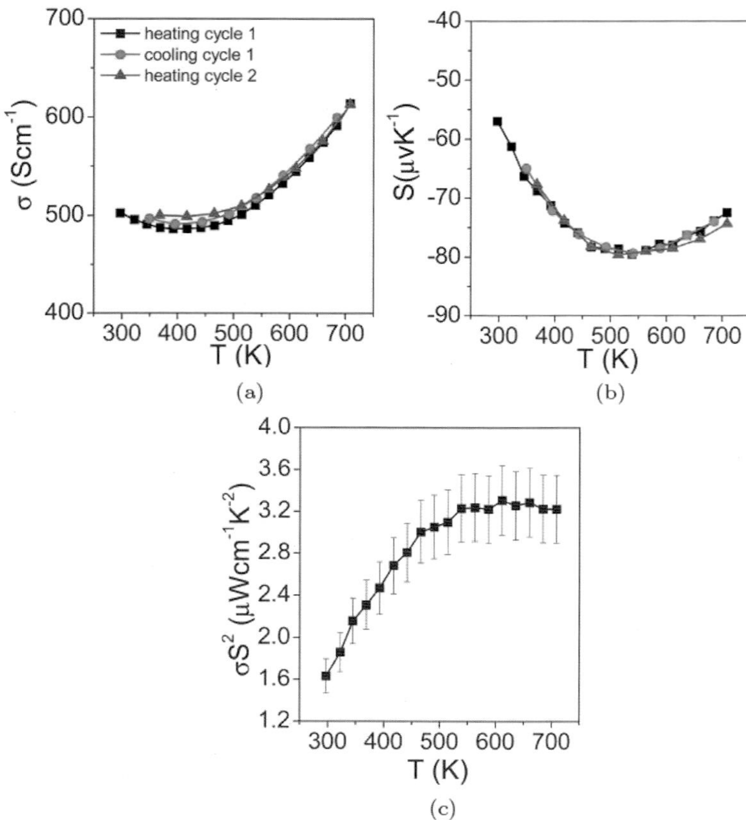

Figure 12. Temperature dependent (a) electrical conductivity (σ); (b) Seebeck coefficient (S) and (c) power factor (σS^2) (with 10% error bars) of the $PbBi_2Se_4$ nanosheet sample. (Adapted with permission from Ref. [71]. Copyright 2014 Royal Society of Chemistry).

(Figure 13) [72–74]. At high temperature, the low symmetry *Pnma*-phase undergoes a displacive phase transition to a high symmetry *Cmcm*-phase [73, 74]. The two-atom-thick SnSe slabs are folded up and create a zigzag accordion-like projection along the crystallographic *b* axis. In both the structures these identical layers are weakly bound (weak van der Waals interaction) and resulting in an anisotropic layered structure (Figure 13) [73, 74]. Zhao and Kanatzidis have extensively investigated the thermoelectric properties of single crystalline SnSe at different crystallographic direction.

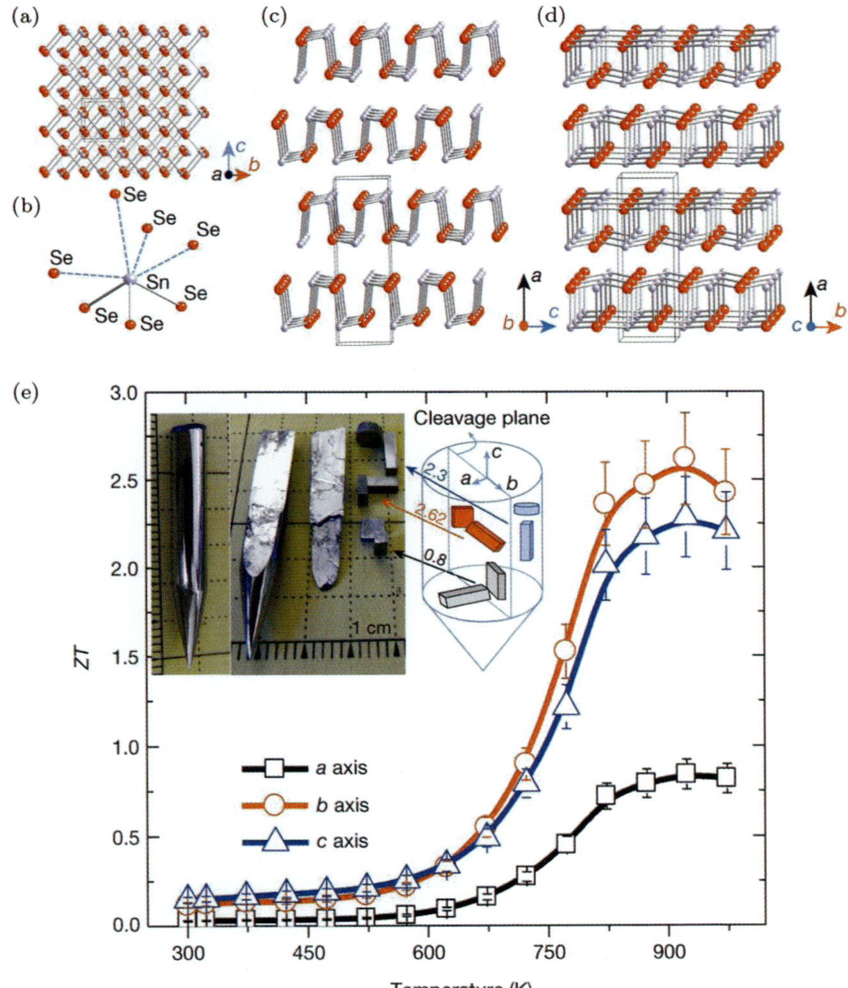

Figure 13. (a) Crystal structure of SnSe along the crystallographic a axis: grey, Sn atoms; red, Se atoms. (b) Highly distorted $SnSe_7$ coordination polyhedron with three short and four long Sn–Se bonds. (c) Structure along the b axis. (d) Structure along the c axis. (e) Main panel, ZT values along different axial directions; the ZT measurement uncertainty is about 15% (error bars). Inset images: left, a typical crystal; right, a crystal cleaved along the (100) plane, and specimens cut along the three axes and corresponding measurement directions. Inset diagram, how crystals were cut for directional measurements; ZT values are shown in the blue, red and grey arrows; colours represent specimens oriented in different directions. (Adapted with permission from Ref. [73]. Copyright 2014 Nature Publishing Group.)

They are able to achieve a record high ZT of 2.6 at 923 K due to moderate power factor and ultra-low thermal conductivity, measured along the b axis in SnSe crystals without addition of any impurities or other optimisation (Figure 13) [73]. Thereafter, investigations on the thermoelectric properties of SnSe gain a significant attention [75–77]. Recently, it has been shown that excellent average ZT value can be obtained in Na-doped p-type SnSe single crystal due to increase in the σ value [78, 79]. The single-crystalline nanosheets of SnSe with four-atomic thickness were synthesised for the first time in 2013 by Li *et al.* [80]. Recently, a theoretical investigation claims that a ZT value as high as 3.3 in single-layered SnSe nanosheet along the armchair and 2.76 along zigzag direction, respectively, with optimal n-type carrier concentration at 700 K [74]. However, detail experimental work is required for practical realisation.

6.8 CsBi$_4$Te$_6$

CsBi$_4$Te$_6$ is a potential material for low-temperature thermoelectric applications. It was first synthesised by Chung *et al.* [81] in the year 2000. CsBi$_4$Te$_6$ crystallises in the space group C2/m, has a layered anisotropic structure composed of anionic [Bi$_4$Te$_6$] slabs alternating with layers of Cs$^+$ ions. The [Bi$_4$Te$_6$] slabs are aligned parallel to b axis and are linked via Bi–Bi bonds. The presence of Bi–Bi bonds in the structure is responsible for a narrow energy gap ($\sim$0.1 eV), nearly half compared to that of Bi$_2$Te$_3$ [81, 82]. The narrower band gap is responsible for the maximum ZT value in CsBi$_4$Te$_6$ being at lower temperature than that of Bi$_2$Te$_3$.

CsBi$_4$Te$_6$ is sensitive to doping, in fact very low doping levels can affect the charge-transport properties. A ZT value of 0.8 at 225 K was obtained for 0.06% SbI$_3$-doped CsBi$_4$Te$_6$ (Figure 14). Various other doping were also been explored in this compound thereafter by the same group [82]. For example, 0.3% BiI$_3$, 0.1% Bi and 0.06% Sb-doped CsBi$_4$Te$_6$ exhibited significantly high power factors at 150 K which is 30–40 K below the temperature compared to that of the 0.05% SbI$_3$-doped CsBi$_4$Te$_6$ [82].

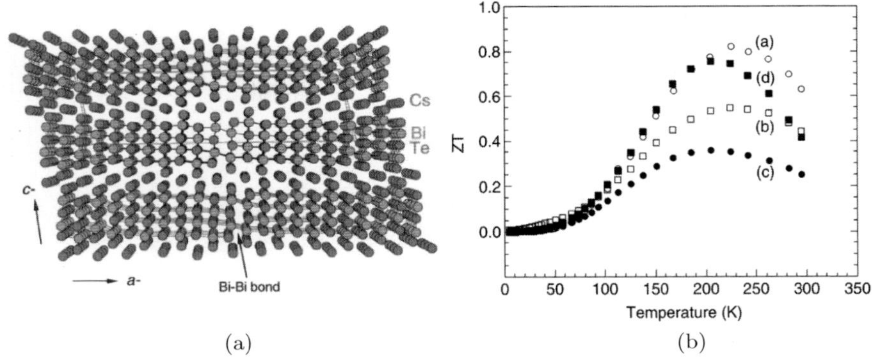

(a) (b)

Figure 14. (a) The structure of $CsBi_4Te_6$ looks down at the crystallographic *b* axis. The Bi–Bi bond is indicated by the arrow. (b) Variable-temperature ZT for (a) 0.05% SbI_3-doped $CsBi_4Te_6$ and for comparison the estimated ZT's for (b) 0.06% Sb (c) 0.3% BiI_3, and (d) 0.1% Bi-doped sample. (Panels (a) and (b) adapted with permission from Ref. [81]. Copyright 2000 AAAS and Ref. [82]. Copyright 2004 American Chemical Society, respectively.)

6.9 BiCuSeO

Layered BiCuSeO, an oxychalcoegnide, is reported to be a promising thermoelectric material by Zhao *et al.* [83]. BiCuSeO crystallises in a layered ZrCuSiAs type tetragonal structure (space group P4/nmm). The structure consists of fluorite like Bi_2O_2 layers which are alternatively stacked with Cu_2Se_2 layers along the crystallographic *c*-axis (Figure 15).

The Bi_2O_2 layer is insulating and acts as charge reservoir, whereas Cu_2Se_2 layer acts as a conduction pathway for carriers [83–85]. BiCuSeO is a multiband material with an indirect band gap of $\sim$0.8 eV. In addition to that, BiCuSeO exhibits extremely low thermal conductivity [83]. The combined effect of strong anharmonicity in bonding arrangements due to the presence of Bi $6s^2$ lone pair, layered structure and the presence of heavy elements in the structure leads to lower phonon group velocity in this compound [83]. Although BiCuSeO shows ultra-low thermal conductivity, but a moderate ZT value has been realised in pristine BiCuSeO ($\sim$0.5 at 923 K) due to poor electrical transport [86]. Thus, to improve the electrical transport properties of BiCuSeO, various divalent metal

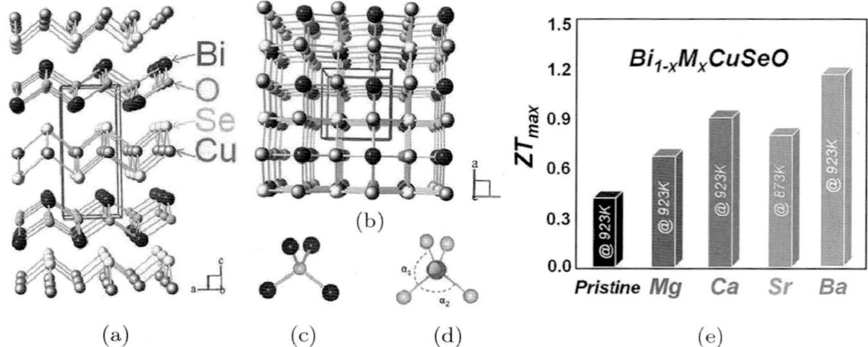

Figure 15. (a) Crystal structure of BiCuSeO along the crystallographic b axis. (b) The structure along the c axis. (c) Bi_4O tetrahedra. (d) $CuSe_4$ tetrahedra. (e) Dimensionless figure-of-merit ZT for the $Bi_{1-x}M_xCuSeO$ (M = Mg, Ca, Sr and Ba) systems. (Adapted with permission from Ref. [83]. Copyright 2014 Royal Society of Chemistry.)

cations (e.g., Mg^{2+}, Ca^{2+}, Sr^{2+} and Ba^{2+}) were used as dopant. The large increase in power factor due to significant improvement in carrier transports and decrease in lattice thermal conductivity due to point defects resulted in an improvement in ZT value. The maximum ZT was improved from $\sim$0.5 for pristine BiCuSeO to $\sim$0.67, $\sim$0.9, $\sim$0.76 and $\sim$1.1 for Mg, Ca, Sr and Ba doped samples, respectively (Figure 15(b)) [87–90]. Apart from optimisation of carrier concentration via uniform doping researchers have also taken strategies to improve the performance such as modulation doping [91], texturation of optimally doped materials, Cu deficiencies and simultaneous Cu/Bi dual deficiencies [86, 92, 93]. Cu deficiencies and simultaneous Cu/Bi dual deficiencies results in a peak ZT value of $\sim$0.81 at 923 K and $\sim$0.84 K at 750 K, respectively [86, 92]. A maximum ZT value of $\sim$1.4 at 923 K was achieved by texturing of $Bi_{0.875}Ba_{0.125}CuSeO$ [93].

6.10   Conclusions and Future Directions

In this chapter, we have discussed about the key concepts, latest development and understanding of the thermoelectric properties of layered chalcogenides. We have learned that layered Bi_2Te_3 and

Sb_2Te_3 are the leading thermoelectric materials for near room temperature applications, while layered SnSe demonstrates remarkably high ZT of ~2.6 at 923 K, which is a potential candidate for the high-temperature power generation. Generally, layered metal chalcogenides show interracially low thermal conductivity due to anisotropic structure and strong lattice anharmonicity. Intergrowth homologous chalcogenides are new candidates for thermoelectric applications as they exhibit low thermal conductivity due to long periodic intergrowth structure. The studies on the thermoelectric properties of layered chalcogenides are progressive and this is an open area of research. Although the thermal conductivity is intrinsically low for layered chalcogenides, attention should be given on improving the Seebeck coefficient by various innovative approaches such as electronic band valley convergence and exploration of resonance level in the electron structure. Many new layered materials with promising thermoelectric properties have been discovered, although a lot of work and progress need to be done for their practical realisation in the form of thermoelectric module. In this context, the combined effort of chemist, physics, materials scientists and engineers will be highly worthy.

Acknowledgements

The authors thank DAE-BRNS project (37(3)20/01/2015/BRNS) for financial support. Kanishka Biswas acknowledges the partial support from the DST Ramanujan Fellowship and Sheikh Saqr Laboratory. Satya N. Guin and Ananya Banik thank JNCASR and inspire programme respectively for the research fellowship.

References

[1] M. Chhowalla, H. S. Shin, G. Eda, L.-J. Li, K. P. Loh and H. Zhang, *Nat. Chem.* **5**, 263 (2013).

[2] M. G. Kanatzidis, *Acc. Chem. Res.* **38**, 361 (2005).

[3] R. Mas-Balles, C. Gomez-Navarro, J. Gomez-Herrero and F. Zamora, *Nanoscale,* **3**, 20 (2011).

[4] C. N. R. Rao, H. S. S. Ramakrishna Matte and U. Maitra, *Angew. Chem. Int. Ed.* **52**, 13162 (2013).

[5]   A. Gupta, T. Sakthivel and S. Seal, *Progr. Mater. Sci.* **73**, 44 (2015).

[6]   Y. Sun, S. Gao and Y. Xie, *Chem. Soc. Rev.* **43**, 530 (2014).

[7]   A. K. Geim, *Nature* **499**, 419 (2013).

[8]   A. K. Geim, *Nat. Mater.* **6**, 183 (2007).

[9]   K. S. Novoselov, D. Jiang, F. Schedin, T. J. Booth, V. V. Khotkevich, S. V. Morozov and A. K. Geim, *Proc. Natl. Acad. Sci. U.S.A.* **102**, 10451 (2005).

[10]   G. J. Snyder and E. S. Toberer, *Nat. Mater.* **7**, 105 (2008).

[11]   J. Sootsman, D. Y. Chung and M. G. Kanatzidis, *Angew. Chem. Int. Ed.* **48**, 8616 (2009).

[12]   M. Zebarjadi, K. Esfarjani, M. S. Dresselhaus, Z. F. Ren and G. Chen, *Energy Environ. Sci.* **5**, 5147 (2012).

[13]   D. M. Rowe. *CRC Handbook of Thermoelectrics*, CRC Press, Boca Raton, FL (1995).

[14]   K. Biswas, J. He, I. D. Blum, C. I. Wu, T. P. Hogan, D. N. Seidman, V. P. Dravid and M. G. Kanatzidis, *Nature* **489**, 414 (2012).

[15]   L.-D. Zhao, S.-H. Lo, Y. Zhang, H. Sun, G. Tan, C. Uher, C. Wolverton, V. P. Dravid and M. G. Kanatzidis, *Nature* **508**, 373 (2014).

[16]   L. D. Hicks and M. S. Dresselhaus, *Phys. Rev. B* **47**, 12727 (1993).

[17]   M. S. Dresselhaus, G. Chen, M. Y. Tang, R. Yang, H. Lee, D. Wang, Z. Ren, J.-P. Fleurial and P. Gogna, *Adv. Mater.* **19**, 1043 (2007).

[18]   J. P. Heremans, B. Wiendlochaac and A. M. Chamoire, *Energy Environ. Sci.* **5**, 5510 (2012).

[19]   M. S. Dresselhaus, G. Dresselhaus, X. Sun, Z. Zhang, S. B. Cronin and T. Koga, *Phys. Solid State* **41**, 1063 (1999).

[20]   T. M. Tritt, *Science* **283**, 804 (1999).

[21]   F. J. DiSalvo, *Science* **285**, 703 (1999).

[22]   G. S. Nolas, J. Sharp and H. J. Goldsmid. *Thermoelectrics Basic Principles and New Materials Development*, Springer, Berlin, Germany (2001).

[23]   H. J. Goldsmid and R. W. Douglas, *Br. J. Appl. Phys.* **5**, 386 (1954).

[24]   D. Teweldebrhan, V. Goyal and A. A. Balandin, *Nano Lett.* **10**, 1209 (2010).

[25]   A. F. Ioffe. *Semiconductor Thermoelements*; Nauka, Moscow, 1956 (in Russian) or Ioffe, A. F. *Semiconductor Thermoelectric and Thermoelectric Cooling*, Info Search, London (1957).

[26]   D. A. Wright, *Nature* **181**, 834 (1958).

[27]   B. Poudel, Q. Hao, Y. Ma, Y. Lan, A. Minnich, B. Yu, X. Yan, D. Wang, A. Muto, D. Vashaee, X. Chen, J. Liu, M. S. Dresselhaus, G. Chen and Z. Ren, *Science* **320**, 634 (2008).

[28] S. II. Kim, K. H. Lee, H. A. Mun, H. S. Kim, S. W. Hwang, J. W. Roh, D. J. Yang, W. H. Shin, X. S. Li, Y. H. Lee, G. J. Snyder and S. W. Kim, *Science* **348**, 109 (2015).

[29] X. Yan, B. Poudel, Y. Ma, W. S. Liu, G. Joshi, H. Wang, Y. Lan, D. Wang, G. Chen and Z. F. Ren, *Nano Lett.* **10**, 3373 (2010).

[30] M. S. Dresselhaus, G. Dresselhaus, X. Sun, Z. Zhang, S. B. Cronin and T. Koga, *Phys. Solid State* **41**, 679 (1999).

[31] R. Venkatasubramanian, E. Siivola, T. Colpitts and B. O'Quinn, *Nature* **413**, 597 (2001).

[32] W. J. Xie, X. F. Tang, Y. G. Yan, Q. J. Zhang and T. M. Tritt, *J. Appl. Phys.* **105**, 113713 (2009).

[33] R. J. Mehta, Y. Zhang, C. Karthik, B. Singh, R. W. Siegel, T. Borca-Tasciuc and G. Ramanath, *Nat. Mater.* **11**, 233 (2012).

[34] (a) J. Fu, S. Song, X. Zhang, F. Cao, L. Zhou, X. Li and H. Zhang, *Cryst. Eng. Comm.* **14**, 2159 (2012). (b) J. Martin, L. Wang, L. Chen and G. S. Nolas, *Phys. Rev. B: Condens. Matter Mater. Phys.* **79**, 115311 (2009) (c) R. Y. Wang, J. P. Feser, X. Gu, K. Yu, R. A. Segalman, A. Majumdar, D. J. Milliron and J. J. Urban, *Chem. Mater.* **22**, 1943 (2010).

[35] J. S. Son, M. K. Choi, M.-K. Han, K. Park, J.-Y. Kim, S. J. Lim, M. Oh, Y. Kuk, C. Park, S.-J. Kim and T. Hyeon, *Nano Lett.* **12**, 640 (2012).

[36] P. Larson, V. A. Greanya, W. C. Tonjes, R. Liu, S. D. Mahanti and C. G. Olson, *Phys. Rev. B* **65**, 085108 (2002).

[37] S. Urazhdin, D. Bilc, S. H. Tessmer, S. D. Mahanti, T. Kyratsi and M. G. Kanatzidis, *Phys. Rev. B* **66**, 161306 (2002).

[38] J. A. Woollam, H. A. Beale and I. L. Spain, *Rev. Sci. Instrum.* **44**, 434 (1973).

[39] Y. Du, S. Z. Shen, K. F. Cai and P. S. Casey, *Prog. Polym. Sci.* **37**, 820 (2012).

[40] Y. Sun, H. Cheng, S. Gao, Q. Liu, Z. Sun, C. Xiao, C. Wu, S. Wei and Y. Xie, *J. Am. Chem. Soc.* 134, 20294 (2012).

[41] M. K. Jana, K. Biswas and C. N. R. Rao, *Chem. Eur. J.* **19**, 9110 (2013).

[42] W. Wangzhong, B. Poudel, J. Yang, D. Z. Wang and Z. F. Ren, *J. Am. Chem. Soc.* **127**, 13792 (2005).

[43] D. M. Rowe, *Thermoelectrics Handbook: Macro to Nano*, Section III, (Chapter 27), p. 16, CRC Press, New York (2006).

[44] W. D. Shi, L. Zhou, S. Y. Song, J. H. Yang and H. J. Zhang, *Adv. Mater.* **20**, 1892 (2008).

[45] G.-H. Dong, Y.-J. Zhu and L.-D. Chen, *J. Mater. Chem.* **20**, 1976 (2010).

[46] G.-H. Dong, Y.-J. Zhu and L.-D. Chen, *Cryst. Eng. Comm,* **13**, 6811 (2011).

[47] R. J. Mehta, C. Karthik, B. Singh, R. Teki, T. Borca-Tasciuc and G. Ramanath, *ACS Nano* **4**, 5055 (2010).

[48] G. D. Mahan, *Solid State Phys.* **51**, 81 (1998).

[49] A. Soni, Z. Yanyuan, Y. Ligen, M. K. K. Aik, M. S. Dresselhaus and Q. Xiong, *Nano Lett.* **12**, 1203 (2012).

[50] Y. Min, J. W. Roh, H. Yang, M. Park, S. I. Kim, S. Hwang, S. M. Lee, K. H. Lee and U. Jeong, *Adv. Mater.* **25**, 1425 (2013).

[51] C.-J. Liu, H. C. Lai, Y.-L. Liu and L.-R. Chen, *J. Mater. Chem.* **22**, 4825 (2012).

[52] I. Lévesque, P.-O. Bertrand, N. Blouin, M. Leclerc, S. Zecchin, G. Zotti, C. I. Ratcliffe, D. D. Klug, X. Gao, F. Gao and J. S. Tse, *Chem. Mater.* **19**, 2128 (2007).

[53] H. Liu, J. Wang, X. Hu, R. I. Boughton, S. Zhao, Q. Li and M. Jiang, *Chem. Phys. Lett.* **352**, 185 (2002).

[54] I. Lévesque, X. Gao, D. D. Klug, S. T. John, C. I. Ratcliffe and M. Leclerc, *React. Funct. Polym.* **65**, 23 (2005).

[55] H. Song, C. Liu, H. Zhu, F. Kong, B. Lu, J. Xu, J. Wang and F. Zhao, *J. Electron. Mater.* **42**, 1268 (2013).

[56] Y. Du, K. F. Cai, S. Chen, P. Cizek and T. Lin, *ACS Appl. Mater. Interfaces* **6**, 5735 (2014).

[57] A. Mrotzek and M. G. Kanatzidis, *Acc. Chem. Res.* **36**, 111 (2003).

[58] M. Ruck and P. F. P. Poudeu, *Z. Anorg. Allg. Chem.* **634**, 482 (2008).

[59] M. Ohta, D. Y. Chung, M. Kuniib and M. G. Kanatzidis, *J. Mater. Chem. A* **2**, 20048 (2014).

[60] P. F. P. Poudeu and M. G. Kanatzidis, *Chem. Commun.* **20**, 2672 (2005).

[61] R. Atkins, M. Dolgos, A. Fiedler, C. Grosse, S. F. Fischer, S. P. Rudin and D. C. Johnson, *Chem. Mater.* **26**, 2862 (2014).

[62] A. Mrotzek and M. G. Kanatzidis, *J. Solid State Chem.* **167**, 299 (2002).

[63] L. Zhang and D. J. Singh, *Phys. Rev. B* **81**, 245119 (2010).

[64] A. Chatterjee and K. Biswas, *Angew. Chem. Int. Ed.* **54**, 5623 (2015).

[65] K. Nakayama, K. Eto, Y. Tanaka, T. Sato, S. Souma, T. Takahashi, K. Segawa and Y. Ando, *Phys. Rev. Lett.* **109**, 236804 (2012).

[66] I. V. Silkin, Y. M. Koroteev, S. V. Eremeev, G. Bihlmayer and E. V. Chulkov, *JETP Lett.* **94**, 217 (2011).

[67] L. E. Shelimova, O. G. Karpinskii, P. P. Konstantinov, E. S. Avilov, M. A. Kretova, I. Y. Nikhezina and V. S. Zemskov, *Inorg. Mater.* **1**, 83 (2010).

[68] K. Yang, W. Setyawan, S. Wang, M. B. Nardelli and S. Curtarolo, *Nat. Mater.* 2012, **11**, 614 (2012).

[69] K. Kuroda, H. Miyahara, M. Ye, S. V. Eremeev, Y. M. Koroteev, E. E. Krasovskii, E. V. Chulkov, S. Hiramoto, C. Moriyoshi, Y. Kuroiwa, K. Miyamoto, T. Okuda, M. Arita, K. Shimada, H. Namatame, M. Taniguchi, Y. Ueda and A. Kimura, *Phys. Rev. Lett.* **108**, 206803 (2012).

[70] O. G. Karpinskii, L. E. Shelimova, E. S. Avilov, M. A. Kretova and V. S. Zemskov, *Inorg. Mater.* **38**, 17 (2002).

[71] A. Chatterjee, S. N. Guin and K. Biswas, *Phys. Chem. Chem. Phys.* **16**, 14635 (2014).

[72] W. J. Baumgardner, J. J. Choi, Y.-F. Lim and T. Hanrath, *J. Am. Chem. Soc.* **132**, 9519 (2010).

[73] L.-D. Zhao, S.-H. Lo, Y. Zhang, H. Sun, G. Tan, C. Uher, C. Wolverton, V. P. Dravid and M. G. Kanatzidis, *Nature* **508**, 373 (2014).

[74] Fancy Qian Wang, Shunhong Zhang, JiabingYuc and Qian Wang, *Nanoscale* **7**, 15962 (2015).

[75] S. Sassi, C. Candolfi, J.-B. Vaney, V. Ohorodniichuk, P. Masschelein, A. Dauscher and B. Lenoir, *Mater. Today* **2**, 690 (2015).

[76] C.-L. Chen, H. Wang, Y.-Y. Chen, T. Day and G. J. Snyder, *J. Mater. Chem. A*, **2**, 11171 (2014).

[77] S. Sassi, C. Candolfi, J.-B. Vaney, V. Ohorodniichuk, P. Masschelein, A. Dauscher and B. Lenoir, *Appl. Phys. Lett.* **104**, 212105 (2014).

[78] L.-D. Zhao, G. J. Tan, S. Q. Hao, J. Q. He, Y. L. Pei, H. Chi, H. Wang, S. K. Gong, H. B. Xu, V. P. Dravid, C. Uher, G. J. Snyder, C. Wolverton and M. G. Kanatzidis, *Science* **351**(6269), 141 (2016).

[79] K. Peng, X. Lu, H. Zhan, S. Hui, X. Tang, G. Wang, J. Dai, C. Uher, G. Wang and X. Zhou, *Energy Environ. Sci.* **9**, 454 (2016).

[80] L. Li, Z. Chen, Y. Hu, X. Wang, T. Zhang, W. Chen and Q. Wang, *J. Am. Chem. Soc.* **135**, 1213 (2013).

[81] D. Y. Chung, T. Hogan, P. Brazis, M. Rocci-Lane, C. Kannewurf, M. Bastea, C. Uher and M. G. Kanatzidis. *Science* **287**, 1024 (2000).

[82] D. Y. Chung, T. Hogan, M. Rocci-Lane, P. Brazis, J. R. Ireland, C. R. Kannewurf, M. Bastea, C. Uher and M. G. Kanatzidis, *J. Am. Chem. Soc.* **126**, 6414 (2004).

[83] L.-D. Zhao, J. He, D. Berardan, Y. Lin, J.-F. Li, C.-W. Nanc and N. Dragoe, *Energy Environ. Sci.* **7**, 2900 (2014).

[84] A. M. Kusainova, P. S. Berdonosov, L. G. Akselrud, L. N. Kholodkovskaya, V. A. Dolgikh and B. A. Popovkin, *J. Solid State Chem.* **112**, 189 (1994).

[85] H. Hiramatsu, H. Kamioka, K. Ueda, M. Hirano and H. Hosono, *J. Ceram. Soc. Jpn.* **113**, 10 (2005).

[86] Y. Liu, L. D. Zhao, Y. C. Liu, J. L. Lan, W. Xu, F. Li, B. P. Zhang, D. Berardan, N. Dragoe, Y. H. Lin, C. W. Nan, J. F. Li and H. M. Zhu, *J. Am. Chem. Soc.* **133**, 20112 (2011).

[87] J. Li, J. H. Sui, C. Barreteau, D. Berardan, N. Dragoe, W. Cai, Y. L. Pei and L. D. Zhao, *J. Alloys Compd.* **551**, 649 (2013).

[88] J. Li, J. H. Sui, Y. L. Pei, C. Barreteau, D. Berardan, N. Dragoe, W. Cai, J. Q. He and L. D. Zhao, *Energy Environ. Sci.* **5**, 8543 (2012).

[89] Y. L. Pei, J. Q. He, J. F. Li, F. Li, Q. J. Liu, W. Pan, C. Barreteau, D. Berardan, N. Dragoe and L. D. Zhao, *NPG Asia Mater.* **5**, e47 (2013).

[90] L. D. Zhao, D. Berardan, Y. L. Pei, C. Byl, L. Pinsard-Gaudart and N. Dragoe, *Appl. Phys. Lett.* **97**, 092118 (2010).

[91] Y. L. Pei, H. Wu, D. Wu, F. Zheng and J. He, *J. Am. Chem. Soc.* **136**, 13902 (2014).

[92] Z. Li, C. Xiao, S. Fan, Y. Deng, W. Zhang, B. Ye and Y. Xie, *J. Am. Chem. Soc.* **137**, 6587 (2015).

[93] J. Sui, J. Li, J. He, Y. L. Pei, D. Berardan, H. Wu, N. Dragoe, W. Cai and L.-D. Zhao, *Energy Environ. Sci.* **6**, 2916 (2013).

Chapter 7

Plasma Chemical and Physical Vapour Deposition Methods and Diagnostics for 2D Materials

Majed A. Alrefae[*], Nicholas R. Glavin[*,†], Andrey A. Voevodin[‡]
and Timothy S. Fisher[§]

[*]*School of Mechanical Engineering
and Birck Nanotechnology Centre, Purdue University,
West Lafayette, IN, USA*

[†]*Air Force Research Laboratory,
Materials & Manufacturing Directorate, WPAFB, OH, USA*

[‡]*Department of Materials Science and Engineering,
University of North Texas, Denton, TX, USA*

[§]*tsfisher@purdue.edu*

Abstract. The growth of 2D materials using plasma chemical and vapour deposition techniques is reviewed in this chapter. We first explain the growth mechanisms of single- and few-layer graphene, and graphene nanopetals in plasma chemical vapour deposition systems. Thereafter, physical vapour depositions of h-BN by pulsed laser deposition and MoS_2 using magnetron sputtering are described. These plasma systems provide many benefits including low growth time, low growth temperatures, high growth rate, and continuous morphology of the deposited films. Moreover, an overview of *in situ* diagnostic methods in plasma systems is covered, including optical emission spectroscopy, plasma imaging, laser absorption spectroscopy, and coherent anti-Stokes Raman spectroscopy. These diagnostics methods are applied to understand, monitor and control the growth of 2D materials in a plasma environment. Overall, plasma systems are projected to be efficient and effective in mass production of 2D materials for existing and future applications.

7.1 Introduction

Since the discovery and isolation of graphene by researchers from the University of Manchester in 2004 [1], the study of two-dimensional (2D) materials has been of great interest. 2D materials have opened a new era for the scientific community upon the discovery that confinement of energy carriers to a plane can enhance physical properties including electron mobility [2], heat transport [3] and inherent material flexibility [4, 5]. Graphene, with a single-layer sp^2-bonded carbon atoms in a honeycomb lattice structure, is also the basis of other carbons structures: fullerenes, carbon nanotubes (CNTs) and graphite. The properties, applications and synthesis of graphene have been reviewed extensively [6–11]. Multiple applications of graphene have been reported including electrochemical sensors and biosensors [12–14], energy and environmental applications [8], carbon capture [15], solar cells [9], silicon-based semiconductor devices [16], corrosion protection [17], oxidation barrier coatings [18], and supercapacitors [19].

The exploration of growth methodologies for atomic layered graphene and other two-dimensional (2D) materials has been a major area of scientific discovery. Recently, interest has spread to the synthesis, properties and applications of 2D materials beyond graphene [20–23]. The initial realisation of these 2D atomic materials was made possible by the mechanical exfoliation of atomic sheets within bulk crystals, including conducting graphite, semiconducting materials including transition metal dichalcogenides (MoS_2, WS_2, black phosphorous, etc.) and insulating layered structures such as hexagonal boron nitride (h-BN) [24]. Alternatively, liquid exfoliation of the layered structures with solvent sonication [25–28] or interstitial intercalation to physically create few-layered grains [29] allows for large dispersions of 2D flakes for deposition. While the mechanical and liquid exfoliation of bulk crystals can lead to high-quality materials at a local scale, researchers have continued to expand and develop direct synthesis methods to grow these materials using techniques that are amenable to larger area device structures. The most popular methodologies for growth include chemical vapour deposition (CVD),

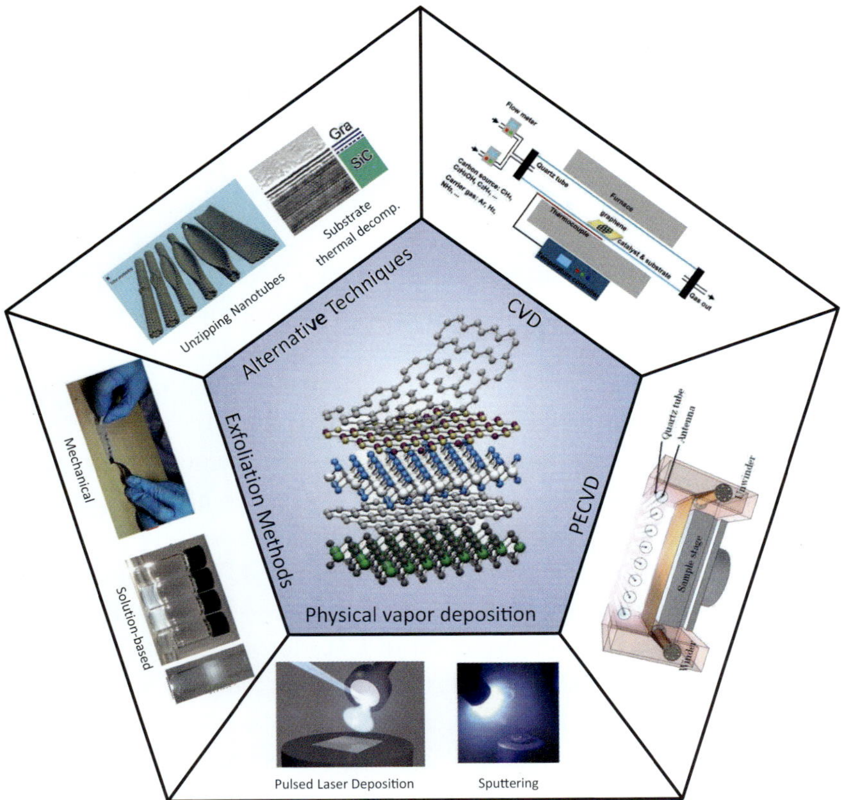

Figure 1. Representation of growth methodologies for 2D materials (inset image reprinted with permission from Ref. [30]; Copyright 2013 Nature Publishing Group), CVD (reprinted with permission from Ref. [31] Copyright 2013 American Chemical Society), PECVD (reprinted with permission from Ref. [32]; Copyright 2012 Elsevier Ltd), physical vapour deposition, exfoliation methods: solution-based. (Reprinted with permission from Ref. [33] Copyright 2015 American Chemical Society) and mechanical method. (Reprinted with permission from Ref. [34] Copyright 2015 Royal Society of Chemistry), and alternative techniques: unzipping nanotubes. (Reprinted with permission from Ref. [35] Copyright 2009 Nature Publishing Group) and substrate thermal composition. (Reprinted with permission from Ref. [36]; Copyright 2013 Royal Society of Chemistry).

plasma enhanced chemical vapour deposition (PECVD), and physical vapour deposition (PVD) as depicted in Figure 1.

CVD is a widely used technique to synthesise thin films from gas-phase reactants [37]. CVD has proven to be the most popular direct growth technique due to the ease of processing, relatively low cost and simple equipment setup, and high-quality material synthesis. Reported in 2008–2009, as one of the first direct growth processes for single-layer graphene [38], CVD has expanded to other materials including transition metal dichalcogenides (MX_2) [39, 40], h-BN [41–45] and other emerging 2D materials such as 2D oxides, nitrides and carbides [46].

Regardless of the success of thermal CVD in growing large-area and high-quality graphene and other 2D films, several factors limit the use of thermal CVD for mass production. Thermal CVD requires heating the substrate to high temperatures, typically near $1000°C$, which then requires the 2D structure to be removed from the substrate and transferred. Despite recent advances in graphene transfer to other substrates for processing [47], the need to eliminate this transfer is important to preserve graphene's unique properties. Furthermore, the deposition time in thermal CVD is longer due to the slow process of heating the feed gases, the chamber walls and the substrate [48]. The portion of CH_4 decomposition to radical species is about 0.0002% in thermal CVD at $900°C$, whereas this percentage increases to more than 80% for plasma CVD (in CNT CVD growth) [49]. Therefore, plasma CVD techniques are likely to be more feasible to overcome challenges inherent in thermal CVD and are discussed here with a comparison to emerging PVD methods for 2D material growth.

7.2 Synthesis of Graphene by Plasma CVD

Plasma CVD can lead to mass production of high-quality graphene at lower substrate temperatures compared to thermal CVD. Furthermore, the deposition time in plasma CVD decreases sharply to a few minutes (or even seconds) for single or few-layer graphene growth. Table 1 provides a summary of plasma CVD systems and deposition parameters that have been used to grow single-, few- and multi-layers graphene (SLG, FLG and MLG, respectively) as well as graphene nanopetals. Different plasma sources have been used to generate the energy required to decompose the feed gas into different radicals.

Table 1. Overview of the plasma CVD systems used to grow graphene.

Layer	Plasma	Power (W)	Pressure (Torr)	Gases	Substrate	Substrate Temperature (°C)	Deposition Time (min)	References
SLG	rf	10–200	0.01	CH_4–H_2–Ar	Cu	700–830	0.2–4	[48]
SLG	r-rf	80	0.05–0.3	CH_4–H_2 (or C_2H_4–H_2)	Sapphire, SiO_2	400–700	20–80	[54]
SLG	MW	1400	20	CH_4–H_2	Ni	450–750	1	[55]
SLG	MW	1000	15–60	CH_4–H_2	Cu	No heating	10	[56]
SLG	MW	10–40	0.5	CH_4–H_2–N_2	Cu	No heating	5–20	[57]
SLG-FLG	DC	170	1.5	CH_4 and H_2	Ni	450	1–12	[58]
SLG-FLG	rf	200	7.5	CH_4–H_2–Ar	Ni	650	0.5	[59]
SLG-FLG	rf	50–600	1	CH_4–H_2–Ar	Cu	950	0.1–60	[60]
SLG-FLG	rf	100–300	0.01-0.02	CH_4–H_2	Cu	500–950	5	[61]
FLG	DC	–	80	CH_4–H_2	Si, Ni	950	5–10	[62]
FLG	DC	500–800	1	C_2H_2–H_2	Ni	1000	10–20	[63]
					Glass	450	6 and 12	[64]
FLG	r-rf	100	0.2	CH_4	many substrates	550	120–240	[65]
FLG	rf	200	1.65	CH_4–H_2–Ar	Co	800	0.25–20	[66]
FLG	rf	0–600	1	CH_4- H_2–Ar	Si, Sapphire	700–1000	1	[67]
FLG	rf	100	0.3	CH_4–Ar	Ni	475	1.7–10	[68]
FLG	MW	1,600–	0.03	CH_4–H_2–Ar	Cu, Al	No heating	0.5–3	[69]
		4,500	0.23				1.7	[32]
FLG	MW	400	10	CH_4–H_2	Cu	No heating	0.5–2	[70]
FLG-MLG	rf	50–150	0.01–0.05	CH_4–H_2–Ar	Ni	200–800	0.2–120	[71]
FLG-MLG	r-rf	550	0.3	CH_4–Ar	quartz, Cu	650	5–15	[72]

(Continued)

Table 1. (*Continued*)

Layer	Plasma	Power (W)	Pressure (Torr)	Gases	Substrate	Substrate Temperature (°C)	Deposition Time (min)	References
MLG	rf	150–250	0.38–0.75	CH_4–Ar	Si	300–380	60–240	[73]
Petals	DC	—	10–200	H_2	Gr anode	No heating	2–3	[74]
Petals	DC	—	3.75	C_2H_2	Cu, Si	750	4	[75]
Petals	DC	400–600	60–150	CH_4–H_2	Si	No heating	25	[76]
Petals	rf	100–500	0.1	CH_4 (or C_2F_6,	many substrates	500	10–4800	[77]
		900		CF_4)–H_2		600–900	5–40	[78]
Petals	rf	900	0.001–	CH_4–H_2	Si	600–1100	20	[79]
			760		Cu, Si	750		[80]
Petals	rf	1000	0.02	C_2H_2–H_2	Ni	620–850	10	[81]
Petals	rf	500	0.15	H_2–"oil"	Si	800	1–4	[82]
Petals	rf	900	1	CH_4–H_2	Cu, Si	680	20 and 60	[83]
Petals	MW	500	1	CH_4–H_2	Si, sapphire	650	5 and 10	[84]
Petals	MW	2,000	40	CH_4–H_2	many substrates	No heating	0.02–50	[85]
							0.2–1.3	[86]
Petals	MW	700–	30	CH_4–O_2–H_2	graphite fibres	No heating	30	[87]
		1000	30–40	CH_4–H_2		900	15	[88]
Petals	MW	300–700	30	CH_4–H_2	SiO_2	No heating	0.5–30	[89]
					Carbon cloth		25	[90]

Notes: SLG: single-layer graphene, FLG: few-layer graphene, MLG: multilayer graphene, petals: graphene nanopetals, DC: direct current plasma, r-rf: remote radio frequency plasma, rf: radio frequency plasma, MW: microwave plasma, CH_4: methane, C_2H_2: acetylene, H_2: hydrogen, Ar: argon, O_2: oxygen Cu: copper, Ni: nickel, Si: silicon, SiO_2: silicon dioxide, Co: cobalt, Gr: graphite anode, Al: aluminium, "oil": oil extracted from a tea tree.

These include direct current (DC), radio frequency (rf) (common frequency of 13.56 MHz) and microwave (MW) (common frequency of 2.45 GHz) plasmas. Further details of common plasma sources used for materials deposition can be found in cited references [50–52]. These plasma sources used for material deposition have typical ionisation fractions of order 10^{-7}–10^{-4} and electron number densities of 10^{9}–10^{11} cm^{-3} [53]. These non-equilibrium plasmas operate at low-gas temperatures and over a range of pressures from 0.01 Torr to 100 Torr.

Energetic electrons in the plasma are able to provide collisional dissociation of the feed gases, which are methane and hydrogen in most cases. For example, the products of electron impact reactions of CH_4 (Reactions R1, R2 and R3 below) can enhance the kinetics of the gas-phase and the generation of CH_3, CH, C_2 radicals, H atom and different ions [91]. These active radicals and ions, among others, reach the substrate surface with high energy, diffuse and provide active nucleation sites for graphene deposition [66, 76]. Plasma modelling, its kinetics and correlation to film deposition are active research areas to understand and optimise growth processes [92–94].

$$\text{Excitation [95]:} \quad CH_4 + e^- \longrightarrow CH_4^* + e^- \tag{R1}$$

$$\text{Ionisation [96]:} \quad CH_4 + e^- \longrightarrow CH_4^+ + 2e^- \tag{R2}$$

$$\text{Dissociation [96]:} \quad CH_4 + e^- \longrightarrow CH_3 + H + e^- \tag{R3}$$

Denysenko *et al.* [94] reported the simulated fluxes of ions and radicals in inductively coupled $Ar/CH_4/H_2$ plasmas for carbon nanostructure deposition [94]. The plasma frequency was 0.46 MHz with power and pressure ranges of 1.8–3 kW and 20–70 mTorr, respectively. The deposited fluxes were defined as the number of radicals and ions deposited on the surface per area per time. These fluxes are represented by the product of the number density, the velocity and either the sticking coefficient for radicals or the ratio between the densities at the wall to the bulk for ions. The deposited flux densities of H, CH, CH_2 and CH_3 radicals at a plasma power near 2 kW were found to be about 2×10^{15}, 2×10^{14}, 2×10^{13} and

$1.5 \times 10^{13}\,\text{cm}^{-2}\,\text{s}^{-1}$, respectively. However, the deposited ion fluxes at the same power were higher, with values of approximately 1×10^{17}, 8×10^{15}, 5×10^{14}, 2×10^{14}, $2 \times 10^{14}\,\text{cm}^{-2}\,\text{s}^{-1}$ for Ar^+, H^+, H_2^+, CH_3^+ and CH_4^+, respectively. These results suggest that ions have more influence on film deposition than neutral species. Furthermore, CH has higher deposition flux than other hydrocarbon radicals, and H (including H^+) has the highest flux among radicals (and ions). The control of these fluxes by optimising the plasma power and the flow rates of CH_4 and Ar can lead to higher quality of 2D films with lower defects and higher deposition rates.

The process parameters for graphene growth by plasma CVD affect the quality and structure of the deposited films [48, 76]. As the plasma power increases, more carbon and hydrogen radicals are produced, directly influencing the quality of the films [76]. More ions with high energy are available to damage the films or create defects [56, 57]. Depending on the plasma source and conditions, the power generally varies from 50 W to 5000 W, as presented in Table 1. However, most plasma discharges used for graphene (and other materials) occur at low pressures to satisfy the breakdown condition, which depends on the product of pressure and the distance between the electrodes known as Paschen's law [97]. This low pressure has the effect of decreasing the density of the plasma, while increasing the kinetic energy of arriving ions, which affects growth rate, morphology and structure of the graphene films [52]. Less dense and higher energy radicals in low-pressure plasmas produce longer surface diffusion times before carbon atoms are locked in the structure, and this process helps to provide a more ordered structure but requires longer deposition times [57].

One recent development in plasma CVD involves atmospheric pressure systems to reduce the use of high-vacuum mechanical equipment and to increase the growth rate by a higher concentration of the carbon-source feed gas [98]. The most common carbon source feed gas used in graphene synthesis is CH_4, but other gases such as C_2H_2, C_2H_4, C_2F_6 and CF_4 may be used, as shown in Table 1. These gases are typically used in low concentrations while H_2 or Ar are added as bath gases to increase the pressure of the system.

Furthermore, H_2 is commonly used due to its etching role as well as to enhance hydrocarbon gas-phase reactions [99].

The growth of graphene by plasma CVD occurs at lower substrate temperatures compared to thermal CVD, and hence a wider range of substrate materials can be used. These substrates range from metals and glasses to carbon fibre. The substrates can be supplementally heated to higher temperatures to enhance surface reactions and diffusion processes, or their heating may solely come from the plasma. For example, thin foils of copper and nickel introduced at room temperature to an MW plasma CVD reactor are heated to 700–900°C in a few seconds [71].

Unlike thermal CVD in which self-limiting graphene growth has been reported [100, 101], the time of deposition in plasma CVD determines the number of layers (or thickness) of graphene films and their morphology [102]. The self-limiting growth mechanism of graphene in thermal CVD processes is due to the highly inert surface of the well-ordered hexagonal structure, preventing nucleation of the subsequent layers. In plasma CVD growth, the surface of graphene has multiple grain boundary and point defects from the higher initial nucleation density and ion bombardment, which promote new nucleation sites on otherwise inert surfaces. For films with reduced surface energy and hence wettability to the surface, a Volmer–Weber growth mechanism [103, 104] leads to film branching and bending away from the initial substrate surface. Thus, the graphene film thickness in plasma CVD processes continues to increase until the few-layer graphene sheets curl to form vertical graphene or graphene nanopetals.

Figure 2 shows atomic force microscopy (AFM) images and Raman spectra of early stages (deposition times of 3, 6, 10, 30, 120 and 150 s) of graphene synthesis on Si substrates using DC plasma CVD of a mixture of CH_4, H_2 and Ar at 1 Pa and 3000 W [102]. At lower times, Figures 2(a)–2(c), scattered islands of graphene form and then enlarge as the deposition time increases. These islands coalesce to form single- and few-layer graphene, Figures 2(d) and 2(e), with some defects. A sharp contrast appears with increasing time (near 150 s as shown in Figure 2(f)) indicating the start of graphene

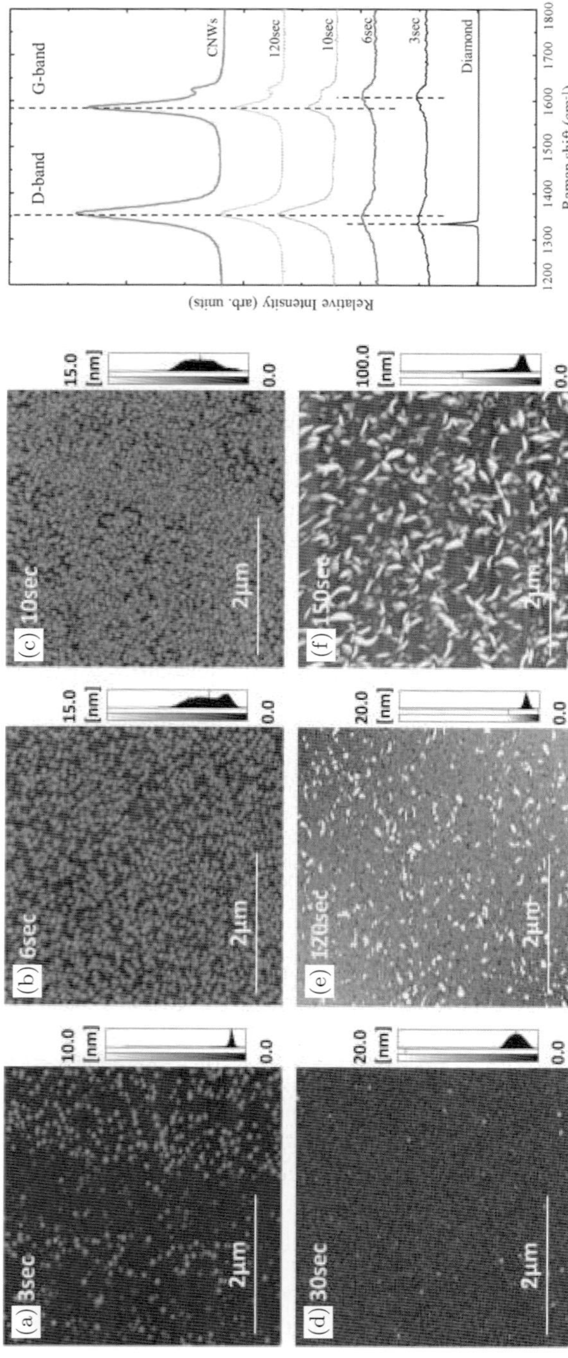

Figure 2. Left: AFM images of graphene deposited on Ni substrates for different times (a) 3, (b) 6, (c) 10, (d) 30, (e) 120, and (f) 150 s. Right: Raman spectra of samples grown in 3, 6, 10 and 120 s deposition time. The laser excitation is 532 nm. The Raman spectra of graphene nanopetals, or carbon nanowall (CNW), after a growth time of 15 min. The Raman spectra of bulk diamond is shown as reference. (Reprinted with permission from Ref. [102] Copyright 2012 Elsevier Ltd).

nanopetals. The Raman spectra of the grown films are shown where the D and G peaks increase with time, indicating more graphitic films with more defects. The growth mechanisms of graphene (single-, few- and many-layers) and nanopetals are discussed in the following subsections.

7.2.1 Growth of single-layer graphene by plasma CVD

The growth of a single-layer graphene with relatively few defects on a copper substrate has been demonstrated using microwave plasma [56, 57]. The process begins with H_2 plasma and a small concentration of cyano radicals, methane and nitrogen to etch the copper oxide and smooth the substrate surface. Nucleation and growth of graphene occur on both the top and the bottom sides of the Cu foil. Single-layer graphene grows on the bottom surface only, whereas the top side of the copper substrate was directly exposed to plasma and thus has more defects. However, the plasma environment is important to etch copper oxide and amorphous carbon and to supply the active radicals for graphene deposition. Also, results from Raman spectroscopy showed a decrease of defects with time due to merging graphene domains to form a single layer. In this recent study, the measured electron mobility at $300\,K$ was as high as $6.0 \times 10^4\,cm^2\,V^{-1}\,s^{-1}$, compared to $3 \times 10^4\,cm^2\,V^{-1}\,s^{-1}$ for graphene grown using thermal CVD [105].

Kim *et al.* [48] used Ar/CH_4 and H_2/CH_4 mixtures in an inductively coupled rf plasma to grow single-layer graphene on copper foil. With the absence of H_2 in the feed gas, the plasma was used to decompose CH_4 to create radicals such as H_2 and H atom. With increasing power, more hydrogen is produced from methane, resulting in increased grain size due to the higher H density as shown in Figure 3. As H atom concentration increased with power from $10\,W$ to $50\,W$, the grain size increased and the nucleation density decreased due to the active role of H as a catalyst for graphene deposition, as shown in Figures 3(a) and 3(b). However, when power increased above $50\,W$, more H was produced causing an apparent saturation effect and providing more etching. Thus, a competition

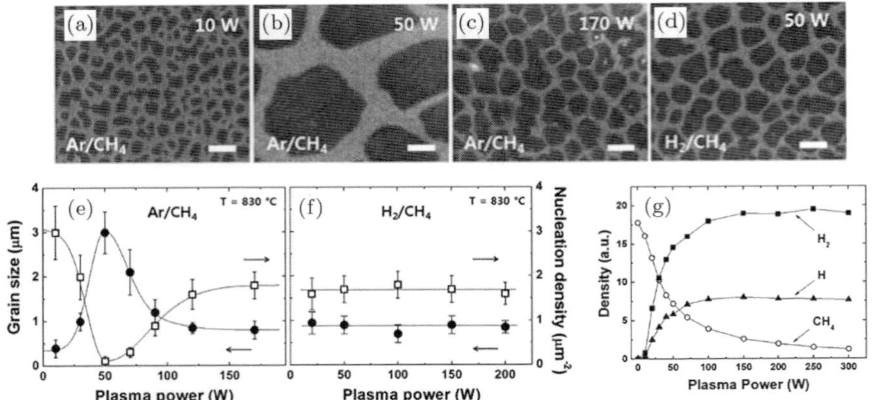

Figure 3. SEM images of graphene after growth at (a) 10 W and 1 min, (b) 50 W and 1 min, (c) 170 W and 20 s in Ar/CH_4 mixture and (d) 50 W and 10 s in H_2/CH_4. The scale bar is $1\,\mu m$. The grain size (filled circles) and the nucleation density (open squares) of graphene as a function of plasma power for (e) Ar/CH_4 and (f) H_2/CH_4 mixtures. (g) The power-dependent H_2, H and CH_4 density in Ar/CH_4 mixture from mass spectroscopy measurements. (Reprinted with permission from Ref. [48]; Copyright 2012 Royal Society of Chemistry).

exists between deposition and etching of graphene due to the roles of H in graphene deposition, as shown in Figure 3(e). For instance, the reaction of H with carbon radicals (CH_x) to produce H_2 and more active radicals such as CH_{x-1}, e.g., $(CH_x + H \rightarrow CH_{x-1} + H_2)$, enhances the deposition process. As the amount of H increases at higher power, etching by H results in decreased graphene grain size (Figure 3(c)) through the reaction: $H + graphene \rightarrow CH + (graphene - C)$. On the other hand, no variation in grain size or nucleation density was observed for H_2/CH_4 plasma due to the presence of hydrogen in the feed gas (Figures 3(d) and 3(f)). Finally, the measured electron mobility was $3200\,cm^2\,V^{-1}\,s^{-1}$ at 300 K with a carrier density of $1 \times 10^{12}\,cm^{-2}$.

The effect of hydrogen on the grain size of single-layer graphene was investigated by Vlassiouk *et al.* [99] in a thermal CVD process. They found that the grain size increases with the H_2 (a surrogate of H) up to a critical pressure of H_2, after which the grain size decreases. The shapes of the grains become more regular with increased H_2 and become hexagonal at the high H_2 pressure that eliminates further increases in grain size.

Microwave plasma was used to grow high-quality SLG on nickel foil by setting H_2:CH_4 ratio to 80:1 [55]. They found that the defects of the graphene increase with decreasing the substrate temperature. For example, the substrate temperature affects the graphene sheet resistance which has a value of 590 $\Omega/\square$ for a graphene grown using a substrate temperature of 750°C.

Wei *et al.* [54] used a remote rf H_2 plasma to grow single-layer graphene on dielectric substrates (sapphire and SiO_2) and found a mobility of the fabricated field-effect transistors (FETs) in the range of 550 to 1600 $cm^2\,V^{-1}\,s^{-1}$. This study demonstrated the use of plasma CVD (with CH_4 or C_2H_4 as a carbon source) to grow high-quality, single layer, micrometre-scale graphene crystals for direct integration into electronic devices without the need for graphene transfer.

With plasma CVD, the growth of single-layer graphene requires lower time depending on the operation conditions [48, 61]. As time increases, more graphene layers are deposited due to the presence of active radicals such as CH, C_2 and CH_3. Also, as the CH_4 concentration increases, more carbon radicals exist, resulting in more layers of graphene [106]. Terasawa and Saiki [61], observed that the number of layers of graphene is higher at lower substrate temperature. For example, they were able to grow single-layer graphene at 900°C but few graphene layers were synthesised at 500°C. Furthermore, as the thickness of the graphene increases or as the CH radical emission intensity increases, the Raman I_D/I_G ratio increases, indicative of a decrease in grain size.

A schematic of the growth mechanism of graphene on Cu substrate provided in Ref. [61] is shown in Figure 4. At lower substrate temperature, the carbon radicals in the plasma (C_2 in this case) start depositing on the Cu substrate to form graphene with small grains. Copper has been shown to decompose hydrocarbons to yield C atoms that can penetrate into the copper and then precipitate to form SLG, as in the case in thermal CVD [101]. Once the substrate

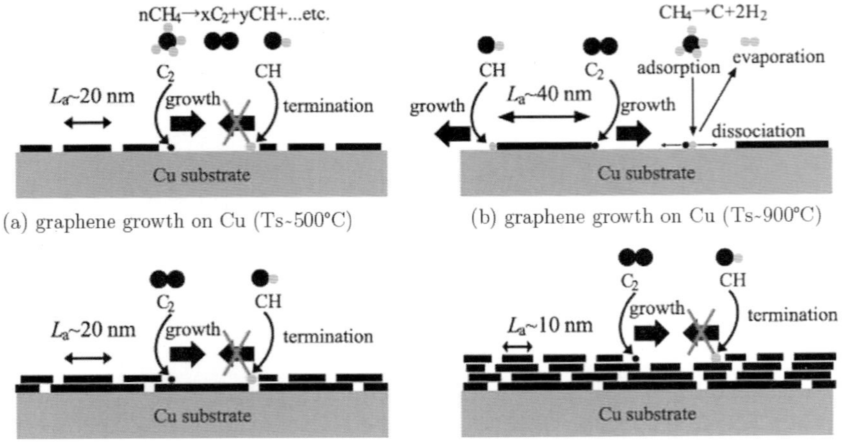

Figure 4. Schematic of different mechanisms to grow graphene in different layers: (a) monolayer at low substrate temperature (T_s), (b) monolayer at higher substrate temperature (T_s), (c) bilayer and (d) multilayer. (Reprinted with permission from Ref. [61]; Copyright 2011 Elsevier Ltd).

temperature is increased to about $900°C$, carbon radicals reach the catalytically active copper surface to grow single-layer graphene with larger grain size. For longer deposition time, a second layer of graphene is deposited due to the presence of active radicals such as C_2. Thus, the rate of this deposition is much faster than for thermal CVD, which depends solely on the Cu to catalyse the deposition. The grain size typically decreases with successive graphene layers until reaching about $10\,$nm depending on the concentrations of C_2 and CH [101].

7.2.2   Growth of few- and many-layer graphene by plasma CVD

The FLG growth mechanism on nickel substrates using plasma CVD is explained in cited references [62, 68]. Carbon radicals originating in the plasma reach the Ni surface where the C atoms dissolve to form NiC if the temperature of the substrate is higher than about $475°C$. This low substrate temperature is enabled in the plasma due to the rich density of carbon radicals. During cooling, the dissolved carbon atoms desorb and precipitate on the Ni surface. Therefore, unlike copper, there is no catalytic effect for graphene growth on Ni substrates as for both Ni and Cu substrates with thermal CVD. However, for substrate temperatures less than 450°C, the solubility of carbon on nickel decreases to 0%; consequently, Cheng *et al.* [71] proposed another growth mechanism for this case, in which nucleation and growth of carbon radicals occurs at defects with no influence of dissolution in the Ni substrate.

The number of graphene layers increases with deposition time [48, 61, 70] and methane concentration in the feed gas [70, 106] and depends on the plasma power [48] as well as the substrate temperature [61, 67]. Furthermore, the substrate material plays a major role in determining the growth time of graphene as well as its crystal size [65]. This may be related to the rate of adsorption of carbon radicals from the plasma based on substrate roughness and lattice orientation [65].

7.2.3 Growth of graphene nanopetals by plasma CVD

Graphene nanopetals were first produced in a hydrogen arc discharge [74] and then a microwave plasma [84] CVD during the synthesis of CNTs. The direction of these nanopetals is predominantly vertical on substrates (thus the pseudonym 'vertical graphene' (VG) due to the presence of a vertical electric field). A recent review by Bo *et al.* [52] summarises the synthesis of nanopetals by plasma CVD in which the effects of plasma sources, feedstock gases, temperature, pressure and substrates are discussed. The growth of graphene nanopetals requires no catalyst and starts with a base of graphene layers that grow parallel to the substrate [79], as in Figure 5(a) [107]. Due to the weak interaction between the graphene monolayers and the forces at the grain boundaries which increase with time, the edges of the top layers are curled upward. These localised vertical edges act as electrical field concentrators, attracting carbon radicals to further increase the height of the nanopetals rather than their width, Figure 5(b) [77]. Such localised electrical field concentration, $\vec{E}$, and its impact on nanostructure growth is an important characteristic of plasma assisted deposition and was reviewed recently for CNT

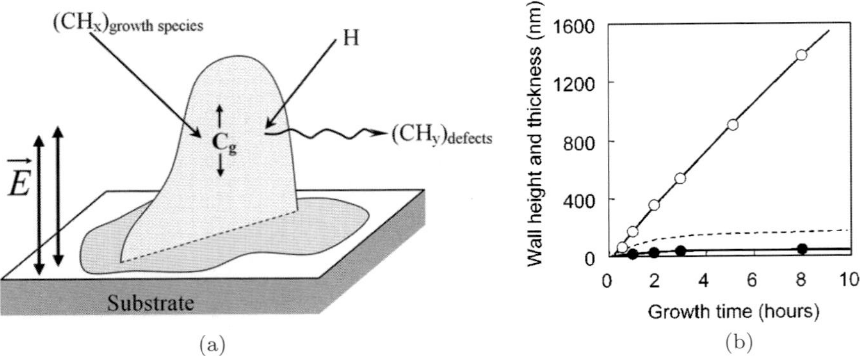

Figure 5. (a) A schematic of a nanopetal growth mechanism on a substrate. $\vec{E}$ is the direction of the electric field near a substrate surface. (Reprinted with permission from Ref. [107] Copyright 2007 Elsevier Ltd) and (b) The height of nanopetals (open circles) and its thickness (closed circles) vs. the growth time. The substrate is Si in C_2F_6/H_2 rf plasma. The dashed line showed the maximum spacing between adjacent nanopetals. (Reprinted with permission from Ref. [77]; Copyright 2004 AIP Publishing LLC).

and 2D material growth [91]. In addition to the electrical filed effect, the presence of atomic hydrogen plays an important role in etching amorphous carbon and to prevent the nucleation and secondary growth on the growing protrusion to yield thin nanopetals. Moreover, the height of these nanopetals increases with time, while the thickness remain relatively constant with time, as shown in Figure 5(b) [77].

To study the effects of different radicals in plasma for nanopetal growth, Mironovich *et al.* [76] varied the plasma conditions in conjunction with a 2D plasma model to trace the variation in species concentrations. Figure 6 shows SEM images of the graphene nanopetals with the corresponding plasma conditions. With increasing current (Figures 6(d)–6(f)), secondary nucleation occurs at the top-side of the nanopetals. Furthermore, with decreasing pressure (Figures 6(g)–6(i)) the height of the nanopetals decreases and more etched holes appear, as indicated by the arrows. If both the current increases and the pressure decreases (at constant temperature), a combination of secondary nucleation and etchant effects is observed due to the increase of C_xH_y concentration with current (Figures 6(j)–6(l)). From the 2D plasma model, increasing the current produces

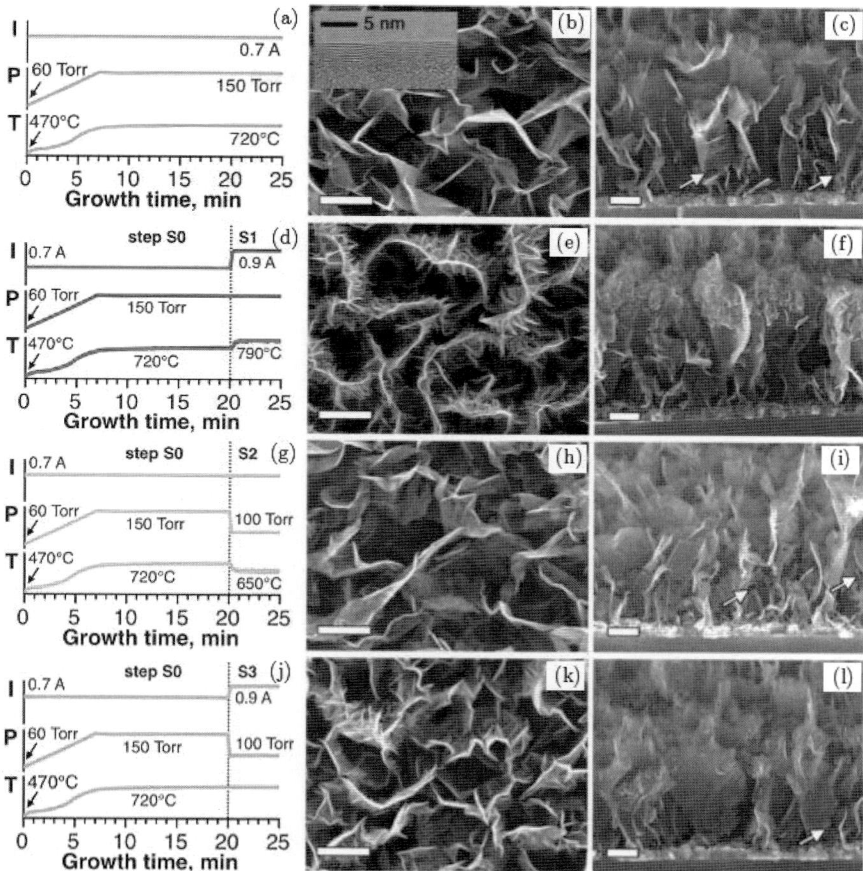

Figure 6. (a, d, g, j) Plasma conditions for graphene nanopetals growth. (b, e, h, k) top-view and (c, f, i, l) side-view of the SEM images of the grown nanopetals at the specified plasma conditions. The scale bar is 1 μm. (Reprinted with permission from Ref. [76] Copyright 2014 Royal Society of Chemistry).

higher concentrations of H and C atoms, and CH, CH$_2$, C$_2$H and C$_3$, C$_3$H radicals. C$_x$H$_y$ radicals are expected to become trapped on the surface of nanopetals to initiate the growth of the secondary nanopetals. However, with decreasing pressure and temperature, fewer C$_x$H$_y$ radicals exist, whereas more hydrogen etching occurs leading to lower growth rates with more defects on the nanopetals.

7.3 Chemical Vapour Deposition vs. Physical Vapour Deposition of 2D Materials: Thermodynamics vs. Kinetics

Variants of chemical vapour deposition to grow 2D materials include metal transformation, chemical vapour transport, powder vapourisation, and reactive chemical vapour deposition [108]. In all of these variations, the mechanism for film growth is driven by thermodynamics, as the heat supplied to the substrate is typically the major energy source required for crystal formation ($\sim$0.1 eV if the substrate is heated to a surface temperature of 1000°C). The thermodynamic process is driven by atomic mobility, or the lateral diffusion rate, of condensed species on the substrate to form thin films starting at specific, discrete nucleation spots. The deposition rate using this technique can be very slow, where some processes require hours to heat and cool the chamber of interest followed by up to an hour of growth time. Furthermore, in many cases, there is an addition of an annealing step. Because of this slow growth rate in comparison to the deposition flux, the condensed atomic species are near equilibrium, where the species have sufficient time to explore their minimal energy representation on the surface before settling into a final configuration [109, 110]. These conditions can lead to incomplete substrate coverage and small single-crystal domains as shown in Figures 7(a) and 7(b) for h-BN growth on a metal foil substrate. The equilibrium-based process can form very uniform layered structures within these domains, as evidenced in Figure 7(c), and even complex patterns such as graphene 2D flower-like structures seen in Figure 7(d). After a critical time, these boundaries will merge, and the physics surrounding these boundaries has been a subject of intense research with regard to thermal transport, electron transport, and other physical phenomena.

In contrast to high equilibrium, thermodynamically driven processes such as CVD, 2D materials have also been shown to nucleate and grow crystalline, highly continuous films by means of PVD. In PVD techniques including magnetron sputtering, pulsed laser deposition, molecular beam epitaxy and others, the plasma generated within a high vacuum chamber can produce ion and electron energies

Figure 7. (a) Growth of monolayer on copper foil. (Reprinted with permission from Ref. [45]. Copyright 2012 American Chemical Society), (b) growth of few-layer h-BN on copper foil. (Reprinted with permission from Ref. [111]. Copyright 2012 IOP Publishing Ltd). (c) A cross-sectional micrograph of high quality h-BN grown by CVD at temperatures higher than 1000°C (Reprinted with permission from Ref. [44]. Copyright 2012 American Chemical Society) and (d) a graphene flower configuration directly grown thermodynamically on a substrate. (Reprinted with permission from Ref. [112]. Copyright 2012 American Chemical Society).

on the order of 1–100 eV. Figure 8(a) schematically represents the different mechanisms, where PVD techniques tend to have a high deposition flux and a relatively low time for atomic surface diffusion [109]. These processes have been known to create metastable and complex structures (including 2D planes), with control of the rate and concentration of species within the plasma [113].

Due to the high-energy plasma species, special provisions are needed to control the energy in order to minimise substrate damage as well as to control the composition of the condensate at the substrate interface. As an example, we consider the use of pulsed laser

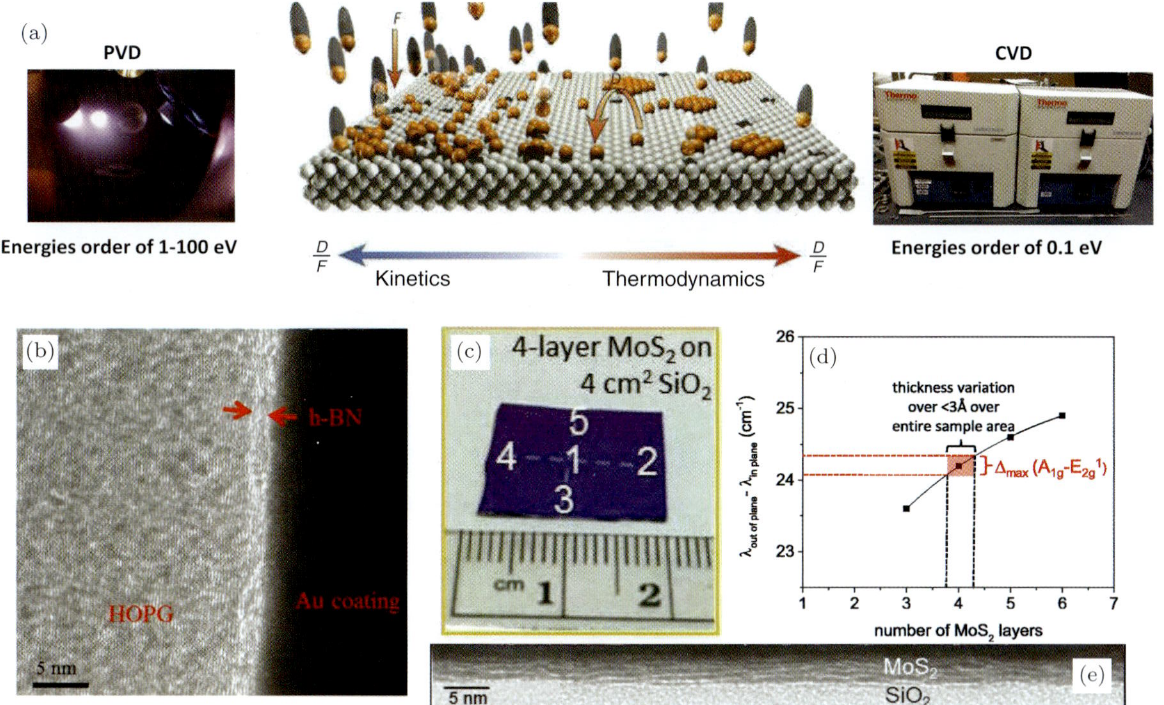

Figure 8. (a) Schematic representing the contract in kinetic (PVD) and thermodynamic (CVD) approaches to film growth where D is the diffusion rate and F is the incoming deposition flux. (Reprinted with permission from Ref. [109]. Copyright 2005 Nature Publishing Group), (b) TEM of few-layer h-BN growth on HOPG by PLD. (Reprinted with permission from Ref. [114] Copyright 2014 Elsevier), (c) large area substrate coverage as well as (d) low thickness variation of MoS$_2$ sputtered films, and (e) TEM cross-section of few-layer MoS$_2$, (c, d and e are reprinted with permission from Ref. [115] Copyright 2014 AIP Publishing LLC).

deposition (PLD) to form few layer continuous h-BN films on HOPG, where a background gas of nitrogen was introduced into the growth chamber to encourage collisions within the plume in the construction of stoichiometric, nanocrystalline h-BN layers in Figure 8(b) [114]. While the CVD-grown h-BN is formed in single crystal domains, the highly kinetically driven PVD process coats the entire substrate uniformly, yet restricts grain growth due to the high deposition flux. This is evidenced by nanocrystalline lateral grain formation on the order of 3–5 nm.

This phenomena is also present in highly uniform MoS_2 growth by magnetron sputtering, where the films exhibit uniform thicknesses over the entire substrate of less than three angstroms, and the incoming energy must be controlled by an exterior magnetic field to steer the ion species away [115, 116]. The drawback to this process is similar to that of h-BN PLD in that the high deposition flux restricts large-area grain growth, as the grains in this case are on the order of 20 nm, in contrast to the $>1\,\mu$m grains from conventional CVD processes. Due to the increased atomic energy of the adsorbate material on the substrate, crystallinity in these 2D films can also be realised at much lower temperatures than conventional thermodynamic approaches. This is evidenced by h-BN growth at $700°$C with PLD, compared to typically $>1000°$C in CVD, as well as crystalline MoS_2 growth at $450°$C with magnetron sputtering, compared to traditional CVD at temperatures $>700°$C. Development of 2D materials synthesis routes through high-energy plasma processing is currently a growing field, as researchers are beginning to expand to other growth techniques besides conventional CVD, and will potentially be an exciting pathway towards commercial 2D device structures.

7.4 Process Diagnostics of Plasma Chemical Vapour Deposition

Different diagnostic techniques have been used to monitor plasma characteristics and their correlation to film growth. Optical emission spectroscopy (OES) can access the upper states of atoms/molecules and measure their relative intensities and temperatures. Plasma imaging is a powerful tool to map the intensity of species at different timescales during growth. These two techniques do not require light sources such as lasers and thus simplify experimental setup. On the other hand, laser absorption techniques can be used to measure the ground state populations of species and molecules. More sophisticated techniques, such as Coherent Anti-Stokes Raman Spectroscopy (CARS) are powerful in accurately measuring species temperatures and concentrations. Other plasma diagnostics techniques such as Langmuir probe [117], mass spectroscopy [48, 118], and Laser Induced Fluorescence (LIF) [119] also provide useful insights of the plasma and its effects on the grown films, but are outside the scope of this work.

7.4.1 Optical emission spectroscopy

OES has been recognised as a simple technique to characterise plasma systems [120–122] and is commonly used to characterise plasma systems for nanomaterial growth. OES is a sensitive and non-intrusive technique to infer gas temperatures and relative densities of emitting species in plasmas. Figure 9 shows a top view of an OES experimental setup where the emission of the plasma is collected through a fibre optic and transmitted to the entrance of the spectrometer. This light is then diffracted through a grating and finally detected by the CCD camera to produce a spectrum as shown in Figure 9.

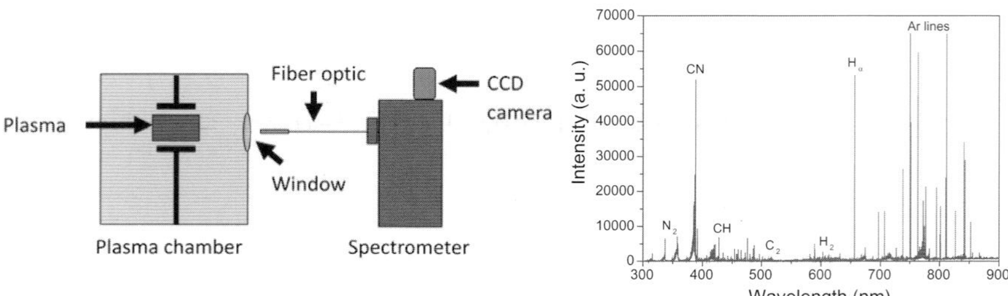

Figure 9. Left: OES experimental setup. Right: an example of the emission spectrum showing important species in the $H_2/CH_4/Ar$ rf plasma.

Plasmas typically have plenty of molecules in electronically excited states that emit photons at different wavelengths. The emission intensity can be expressed as:

$$I = N_u A_{ul} h v_{ul},\tag{7.1}$$

where I is the intensity of the emission corresponding to a transition from an upper energy level (u) to a lower energy level (l), N_u is the number density in the upper level, A_{ul} is the Einstein coefficient for spontaneous emission, h the Planck's constant, and v_{ul} is the frequency of the transition. The population N_u can be expressed assuming a Boltzmann distribution:

$$N_u = N_o \frac{g_u}{Z} \exp\left[-\frac{E_u}{k_B T}\right],\tag{7.2}$$

where N_o is the population at the ground level, g_u is the degeneracy of the upper state, Z is the partition function, E_u is the energy of the upper state, k_B is Boltzmann's constant, and T is the temperature. Knowledge of the molecule spectroscopic model aids in determining the temperature and the relative densities from the emission spectrum [123].

7.4.1.1 *Temperature measurements*

The system temperatures (rotational, vibrational and excitation, or 'electronic') can be found either by fitting Equations (7.1) and (7.2) to the measured intensity or using the Boltzmann plot of the specified transitions. For non-equilibrium plasma, the electron temperature is higher than the translational (gas) temperature, and in general, the following inequality holds: $T_{\text{electron}} \geq T_{\text{vibration}} \geq T_{\text{rotation}} \geq T_{\text{translation}}$. OES can be used to access all of these modes to estimate, using simple experimental setup, the temperature of each level [121, 124–126]. In general, a Boltzmann plot is used to determine the corresponding temperature. Alternatively, fitting the measured spectrum allows the estimation of the rotational and/or vibrational temperatures.

A recent review by Brüggeman *et al.* [127], surveyed different methods of finding gas temperatures from rotational lines of non-equilibrium plasmas using OES. The measured rotational temperature can be used to represent the translational (gas) temperature if both translation and rotational levels are in thermal equilibrium. In such cases, the rotational energy transfer must be fast so that the rotational level becomes thermally distributed. In most cases of plasmas for CVD processes, the excitation mechanisms are dominated by the electron excitation. Consequently, the rotational distribution of the excited state is a projection of that of the ground state, and their temperatures are related. The gas temperature can be determined from the rotational temperature from several species, such as N_2, OH, N_2^+, H_2, O_2, CN, C_2 and CH. Table 2 lists the species that have been used to measure the gas temperature in non-equilibrium plasmas that have similar behaviour as in the plasma CVD systems.

7.4.1.2 *OES as an in situ growth characterisation tool*

The emission intensities of H, CH and C_2 from OES in a $CH_4/H_2/Ar$ plasma for diamond growth in a microwave plasma CVD reactor

Table 2. Commonly used species with their transitions to measure rotational temperature in plasma systems.

Species	Transition	Wavelength (nm)	Method	References
N_2	C–B (2nd positive system)	360–382	Spectrum fit	[128–130]
N_2^+	B–X (1st negative system)	380–395	Spectrum fit	[130]
H_2	$d^3\Pi - a^3\Sigma$ (Fulcher α band)	590–620	Boltzmann plot Spectrum fit	[129, 131]
C_2	$d^3\Pi_g - a^3\Pi_u$	510–520 520–570	Spectrum fit Boltzmann plot	[132, 133]
CN	$B - a$	382–389	Spectrum fit	[132]
CH	$A^2\Delta - X^2\Pi$	420–440	Spectrum fit	[134, 135]
OH	A–X	306–314	Spectrum fit	[128, 130, 136]
O_2	$B - X$	758–772	Spectrum fit	[136]
NO	A–X	240–248	Spectrum fit	[128]

were compared to the results from Cavity Ring Down Spectroscopy (CRDS) [137]. Using a scaling factor, the results from OES showed similar trends as a function of input parameters when compared to CRDS. Hence, OES is commonly used as an *in situ* diagnostic tool to correlate the quality of deposited films during plasma CVD. The growth of diamond has been extensively studied using OES in microwave plasma assisted CVD systems [138–144]. It has also been used to characterise the growth of graphene nanopetals in plasma CVD systems [98, 146–149]. For example, Ma *et al.* [150] applied OES to characterise graphene growth as a function of plasma parameters (power, flow rates of C_2H_2 and H_2) on Cu substrates. C_2 and H emission intensities were used as signatures for the quality of graphene growth through correlation to sheet resistance and the (I_D/I_G) ratio from Raman scattering spectroscopy, respectively. Similarly, the emission intensity of CH was correlated with I_D/I_G during the growth of graphene on Cu foil [61]. This linear relationship between CH and I_D/I_G confirms the decrease in the grain size for multilayer graphene with increasing CH radical concentration.

7.4.2 Plasma imaging diagnostics

In addition to the optical emission spectroscopy, CCD camera imaging of transient plume dynamics using intensity mapping can give insight into spatial distributions of the plasma at specific timescales within the growth process. With the use of a gate pulsar, small windows as narrow as 100 ns can be imaged and intensity maps created for that plasma at that particular moment in time. The inset of Figure 10(a) displays three of these captured images of a boron nitride plume during PLD of 2D h-BN [151]. Plume confinement is evident at 1000 ns, as the 50 mTorr nitrogen background gas creates collisions within the plume, where at these pressures, the plume then abides in the shock-wave regime characterised by a hemispherical shock front [152].

Broadband spatial distributions can be very helpful in the growth of 2D materials, as a target to substrate working distance as well as plasma energies and energy distributions are key to the formation of continuous, defect-mitigated 2D layered materials. With the use of

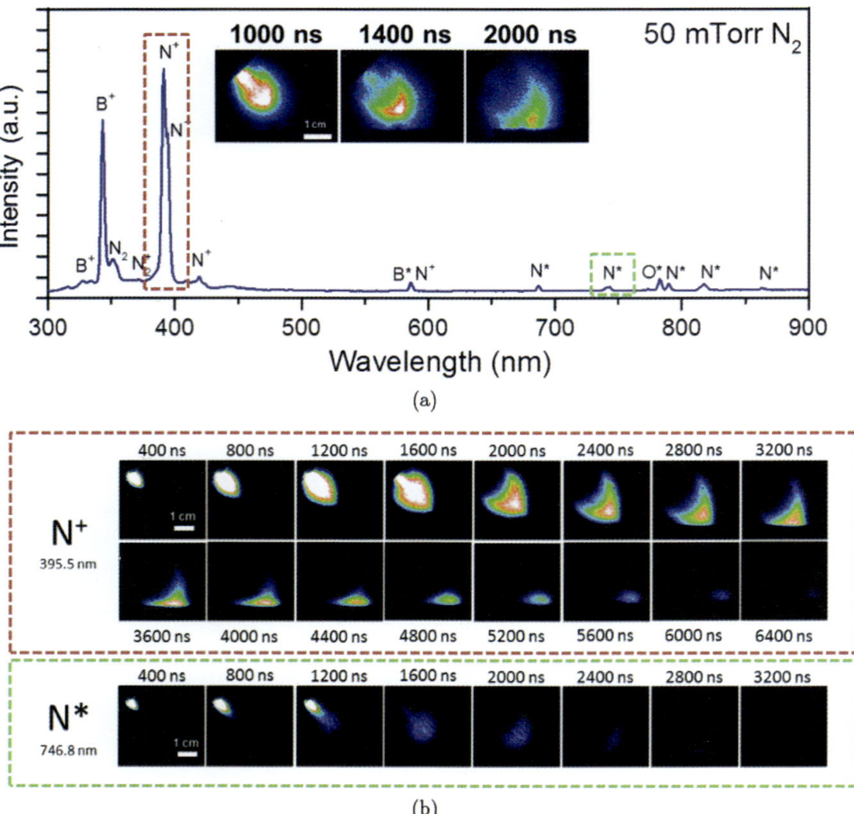

(a)

(b)

Figure 10. (a) OES spectrum of BN plume from PLD during few-layer h-BN growth and insight is plasma imaging at 1000 ns, 1400 ns and 2000 ns after the laser impact with the target and (b) plasma imaging through narrow bandpass filters for N^+ and N^* at 395.5 nm with a 400 nm filter and 746.8 nm with a 750 nm filter.

narrow notch filters, information on spatial distribution of individual excited and ions species can be realised. As shown in Figures 10(a) and 10(b), notch filtering of the ionised N^+ and excited N^* reveals that the N^+ species is delayed in towards the front of the plume, where N^* reaches the condensate interface first. Through OEM and the notch filter analysis, it was determined that N^* was more critical for BN film formation, and N^+, as it lags behind the main plume and plays a much lesser role in film formation. This technique was also used to track and identify critical ionised and neutral species with regards to growth of DC magnetron sputtered MoS_2 2D films [116]. In this case, a magnetic field was used to steer the ionised Ar species from created defects in the MoS_2 film, and notch filtering helped to identify the velocity and relative magnitude at which the ions were deflected. The use of spatial plasma imaging can be a tool to aid in the understanding of energy distributions, spatial concentration gradients, and with the use of a notch filter, species concentration gradients, to optimise the growth conditions required in 2D materials.

7.4.3   Laser absorption spectroscopy

Absorption spectroscopy for plasma CVD reactors has been widely used as a non-intrusive and real-time technique to optimise the growth conditions of nanoscale films. Different laser systems have been implemented to probe the concentrations of important neutral species such as CH_4 and C_2H_2, and radicals such as CH_3 and C_2. In absorption experiments, a light source is used to emit a beam that passes through the plasma to be measured by a detector. The Beer-Lambert law represents the decay of a monochromatic beam as it passes through a homogenous medium:

$$\frac{I(v)}{I_o(v)} = \exp(-k_v L), \tag{7.3}$$

where $I(v)$ is the transmitted beam intensity, $I_o(v)$ is the reference beam intensity, k_v is the spectral absorption coefficient in cm^{-1}, and L is the path length of the medium in cm. The absorption coefficient is related to the concentration of the species (in cm^{-3}) through the absorption cross-section (in cm^2) which is known for many species from databases such as HITRAN [153].

The direct absorption (DA) technique is a common approach to measure high-concentration species with high absorption coefficients. For low concentrations, a multi-pass setup is implemented to increase the optical pathlength (L) hence increasing the absorption. Furthermore, wavelength modulation spectroscopy (WMS) can be used to increase the detection limit with higher signal-to-noise ratio. Also, cavity ring down spectroscopy (CRDS) is used by mounting highly reflective mirrors (more than 99.999%) in which the transmitted signal is measured to determine the ringdown time, which is related to concentration. Table 3 lists common species detected by absorption spectroscopy in plasma systems for material deposition.

Table 3. Summary of measured species concentration in plasma CVD systems with wavelength and laser technique used.

Species	Wavelength (nm)	Method	References
CH_4	1,648.5	Multi-pass WMS using TDLAS	[154]
	3,358.4	DA and WMS using DFG laser	[155]
	7,855	DA using QCL	[156]
C_2H_2	1,510.3	CRDS using diode laser	[157]
	3,113.3	DA and WMS using DFG laser	[155]
	7,449.4	DA using QCL	[158–160]
	7,677.6	DA using TDLAS	[161, 162]
	12,500.0	DA in multi-pass cell using TDLAS	[163]
C_2H_6	3,333.3	DA using TDLAS	[161, 162, 164]
	12,500.0	DA in multi-pass cell using TDLAS	[163]
C_2H_4	3,108.5	WMS using DFG laser	[155]
	3,340.3	DA using TDLAS	[162]
	10,548.5	Multi-pass DA using TDLAS	[165]
CH_3	216.4	DA using deuterium lamp	[166]
	216.4	DA using TDLAS	[166]
	16,501	DA using TDLAS	[162]
CH	431.1	CRDS using a tunable dye laser	[167–169]
C_2	231.3	DA using deuterium lamp	[164]
	516.5	DA using mercury arc lamp	
CN	388.3	DA using deuterium lamp	[164]
C_2N_2	4,651.3	Multi-pass DA using TDLAS	[165]
H	656.3	DA using diode laser	[169]
	486.1	CRDS using a tunable dye laser	[169]
	121.6	DA using MHCL	[170]

Notes: DA: Direct Absorption, WMS: Wavelength Modulation Spectroscopy, CRDS: Cavity Ring Down Spectroscopy, DFG: Difference Frequency Generation laser, QCL: Quantum Cascade Laser, TDLAS: Tunable Diode Laser Absorption Spectroscopy and MHCL: Micro-Discharge Hollow-Cathode lamp.

7.4.4 Coherent Anti-Stokes Raman Spectroscopy

CARS is a nonlinear optical process in which two or three beams are used to generate a CARS signal. A typical CARS energy level diagram shows three laser beams: probe, pump and stokes focused in a probe volume. Phase matching is maximised when the wave vector relation $k_4 = k_1 - k_2 + k_3$ is satisfied and resulted in the generation of a coherent anti-stokes laser-like signal with a frequency of $\omega_4 = \omega_1 - \omega_2 + \omega_3$.

The simultaneous process of CARS begins by exciting the Raman transition of a molecule with the pump beam ω_1 to a virtual state which then de-excites at a frequency similar to the Stokes beam ω_2. The probe beam with frequency of ω_3 then excites the molecule to another virtual state that results in the generation of the anti-Stokes beam. The selection of frequencies for the pump, Stokes and probe laser beams is important to match the molecular oscillation frequency. Figure 11 shows an experimental setup of a CARS experiment to measure the temperature and the concentration of H_2 in H_2/CH_4 microwave plasma used for graphene growth. The CARSFT code [171] can be used to fit the CARS signal, thus providing the temperature and concentration of H_2.

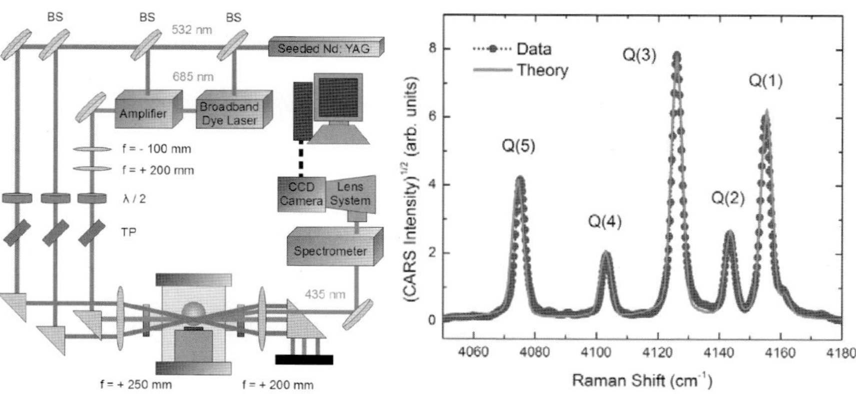

Figure 11. Left: CARS experimental setup in microwave plasma used for graphene deposition. Right: Experimental CARS signal of H_2 and fitted spectrum using CARSFT to give temperature of 1341 K. (Reprinted with permission from Ref. [172]. Copyright 2014 ASME).

Table 4. Summary of CARS measurements in plasma CVD systems.

Species	Measured parameter	Raman shift (cm^{-1})	References
H_2	Concentration and temperature	4,000–4,200	[174, 175]
	Temperature	4,000–4,200	[173, 176–179]
N_2	Temperature	2,290–2,340	[178]
C_2H_2	Concentration	1,860–2,200	[180]
		3,374–3,368	[174]
CH_4	Concentration	2,912–2,922	[174]
GeH_4	Concentration	2,111	[181]
SiH_4	Concentration	2,187	[181, 182]
Si_2H_6	Concentration	2,163	[182]

CARS experiments provide uniquely high spatial resolution in the measurements of temperature and concentration. Furthermore, its signal is coherent, and therefore the emission background from plasmas can be easily eliminated. Hence, this *in situ* technique can precisely study the process of plasma CVD, as summarised in Table 4. Recent developments of CARS using nanosecond, picosecond and femtosecond lasers have been reviewed by Roy *et al.* [173] for applications in reacting flows. CARS experiments in plasma CVD systems have great potential for simultaneous detection of multiple species and temperature that results in better understanding of the deposition process of graphene and other 2D materials.

7.5   Conclusion

Plasma-assisted chemical and physical deposition methods for the growth of graphene, 2D transition metal dichalcogenides and BN ultra-thin films provide new possibilities for innovation in the synthesis of these emerging 2D materials. Plasma techniques greatly reduce production cost at larger scales by decreasing growth time and substrate temperatures. Furthermore, due to this lower temperature environment, additional materials can be used as substrates, thus eliminating the transfer procedures that might damage the 2D films or create defects. The processes in plasma chemical and physical deposition methods rely on the presence of highly chemically active and energetic ions, atoms and radicals that promote reactivity and assembly on the substrate surfaces. The advantages are clearly apparent in the case of the graphene growth by plasma CVD in which the use of direct current (DC), radio frequency (rf) and microwave (MW) plasma sources for precursor gas dissociation and excitation help in reducing the growth time and minimising (and eliminating, in many cases) substrate heating requirements. Similar results have been reported for PVD synthesis of 2D MoS_2 by magnetron sputtering and few nanometre thick BN film growth by pulsed laser deposition.

The growth process of the plasma-assisted deposited 2D films is continuous and results in a gap-free morphology that overcomes the limitations of the island growth typical of thermal CVD. Furthermore, unique 2D materials such as graphene nanopetals and ultra-thin dense amorphous BN layers can be grown solely in the plasma assisted processes. The growth mechanisms of these unique materials are linked to the presence of a vertical electric field and the ionised energetic species that are only attainable with the use of plasma.

The main challenge of the plasma processes involves controlling the density and energy of the ionised species to avoid the damage to the growing film by collisional energy transfer and ions bombardment. To overcome this issue, deeper understanding of the plasma environment is needed. Plasma modelling and plasma diagnostics techniques should be developed and used to optimise the wide range of plasma processes. Specifically, methods using spectroscopic and plasma optical emission approaches are attractive for *in situ* process monitoring and control for 2D materials growth and for future scaled-up manufacturing of these materials. These diagnostics techniques include OES, plasma optical imaging, laser absorption and CARS. These techniques have been reviewed here to provide the necessary foundation for further advances.

References

[1] K. S. Novoselov, A. K. Geim, S. V. Morozov, D. Jiang, Y. Zhang, S. V. Dubonos, I. V. Grigorieva and A. A. Firsov, *Science* **306**(5696), 666–669 (2004).

[2] K. S. Novoselov, A. K. Geim, S. V. Morozov, D. Jiang, M. I. Katsnelson, I. V. Grigorieva, S. V. Dubonos and A. A. Firsov, *Nature* **438**(7065), 197 (2005).

[3] E. Pop, V. Varshney and A. K. Roy, *MRS Bull.* **37**(12), 1273 (2012).

[4] D. Akinwande, *Proceedings of SPIE — The International Society for Optical Engineering* **9083**, 9083 (2014).

[5] S. J. Kim, K. Choi, B. Lee, Y. Kim and B. H. Hong, *Annu. Rev. Mater. Res.* **45**(1), 63 (2015).

[6] C. N. R. Rao, A. K. Sood, K. S. Subrahmanyam and A. Govindaraj, *Angew. Chem. Int. Ed. Engl.* **48**(42), 7752 (2009).

[7] W. Choi, I. Lahiri, R. Seelaboyina and Y. S. Kang, *Crit. Rev. Solid State Mater. Sci.* **35**(1), 52–71 (2010).

[8] G. Zhao, T. Wen, C. Chen and X. Wang, *RSC Adv.* **2**(25), 9286 (2012).

[9] S. Das, P. Sudhagar, Y. S. Kang and W. Choi, *J. Mater. Res.* **29**(03), 299 (2014).

[10] X. Huang, Z. Yin, S. Wu, X. Qi, Q. He, Q. Zhang, Q. Yan, F. Boey and H. Zhang, *Small* **7**(14), 1876 (2011).

[11] Y. H. Wu, T. Yu and Z. X. Shen, *J. Appl. Phys.* **108**(7), 071301 (2010).

[12] Z. Zhu, L. Garcia-Gancedo, A. J. Flewitt, H. Xie, F. Moussy and W. I. Milne, *Sensors (Basel).* **12**(5), 5996 (2012).

[13] S. Kochmann, T. Hirsch and O. S. Wolfbeis, *TrAC Trends Anal. Chem.* **39**, 87 (2012).

[14] M. S. Artiles, C. S. Rout and T. S. Fisher, *Adv. Drug Deliv. Rev.* **63**(14–15), 1352–1360 (2011).

[15] A. T. Najafabadi, *Renew. Sustain. Energy Rev.* **41**, 1515 (2015).

[16] K. Kim, J.-Y. Choi, T. Kim, S.-H. Cho and H.-J. Chung, *Nature* **479**(7373), 338–344 (2011).

[17] A. C. Stoot, L. Camilli, S.-A. Spiegelhauer, F. Yu and P. Bøggild, *J. Power Sources* **293**, 846 (2015).

[18] A. S. Kousalya, A. Kumar, R. Paul, D. Zemlyanov and T. S. Fisher, *Corros. Sci.* **69**, 5–10 (2013).

[19] G. Xiong, C. Meng, R. G. Reifenberger, P. P. Irazoqui and T. S. Fisher, *Electroanalysis* **26**(1), 30 (2014).

[20] M. Chhowalla, H. S. Shin, G. Eda, L.-J. Li, K. P. Loh and H. Zhang, *Nat. Chem.* **5**(4), 263–275 (2013).

[21] S. Z. Butler, S. M. Hollen, L. Cao, Y. Cui, J. A. Gupta, H. R. Gutiérrez, T. F. Heinz, S. S. Hong, J. Huang, A. F. Ismach, E. Johnston-Halperin, M. Kuno, V. V Plashnitsa, R. D. Robinson, R. S. Ruoff, S. Salahuddin, J. Shan, L. Shi, M. G. Spencer,

M. Terrones, W. Windl and J. E. Goldberger, *ACS Nano.* **7**(4), 2898 (2013).

[22] R. Lv, J. A. Robinson, R. E. Schaak, D. Sun, Y. Sun, T. E. Mallouk and M. Terrones, *Acc. Chem. Res.* **48**(1), 56 (2015).

[23] Q. Peng, A. K. Dearden, J. Crean, L. Han, S. Liu, X. Wen and S. De, *Nanotechnol. Sci. Appl.* **7**(2), 1 (2014).

[24] K. S. Novoselov, D. Jiang, F. Schedin, T. J. Booth, V. V. Khotkevich, S. V. Morozov and A. K. Geim, *Proc. Natl. Acad. Sci. U.S.A.* **102**(30), 10451 (2005).

[25] G. Gao, W. Gao, E. Cannuccia, J. Taha-Tijerina, L. Balicas, A. Mathkar, T. N. Narayanan, Z. Liu, B. K. Gupta, J. Peng, Y. Yin, A. Rubio and P. M. Ajayan, *Nano Lett.* **12**(7), 3518 (2012).

[26] S. Gayathri, P. Jayabal, M. Kottaisamy and V. Ramakrishnan, *AIP Adv.* **4**(2), 027116 (2014).

[27] E. Varrla, C. Backes, K. R. Paton, A. Harvey, Z. Gholamvand, J. McCauley and J. N. Coleman, *Chem. Mater.* **27**(3), 1129 (2015).

[28] J. Kang, J. D. Wood, S. A. Wells, J.-H. Lee, X. Liu, K.-S. Chen and M. C. Hersam, *ACS Nano.* **9**(4), 3596 (2015).

[29] X. Fan, P. Xu, D. Zhou, Y. Sun, Y. C. Li, M. A. T. Nguyen, M. Terrones and T. E. Mallouk, *Nano Lett.* **15**(9), 5956 (2015).

[30] A. K. Geim and I. V Grigorieva, *Nature* **499**(7459), 419 (2013).

[31] D. Wei, B. Wu, Y. Guo, G. Yu and Y. Liu, *Acc. Chem. Res.* **46**(1), 106 (2013).

[32] T. Yamada, M. Ishihara, J. Kim, M. Hasegawa and S. Iijima, *Carbon N. Y.* **50**(7), 2615 (2012).

[33] J. Zhu, J. Kang, J. Kang, D. Jariwala, J. D. Wood, J.-W. T. Seo, K.-S. Chen, T. J. Marks and M. C. Hersam, *Nano Lett.* **15**(10), 7029 (2015).

[34] M. Yi and Z. Shen, *J. Mater. Chem. A* **3**(22), 11700 (2015).

[35] D. V. Kosynkin, A. L. Higginbotham, A. Sinitskii, J. R. Lomeda, A. Dimiev, B. K. Price and J. M. Tour, *Nature* **458**(7240), 872 (2009).

[36] W. Norimatsu and M. Kusunoki, *Phys. Chem. Chem. Phys.* **16**(8), 3501 (2014).

[37] M. L. Hitchman and K. F. Jensen, *Chemical Vapor Deposition: Principles and Applications.* Academic Press Inc. (1993).

[38] Q. Yu, J. Lian, S. Siriponglert, H. Li, Y. P. Chen and S.-S. Pei, *Appl. Phys. Lett.* **93**(11), 113103 (2008).

[39] A. M. van der Zande, P. Y. Huang, D. A. Chenet, T. C. Berkelbach, Y. You, G.-H. Lee, T. F. Heinz, D. R. Reichman, D. A. Muller and J. C. Hone, *Nat. Mater.* **12**(6), 554 (2013).

[40] B. Liu, M. Fathi, L. Chen, A. Abbas, Y. Ma and C. Zhou, *ACS Nano.* **9**(6), 6119 (2015).

[41] Y. Shi, C. Hamsen, X. Jia, K. K. Kim, A. Reina, M. Hofmann, A. L. Hsu, K. Zhang, H. Li, Z.-Y. Juang, M. S. Dresselhaus, L.-J. Li and J. Kong, *Nano Lett.* **10**(10), 4134 (2010).

[42] G. E. Wood, A. J. Marsden, J. J. Mudd, M. Walker, M. Asensio, J. Avila, K. Chen, G. R. Bell and N. R. Wilson, *2D Mater.* **2**(2), 025003 (2015).

[43] R. Y. Tay, M. H. Griep, G. Mallick, S. H. Tsang, R. S. Singh, T. Tumlin, E. H. T. Teo and S. P. Karna, *Nano Lett.* **14**(2), 839 (2014).

[44] K. H. Lee, H.-J. Shin, J. Lee, I. Lee, G.-H. Kim, J.-Y. Choi and S.-W. Kim, *Nano Lett.* **12**(2), 714 (2012).

[45] K. K. Kim, A. Hsu, X. Jia, S. M. Kim, Y. Shi, M. Hofmann, D. Nezich, J. F. Rodriguez-Nieva, M. Dresselhaus, T. Palacios and J. Kong, *Nano Lett.* **12**(1), 161 (2012).

[46] X. He, F. Liu, P. Hu, W. Fu, X. Wang, Q. Zeng, W. Zhao and Z. Liu, *Small* **11**(40), 5423 (2015).

[47] J. Kang, D. Shin, S. Bae and B. H. Hong, *Nanoscale* **4**(18), 5527 (2012).

[48] Y. S. Kim, J. H. Lee, Y. D. Kim, S.-K. Jerng, K. Joo, E. Kim, J. Jung, E. Yoon, Y. D. Park, S. Seo and S.-H. Chun, *Nanoscale* **5**(3), 1221 (2013).

[49] D. B. Hash and M. Meyyappan, *J. Appl. Phys.* **93**(1), 750 (2003).

[50] H. Conrads and M. Schmidt, *Plasma Sources Sci. Technol.* **9**(4), 441 (2000).

[51] A. Bogaerts, E. Neyts, R. Gijbels and J. van der Mullen, *Spectrochim. Acta Part B At. Spectrosc.* **57**(4), 609 (2002).

[52] Z. Bo, Y. Yang, J. Chen, K. Yu, J. Yan and K. Cen, *Nanoscale* **5**(12), 5180 (2013).

[53] M. A. Lieberman and A. J. Lichtenberg, *Principles of Plasma Discharges and Materials Processing.* Hoboken, NJ, USA: John Wiley & Sons, Inc. (2005).

[54] D. Wei, Y. Lu, C. Han, T. Niu, W. Chen and A. T. S. Wee, *Angew. Chem. Int. Ed. Engl.* **52**(52), 14121 (2013).

[55] Y. Kim, W. Song, S. Y. Lee, C. Jeon, W. Jung, M. Kim and C.-Y. Park, *Appl. Phys. Lett.* **98**(26), 263106 (2011).

[56] N. Woehrl, O. Ochedowski, S. Gottlieb, K. Shibasaki and S. Schulz, *AIP Adv.* **4**(4), 047128 (2014).

[57] D. A. Boyd, W.-H. Lin, C.-C. Hsu, M. L. Teague, C.-C. Chen, Y.-Y. Lo, W.-Y. Chan, W.-B. Su, T.-C. Cheng, C.-S. Chang, C.-I. Wu and N.-C. Yeh, *Nat. Commun.* **6**, 6620 (2015).

[58] C. S. Lee, L. Baraton, Z. He, J.-L. Maurice, M. Chaigneau, D. Pribat and C. S. Cojocaru, *Proc. SPIE* **7761**, 77610 (2010).

[59] J. L. Qi, W. T. Zheng, X. H. Zheng, X. Wang and H. W. Tian, *Appl. Surf. Sci.* **257**(15), 6531 (2011).

[60] L. Van Nang and E.-T. Kim, *J. Electrochem. Soc.* **159**(4), K93 (2012).

[61] T. Terasawa and K. Saiki, *Carbon N. Y.* **50**(3), 869 (2012).

[62] A. N. Obraztsov, E. A. Obraztsova, A. V. Tyurnina and A. A. Zolotukhin, *Carbon N. Y.* **45**(10), 2017 (2007).

[63] H. K. Jeong, J. D. C. Edward, G. H. Yong and L. Choong Hun, *J. Korean Phys. Soc.* **58**(1), 53 (2011).

[64] C. S. Lee, C. S. Cojocaru, W. Moujahid, B. Lebental, M. Chaigneau, M. Châtelet, F. Le Normand and J.-L. Maurice, *Nanotechnology* **23**(26), 265603 (2012).

[65] L. Zhang, Z. Shi, Y. Wang, R. Yang, D. Shi and G. Zhang, *Nano Res.* **4**(3), 315 (2010).

[66] S. Wang, L. Qiao, C. Zhao, X. Zhang, J. Chen, H. Tian, W. Zheng and Z. Han, *New J. Chem.* **37**(5), 1616 (2013).

[67] L. Van Nang and E.-T. Kim, *Mater. Lett.* **92**, 437 (2013).

[68] K.-J. Peng, C.-L. Wu, Y.-H. Lin, Y.-J. Liu, D.-P. Tsai, Y.-H. Pai and G.-R. Lin, *J. Mater. Chem. C* **1**(24), 3862 (2013).

[69] J. Kim, M. Ishihara, Y. Koga, K. Tsugawa, M. Hasegawa and S. Iijima, *Appl. Phys. Lett.* **98**(9), 091502 (2011).

[70] A. Kumar, A. A. Voevodin, D. Zemlyanov, D. N. Zakharov and T. S. Fisher, *Carbon N. Y.* **50**(4), 1546 (2012).

[71] L. Cheng, K. Yun, A. Lucero, J. Huang, X. Meng, G. Lian, H.-S. Nam, R. M. Wallace, M. Kim, A. Venugopal, L. Colombo and J. Kim, *J. Mater. Chem. C* **3**(20), 5192 (2015).

[72] S. Chugh, R. Mehta, N. Lu, F. D. Dios, M. J. Kim and Z. Chen, *Carbon N. Y.* **93**, 393 (2015).

[73] S. Chen, M. Gao, R. Cao, H. Du, J. Yang, L. Zhao and Z. Ma, *J. Mater. Sci. Mater. Electron.* **26**(3), 1485 (2014).

[74] Y. Ando, X. Zhao and M. Ohkohchi, *Carbon N. Y.* **35**(1), 153 (1997).

[75] D. Banerjee, S. Mukherjee and K. K. Chattopadhyay, *Appl. Surf. Sci.* **257**(8), 3717 (2011).

[76] K. V. Mironovich, D. M. Itkis, D. A. Semenenko, S. A. Dagesian, L. V. Yashina, E. Y. Kataev, Y. A. Mankelevich, N. V. Suetin and V. A. Krivchenko, *Phys. Chem. Chem. Phys.* **16**(46), 25621 (2014).

[77] M. Hiramatsu, K. Shiji, H. Amano and M. Hori, *Appl. Phys. Lett.* **84**(23), 4708 (2004).

[78] J. Wang, M. Zhu, R. A. Outlaw, X. Zhao, D. M. Manos and B. C. Holloway, *Carbon N. Y.* **42**(14), 2867 (2004).

[79] J. Zhao, M. Shaygan, J. Eckert, M. Meyyappan and M. H. Rümmeli, *Nano Lett.* **14**(6), 3064 (2014).

[80] K. Davami, M. Shaygan, N. Kheirabi, J. Zhao, D. A. Kovalenko, M. H. Rümmeli, J. Opitz, G. Cuniberti, J.-S. Lee and M. Meyyappan, *Carbon N. Y.* **72**, 372 (2014).

[81] M. Cai, R. A. Outlaw, R. A. Quinlan, D. Premathilake, S. M. Butler and J. R. Miller, *ACS Nano.* **8**(6), 5873 (2014).

[82] M. V. Jacob, R. S. Rawat, B. Ouyang, K. Bazaka, D. S. Kumar, D. Taguchi, M. Iwamoto, R. Neupane and O. K. Varghese, *Nano Lett.* **15**(9), 5702 (2015).

[83] K. Davami, M. Shaygan and I. Bargatin, *J. Mater. Res.* **30**(05), 617 (2015).

[84] Y. Wu, P. Qiao, T. Chong and Z. Shen, *Adv. Mater.* **14**(1), 64 (2002).

[85] A. Malesevic, R. Vitchev, K. Schouteden, A. Volodin, L. Zhang, G. Van Tendeloo, A. Vanhulsel and C. Van Haesendonck, *Nanotechnology* **19**(30), 305604 (2008).

[86] R. Vitchev, A. Malesevic, R. H. Petrov, R. Kemps, M. Mertens, A. Vanhulsel and C. Van Haesendonck, *Nanotechnology* **21**(9), 095602 (2010).

[87] T. Bhuvana, A. Kumar, A. Sood, R. H. Gerzeski, J. Hu, V. S. Bhadram, C. Narayana and T. S. Fisher, *ACS Appl. Mater. Interfaces* **2**(3), 644 (2010).

[88] R. H. Gerzeski, A. Sprague, J. Hu and T. S. Fisher, *Carbon N. Y.* **69**, 424 (2014).

[89] G. Xiong, K. P. S. S. Hembram, D. N. Zakharov, R. G. Reifenberger and T. S. Fisher, *Diam. Relat. Mater.*, 27, 1 (2012).

[90] G. Xiong, C. Meng, R. G. Reifenberger, P. P. Irazoqui and T. S. Fisher, *Adv. Energy Mater.* **4**(3), 1300515 (2014).

[91] K. Ostrikov, E. C. Neyts and M. Meyyappan, *Adv. Phys.* **62**(2), 113 (2013).

[92] Y. A. Mankelevich, M. N. R. Ashfold and J. Ma, *J. Appl. Phys.* **104**(11), 113304 (2008).

[93] G. Lombardi, K. Hassouni, G.-D. Stancu, L. Mechold, J. Röpcke and A. Gicquel, *J. Appl. Phys.* **98**(5), 053303 (2005).

[94] I. B. Denysenko, B. S. Xu, J. D. Long, P. P. Rutkevych, N. A. Azarenkov and K. Ostrikov, *J. Appl. Phys.* **95**(5), 2713 (2004).

[95] D. K. Davies, L. E. Kline and W. E. Bies, *J. Appl. Phys.* **65**(9), 3311 (1989).

[96] T. Nakano, H. Toyoda and H. Sugai, *Japanese J. Appl. Physics, Part 1 Regul. Pap. Short Notes Rev. Pap.* **30**(11 A), 2912 (1991).

[97] F. Paschen, *Ann. of Physics*, **273**(5), 69–96 (1889).

[98] Z. Bo, K. Yu, G. Lu, P. Wang, S. Mao and J. Chen, *Carbon N. Y.* **49**(6), 1849 (2011).

[99] I. Vlassiouk, M. Regmi, P. Fulvio, S. Dai, P. Datskos, G. Eres and S. Smirnov, *ACS Nano.* **5**(7), 6069 (2011).

[100] X. Li, W. Cai, J. An, S. Kim, J. Nah, D. Yang, R. Piner, A. Velamakanni, I. Jung, E. Tutuc, S. K. Banerjee, L. Colombo and R. S. Ruoff, *Science* **324**(5932), 1312 (2009).

[101] X. Li, W. Cai, L. Colombo and R. S. Ruoff, *Nano Lett.* **9**(12), 4268 (2009).

[102] A. Yoshimura, H. Yoshimura, S. C. Shin, K. Kobayashi, M. Tanimura and M. Tachibana, *Carbon N. Y.* **50**(8), 2698 (2012).

[103] R. Koch, D. Hu and A. K. Das, *Phys. Rev. Lett.* **94**(14), 146101 (2005).

[104] X. Song, J. Liu, L. Yu, J. Yang, L. Fang, H. Shi, C. Du and D. Wei, *Mater. Lett.* **137**, 25 (2014).

[105] Y. Hao, M. S. Bharathi, L. Wang, Y. Liu, H. Chen, S. Nie, X. Wang, H. Chou, C. Tan, B. Fallahazad, H. Ramanarayan, C. W. Magnuson, E. Tutuc, B. I. Yakobson, K. F. McCarty, Y.-W. Zhang, P. Kim, J. Hone, L. Colombo and R. S. Ruoff, *Science* **342**(6159), 720 (2013).

[106] J. Qi, L. Zhang, J. Cao, W. Zheng, X. Wang and J. Feng, *Chinese Sci. Bull.* **57**(23), 3040 (2012).

[107] M. Zhu, J. Wang, B. C. Holloway, R. A. Outlaw, X. Zhao, K. Hou, V. Shutthanandan and D. M. Manos, *Carbon N. Y.* **45**(11), 2229 (2007).

[108] S. Das, J. A. Robinson, M. Dubey, H. Terrones and M. Terrones, *Annu. Rev. Mater. Res.* **45**(1), 1 (2015).

[109] J. V. Barth, G. Costantini and K. Kern, *Nature* **437**(7059), 671 (2005).

[110] J. Barth, *Surf. Sci. Rep.* **40**(3–5), 75–149 (2000).

[111] N. Guo, J. Wei, L. Fan, Y. Jia, D. Liang, H. Zhu, K. Wang and D. Wu, *Nanotechnology* **23**(41), 415605 (2012).

[112] Y. Zhang, L. Zhang, P. Kim, M. Ge, Z. Li and C. Zhou, *Nano Lett.* **12**(6), 2810 (2012).

[113] C. Muratore, V. Varshney, J. J. Gengler, J. Hu, J. E. Bultman, A. K. Roy, B. L. Farmer and A. A. Voevodin, *Phys. Chem. Chem. Phys.* **16**(3), 1008 (2014).

[114] N. R. Glavin, M. L. Jespersen, M. H. Check, J. Hu, A. M. Hilton, T. S. Fisher and A. A. Voevodin, *Thin Solid Films* **572**, 245 (2014).

[115] C. Muratore, J. J. Hu, B. Wang, M. A. Haque, J. E. Bultman, M. L. Jespersen, P. J. Shamberger, M. E. McConney, R. D. Naguy and A. A. Voevodin, *Appl. Phys. Lett.* **104**(26), 261604 (2014).

[116] A. A. Voevodin, A. R. Waite, J. E. Bultman, J. Hu and C. Muratore, *Surf. Coatings Technol.* **280**, 260 (2015).

[117] H. J. Park, B. W. Ahn, T. Y. Kim, J. W. Lee, Y. H. Jung, Y. S. Choi, Y. Il Song and S. J. Suh, *Thin Solid Films* **587**, 8 (2015).

[118] M. C. McMaster, W. L. Hsu, M. E. Coltrin, D. S. Dandy and C. Fox, *Diam. Relat. Mater.* **4**(7), 1000 (1995).

[119] J. Luque, W. Juchmann and J. B. Jeffries, *Appl. Opt.* **36**(15), 3261 (1997).

[120] W. R. Harshbarger, R. A. Porter and P. Norton, *J. Electron. Mater.* **7**(3), 429 (1978).

[121] J. W. Coburn and M. Chen, *J. Appl. Phys.* **51**(6), 3134 (1980).

[122] C. J. Mogab, A. C. Adams and D. L. Flamm, *J. Appl. Phys.* **49**(7), 3796 (1978).

[123] N. M. Laurendeau, *Statistical thermodynamics: fundamentals and applications.* Cambridge University Press, Cambridge (2005).

[124] N. Britun, M. Gaillard, A. Ricard, Y. M. Kim, K. S. Kim and J. G. Han, *J. Phys. D. Appl. Phys.* **40**(4), 1022–1029 (2007).

[125] D. L. Crintea, U. Czarnetzki, S. Iordanova, I. Koleva and D. Luggenhölscher, *J. Phys. D. Appl. Phys.* **42**(4), 045208 (2009).

[126] A. Gicquel, M. Chenevier, K. Hassouni, A. Tserepi and M. Dubus, *J. Appl. Phys.* **83**(12), 7504 (1998).

[127] P. J. Brüggeman, N. Sadeghi, D. C. Schram and V. Linss, *Plasma Sources Sci. Technol.* **23**(2), 023001 (2014).

[128] C. O. Laux, T. G. Spence, C. H. Kruger and R. N. Zare, *Plasma Sources Sci. Technol.* **12**(2), 125 (2003).

[129] R. K. Garg, T. N. Anderson, R. P. Lucht, T. S. Fisher and J. P. Gore, *J. Phys. D. Appl. Phys.* **41**(9), 095206 (2008).

[130] T.-L. Zhao, Y. Xu, Y.-H. Song, X.-S. Li, J.-L. Liu, J.-B. Liu and A.-M. Zhu, *J. Phys. D. Appl. Phys.* **46**(34), 345201 (2013).

[131] Z. Qing, D. K. Otorbaev, G. J. H. Brussaard, M. C. M. van de Sanden and D. C. Schram, *J. Appl. Phys.* **80**(3), 1312 (1996).

[132] B. A. Cruden, M. V. V. S. Rao, S. P. Sharma and M. Meyyappan, *J. Appl. Phys.* **91**(11), 8955 (2002).

[133] H. N. Chu, E. A. Den Hartog, A. R. Lefkow, J. Jacobs, L. W. Anderson, M. G. Lagally and J. E. Lawler, *Phys. Rev. A* **44**(6), 3796 (1991).

[134] J. Luque, M. Kraus, A. Wokaun, K. Haffner, U. Kogelschatz and B. Eliasson, *J. Appl. Phys.* **93**(8), 4432 (2003).

[135] S. W. Reeve, *J. Vac. Sci. Technol. A Vacuum, Surfaces, Film.* **13**(2), 359 (1995).

[136] S. Y. Moon and W. Choe, *Spectrochim. Acta Part B At. Spectrosc.* **58**(2), 249 (2003).

[137] J. Ma, M. N. R. Ashfold and Y. A. Mankelevich, *J. Appl. Phys.* **105**(4), 043302 (2009).

[138] B. Marcus, M. Mermoux, F. Vinet, A. Campargue and M. Chenevier, *Surf. Coatings Technol.* **47**(1–3), 608 (1991).

[139] T. Lang, J. Stiegler, Y. von Kaenel and E. Blank, *Diam. Relat. Mater.* **5**(10), 1171 (1996).

[140] M. A. Elliott, P. W. May, J. Petherbridge, S. M. Leeds, M. N. R. Ashfold and W. N. Wang, *Diam. Relat. Mater.* **9**(3–6), 311 (2000).

[141] T. Vandevelde, M. Nesladek, C. Quaeyhaegens and L. Stals, *Thin Solid Films* **290**, **291**, 143 (1996).

[142] J. C. Richley, M. W. Kelly, M. N. R. Ashfold and Y. A. Mankelevich, *J. Phys. Chem. A* **116**(38), 9447 (2012).

[143] A. N. Obraztsov, A. A. Zolotukhin, A. O. Ustinov, A. P. Volkov, Y. Svirko and K. Jefimovs, *Diam. Relat. Mater.* **12**(3–7), 917 (2003).

[144] K. W. Hemawan and R. J. Hemley, *J. Vac. Sci. Technol. A Vacuum, Surfaces, Film.* **33**(6), 061302 (2015).

[145] E. Sandoz-Rosado, W. Page, D. O'Brien, J. Przepioski, D. Mo, B. Wang, T.-T. Ngo-Duc, J. Gacusan, M. W. Winter, M. Meyyappan, R. D. Cormia, S. Takahashi and M. M. Oye, *J. Mater. Res.* **29**(03), 417 (2013).

[146] Y. Ma, H. Jang, S. J. Kim, C. Pang and H. Chae, *Nanoscale Res. Lett.* **10**(1), 1019 (2015).

[147] S. Vizireanu, S. D. Stoica, C. Luculescu, L. C. Nistor, B. Mitu and G. Dinescu, *Plasma Sources Sci. Technol.* **19**(3), 034016 (2010).

[148] S. Mori, T. Ueno and M. Suzuki, *Diam. Relat. Mater.* **20**(8), 1129 (2011).

[149] M. Zhu, J. Wang, R. A. Outlaw, K. Hou, D. M. Manos and B. C. Holloway, *Diam. Relat. Mater.* **16**(2), 196 (2007).

[150] Y. Ma, D. Kim, H. Jang, S. M. Cho and H. Chae, *J. Nanosci. Nanotechnol.* **14**(12), pp. 9065–9072, Dec. (2014).

[151] N. R. Glavin, C. Muratore, M. L. Jespersen, J. Hu, T. S. Fisher and A. A. Voevodin, *J. Appl. Phys.* **117**(16), 165305 (2015).

[152] D. B. Geohegan, *Pulsed Laser Deposition of Thin Films*, pp. 115–165 Wiley & Sons, Inc., Newyork (1994).

[153] L. S. Rothman, I. E. Gordon, Y. Babikov, A. Barbe, D. Chris Benner, P. F. Bernath, M. Birk, L. Bizzocchi, V. Boudon, L. R. Brown, A. Campargue, K. Chance, E. A. Cohen, L. H. Coudert, V. M. Devi, B. J. Drouin, A. Fayt, J.-M. Flaud, R. R. Gamache, J. J. Harrison, J.-M. Hartmann, C. Hill, J. T. Hodges, D. Jacquemart, A. Jolly, J. Lamouroux, R. J. Le Roy, G. Li, D. A. Long, O. M. Lyulin, C. J. Mackie, S. T. Massie, S. Mikhailenko, H. S. P. Müller, O. V. Naumenko, A. V. Nikitin, J. Orphal, V. Perevalov, A. Perrin,

314 *M. A. Alrefae et al.*

E. R. Polovtseva, C. Richard, M. A. H. Smith, E. Starikova, K. Sung, S. Tashkun, J. Tennyson, G. C. Toon, V. G. Tyuterev and G. Wagner, *J. Quant. Spectrosc. Radiat. Transf.* **130**, 4 (2013).

[154] C. Busch, I. Möller and H. Soltwisch, *Plasma Sources Sci. Technol.* **10**(2), 259 (2001).

[155] J. H. van Helden, G. Hancock, R. Peverall and G. A. D. Ritchie, *J. Phys. D. Appl. Phys.* **44**(12), 125202 (2011).

[156] A. Cheesman, J. A. Smith, M. N. R. Ashfold, N. Langford, S. Wright and G. Duxbury, *J. Phys. Chem. A* **110**(8), 2821 (2006).

[157] J. B. Wills, M. N. R. Ashfold, A. J. Orr-Ewing, Y. A. Mankelevich and N. V. Suetin, *Diam. Relat. Mater.* **12**(8), 1346 (2003).

[158] J. Ma, A. Cheesman, M. N. R. Ashfold, K. G. Hay, S. Wright, N. Langford, G. Duxbury and Y. A. Mankelevich, *J. Appl. Phys.* **106**(3), 033305 (2009).

[159] M. Hundt, P. Sadler, I. Levchenko, M. Wolter, H. Kersten and K. (Ken) Ostrikov, *J. Appl. Phys.* **109**(12), 123305 (2011).

[160] D. Lopatik, N. Lang, U. Macherius, H. Zimmermann and J. Röpcke, *Meas. Sci. Technol.* **23**(11), 115501 (2012).

[161] V. S. der Gathen, J. Röpcke, T. Gans, M. Käning, C. Lukas and H. F. Döbele, *Plasma Sources Sci. Technol.* **10**(3), 530 (2001).

[162] C. Rond, S. Hamann, M. Wartel, G. Lombardi, A. Gicquel and J. Röpcke, *J. Appl. Phys.* **116**(9), 093301 (2014).

[163] A. Serdioutchenko, I. Möller and H. Soltwisch, *Spectrochim. Acta. A. Mol. Biomol. Spectrosc.* **60**(14), 3311 (2004).

[164] G. Lombardi, K. Hassouni, F. Bénédic, F. Mohasseb, J. Röpcke and A. Gicquel, *J. Appl. Phys.* **96**(11), 6739 (2004).

[165] F. Hempel, P. B. Davies, D. Loffhagen, L. Mechold and J. Röpcke, *Plasma Sources Sci. Technol.* **12**(4), S98 (2003).

[166] G. Lombardi, G. D. Stancu, F. Hempel, A. Gicquel and J. Röpcke, *Plasma Sources Sci. Technol.* **13**(1), 27 (2004).

[167] X. Wu, C. Li, Y. Wang, Z. Wang, C. Feng and H. Ding, *Appl. Phys. B* **120**(4), 659 (2015).

[168] R. Engeln, K. G. Y. Letourneur, M. G. H. Boogaarts, M. C. M. Van De Sanden and D. C. Schram, *Chem. Phys. Lett.*, **310**(5–6), 405 (1999).

[169] C. J. Rennick, J. Ma, J. J. Henney, J. B. Wills, M. N. R. Ashfold, A. J. Orr-Ewing and Y. A. Mankelevich, *J. Appl. Phys.* **102**(6), 063309 (2007).

[170] M. Hiramatsu and M. Hori, *Jpn. J. Appl. Phys.*, **45**(6B), 5522 (2006).

[171] R. E. Palmer, *The CARSFT Computer Code for Calculating Coherent Anti-Stokes Raman Spectra: User and Programmer Information*, Sandia National Laboratories Report, SAND89-8206 (1989).

[172] A. D. Tuesta, A. Bhuiyan, R. P. Lucht and T. S. Fisher, *J. Micro Nano-Manufacturing* **2**(3), 31002 (2014).

[173] S. Roy, J. R. Gord and A. K. Patnaik, *Prog. Energy Combust. Sci.* **36**(2), 280 (2010).

[174] S. O. Hay, W. C. Roman and M. B. Colket, *J. Mater. Res.* **5**(11), 2387 (1990).

[175] A. D. Tuesta, A. Bhuiyan, R. P. Lucht and T. S. Fisher, *J. Micro Nano-Manufacturing* **4**(1), 11005 (2015).

[176] A. Gicquel, K. Hassouni, S. Farhat, Y. Breton, C. D. Scott, M. Lefebvre and M. Pealat, *Diam. Relat. Mater.* **3**(4–6), 581 (1994).

[177] C. F. Kaminski and P. Ewart, *Appl. Phys. B Lasers Opt.* **64**(1), 103 (1997).

[178] O. Leroy, J. Perrin, J. Jolly, M. Péalat and M. Lefebvre, *J. Phys. D. Appl. Phys.* **30**(4), 499 (1997).

[179] V. Kornas, V. Schulz-von der Gathen, T. Bornemann, H. F. Döbele and G. Prosz, *Plasma Chem. Plasma Process.* **11**(2), 171 (1991).

[180] V. Kornas, A. Roth, H. F. Döbele and G. Proß, *Plasma Chem. Plasma Process.* **15**(1), 71 (1995).

[181] Y. H. Shing, J. W. Perry and C. E. Allevato, *Sol. Cells* **24**(3–4), 353 (1988).

[182] N. Hata, A. Matsuda and K. Tanaka, *Japanese J. Appl. Physics, Part 1 Regul. Pap. Short Notes* **25**(1), 108 (1986).

Chapter 8

Metal Contacts to MoS$_2$

Naveen Kaushik[*], Sameer Grover[†], Mandar M. Deshmukh[†,‡]
and Saurabh Lodha[*,§]

[*]*Indian Institute of Technology Bombay, Mumbai 400076, India*

[†]*Tata Institute of Fundamental Research, Mumbai 400005, India*
[‡]*deshmukh@tifr.res.in*
[§]*slodha@iitb.ac.in*

Abstract. The performance of transistors based on 2D transition metal dichalcogenides such as MoS$_2$ is significantly limited by large contact resistance at the metal-MoS$_2$ interface. In this chapter, we present a comprehensive review of metal-MoS$_2$ contact interfaces and recent progress made towards realising low contact resistance on MoS$_2$. Review of metal-semiconductor contact and Fermi-level pinning theory is followed by an analysis of contact resistance and Schottky barrier height data, along with methods used for their extraction, for different work function metals on MoS$_2$ as collated from several experimental and simulation studies. Reduction in contact resistance towards its quantum limit using conventional (doping and interfacial layers) and unconventional (different contact geometries) methods, and different Fermi-level pinning mechanisms responsible for n-type conduction in MoS$_2$ have also been discussed and compared in detail.

8.1 Introduction

Intrinsic electrical and physical properties of MoS$_2$ such as its energy band gap, ultra-thin nature, thermal and chemical stability, etc. make it a promising material for future electronic, optoelectronic, sensing and spintronic applications [1]. However, poor connection (interface) with external metallic circuits limits its potential in

317

future device technologies to a great extent [2]. The performance of MoS_2-based devices has been found to be strongly affected by the metals used to form contacts for interfacing it with external circuits and systems [3]. Unlike conventional silicon field-effect transistors (FETs), carrier injection in a large variety of MoS_2 devices is governed by Schottky barriers at the contact interfaces [4]. In view of the strong influence of metal contact Schottky barriers on device performance, a comprehensive picture of current understanding of contacts to MoS_2, different mechanisms for Fermi-level pinning (FLP) and recent progress towards realising low-resistance contacts to MoS_2 are presented in this chapter.

8.2   Review of Schottky Barrier Height Theory

Rectifying contacts at metal–semiconductor interfaces were first reported by Braun in 1874 [5]. However, major breakthroughs came in 1929 and 1938, when Schottky *et al.* [6] gave the first insights into metal–semiconductor contacts by measuring the differential capacitance at the interface. Analogous to a parallel plate capacitor, rectifying behaviour at the metal–semiconductor interface was attributed to a space charge region in the semiconductor. Hence, these metal–semiconductor contacts are also referred to as Schottky contacts. However, the theory given by Schottky–Mott could not completely explain the experimental results. This led to the development of various theories involving interface states, metal induced gap states (MIGS), charge neutrality level and interface dipoles to explain the experimental behaviour of metal–semiconductor contacts.

8.2.1   Schottky–Mott rule

A discussion on metal–semiconductor contacts always begins with the Schottky–Mott rule [6]. It signifies the critical role played by the electrostatics in energy band alignment. When a metal is brought in contact with a semiconductor, the potential (ΔV) developed between them is equal to the difference in their work functions,

$$q\Delta V = \Phi_m - \Phi_{sc}, \tag{8.1}$$

where Φ_m and Φ_{sc} are the work functions of the metal and the semiconductor, respectively. Analogous to a parallel plate capacitor, the potential difference between the metal and the semiconductor can be expressed as,

$$q\Delta V = \Phi_m - \Phi_{sc} = q\frac{Q_{sc}}{C} = q\frac{Q_{sc}}{\varepsilon_0}d, \qquad (8.2)$$

where capacitance per unit area (C) is modelled using a parallel plate capacitor geometry, q is the electron charge and ε_0 is the dielectric constant of vacuum. For intimate contacts $d = 0$ (vacuum gap width), and hence Eq. (8.2) can be written as,

$$\Phi_m - \Phi_{sc} = 0. \qquad (8.3)$$

It can be seen from Figure 1(a) that, $\Phi_{sc} = \chi_{sc} + E_c - E_F$ for an n-type semiconductor. Here, χ_{sc} is the electron affinity and E_F is the Fermi energy level of the semiconductor.

Equation (8.3) can be written as,

$$\Phi_m - \chi_{sc} - (E_C - E_{Fm}) = 0, \qquad (8.4)$$

$$\Phi_B = E_C - E_{Fm} = \Phi_m - \chi_{sc}. \qquad (8.5)$$

The difference between E_C of the semiconductor and the metal Fermi-level at the interface can be defined as Schottky barrier height (Φ_B) shown in Figure 1(b).

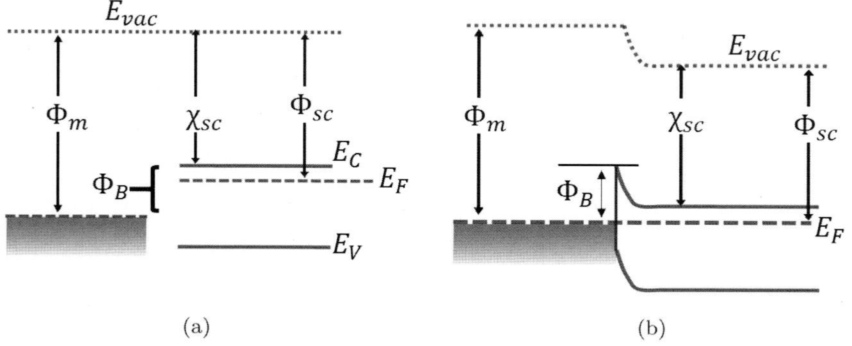

Figure 1. (a) Energy levels of metal and n-type semiconductor when not in contact, (b) energy band alignment of the metal–semiconductor contact in thermal equilibrium.

8.2.2 Fermi-level pinning

As seen from Eq. (8.5) (Schottky–Mott rule), the contact barrier height should vary in linear proportion (slope = 1) with the metal work function. However, most experimental data does not strictly follow the Schottky–Mott rule and in some cases, Φ_B is seen to be completely insensitive to variation in metal work function. Barrier height insensitivity to metal work function is also known as FLP. The inconsistency of experimental results with the Schottky–Mott rule, may be attributed to the fact that it does not consider chemical reactions or charge redistribution at the metal–semiconductor interface. Whereas in practice, interface dipoles arise due to charge redistribution and affect the measured Schottky barrier height. Therefore, Φ_B also depends on other interface parameters besides the metal work function.

8.2.2.1 *Interface states*

The discrepancy between the Schottky–Mott model and measured experimental Schottky barrier height data was first addressed by Bardeen in 1947 [7]. He showed that electronic states localised at the interface within the semiconductor band gap also play a role in determining the Schottky barrier height. Therefore, the overall charge neutrality equation can be written as,

$$Q_m + Q_{sc} + Q_{is} = 0, \tag{8.6}$$

where Q_m is charge on metal, Q_{sc} is charge on semiconductor and Q_{is} is charge at the interface.

Cowley and Sze [8], further developed Bardeen's idea of interface states assuming a uniform distribution across the energy band gap of the semiconductor as shown in Figure 2. These interface states can create a dipole and alter the semiconductor band bending significantly. The barrier height can now be shown to depend on the charge neutrality level Φ_{CNL} — an energy level (measured from conduction band) that corresponds to net charge neutrality at the semiconductor surface with interface states. Equation (8.5) can be written as,

$$\Phi_B = S(\Phi_m - \chi_s) + (1 - S)(\Phi_{\mathrm{CNL}}), \tag{8.7}$$

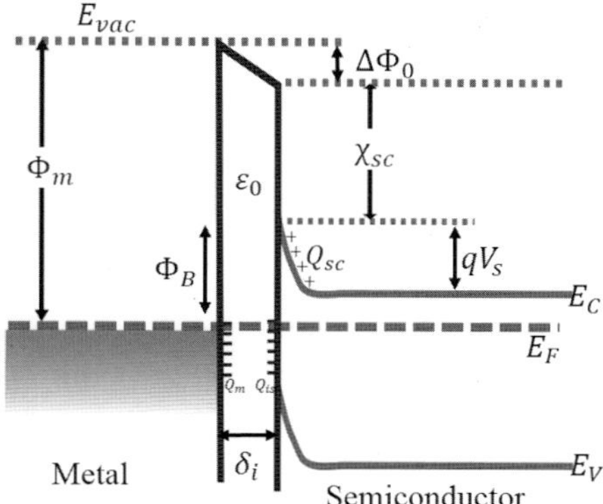

Figure 2. Energy band alignment of a metal and n-type semiconductor with interface states.

where S is the FLP factor defined as,

$$S = \frac{\partial \Phi_B}{\partial \Phi_m}. \tag{8.8}$$

According to Cowley and Sze's model, the surface state density is given by,

$$D_S = \frac{(1 - S)\varepsilon_i \varepsilon_0}{S q^2 \delta_i}, \tag{8.9}$$

where δ_i is the interface layer thickness, usually less than the lattice constant and ε_i is the dielectric constant at the metal–semiconductor interface [7, 8]. For the case of complete lack of FLP due to zero surface state density, i.e., $S = 1$, Eq. (8.7) results into the Schottky–Mott model as expected.

$$\Phi_B = \Phi_m - \chi_s. \tag{8.10}$$

For $S = 0$, the Fermi-level is pinned at Φ_{CNL} and Φ_B becomes completely insensitive to the metal work function. In other words, Φ_B is now entirely dependent on the surface properties of the semiconductor as given by Eq. (8.11).

$$\Phi_B = \Phi_{\mathrm{CNL}}. \tag{8.11}$$

The interface states proposed in this model are determined entirely by the surface properties of the semiconductor and do not depend on the contact metal. Hence, barrier height in Eq. (8.7) is found to vary linearly with metal work function with slope S that can vary from 0 (FLP) to 1 (Schottky–Mott).

8.2.2.2   *Metal induced gap states*

Unlike the surface states proposed by Bardeen, Heine attributed the origin of interface states at the metal–semiconductor interface to the decaying tails of the electron wave-functions in the semiconductor band gap [9]. These interface states are called MIGS. Decay length of these virtual MIGS is proportional to the inverse of the band gap, which in turn depends on the dielectric constant of the semiconductor. In MoS_2, FLP has been attributed to MIGS at the metal-MoS_2 interface which will be discussed in detail in Section 8.8.

8.2.2.3   *Charge transfer at interfaces*

MIGS, interface states and Schottky–Mott theory assume a constant work function for the contact metal. However, work function is not constant after the metal makes contact to the semiconductor. Two components contribute towards the work function: (1) surface dipoles which can change according to the surface orientation of the metal, and, (2) electronegativity of the metal which is a bulk property. The dipole due to charge redistribution at the metal–semiconductor interface can alter the metal work function and atomic bonding significantly, resulting in a notably different metal–semiconductor energy band alignment than the classical Schottky–Mott theory.

8.3   Experimental Studies of Schottky Contacts on MoS_2

There have been extensive investigations on electrical contacts to two-dimensional (2D) materials such as MoS_2 in the recent past. MoS_2-based devices were found to be n-type in most experimental reports. Even high work function metals like Au and Pt (except Pd which shows both p [10] and n-type [11] transport) have been found to pin above midgap of MoS_2. A summary of Schottky barrier heights

Table 1. Summary of Schottky barrier height (Φ_B) data for a range of metals deposited on MoS$_2$.

Metal	Work Function	Contact Type	Low Temperature $I_{DS} - V_{GS}$ Measurement	TLM	DFT Simulation
Sc	3.6	n	0.02–0.04 [12]		−0.1 [13]
In	4.2	n			0.47 [14]
Ti	4.3	n	0.05 [12], 0.03–0.07 [15]		0.35, 0.3 [14]
Ni	4.5	n	0.12 [12]		0.45 [13]
Mo	4.5	n			0.1 [16], 0.13 [14]
Au	5.1–5.4	n	0.06–0.16 [3]	0.126 [3]	0.62 [14]
Pd	5.2–5.6	n/p	0.38–0.5 [3], 0.25 [17]	0.4 [3]	0.55 [13], 0.9 [14]
Pt	5.6	n	0.22 [12]		0.6 [13]

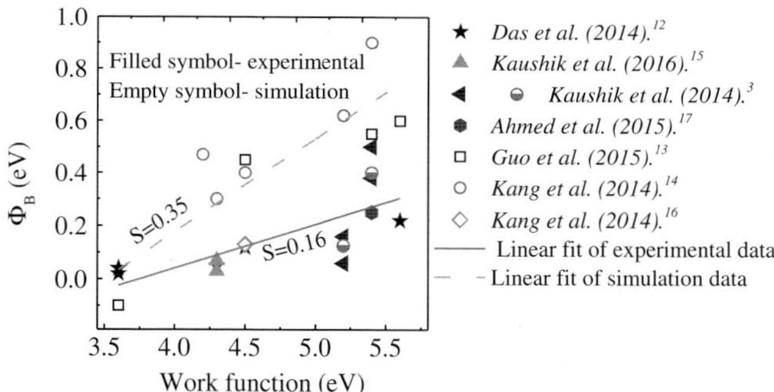

Figure 3. Summary of Schottky barrier heights for different work function metals on MoS$_2$. The dotted and solid lines are linear fits to simulated and experimental Φ_B, respectively.

(Φ_B), for different metals on exfoliated MoS$_2$ flakes extracted using low temperature $I_{DS} - V_{GS}$ measurements, transmission line measurements (TLM) and density functional theory (DFT) simulations is shown in Table 1.

The variation in reported Φ_B for a single metal could be due to variations in MoS$_2$ flake defect densities, metal deposition techniques, etc. Schottky barrier height vs. work function data confirms partial FLP at the metal-MoS$_2$ interface. The FLP factor for experimental ($S = 0.16$) and simulation studies ($S = 0.35$) are shown in Figure 3. The different S values for experimental and simulation data can be attributed to the fact that simulations are usually done on pristine or defect free MoS_2, whereas exfoliated or chemical vapour deposition (CVD) grown MoS_2 used in most experimental studies is not defect free. However, FLP in MoS_2 is quite different from pinning in conventional semiconductors [18]. Different mechanisms of FLP and reasons behind n-type conduction in MoS_2 are discussed in detail later in this chapter.

8.4 Computational Study of Metal Contacts to MoS_2

Apart from experimental reports, there have been numerous reports published on computational studies of metal contacts to MoS_2 (Table 1). Similar to the experimental reports, most of the metals were found to be pinned close to the conduction band resulting in n-type conduction. An important point to note here is that the Φ_B values were calculated from first principle calculations on defect free MoS_2, hence, the n-type behaviour cannot be attributed to defects alone [13].

8.4.1 Selection of contact metals

Selection of contact metals to 2D transition metal dichalcogenides (TMDs) such as MoS_2 is a very crucial step in realising low-resistance contacts. Physical and chemical properties such as high melting point, high electrical conductance and chemical stability make Al, Ti, Cr, Ni, Cu, Pd, Ag, In, Pt and Au promising candidates for contact metals. Kang *et al.* [14] have highlighted the parameters that play a key role in achieving efficient electron injection to/from the Schottky contacts. First is the presence of d-orbitals in the contact metal for mixing/hybridising with the d-orbitals of the transition metal in the TMDs and the second is small lattice mismatch to maximise orbital overlap [14]. Therefore, the large lattice mismatch of Cr and Ni with MoS_2 and the absence of d-orbitals in Al make them unsuitable for forming low-resistance contacts on MoS_2. Partial density of states data in Figure 4 indicates that Ti–MoS_2 contact has the highest density of states close to E_F. Hence, Ti is the most appropriate contact metal for MoS_2.

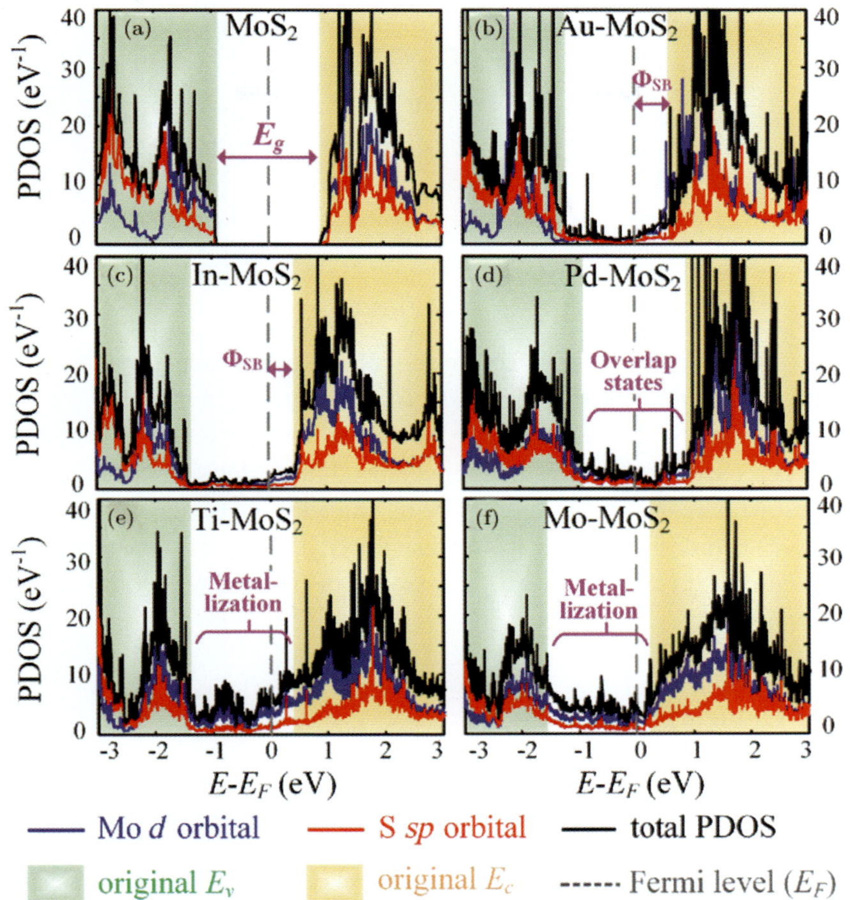

Figure 4. Partial density of states for different contact metals on MoS$_2$: (a) MoS$_2$, (b) Au–MoS$_2$, (c) Pd–MoS$_2$, (d) In–MoS$_2$, (e) Ti–MoS$_2$ and (f) Mo–MoS$_2$. (Adapted from Ref. [14].)

8.4.2   Effect of contact interface on Schottky barrier height

Different contact interface geometries (end-contacts and side-contacts shown in Figure 5) of metal-MoS$_2$ contacts have been studied to investigate their effect on Schottky barrier height [14]. Using *ab initio* DFT calculations, Kang *et al.* [14] concluded that strong overlap of electron orbitals in end-contacts shows better

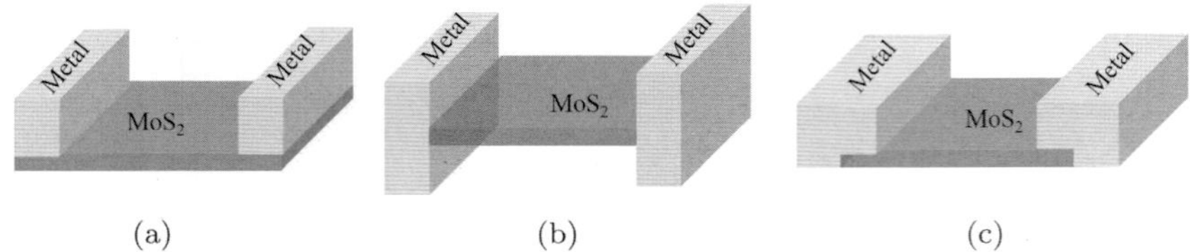

Figure 5. Different contact interface geometries for metal-MoS_2: (a) side-contact, (b) end-contact and (c) side + end-contact.

electron injection when compared to side-contacts resulting in smaller Schottky/tunnel barriers. In back gated multilayer transistors, the MoS_2 layers that are closer to the gate are inverted to a larger extent and have higher sheet charge density compared to layers which are farther away from the gate. Therefore, the side-contact geometry suffers from very high inter-layer resistance [19] and cannot provide efficient electron injection when compared to end-contacts.

8.5   Device Fabrication

Back gate MoS_2 transistor fabrication starts with 90 nm or 280 nm (for flake visibility) thermally grown SiO_2 on a degenerately doped p-type Si substrate. Oxygen plasma can be used to improve MoS_2 adhesion on the SiO_2 surface. MoS_2 flakes can either be exfoliated [1, 20] from a molybdenite crystal using scotch tape method similar to graphene [21, 22] or can be grown by CVD [23]. Optical micrograph of mechanically exfoliated bilayer and multilayer MoS_2 flakes is shown in Figure 6(a). Scotch tape residue is removed by soaking samples in acetone for 3–6 h [24]. The flakes are then identified to be multi/monolayer using an optical microscope, atomic force microscopy (AFM) (Figure 6(b)) and Raman spectroscopy (Figure 7(c)). After flake identification, alignment markers are made by standard electron beam lithography followed by electron beam source-drain patterning on electron beam resist (first EL9 is spin coated to get a thickness of 300 nm and then polymethylmethacrylate (PMMA) is coated over it to get a thickness of 100 nm). Contact

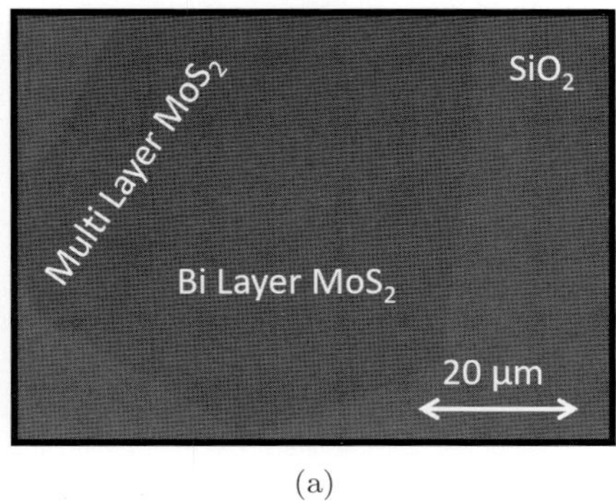

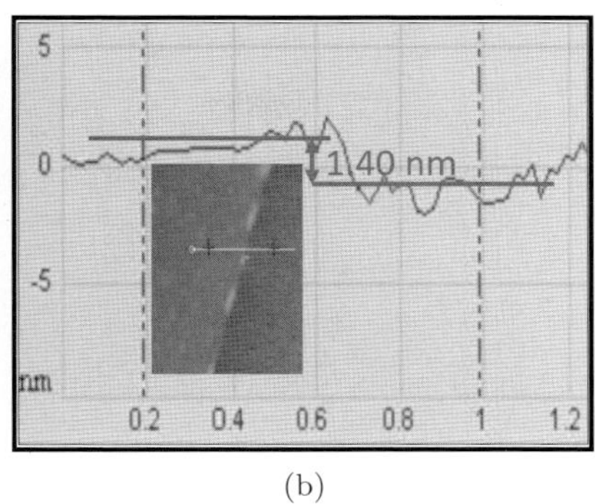

(a)   (b)

Figure 6. (a) Optical micrograph showing bilayer and multilayer MoS$_2$, and (b) AFM linescan data for a bilayer MoS$_2$ flake.

metals (Ti, Ni, Au, Pd, etc.) are then deposited using an electron beam evaporator under high vacuum conditions on the patterned resist. Finally, the underlying resist can be removed by soaking the sample in acetone for 4–5 h, so that only patterned metal is left. Schematic of a back-gated MoS$_2$ transistor and the corresponding process flow are shown in Figures 7(a) and 7(b). Scanning electron microscope (SEM) image with Au and Pd contacts on the same flake is shown in Figure 7(d).

8.6 Schottky Barrier Height Extraction Using Low-temperature Measurements

Φ_B for metal-MoS$_2$ contacts can be extracted using low-temperature I–V measurements. Different current transport regimes such as thermionic emission (TE), thermionic field emission (TFE) and field emission (FE) can be seen in the variable temperature transfer characteristics of an Au$-$MoS$_2$ device shown in Figure 8(a). The strong dependence of drain current on temperature for small gate voltages indicates TE dominated transport regime. At large V_{GS}, larger band bending increases tunnelling current significantly, hence, drain current does not depend on temperature which indicates the transistor is operating in FE regime. Effective Φ_B values were calculated from the slopes of Arrhenius plots for various V_{GS} as

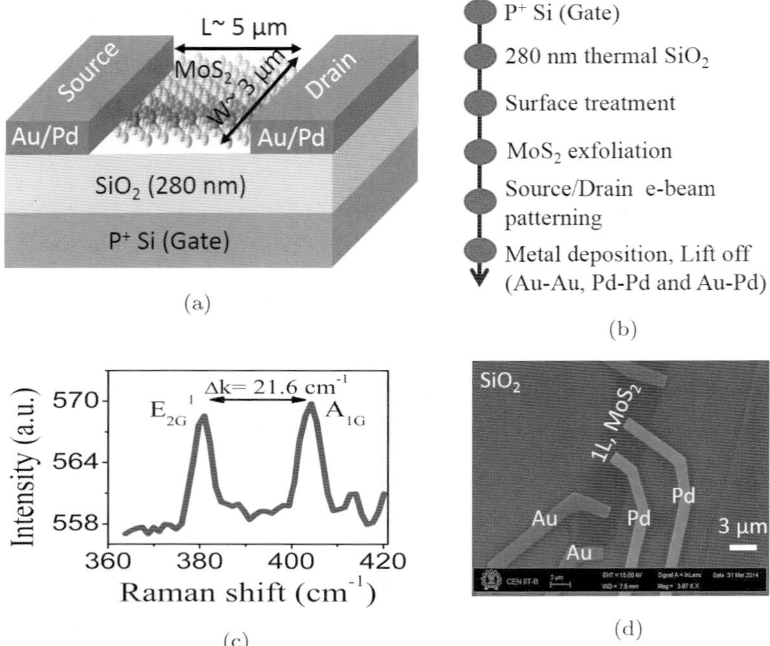

Figure 7. (a) 3D cross-section schematic of fabricated metal-MoS₂–metal (metal = Au, Pd) transistor devices, (b) Process flow for device fabrication, (c) Raman spectrum of a bilayer flake indicating pristine MoS₂ quality, and (d) Scanning electron micrograph of a fabricated MoS₂ transistor with Au and Pd source-drain contacts. (Adapted from Ref. [3].)

shown in Figure 8(b) based on standard TE theory relating I_{DS} to Φ_B [12],

$$I_{DS} = AA^*T^2 e^{(-q\Phi_B/k_B T)}[1 - e^{qV_{DS}/k_B T}], \qquad (8.12)$$

$$A^* = \frac{4qk_B^2 m^*}{h^3}, \qquad (8.13)$$

where A is the contact area, A^* is the effective Richardson's constant, k_B is Boltzmann's constant, h is Planck's constant, T is temperature and $m^* = 0.45m_o$ [25] is the effective mass for electrons in MoS₂. Barrier height extraction for large negative gate voltages will cause overestimation of Φ_B due to upward band bending as shown in Figure 8(c). Similarly, large positive gate voltages will underestimate

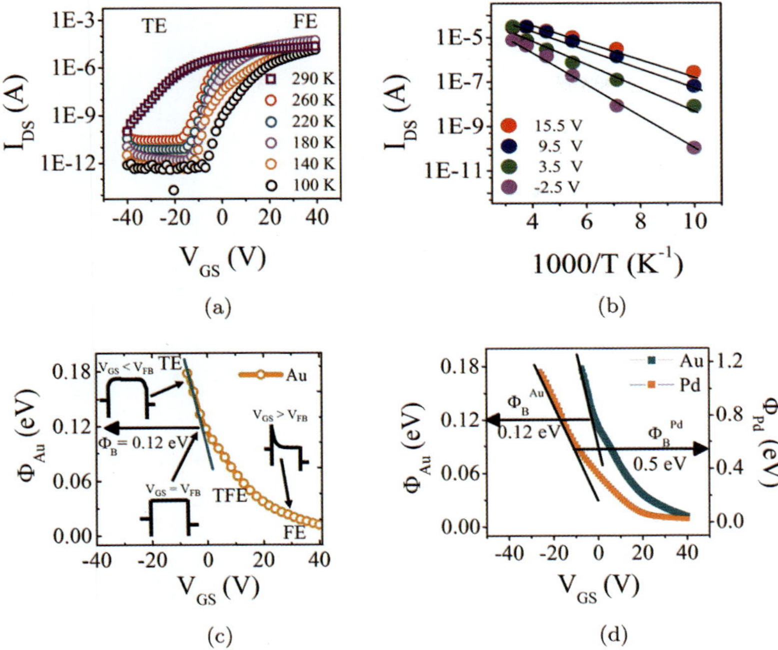

Figure 8. (a) Variable temperature $I_{DS}-V_{GS}$ transfer characteristics for an Au$-$MoS$_2$ transistor, (b) Arrhenius plots of I_{DS} at different V_{GS} values, (c) True Schottky barrier height extraction at flat band for an Au$-$MoS$_2$ device, and (d) True Schottky barrier height values extracted for Au$-$MoS$_2$ and Pd$-$MoS$_2$ contact interfaces. (Adapted from Ref. [3].)

Φ_B and show a nonlinear behaviour in gate tunability of Φ_B. Therefore, at flat band condition Φ_B changes from a linear to a nonlinear regime and gives the true barrier height for Au and Pd contacts to MoS$_2$ as shown in Figure 8(d).

8.7   Contact Resistance

8.7.1   Contact resistance extraction using transmission line measurement

TLM is widely used to extract the contact resistance (R_C) for metal–semiconductor contacts. Total resistance of a metal–semiconductor contact is given by

$$R_{\text{Total}} = 2R_C + R_{sh} \times L, \tag{8.14}$$

Figure 9. Variable channel length MoS$_2$ transistor devices on a rectangular MoS$_2$ flake that also act as a TLM structure for estimation of R_C and R_{sh}.

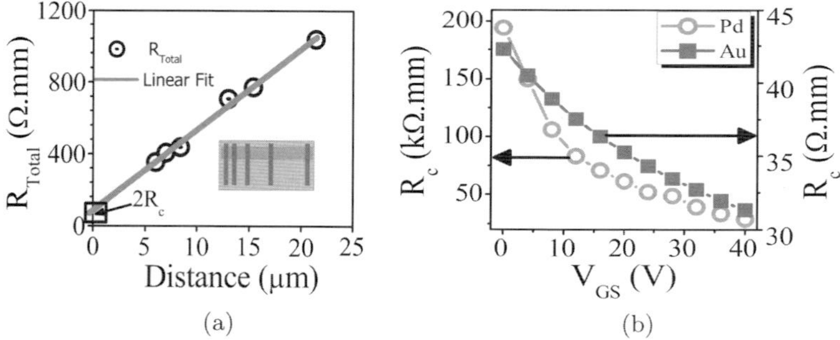

(a)

(b)

Figure 10. (a) Total resistance (R_{Total}) vs. distance L for Au–MoS$_2$ contacts, (b) R_C vs. V_{GS} for Au–MoS$_2$ and Pd–MoS$_2$ contacts. (Adapted from Ref. [3].)

where R_{Total} is the total resistance measured between two metal contacts separated by distance L on a semiconductor with sheet resistance R_{sh} and contact resistance R_C. Transistors with varying channel length shown in Figure 9 enable the extraction of R_C as well as R_{sh}.

Figure 10(a) depicts R_{Total} for different L measured between Au–MoS$_2$ contacts. An extrapolation of linear fits for different V_{GS} values gives the y-intercepts as well as slopes which represent $2R_C$ and R_{sh}, respectively. The change in R_C with varying V_{GS} shown in Figure 10(b) can be attributed to modulation of the Schottky contact barrier with applied back gate voltage. Increasing V_{GS} increases the downward band bending at the metal-MoS$_2$ interface which in turn increases the tunnelling component of the current and lowers the R_C.

8.7.2 Contact resistance using $I_{DS} - V_{DS}$

An approximate value of the contact resistance can also be estimated using $I_{DS} - V_{DS}$ output characteristics of MoS$_2$ transistors. Total transistor resistance for an applied V_{GS} is given by,

$$R_{\text{Total}} = 2R_C + R_{\text{channel}}, \qquad (8.15)$$

where R_{Total} is the total resistance and R_{channel} is the resistance of the MoS$_2$ channel. At high gate voltages, the channel is populated with a large density of electrons making the channel resistance very low. Hence, at large V_{GS}, $R_{\text{Total}} \sim 2R_C$ limits the drain current [26]. The value of R_{Total}, and hence R_C, can be extracted from $I_{DS} - V_{DS}$ data at large V_{GS}. It is important to note that this method of R_C estimation can only be applied for contact resistance limited devices.

8.7.3 Schottky barrier height extraction from contact resistance or vice versa

An analytical model [27] can be used to validate the experimental Φ_B and R_C data. Contact resistance vs. sheet charge density plots (shown in Figure 11) for varying Φ_B have been used to validate/calculate the Φ_B values [3]. Sheet charge density in MoS$_2$ transistors can be expressed as,

$$n_{2d} = \frac{LI_{DS}}{q\mu W V_{DS}}, \qquad (8.16)$$

where I_{DS} is the drain to source current at $V_{GS} = 0$ V, μ is the mobility and L and W are the length and width of the device, respectively. The n_{2d} and R_C values extracted from TLM measurements have been used to estimate the Φ_B values as shown in Figure 10. These values nearly match the Φ_B values extracted from low-temperature measurements as shown in Table 2.

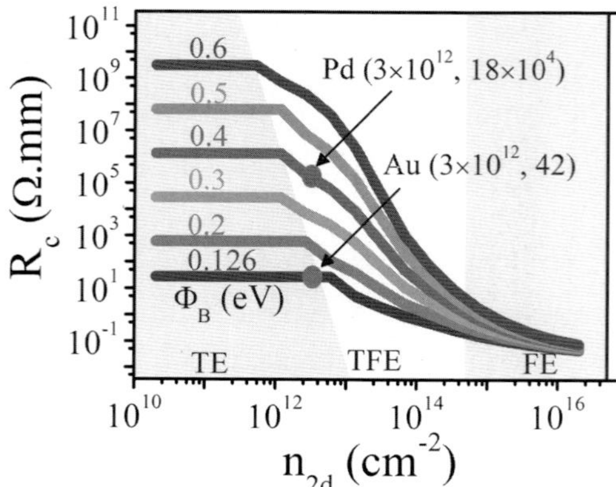

Figure 11. R_C vs. sheet electron charge density for varying Φ_B.

Table 2. Experimental and analytical Φ_B values for Au and Pd contacts on MoS_2.

Metal Contact	R_C Ω mm	Low Temp. Φ_B	Simulated Φ_B
Au	42	0.06–0.16	0.126
Pd	18×10^4	0.38–0.5	0.4

Note: Adapted from Ref. [3].

8.7.4 Schottky barrier height and contact resistance validation from photocurrent measurements

Scanning photocurrent microscopy can also be used to study and validate the band alignment of metal-MoS_2 interfaces. Spatial photocurrent maps for an Au$-MoS_2$(1L)$-$Pd device are shown in Figure 12(a). Asymmetry in current magnitude for positive and negative V_{DS} as seen in Figure 12(b) indicates the asymmetry in Φ_B for Au and Pd contacts to MoS_2. Photoresponse in TMDs is a combination of photothermoelectric effects and separation of photoexcited electron–hole pairs. However, photoresponse in monolayer MoS_2 is mainly due to the photothermoelectric effect and not by separation of electron–hole pairs [28]. Therefore, photothermal contribution to the total photocurrent is negligible at the Pd$-MoS_2$ contact

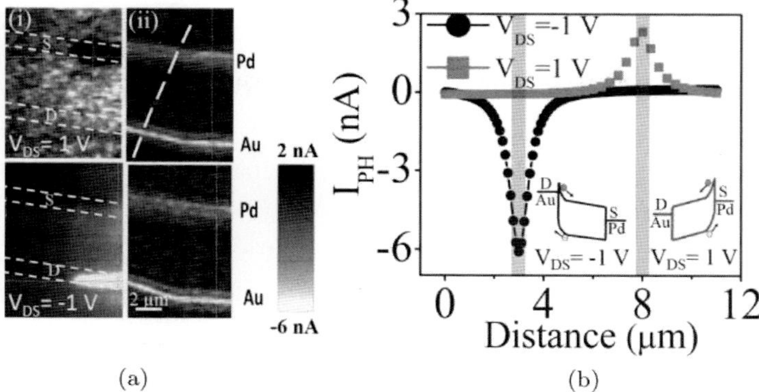

Figure 12. (a) Photocurrent maps of Au$-$MoS$_2$(1L)$-$Pd device at $V_{DS} = \pm 1$ V, $V_{GS} = 0$ V, (b) Line profile of photocurrent from drain to source (shown with a white line in Figure (a,i) and band diagram of Au$-$MoS$_2$$-$Pd device at $V_{DS} = \pm 1$ V, $V_{GS} = 0$ V. (Adapted from Ref. [3].)

due to a large Φ_B. Hence, only photogeneration will contribute to I_{PH} for positive V_{DS}. On the other hand, small Φ_B at Au$-$MoS$_2$ contact interface causes both photogeneration and photothermal mechanisms to contribute to the photocurrent for negative V_{DS} as shown in Figure 12(b). Hence, a large I_{PH} is observed for negative V_{DS}.

8.8   Why does MoS$_2$ Show *n*-type Conduction?

8.8.1   Fermi-level pinning

Wallace *et al.* [18] have studied FLP in MoS$_2$ and concluded that it is intrinsically different from traditional pinning such as Bardeen pinning (interface states), MIGS and defect induced pinning. Partial FLP at metal-MoS$_2$ interfaces has been attributed to two reasons, first is a modification in contact metal work function due to charge redistribution at the metal-MoS$_2$ interface, and second is the formation of gap states mainly of Mo *d*-orbital character [18]. Unlike conventional semiconductors, the formation of these gap states is due to weakened intra-layer S$-$Mo bonding at the metal-MoS$_2$ interface. It is important to note that these studies were done on pristine MoS$_2$, therefore this partial pinning cannot be attributed to any surface states and/or intrinsic defects. Some of the difference in Φ_B values reported in experimental work can be attributed to the fact that the exfoliated MoS_2 (used in most of the experimental reports) is not free of defects and/or interface states. Therefore, partial FLP exists even in pristine MoS_2 due to metal-MoS_2 interaction. However, Guo *et al.* [13] have attributed the FLP in MoS_2 to MIGS and concluded that gap states arise from metal-chalcogen atom (S in MoS_2) bonding without disturbing the weak intra-layer (S−Mo) bonding.

8.8.2   Chalcogen vacancies

N-type conduction in TMDs has also been attributed to the presence of chalcogen vacancies. Liu *et al.* [29] have shown that for all the various possible process conditions for formation of MoS_2, sulphur vacancies have lower formation energy than Mo vacancies and they give rise to a defect level in the upper half of the band gap of MoS_2. Sulphur vacancy has a formation energy of 2.35 eV in S-rich conditions, whereas Mo vacancy needs 8.02 eV in Mo-rich conditions. Therefore, reactive metal electrodes create S defects in MoS_2 and introduce transition states close to the conduction band. The formation energy for S vacancy is lowered further or can even become negative if E_F is close to the conduction band in the S-poor limit. Therefore, most metals are usually pinned in the upper half of the band gap resulting in n-type behaviour of MoS_2 transistors.

8.8.3   Positive charges in the gate oxide

N-type characteristics of MoS_2 back gate transistors have also been attributed to positive fixed charges in the gate oxide which can induce electrons in the MoS_2 channel [30, 31]. However, n-type behaviour of suspended MoS_2 transistors cannot be explained through the fixed charge theory. Therefore, the origin of n-type conduction in MoS_2 cannot be attributed to oxide fixed charge alone. The oxide substrate has the potential to reduce the threshold voltage (V_T) by a large amount leading to a negative V_T in most reported MoS_2 back gate transistors.

8.9 Approaches to Reduce Contact Resistance and Schottky Barrier Height

8.9.1 Interfacial layer

8.9.1.1 *TiO$_2$ as an interfacial layer*

Reduction in Φ_B and contact resistance has been reported using ultra-thin TiO$_2$ as an interfacial layer (IL) between metal and MoS$_2$ [15]. Unlike the case of Ge, where TiO$_2$ unpins the Fermi-level [32], a reduced and constant ($\sim$40 meV) Φ_B was observed for MoS$_2$, irrespective of the metal work function, which indicates an increase in n-type doping at the TiO$_2-$MoS$_2$ interface. Interfacial doping at TiO$_2-$MoS$_2$ contact interface was validated using DFT calculations and, X-ray and ultraviolet photoelectron spectroscopy (XPS and UPS) measurements. Schematics of back gated MoS$_2$ transistors without and with TiO$_2$ are shown in Figures 13(a) and 13(b), respectively. Optical micrograph and SEM image of fabricated back gated transistors with and without TiO$_2$ on the same MoS$_2$ flake are shown in Figures 14(a) and 14(b), respectively.

Low temperature $I_{DS} - V_{GS}$ measurements were carried out to extract the metal-MoS$_2$ and metal-TiO$_2$-MoS$_2$ interface Φ_B values. Reduction in Φ_B from 0.12 eV for Au$-$MoS$_2$ to 0.028 eV for Au$-$TiO$_2-$MoS$_2$ contacts was observed (shown in Figure 15(a)).

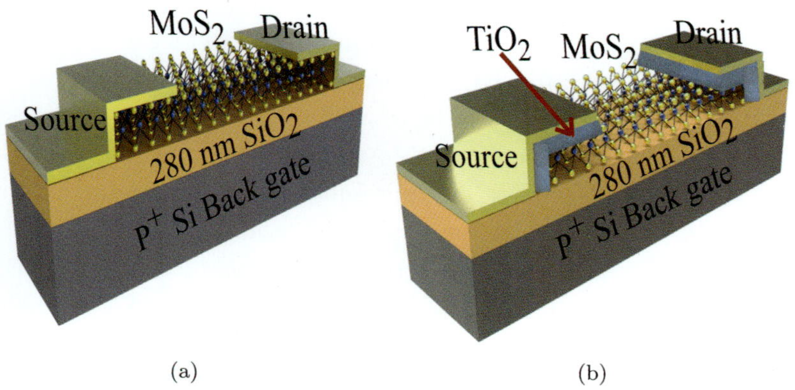

Figure 13. 3D schematics of MoS$_2$ transistors (a) without TiO$_2$ IL (control), and (b) with TiO$_2$ as an IL between the contact metal and MoS$_2$.

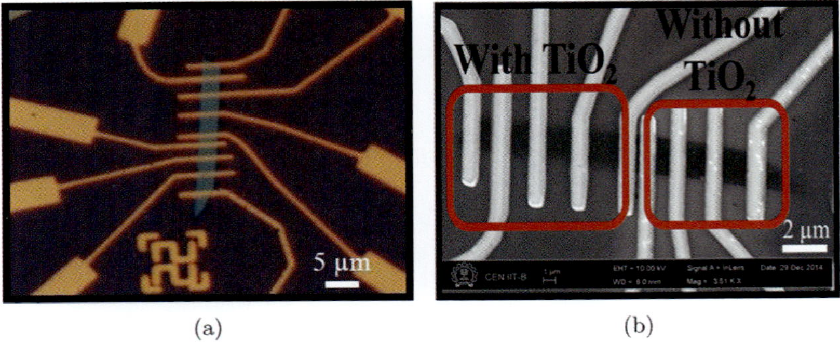

(a)   (b)

Figure 14. (a) Optical micrograph, (b) SEM image of MoS_2 transistors with varying channel length (TLM structure with varying contact separation) with and without TiO_2 IL on the same flake.

Constant Schottky barrier height (SBH) for various contact metals indicates increased n-type doping of MoS_2 at the TiO_2-MoS_2 interface, which is shown in Figure 15(c). The increased doping at the interface causes tunnelling dominated transport and makes the SBH insensitive to the metal work function.

R_C and R_{sh} values were extracted using TLM measurements as shown in Figures 16(a) and 16(b). Reduction ($24\times$) in R_C observed with TiO_2 IL is in good agreement with the substantial reduction in effective SBH shown in Figure 15(a). Strong dependence of contact resistance on gate voltage is due to the modulation of barriers at the metal-MoS_2 junction. The interfacial doping model with TiO_2 IL was validated using XPS and UPS measurements.

To confirm interfacial doping of MoS_2, XPS was used to measure the shift in binding energy of Mo and S peaks. Mo-$3d$ and S-$2p$ peaks shift towards higher binding energies by $0.5\,\mathrm{eV}$ and $0.54\,\mathrm{eV}$, respectively as shown in Figures 17(a) and 17(b). This can be interpreted as a Fermi-level shift towards the conduction band due to increased n-type doping of MoS_2. UPS measurements of TiO_2-MoS_2 also show a reduction in work function from $4.6\,\mathrm{eV}$ to $4.3\,\mathrm{eV}$ for MoS_2 and TiO_2-MoS_2, respectively, indicating that TiO_2 dopes MoS_2 n-type at the interface, thereby causing a reduction in Φ_B and R_C.

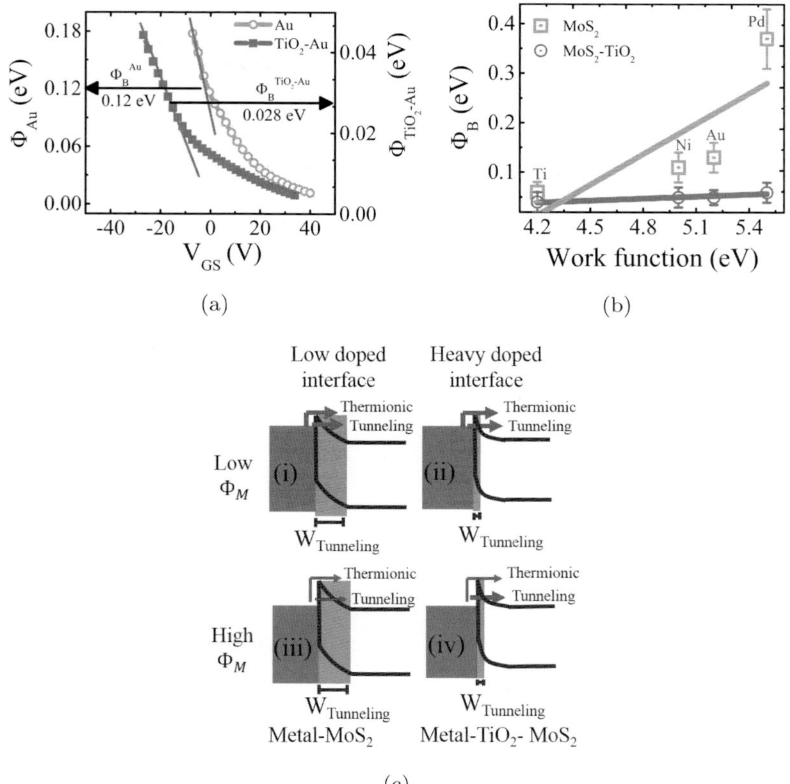

Figure 15. (a) Effective Φ_B values extracted with and without TiO$_2$ as an IL, (b) effective Φ_B vs. metal work function with and without TiO$_2$ showing reduced and nearly constant SBH with TiO$_2$ IL, (c) Energy band diagrams showing the impact of increased interfacial doping on current transport for metal-MoS$_2$ contacts for low as well as high work function (Φ_m) metals. (Adapted from Ref. [15].)

8.9.1.2 *Barrier free contact using graphene electrodes*

Liu *et al.* [33] have explored graphene as a contact electrode to circumvent the high contact resistance problem and demonstrated barrier free contacts to MoS$_2$ as shown in Figure 18. As discussed earlier, FLP and Φ_B in MoS$_2$ depend on the metal work function and how the contact metal affects intra-layer S–Mo bonding of MoS$_2$. Unlike conventional contact metals, graphene has two important properties, (i) tunability of the Fermi-level of graphene with gate

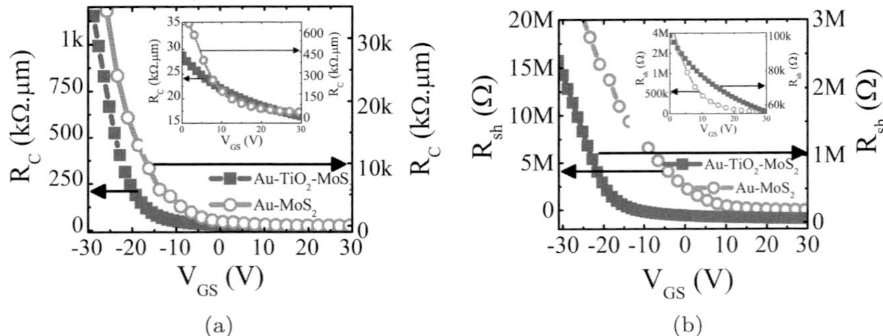

Figure 16. (a) Contact resistance data extracted from TLM devices without and with TiO_2 IL, (b) reduction in sheet resistance with TiO_2 as an IL between the contact metal and MoS_2 confirms the reduction in Φ_B. (Adapted from Ref. [15].)

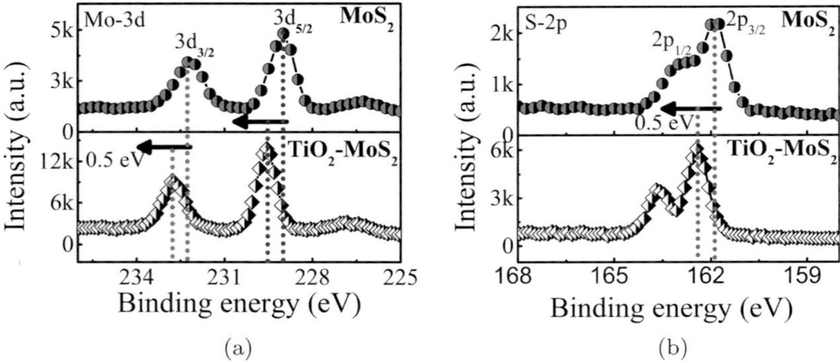

Figure 17. (a) Mo-$3d$ and, (b) S-$2p$ XPS core level shifts (0.5 eV) to higher binding energies for TiO_2-MoS_2 vs. bare MoS_2 indicate Fermi-level movement towards the MoS_2 conduction band in the presence of TiO_2. (Adapted from Ref. [15].)

potential that allows perfect band alignment with MoS_2 and (ii) its highly inert and stable nature that inhibits any reaction with MoS_2. Hence, it does not damage or weaken the intra-layer S$-$Mo bonding in MoS_2, which is known to be responsible for FLP in metal-MoS_2 contacts [33]. Therefore, tunability of E_F with V_{GS} and its inert nature make graphene an attractive choice as a contact electrode for TMDs. Graphene contacts to MoS_2 exhibit linear $I_{DS} - V_{DS}$ characteristics (shown in Figure 18(d)) at cryogenic temperatures down to $1.9\,\mathrm{K}$ thereby confirming barrier free contacts at the graphene–MoS_2 interface.

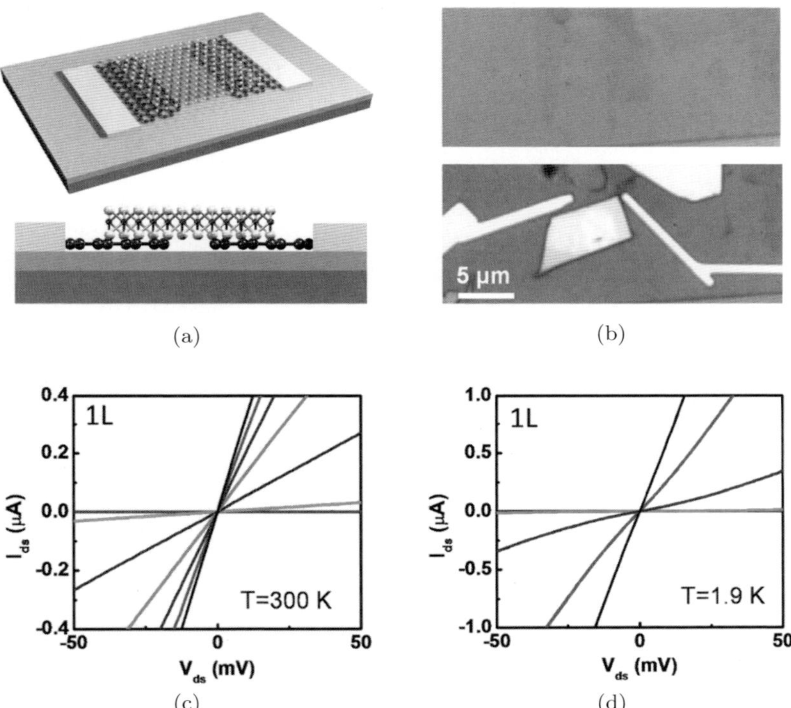

Figure 18. (a) Schematics of MoS$_2$ FETs with bottom-graphene electrodes. (b) Optical micrograph showing two graphene flakes placed close to each other (top panel) and the final device after MoS$_2$ transfer on graphene strips (bottom panel). (c) and (d) Output characteristics of a monolayer MoS$_2$ device at (c) room temperature and (d) low temperature (1.9 K) for varying gate voltages. (Adapted from Ref. [33].)

8.9.2 MoO$_x$ as an interfacial layer for p-type conduction

P-type (hole injecting) contacts to MoS$_2$ have been difficult to realise as most metals pin close to the conduction band resulting in n-type conduction. However, p-type FETs based on p-type contacts are highly essential for complementary MOS (CMOS) technologies. Chuang *et al.* [34] have demonstrated efficient hole injection through contacts to MoS$_2$ using molybdenum trioxide (MoO$_x$, $x < 3$). High work function MoO$_x$ was used as an interlayer between metal and

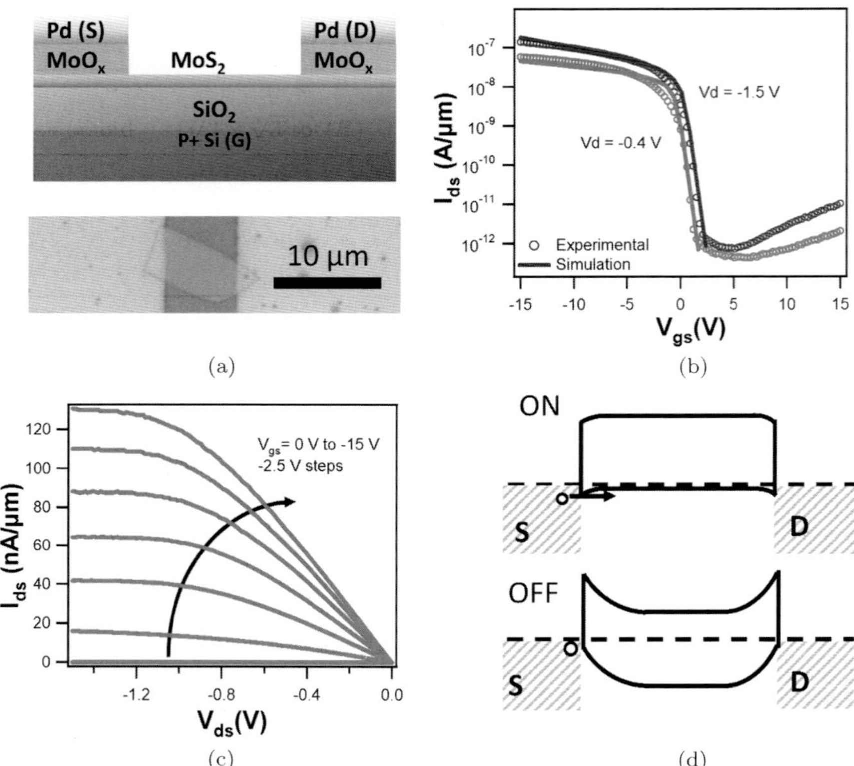

Figure 19. (a) 3D schematic and optical micrograph of MoS$_2$ back gate transistor with MoO$_x$ as an IL, (b) I_{DS} vs. V_{GS} characteristics showing p-type behaviour of the MoS$_2$ transistor, and (c) I_{DS} vs. V_{DS} characteristics for a MoS$_2$ p-FET with MoO$_x$ IL, (d) Energy band diagrams for positive and negative gate voltages. (Adapted from Ref. [34].)

MoS$_2$ as shown in Figure 19. WSe$_2$ FETs, that are intrinsically p-type, have also shown improved performance using MoO$_x$ contacts indicating a reduction in the hole contact barrier height.

8.9.3   Contact resistance reduction using chloride doping

Yang *et al.* [35] have demonstrated contact resistance reduction in MoS$_2$ by using chemical doping (Figure 20). Dichloroethane (DCE) was used as a doping reagent to alleviate the contact resistance

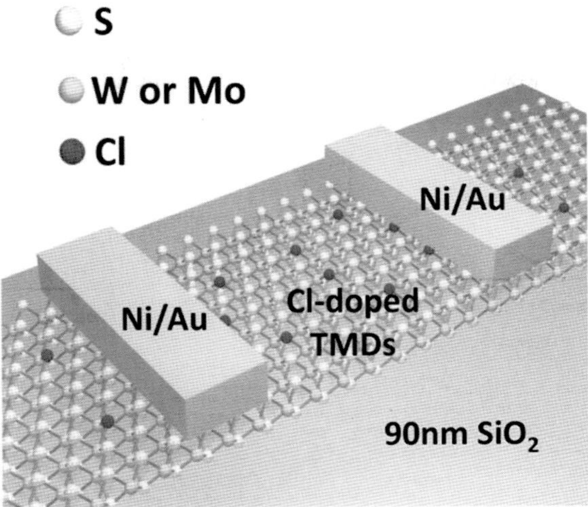

Figure 20. Schematic of the MoS$_2$ back gate FET with Cl doping. (Adapted from Ref. [35].)

problem in few-layer MoS$_2$ flakes. Improvement in transistor performance was attributed to increased n-type doping ($2 \times 10^{19}\,\text{cm}^{-3}$) at the metal-MoS$_2$ interface. This results in thinning of the tunnelling barrier width and reduction in R_C. The contact resistance reduces from $5.4\,\text{k}\Omega\,\mu\text{m}$ without Cl doping to $0.5\,\text{k}\Omega\,\mu\text{m}$ with Cl doping. Improvement in transistor performance was also observed with a high drain current value of $460\,\mu\text{A}/\mu\text{m}$ for 100 nm channel length at $V_{DS} = 1.6\,\text{V}$. Drain current insensitivity to temperature also indicates thinning of the tunnelling barrier (shown in Figure 21(b)), which increases the tunnelling current to/from the contacts significantly.

8.9.4 Phase-engineered low-resistance contacts for ultra-thin MoS$_2$ transistors

Conventional techniques to reduce contact resistance use an IL between the metal and the semiconductor or increase the semiconductor doping underneath the contact. IL lowers the contact resistance by unpinning the Fermi-level whereas increased channel doping enables increased tunnelling by thinning the barrier. However, the IL needs to be ultra-thin to avoid addition of series tunnelling

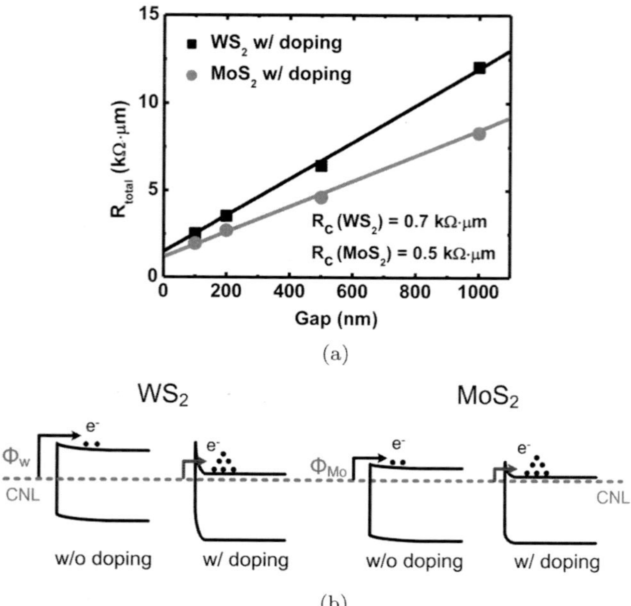

Figure 21. (a) Reduction in MoS_2 contact resistance from $5.4\,k\Omega\,\mu m$ to $0.5\,k\Omega\,\mu m$ with Cl doping. (b) Energy-band diagram of the metal and Cl doped MoS_2 showing reduction in the effective Schottky barrier. (Adapted from Ref. [35].)

resistance, whereas chemical doping of MoS_2 is difficult because of its ultra-thin nature. Yalcin *et al.* [36] have demonstrated the lowest contact resistance value on MoS_2 by exploiting its metallic 1T-phase (shown in Figure 22(b)). MoS_2 can exist in two phases, the more common, stable and semiconducting 2H-phase and the metastable 1T-phase which is metallic in nature due to half filled *d*-orbital bands. Phase conversion from 2H to 1T MoS_2 was demonstrated through treatment of MoS_2 with an organometallic solution containing *n*-butyl lithium [36]. In this chemical process, MoS_2 receives electrons from lithium that convert it from the 2H- to 1T-phase [37]. Reduction in contact resistance was observed from ~ 1–$10\,k\Omega\,\mu m$ for the 2H-phase to ~ 0.2–$0.3\,k\Omega\,\mu m$ for the 1T MoS_2 phase also resulting in significant improvement in transistor performance. Contacts made on 1T MoS_2 were also found to be more reproducible when compared

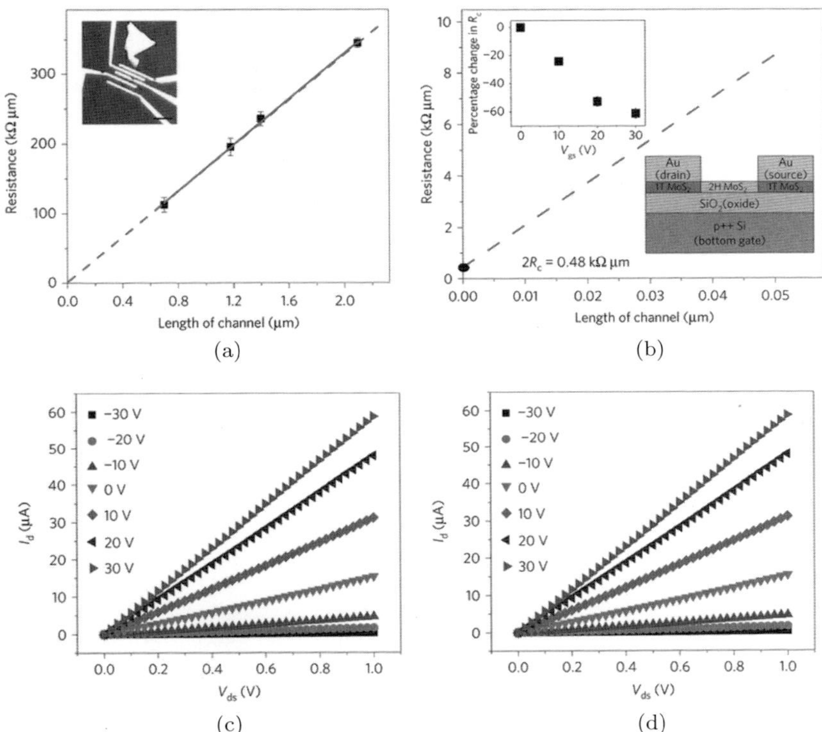

Figure 22. Contact resistance values extracted using TLM on (a) 2H-phase of MoS$_2$, and (b) 1T MoS$_2$, (c) and (d) $I_{DS} - V_{DS}$ data of back gated FETs for different gate voltages: (c) nonlinear behaviour of output characteristics confirms large Schottky barrier metal-2H MoS$_2$ contacts, and (d) linear behaviour indicating low SBH metal-1T MoS$_2$ contacts. (Adapted from Ref. [36].)

with contacts made on 2H MoS$_2$. Transistor drain current with 1T contacts was found to be insensitive to variation in the work function of contact metals suggesting tunnelling dominated transport at the 1T MoS$_2$ contact interface.

8.10 Summary and Theoretical Limit of Contact Resistance

Table 3 summarises the data from various reports of approaches to reduce contact resistance on MoS$_2$.

Table 3. Summary of approaches used for R_C reduction in MoS$_2$ transistors.

Ref.	Approaches to Reduce R_C	Metal Contact	V_{GS} (V)	t_{OX} (nm)	No. of layers	R_C (k$\Omega\,\mu$m)
[36]	1T MoS$_2$	Au	0	100	1–3	$\sim$0.48
[35]	Cl doping	Au	50	90	5–7	0.5
[38]	PEI doping	Ti/Au	10	90	5–7	12.5
[39]	1 nm TiO$_2$ IL	Co	0	285	14–18	2.4
[40]	By optimising metal deposition pressure	Au	10	90	3–7	$\sim$2
[15]	Interfacial doping using TiO$_2$	Au	30	290	10–15	2.4–15
[41]	Nickel etched graphene contacts	Ni	50	285	2	0.2
[42]	Atomic passivation of Au surface	Au	NA	NA	1	NA
[43]	Boron nitride IL	Au	NA	NA	1	NA
[44]	Sulphur treatment	Ni, Pd	NA	30	multi	$\sim$4–6

The quantum limit for minimum contact resistance is given by [41, 42],

$$R_C = \frac{h}{(2q^2 M)},\qquad(8.17)$$

where k_F is the Fermi wave vector and $M \simeq k_F W$ is the number of electron modes whose wavelength fit the conductor of width W [37]. Sheet density of electrons in a 2D channel is defined as,

$$n_{2D} = \frac{k_F{}^2}{2\pi}\qquad(8.18)$$

Using these equations, the minimum contact resistance normalised to width can be expressed as

$$R_C W \simeq h/2q^2 k_F \simeq \frac{0.026}{\sqrt{n_{2D}}}k\Omega\,\mu\text{m}.\qquad(8.19)$$

n_{2D} in 2D TMDs varies typically from 10^{10} to $10^{13}\,\text{cm}^{-2}$. The lowest reported contact resistance values for TMDs and conventional semiconductors along with the quantum limit are shown in Figure 23. Contact resistance values obtained for 1T and doped MoS$_2$ are still

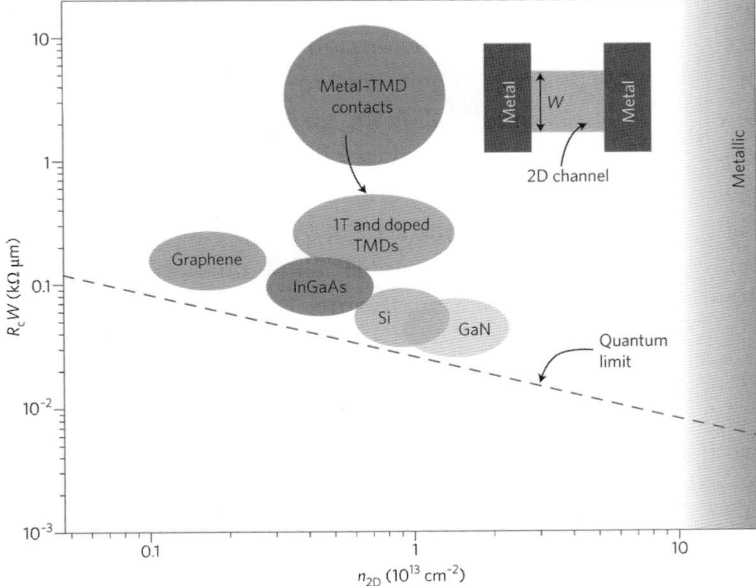

Figure 23. Theoretical quantum limit of contact resistance along with lowest reported contact resistance values for 2D TMDs, graphene and conventional semiconductor materials. (Adapted from Ref. [37].)

high compared to conventional semiconductors such as Si and GaN. Significant efforts are therefore required to reduce R_C as close to the theoretical limit as possible.

Acknowledgements

We acknowledge the help received from Shruti Karande, John Mathew, Natasha Goyal and Kartikey Thakar towards proofreading this chapter and financial support from the Department of Science and Technology, Government of India.

References

[1] R. Ganatra and Q. Zhang, *ACS Nano.* **8**(5), 4074 (2014).

[2] G. H. Xing, D. Jena and K. Banerjee, *Nat. Mater.* **13**, 1076 (2014).

[3] N. Kaushik, A. Nipane, F. Basheer, S. Dubey, S. Grover, M. M. Deshmukh and S. Lodha, *Appl. Phys. Lett.* **105**(11), 113505 (2014).

[4] H. Liu, M. Si, Y. Deng, A. T. Neal, Y. Du, S. Najmaei, P. M. Ajayan, J. Lou and P. D. Ye, *ACS Nano.* **8**(1), 1031 (2014).

[5] W. Mönch, *Electronic structure of metal–semiconductor contacts*, volume 4. Springer Science & Business Media, New York (2012).

[6] W. Schottky, *Naturwissenschaften* **26**(52) 843 (1938).

[7] J. Bardeen, *Phys. Rev.* **71**, 717 (1947).

[8] A. M. Cowley and S. M. Sze, *J. Appl. Phys.* **36**(10), (1965).

[9] Volker Heine, *Phys. Rev.* **138**, A1689 (1965).

[10] A. K. Boyd, M. Rinzan, A. Y. Liu, M. Paranjape, M. Fontana, T. Deppe and P. Barbara, *Sci. Rep.* **3**, 1634 (2013).

[11] A. T. Neal, H. Liu, J. J. Gu and P. D. Ye, *Device Research Conference (DRC), 2012 70th Annual*, p. 65 (2012).

[12] S. Das, H.-Y. Chen, A. V. Penumatcha and J. Appenzeller, *Nano Lett.* **13**(1), 100 (2013).

[13] Y. Guo, D. Liu and J. Robertson, *ACS Appl. Mater. Interfaces* **7**(46), 25709–25715 (2015). PMID: 26523332.

[14] J. Kang, W. Liu, D. Sarkar, D. Jena and K. Banerjee, *Phys. Rev. X* **4**, 031005 (2014).

[15] N. Kaushik, D. Karmakar, A. Nipane, S. Karande and S. Lodha, *ACS Appl. Mater. Interfaces* **8**(1), 256 (2016). PMID: 26649572.

[16] J. Kang, W. Liu and K. Banerjee, *Appl. Phys. Lett.*, **104**(9), (2014).

[17] F. Ahmed, M. S. Choi, X. Liu and W. J. Yoo, *Nanoscale* **7**, 9222 (2015).

[18] R. M. Wallace, K. Cho, C. Gong and L. Colombo, *Nano Lett.* **14**, 1714 (2014).

[19] S. Das and J. Appenzeller, *Physica Status Solidi (RRL) Rapid Res. Lett.* **7**(4), 268–273 (2013).

[20] B. Radisavljevic, A. Radenovic, J. Brivio, V. Giacomett and A. Kis, *Nat. Nanotechnol.* **6**, 147 (2011).

[21] A. K. Geim, *Rev. Mod. Phys.* **83**, 851 (2011).

[22] A. K. Geim and K. S. Novoselov, *Nat. Mater.* **6**, 183 (2007).

[23] H. F. Liu, S. L. Wong and D. Z. Chi, *Chem. Vapor Deposition* **21**(10–12), 241 (2015).

[24] Y. Huang, J. Wu, X. Xu, Y. Ho, G. Ni, Q. Zou, G. K. W. Koon, W. Zhao, A. H. C. Neto, G. Eda, C. Shen and B. Özyilmaz, *Nano Res.* **6**(3), 200 (2013).

[25] Y. Yoon, K. Ganapathi and S. Salahuddin, *Nano Lett.* **11**(9), 3768 (2011).

[26] W. S. Hwang, J. H. Lee, J. Lee, J. Yang, C. Jung, H. Kim, J. B. Yoo, S. Kim, A. Konar and J. Y. Choi, *Nat. Commun.* **3**, 1011 (2012).

[27] K. Varahramyan and E. J. Verret, *Solid-State Electron.* **39**(11), 1601 (1996).

[28] M. Buscema, M. Barkelid, V. Zwiller, H. S. J. van der Zant, G. A. Steele and A. Castellanos-Gomez, *Nano Lett.* **13**(2), 358 (2013). PMID: 23301811.

[29] D. Liu, Y. Guo, L. Fang and J. Robertson, *Appl. Phys. Lett.* **103**(18) (2013).

[30] H. Liu. *IEEE Electron Dev. Lett.* **33**, 546 (2012).

[31] D. Kufer and G. Konstantatos, *Nano Lett.* **15**(11), 7307 (2015). PMID: 26501356.

[32] A. Agrawal, J. Lin, B. Zheng, S. Sharma, S. Chopra, K. Wang, A. Gelatos, S. Mohney and S. Datta, In *VLSI Technology VLSIT, 2013 Symposium on T200* (2013).

[33] Y. Liu, H. Wu, H.-C. Cheng, S. Yang, E. Zhu, Q. He, M. Ding, D. Li, J. Guo, N. O. Weiss, Y. Huang and X. Duan, *Nano Lett.* **15**(5), 3030 (2015).

[34] S. Chuang, C. Battaglia, A. Azcatl, S. McDonnell, J. S. Kang, X. Yin, M. Tosun, R. Kapadia, H. Fang, R. M. Wallace and A. Javey, *Nano Lett.* **14**(3), 1337 (2014). PMID: 24568656.

[35] L. Yang, K. Majumdar, H. Liu, Y. Du, H. Wu, M. Hatzistergos, P. Y. Hung, R. Tieckelmann, W. Tsai, C. Hobbs and P. D. Ye, *Nano Lett.* **14**(11), 6275 (2014). PMID: 25310177.

[36] S. E. Yalcin, B. Branch, G. Gupta, A. D. Mohite, R. Kappera, D. Voiry and M. Chhowalla, *Nat. Mater.* **13**, 1128 (2014).

[37] K. Banerjee, D. Jena and G. H. Xing, *Nat. Mater.* **13**, 1076 (2014).

[38] Y. Du, H. Liu, A. T. Neal, M. Si and P. D. Ye, *IEEE Electron Dev. Lett.* **34**(10), 1328 (2013).

[39] A. Dankert, L. Langouche, M. V. Kamalakar and S. P. Dash, *ACS Nano* **8**(1), 476 (2014).

[40] C. D. English, G. Shine, V. E. Dorgan, K. C. Saraswat and E. Pop, In *72nd Device Research Conference* **193** (2014).

[41] R. Landauer, *IBM J. Res. Develop.* **1**(3), 223 (1957). DOI: 10.1021/nn506567r

[42] Y. V. Sharvin, *Zh. Eksperim. i Teor. Fiz.* **48** (1965). https://doi.org/10.1088/2053-1583/4/1/015019

[43] DOI: 10.1039/C6CP02132H

[44] DOI: 10.1109/TED.2016.2554149

Chapter 9

Strain Dependent Properties of 2D MX_2 (M = Mo and W; X = S, Se and Te)

Tribhuwan Pandey[*], Swastibrata Bhattacharyya[*,†]
and Abhishek K. Singh[*,‡]

*Materials Research Centre, Indian Institute of Science,
Bangalore 560012, India

†Department of Physics, Yokohama National University,
79-5 Tokiwadai, Hodogaya-ku, 240-8501,
Yokohama, Japan

‡abhishek@mrc.iisc.ernet.in

Abstract. One of the fascinating properties of the new families of two-dimensional (2D) crystals is their high stretchability, which gives rise to possibility of using external strain to manipulate their optical and electronic properties in a controlled manner. The objective of this chapter is to give an overview of the recent progresses made in controlling the electronic, optical and vibrational properties of 2D transition metal dichalcogenides (MX_2, where M = Mo, W and X = S, Se, Te), by means of strain engineering. We will discuss the electronic origin of strain induced semiconductor-metal transition in these 2D materials. We will present the role played by Raman spectroscopy in mapping out the changes in vibrational properties of these 2D materials under applied strain. The effect of strain induced changes in electronic structure on controlling the transport properties of these materials will also be covered. Finally, we will review the recent experimental results, which verifies some of the theoretical predictions and show the immense potential of strain engineering in the range of applications.

9.1 Introduction

Transition metal dichalcogenides (TMDs) have the similar layered structure as that of graphite and are well known for their application as solid lubricant [1], catalyst [2] and photovoltaic [3, 4] material. Free-standing two-dimensional (2D) sheets of single to few layers of TMDs have been obtained experimentally [5–14] and they show excellent optical absorption and photoconductivity [5, 15]. TMDs have a stoichiometry of MX_2, where M stands for metal and X stands for the chalcogen atom. Depending on the combination of metal and chalcogens, TMDs offer a wide range of 2D materials including metals [16, 17], superconductors [18, 19], charge density wave systems [20, 21], Mott insulators [22] and semiconductors [23, 24]. The type of metal as well as chalcogen atoms determine the electronic, structural, vibrational and mechanical properties [16] of MX_2. For example, 2D sheets of NbX_2 and TaX_2 are metallic while semiconducting TMDs include MoX_2 and WX_2 with X = Te, S and Se. Semiconducting TMDs have attracted recent research interest and have emerged as promising materials for a range of applications such as nanoelectronics and nanophotonics. The presence of intrinsic band gap in the range of $1 \sim 2\,\mathrm{eV}$, is an advantage of these materials over other similar stable materials such as gapless graphene and insulating BN.

A single layer of MX_2 is a quasi 2D sheet consisting of three hexagonally arranged planes: one plane of transition metal (M = Ti, Nb, Ta, Mo and W) sandwiched between two planes of chalcogen atoms (X = S, Se and Te). The metals and chalcogens form strong ionic-covalent bonds within the layer. In bulk and few-layered TMDs, the layers are held together by weaker van der Waals (vdW) bonding. This leads to an anisotropic structure which contributes to various directional dependent properties. The bulk semiconducting TMDs are indirect band gap semiconductors having band gaps in the range of 1.0–$1.35\,\mathrm{eV}$ [23]. With the reduction in number of layers the band gap of TMDs increases [24] and a transition from indirect to direct gap occurs for a monolayer [25–27]. The TMD based field effect transistors with high room-temperature current on/off ratios [28] and higher on-current density [29, 30] as well as integrated circuits [31] have been successfully fabricated. For practical utility of these materials in electronic and photonic applications, varying the electronic, structural, optical and other properties is very essential. The tuning of electronic properties has been achieved by doping [16, 32], intercalation [33, 34] and applying electric field [35]. Change in electronic and optical properties was reported for MoS$_2$ monolayer via molecular charge transfer [36]. Tuning of thermoelectric properties has been achieved by intercalating SnS layer into the van der Waals gap of TiS$_2$. However, most of these techniques have limitations in practical reversible applications. There is a lack of an effective method of doping or functionalisation in a controlled fashion on the other hand growth of specific sized nanostructures (e.g., nanoribbons and quantum dots), free standing or embedded is yet to be achieved experimentally. The requirement of high electric field as well as the need for an extra electrode is another disadvantage in electric field induced tuning of the band gap in low dimensional materials. Development of simpler yet effective methods is thus necessary to achieve this goal experimentally for the potential application of these materials in various nanodevices.

Application of strain is known to be an effective way to manipulate electronic, transport, structural and optical properties of materials and has also been applied for MX$_2$. Various kind of strains such as uniaxial, biaxial or hydrostatic can be applied offering a multidimensional approach in achieving tunability of the properties. While biaxial strains are planar (along with **a** and **b** axis), the uniaxial strain can be either planar or normal (along **c** axis). Because of the anisotropy in the structure, the influence of different types of strain on its properties varies. Moreover, a comparatively larger amount of force is required to strain the sheet along the planar direction than along the normal direction. Thus, while applying the hydrostatic pressure, the amount of strain along the normal direction is always larger than that of along the planar direction. At very high strain, the vdW type of bonding along the normal direction changes to the covalent type as the distance between the layers decreases [37, 38]. Change in electronic structure and band gap with the application of strain has been investigated theoretically for bulk, 2D sheets, 1D nanoribbons, and nanotubes of MX$_2$. Bulk to few-layers of MX$_2$ along with the monolayers show semiconductor to metal transition under biaxial tensile and compressive strain [39–44] as well as normal compressive strain [37, 44]. In experiments also application of hydrostatic pressure has shown to reduce the resistivity for bulk MX_2 [45] and a complete semiconductor to metal transition was observed for few-layered MoS_2 [38]. The effect of pressure on electronics, thermoelectric, optical and vibrational properties of MX_2 mono and multilayer has been reported experimentally. An increase in direct optical band gap of the monolayers has been found under the application of hydrostatic pressure [46]. Most importantly, the structure of 2D MX_2 is robust upto a very high pressure without any permanent deformation or structural phase transition. However, the nature of inter-layer interaction changes from weak vdW type to a strong covalent one [38]. The strain has an effect on the thermoelectric properties of these materials as well. Strain modifies the dispersion of bands, which influences the transport properties thereby modifying the thermoelectric performance of the materials [44].

In this chapter, we will present pressure/strain induced changes in electronic, optical, vibrational and transport properties of 2D materials. We will discuss briefly various computational methods, which are successfully applied to study the strain induced changes in properties of these materials. We arrange this chapter as follows. In Section 9.2, we discuss the most common theoretical methodology adopted to calculate electronic structure, transport and phonon spectra of 2D materials. A general information about the geometry and electronic structure of MX_2 has been provided in Section 9.3. In Section 9.4, we review some theoretical work done on the effect of strain on electronic, transport and vibrational properties. In Section 9.5, we review some experimental progresses on the application of strain on these materials. Finally, in Section 9.6, a general conclusion and future perspective will be given.

9.2 Computational Methodology

Theoretical results, described in this chapter have been mostly obtained using *ab initio* electronic structure calculations based on the solution of the Kohn–Sham equation within the density functional theory (DFT) [47–49]. DFT has been very successful in predicting materials properties using the local density appropriation (LDA) and generalised gradient approximation (GGA). There are many different ways to solve Kohn–Sham equations, a plane wave approach using pseudo potential [47–49] is quite effective for understanding the bonding and associated properties of a material. This method provides the possibility to understand the evolution of properties of materials from a microscopic point. The results obtained by such methods often agree with experiments. These calculations are very advantageous in exploring the properties of both known and unknown materials. Better methods have also been developed and used to improve the results calculated using GGA/LDA or PBE/GGA, such as hybrid Heyd–Scuseria–Ernzerhof (HSE) [50, 51] functional and GW approximation [52]. In HSE approach, the exchange potential is split into a long- and a short- range part. 1/4th of the PBE exchange is replaced by the Hartree-Fock (HF) exact exchange and the full PBE correlation energy is added. GW approximation is based on many-body perturbation theory. Here, G denotes the electron Greens function defined by the Kohn–Sham eigenstates $\phi_{n\mathbf{k}}(r)$ and eigenvalues $\epsilon_{n\mathbf{k}}$ corresponding to the band index n and the wave vector k. The W represents the screened Coulomb interaction. GW0 is a partially self-consistent method, in which the G was iterated but the W was kept fixed to the initial DFT (PBE) W0. Such methods improve the band gap of materials as compared with the GGA however, are computationally very expensive. The weak van der Waals interaction between the layers has an effect in determining the inter-layer distance for the bilayers as well as for the bulk MX$_2$. The van der Waals interaction originates from dynamical correlations between fluctuating charge distributions and cannot be described by the PBE functional. Various methods have been developed to incorporate the effect of van der Waals interactions within DFT among which Grimme's DFT-D2 method [53] is widely used. In this method, the van der Waals interactions were incorporated through a pair-wise force field, by adding a semi-empirical dispersion potential (D) to the conventional Kohn–Sham DFT energy. Other such methods, include non-local van der Waals density functional (vdW-DF) [54] and the second version of the van der Waals density functional (vdW-DF2) [55].

Electrical conductivity of MX_2 compounds from the electronic structure is usually calculated by using Boltzmann transport theory (BTT) within constant scattering time approximation (CSTA) [56, 57]. In CSTA, it is assumed that the scattering time does not vary strongly with energy as well as with the temperature or the doping-level. In the BTT method, group velocities are calculated by Fourier interpolation [58] of the band energies, obtained using electronic structure calculations under DFT formalism, as a function of $\mathbf{k}$ with a subsequent computation of transport properties. σ/τ as a function of temperature (T) and carrier concentration (n) can be calculated from the electronic structure. However, to calculate σ one must approximate the relaxation time, e.g., from the experimental data. The more details regarding transport calculation can be found in Refs. [56, 59].

Various procedures of calculating the vibrational spectrum includes frozen phonon method, molecular dynamics spectral analysis and density functional perturbation theory (DFPT) [60] to name a few. Among these methods, linear response DFPT has been implemented in various DFT codes such as Vienna Ab initio Simulation Package (VASP) [61] and is widely used to calculate phonon dispersion of various materials. Using an additional tool, such as Phonopy [62], which supports VASP interface, the phonon frequencies and electron–phonon coupling constants can be extracted. Interested reader can refer to Refs. [62, 63], for more details regarding phonon calculations.

9.3 Properties of MX_2

9.3.1 Crystal structure

TMDs are a class of layered materials with the chemical formula MX_2, where M is a transition metal atom and X is a chalcogen atom. Under ambient conditions, group VI bulk TMDs are reported to exist in a layered hexagonal crystal structure. This structure is composed of monolayers wherein the X atoms are in trigonal

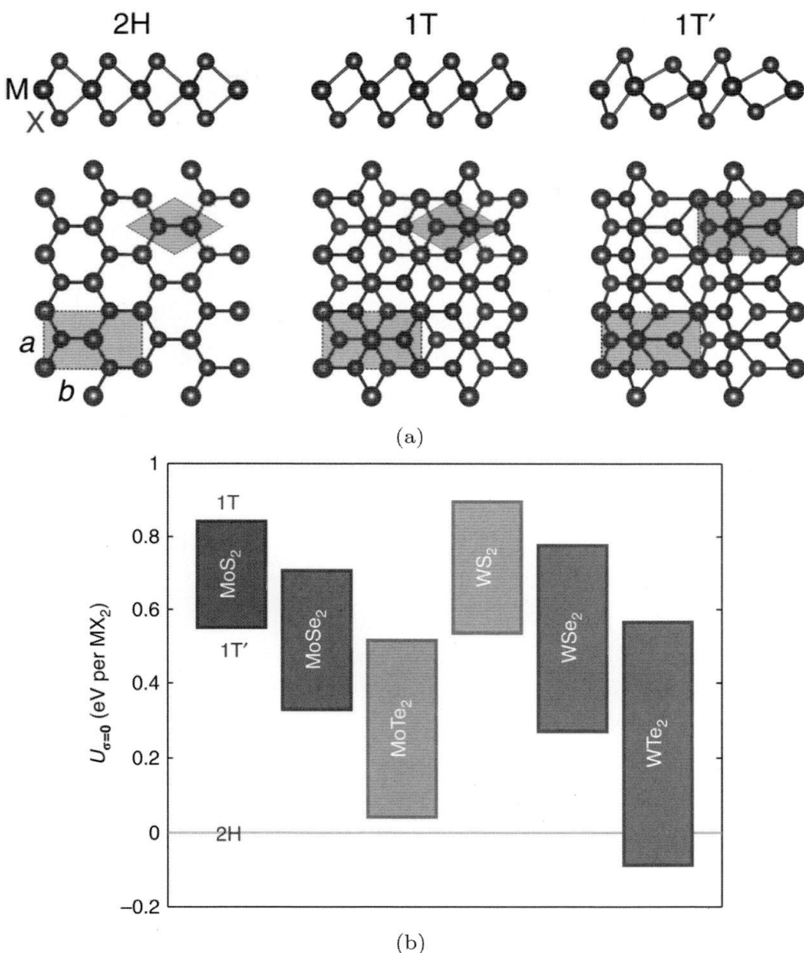

Figure 1. (a) Schematic atomic arrangements of the monolayer MS$_2$ (M = Mo, and W) showing the in-plane and out-of-plane view in 2H-, 1T- and 1T′-phase, (b) Ground-state energy differences between monolayer phases. The energy U is given per formula unit of MX$_2$ for the 2H-, 1T′- and 1T-phases. (The part (a) and (b) are reprinted from Refs. [64, 65], respectively, with permission from Macmillan Publishers Ltd.)

prismatic coordination around the M atoms as shown in Figure 1(a). The symmetry space group of bulk and single-layer MoS$_2$ are $P3m1$ (point group D_{6h}) and $P6m2$ (point group D_{3h}), respectively. Consequently, systems with an even number of layers belong to the space group $P3m1$ (with inversion symmetry), and systems with an odd number of layers to the $P6m2$ space group (without inversion symmetry). Each monolayer is composed of a metal layer sandwiched between two chalcogen layers, forming a three atom thick X–M–X structure. The weak inter-layer attraction of TMDs allows exfoliation of these stable three-atom-thick layers. This phase is commonly known as 2H-phase. In 1T-phase, one of the X layers in a 2H structure is shifted resulting in octahedral coordination of X atoms around the M atoms (as shown in Figure 1(a)). Due to this change in coordination the crystal becomes metallic. This phase is observed in group IV and group V TMD compounds [66]. However, 1T-phase is not stable for group VI TMDs. The group VI TMDs do have a stable metallic structure with distorted octahedral-like M–X coordination as shown in Figure 1(a). This lower-symmetry phase, often referred as 1T$'$ and is considered as a distorted version of the 1T structure [66–68].

The relative stability of these phases has been discussed in Refs. [65, 69]. Duerloo *et al.* [65], have shown that at ambient condition 2H-phase is the most stable for all the group VI monolayer TMDs. It is also evident from their study that the energy associated with the 1T to 1T$'$ transformation is of the order of several tenths of an eV per MX_2 formula unit as shown in Figure 1(b). Similar results were shown by Calandra [69] for monolayer MoS_2. The dynamical stability of various group VI polytypes is discussed by many theoretical studies [46, 65, 69, 70]. The 2H- and 1T$'$-phases do not show any imaginary vibrational frequency in the phonon dispersion, indicating their dynamical stability as shown in Figures 2(a)–2(c). The in-plane acoustical modes (longitudinal acoustic, LA, and transverse acoustic, TA) have linear dispersions and higher energy than the out-of-plane acoustic (ZA) mode. The latter displays a q^2-dependence analogous to that of the ZA mode in graphene, which is a consequence of the point-group symmetry [71]. In the case of monolayer MoS_2, the symmetry is reduced from D_{6h} to D_{3h} and the number of phonon branches reduced to nine [70, 72, 73]. On the other hand, for 1T-phase the optical phonon modes have imaginary vibrational frequency

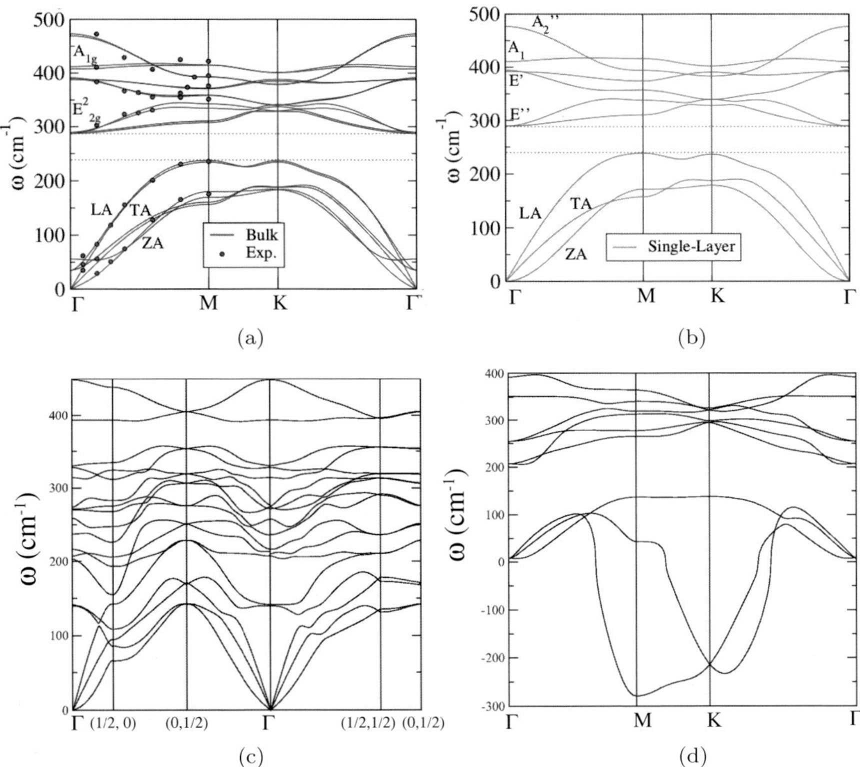

Figure 2. Phonon dispersion curves for (a) 2H-bulk, (b) 2H-monolayer, (c) 1T′-monolayer, and (d) 1T-monolayer MoS$_2$. Part (a) and (b) are reprinted from Ref. [70] with permission of American Physical Society. Part (c) and (d) are reprinted from Ref. [69] with permission of American Physical Society.

[46, 65, 69, 70] as shown in Figure 2(d). This result asserts that the high-symmetry 1T structure is unstable for group VI chalcogenides.

9.3.2 Electronic structure

The electronic structure of bulk and multilayer TMDs has been described in Refs. [37, 40, 74–77]. The orbital character of the electronic bands of TMDs is discussed in details in Refs. [74, 75]. The band structures and partial density of states (PDOS) of pristine monolayer MoS$_2$, MoSe$_2$, WS$_2$ and WSe$_2$ are shown in Figures 3(a)–3(d) [75]. All of the four TMDs are direct band-gap semiconductors

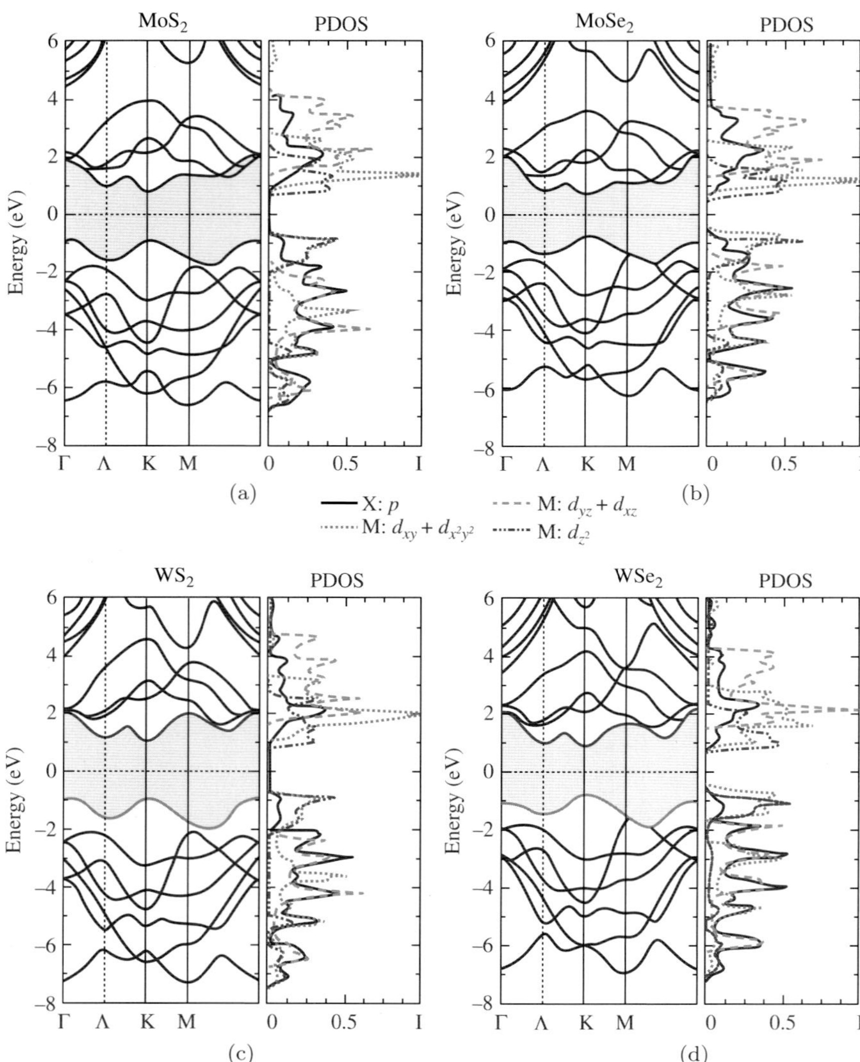

Figure 3. The band structures and partial density of states of different molecular orbitals for (a) MoS$_2$, (b) MoSe$_2$, (c) WS$_2$, and (d) WSe$_2$. (Reprinted from Ref. [75] with permission of American Physical Society).

with both valence band maxima (VBM) and conduction band minima (CBM) located at the K point of the first Brillouin zone. The PDOS shows that the electronic states near the CBM and VBM originate mainly from d_{z^2}, $d_{x^2-y^2}$ and d_{xy} orbitals of the M atom and the p orbitals of the X atom. The strong coupling between p orbitals of the X atom and $d_{xz} + d_{yz}$ orbitals of the M atom leads to a large splitting between their bonding and antibonding states. As discussed by Kośmider *et al.* [78], one of the main characteristics of TMDs is the strong spin-orbit coupling (SOC), which leads to a splitting of the conduction and valence bands. The valence band of MoS$_2$ and MoSe$_2$ is split by nearly 150 meV, whereas due to heavier mass of W the splitting for WS$_2$ and WSe$_2$ is considerably larger (of the order of 400 meV).

It is well known that due to the presence of artificial self-interaction and the absence of the derivative discontinuity in the exchange-correlation potential in the PBE/GGA methods, DFT underestimates the band gap. In low-dimensional materials, the strong exciton binding due to the weak screening of Coulomb interaction is also important, and more sophisticated many-body methods, such as GW usually provides better descriptions of electronic structures [75, 77]. In order to get an accurate description of its electronic structure, several researchers have preformed GW calculations [77, 79–82]. As shown by Shi *et al.* [77], the optical gap obtained by solving the Bethe–Salpeter equation (BSE) including excitonic effects on top of the partially self-consistent GW0, the predicted optical gap magnitude is in good agreement with the experimental value of 1.9 eV [15, 26, 83].

9.3.3 Effect of number of layers on electronic structure

Recent theoretical studies show that the electronic structure of TMDs is strongly dependent on the number of layers. [26, 44, 77] Unstrained few layers to bulk MX$_2$ have indirect band gaps with VBM and CBM at Γ and K-points of the Brillouin zone, respectively [37, 40, 44]. As the electronic properties of the family of group-VI TMDs are similar, herein we take MoS$_2$, as a typical example to discuss the layer dependent electronic structure. Figure 4(a) shows the calculated electronic structures of bilayer, trilayer, quadlayer and pentalayer MoS$_2$. The band gaps sensitively depend on the number of layers of MoS$_2$, it changes from 1.2 eV for bulk to

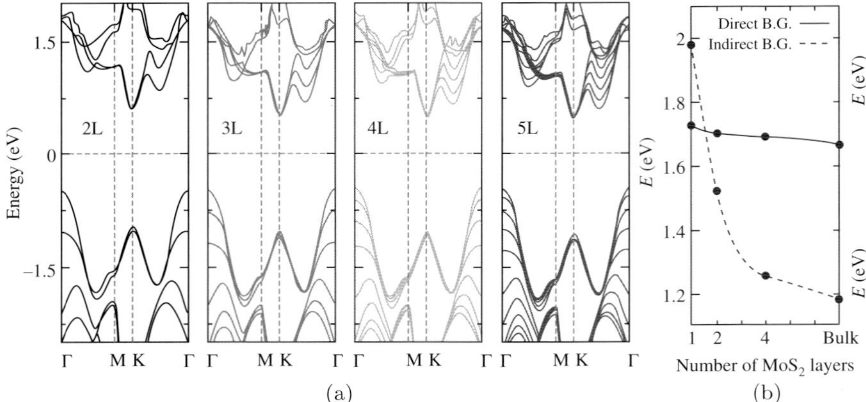

Figure 4. (a) Band structures of nL-MoS$_2$ for $n = 1$, 2, 3, 4 and 5 are shown along the high-symmetric points. The horizontal dotted line represents the Fermi level, (b) Variation of the band gap with the thickness of few-layer MoS$_2$. Part (a) is reprinted from Ref. [44] with permission of the IOP Publishing and part (b) is reprinted from Ref. [40] with permission of Springer.

1.9 eV for the monolayer MoS$_2$, as shown in Figure 4(b). Unlike bulk and few layered TMDs, monolayer TMDs are direct band-gap semiconductors as both the CBM and VBM shift to K points due to the quantum confinement effect [26, 77, 83]. Furthermore, the spacing between the VBM and VBM-1 at the Γ-point decreases as the number of layers increases. This nature of the band structure has important implications under the applied strain. This change in electronic properties can be monitored by the energy evolution of the absorption/reflection spectra and the band-edge emissions in the PL measurements [15, 26, 83]. Hence, PL measurements can serve as a technique to measure the layer thickness in the TMDs.

9.4   Effect of Strain on Properties of MX$_2$

In the last decade, the area of strain engineering has become very popular in both experimental and theoretical research community. In the following section of this chapter, we will give an overview of the recent progress to control the optical and electronics properties of TMDs, by means of strain engineering. We will concentrate on semiconducting TMDs (MoS$_2$, WS$_2$, MoSe$_2$ and WSe$_2$). The benefits of strain engineering in 2D crystals for applications in nanoelectronics and optoelectronics will also be discussed.

9.4.1 Electronic properties

In this section, the recent theoretical progress on engineering the band structure of semiconducting 2D-TMDs beyond graphene, by normal compressive and biaxial strains will be reviewed.

9.4.1.1 *Normal compressive strain*

The effect of normal compressive strain in the band structure of 2D semiconducting crystals has been the focus of many theoretical studies [40, 41, 76, 77, 84–90]. There are a number of studies which investigate the effect of normal compressive strain on the electronic structure of TMDs. Since all the TMDs considered in this chapter exhibit the same electronic structure and similar variation under strain, here, we will discuss the electronic structure of MoS$_2$ under normal compressive strain in details. The normal compressive strain has been simulated by reducing the inter-layer distance (d). The applied strain can be calculated as $\epsilon = (d_0 - d)/d_0$, where d_0 and d are equilibrium and instantaneous inter-layer distances, respectively.

The bulk MoS$_2$ has an indirect band gap between Γ and Λ point [39, 76, 77]. On the other hand, the monolayer MoS$_2$, has a direct band gap, which is located at the K point [77]. To illustrate the effect of normal compressive strain, we take an example of bilayer MoS$_2$ as discussed in Ref. [37]. As the inter-layer separation decreases, the layers start to interact chemically, which leads to the lifting of double degeneracy of the bands, Figure 5. This splitting increases with the increasing NCS. The valence band maxima (VBM) moves away (towards) the Fermi level at $K(\Gamma)$-point. The semiconductor to metal (S–M) transition occurs when VBM crosses the Fermi level at the Γ point after a critical applied NCS. Likewise, the conduction band minima (CBM) also moves towards the Fermi level with increasing NCS. The band structures of the other MX$_2$ materials undergo similar changes S–M transition under applied NCS [37, 75].

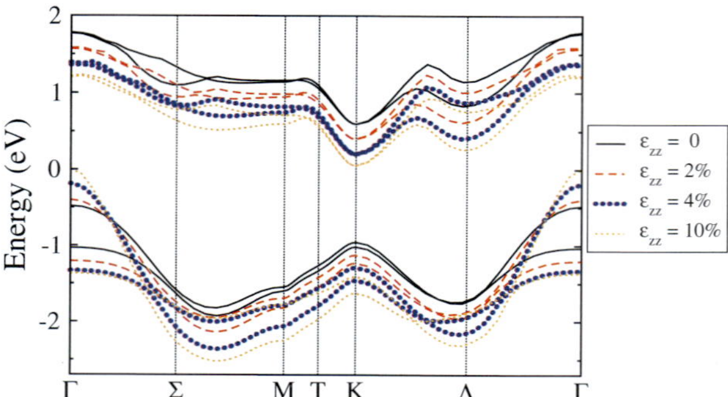

Figure 5. Band structure of bilayer MoS_2 under normal compressive strain (along with the **c** axis), plotted along high-symmetry lines in the Brillouin zone. For clarity, only the highest two valence bands and lowest two conduction bands are shown.

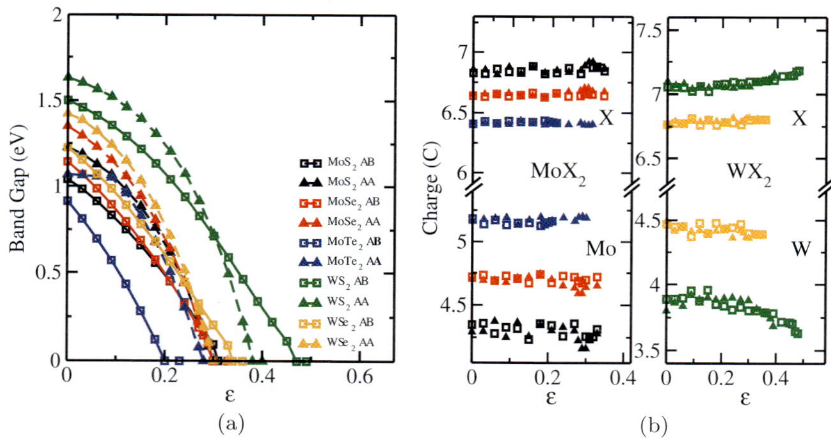

Figure 6. (a) Change in band gap with applied strain for MX_2 bilayers for both AA and AB stacking using PBE+vdW method, (b) Average Bader charge of M and X for all the bilayers as a function of strain. The left and right panels correspond to M = Mo and W, respectively. (Reprinted from Ref. [37] with permission of the American Physical Society).

Bhattacharyya and Singh [37] have discussed the S–M transition in various bilayer TMDs, under normal compressive strains as shown in Figure 6(a). Their results showed that an S–M transition can be achieved in all the TMDs irrespective of their stacking (AB or AA). For a given metal atom, the S–M transition strain decreases as one goes down the column of X atoms in the periodic table. This is caused by increased delocalisation of the atomic orbitals, which leads to reduced interaction between M and X atoms, resulting in an S–M transition at lower pressure. Such behaviour is consistent with the trend of the band gap, which also decreases from S to Te. Similar evidence also comes from a Bader charge analysis. The average charge on M and X as a function of applied strain is shown in Figure 6(b). The charge transfer from M to X increases as a function of NCS, a change which is more prominent for WX$_2$. Furthermore, for a given M atom, the value of the average charge on the X atom decreases as we go down the periodic table, i.e., from S to Te, replicating loosely the trend of electron affinity of the X atoms. This, essentially, results in a weaker interaction between the M and X atoms, which leads to the S–M transition pressure to be lowest for Te and highest for S. Furthermore, in comparison to W, Mo has a lower ionisation potential, which measures the ability to donate charge. Mo can donate charges relatively easily and thereby facilitate the X–X interaction, which results in an S–M transition in MoX$_2$ at a lower pressure than in WX$_2$.

Similar to the bilayer and bulk TMDs, monolayer TMDs are also shown to undergo S–M transition under strain. In the case of monolayer TMDs, the normal compressive strain can be generated by compressing the sulphur–sulphur distance within the same layer [46, 91]. As discussed by Peña-Álvarez *et al.* [91] with application of strain, there is a direct to indirect band gap transition around 0.6 GPa Figure 7. This observation is in excellent agreement with the experimental study, where the photo luminescence (PL) disappears around 0.5 GPa normal compression [91]. From this value onwards, the VBM and CBM continue to converge with compression. At 2.8 GPa, both the VBM at the Γ point and the CBM at the K point cross the Fermi level, leading to the transition to a metallic phase.

9.4.1.2 *Biaxial strain*

For TMDs, a number of studies have investigated the dependence of the band structure and phonon modes on external strain [40, 41, 76,

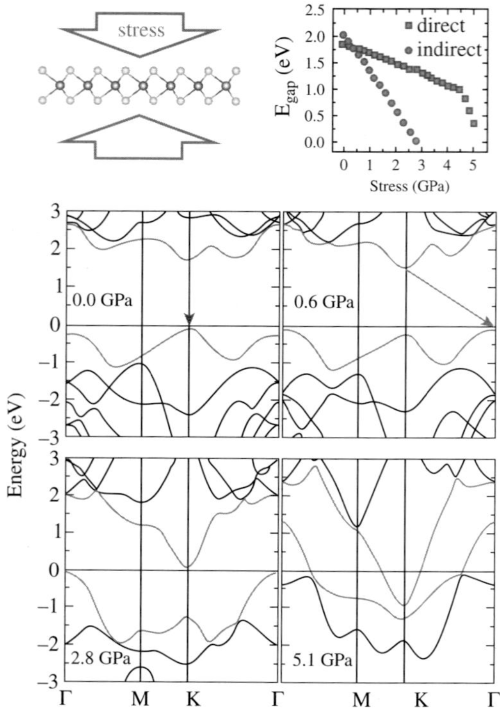

Figure 7. The electronic structure of monolayer MoS_2 under normal compression. Evolution of the direct and indirect band gap energies with stress. (Reprinted from Ref. [91] with permission of the American Chemical Society.)

77, 84–90], and its consequences on different physical properties [44, 92, 93]. In Figures 8(a) and 8(b) the evolution of the band gap with uniform biaxial strain is shown for single-layer MoS_2 and WS_2, respectively. In the absence of strain, monolayer TMDs are direct band gap semiconductors. Uniform biaxial tensile strain leads to a linear decrease in the band gap, and a direct-to-indirect gap transition is reached, with the valence band edge being now situated at the Γ-point [94]. Various other groups have recently reported a similar DFT band gap reduction under biaxial tensile strain [40, 75]. The calculations also find that the band gap decreases with the biaxial tensile strain and the monolayer may become a metal under a larger strain of 8–10% [40, 76, 95]. Johari and Shenoy [76] have discussed the effect of both tensile and shear (when the monolayer

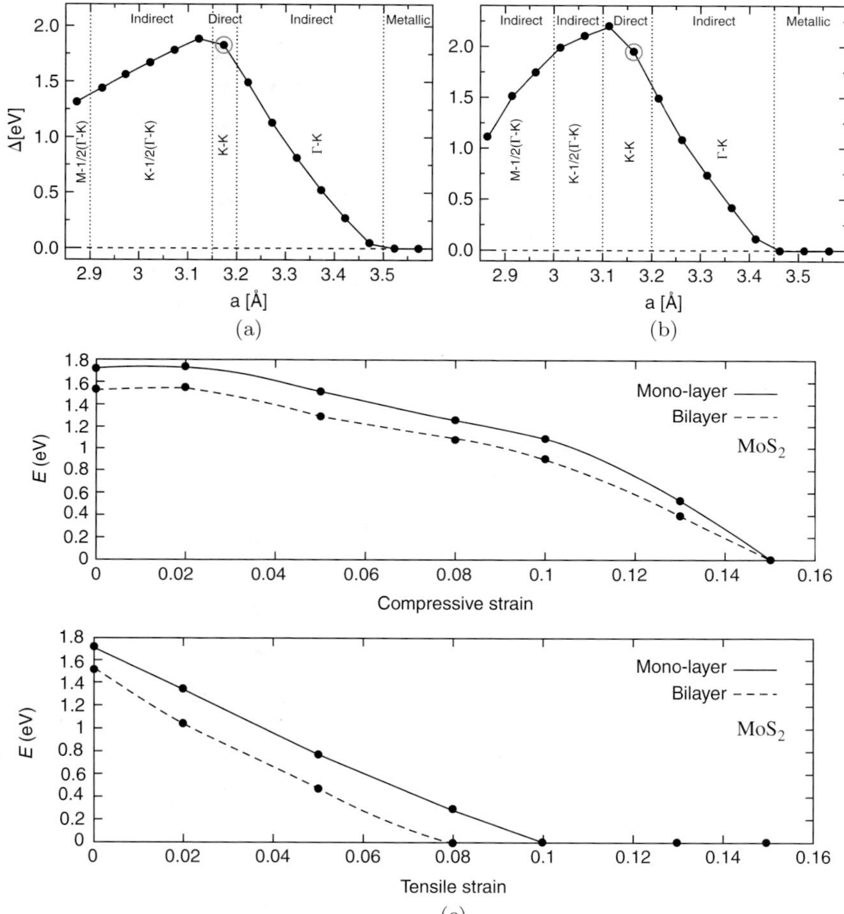

(a)

(b)

(c)

Figure 8. Evolution of the band gap of a single layer (a) MoS$_2$, and (b) WS$_2$ under uniform isotropic tensile and compressive strain. The valence and conduction bands edges evolve with strain and corresponding high symmetric points are marked, (c) Evolution of the band gap with biaxial compressive (top) and tensile (bottom) strain for single layer and bilayer MoS$_2$. Part (a) and (b) are reprinted from Ref. [84] with permission of the American Physical Society. Part (c) is reprinted from Ref. [40] with permission of Springer.

is expanded in one direction and compressed in the other) strains on electronic properties of monolayer TMDs. A direct to indirect and S–M transition was achieved for all types of tensile and shear strains. The S–M transition, however, significantly depends on the type of applied strain and the type of chalcogenide atoms. The heavier chalcogenides require relatively more tensile and less shear strain to attain a direct-to-indirect band-gap transition. In addition, under 10% biaxial tensile strain an S–M transition can be achieved in all semiconducting TMDs, while through pure shear strain this transition can only be achieved by expanding and compressing the monolayer of MTe_2 (M = Mo and W) in the y- and x-directions, respectively [76].

The biaxial tensile strain also modifies the absorption spectra and an optical gap of the monolayer. As shown in the Figure 9(a), the strain-dependent optical absorption spectra indicates that the direct-to-indirect transition of the band gap with biaxial tensile strain does not affect the strength of the vertical optical absorption. The lowest optical excitation in Figure 9(a) originates from the doubly degenerate singlet excitons around the K points [96]. As shown by Feng *et al.* [97], the exciton energy of 2.0 eV at zero strain is in good agreement with the energy (A peak at 1.9 eV) obtained from photoluminescence spectroscopy [83]. The excitation energy reduces substantially when the strain is imposed, and it decreases to 1.1 eV at 9% strain. This large constant exciton binding energy together

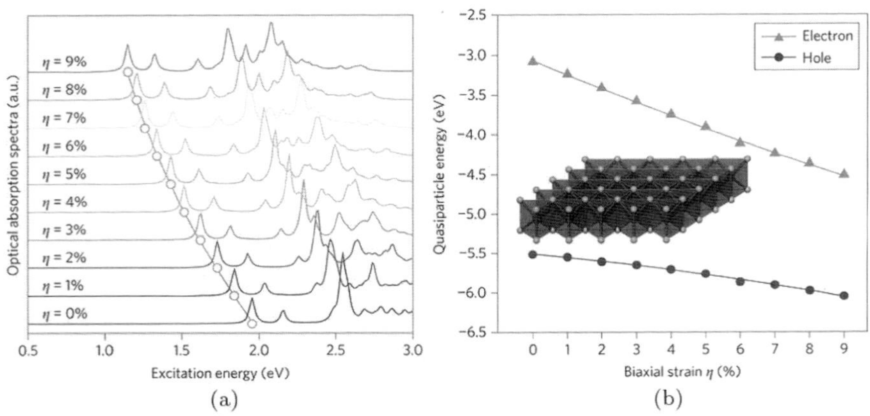

Figure 9. (a) Biaxial strain-dependent optical absorption spectra calculated by BSE, (b) Biaxial strain-dependent GW quasiparticle energies for electrons and holes at the K point. (Reprinted from Ref. [97] with permission of Macmillan Publishers Ltd).

with the reduced quasiparticle energies (Figure 9(b)) indicates that MoS$_2$ can be a good solar cell material under biaxial tensile strain [97].

Similar to the normal compressive and biaxial tensile strain, an S–M transition can also be achieved under biaxial compressive strain [44]. In contrast to biaxial tensile strain, where band gap smoothly reduces, the band gap increases with compressive strain up to a ~2% strain [41, 44, 75, 98] followed by a decrease with larger strain as shown in Figure 8(c) [40, 44]. The calculations find that biaxial compressive strain in an MoS$_2$ monolayer lowers the energies of valence and conduction band at Γ and Q points, respectively. The band gap also changes to an indirect one. Monolayers of other TMDs demonstrate qualitatively the same behaviours under biaxial strain [40, 41, 76, 99]. Dong *et al.* [100] have investigated the effect of biaxial compressive strain on bilayer MoS$_2$. They report an indirect band gap for all values of strain along the basal plane. Several transitions for the indirect band gap were observed at different strain values [100]. The nature of the dispersion changes near the Fermi level with the strain as shown in Figure 10 for 2L-MoS$_2$ [44, 99]. The change in band structure is also observed to be very different for biaxial compressive strain compared to normal and biaxial tensile strain. This indicates that distinct transport properties (carrier effective masses, mobilities, etc.) can be realised in TMDs under different type of strains [39, 44, 100–102]. Important point to

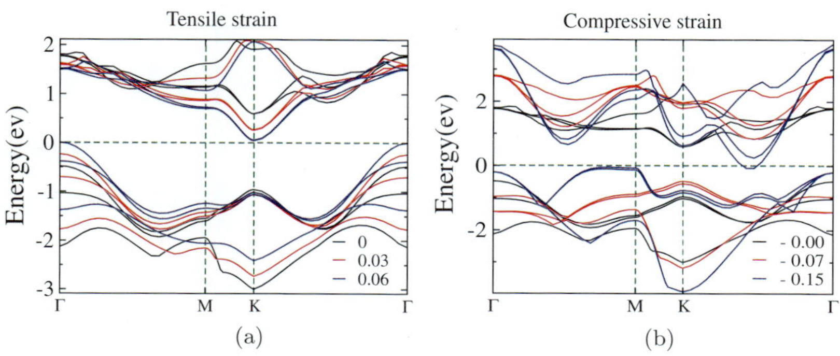

Figure 10. Band structure of bilayer MoS$_2$ under (a) biaxial tensile and (b) biaxial compressive strain.

note that, due to the relatively small bending rigidity, compressive strain is difficult to apply on suspended TMDs, although substrate-induced compression is possible by depositing layers of TMDs on an elastomeric substrate.

9.4.1.3 *Mechanism of S–M transition*

As discussed in the previous sections, a semiconductor to metal transition can be achieved in all the TMDs under different type of strains. The mechanism for this S–M transition under NC, BC and BT strain, can be understood by analysing the contribution from different molecular orbitals by performing LDOS and band-decomposed charge density calculations [37, 40, 44]. The results are shown in Figure 11 for 3L-MoS$_2$ under NC strain and in Figure 12 for 2L-MoS$_2$ under biaxial strain. For all unstrained MoS$_2$ multilayers, the Mo-d and S-p orbitals have the largest contribution to the VB as well as the CB as shown in Figure 11(a) and 12(a)

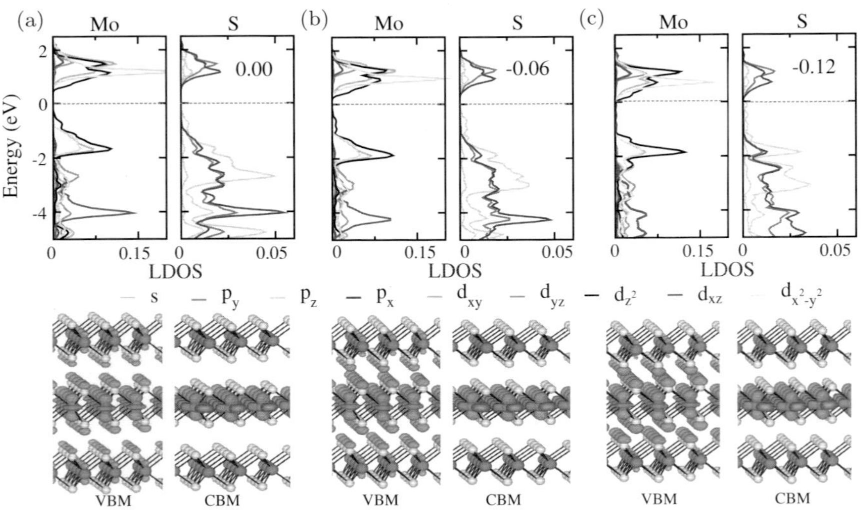

Figure 11. Angular momentum projected density of states (LDOS) of Mo and S and corresponding isosurfaces (value of 0.09 e/Å^3) of the band-decomposed charge density (VBM and CBM) for 3L-MoS$_2$ under NC strains of (a) 0.00 (b) −0.06 and (c) −0.12, respectively. The Fermi level is shown by dotted lines. (Reprinted from Ref. [44] with permission of the IOP Publishing.)

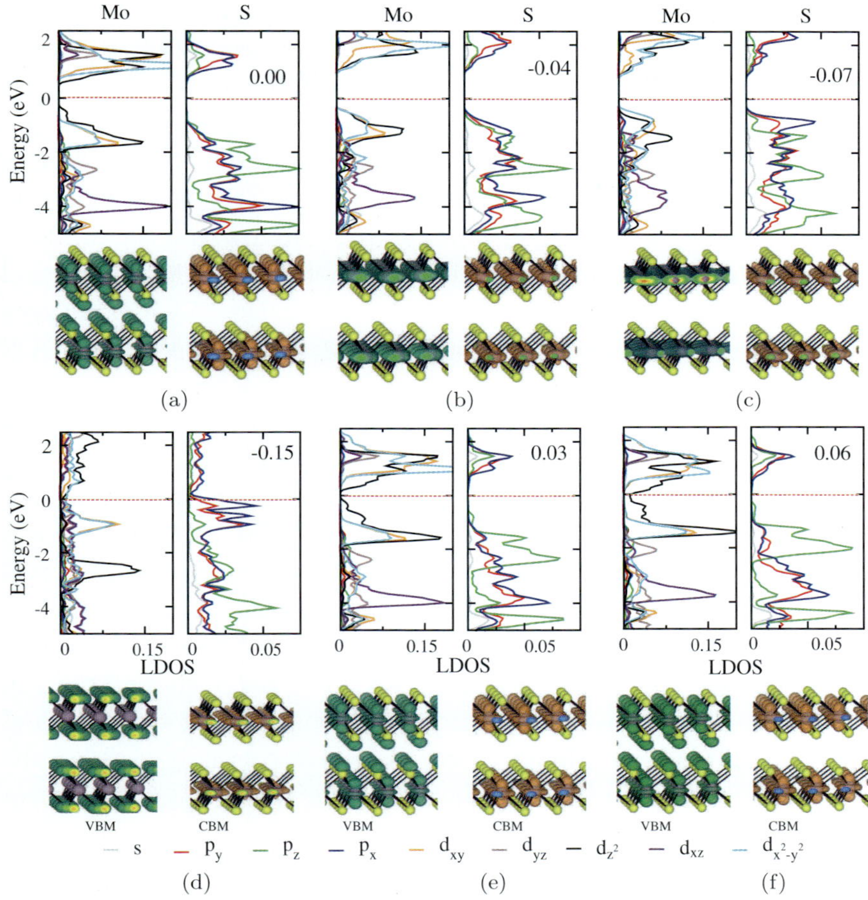

Figure 12. Angular momentum projected density of states (LDOS) of Mo and S and corresponding isosurface (value of $0.09\,\mathrm{e/\mathring{A}^3}$) of band-decomposed charge density (VBM and CBM) for 2L-MoS$_2$ under BC and BT strains. (a) For unstrained case. (b)–(d) Under -0.04, -0.07 and -0.15 BC strains, respectively. (e)–(f) under 0.03 and 0.06 BT strains. The Fermi level is shown by red dotted lines. (Reprinted from Ref. [44] with permission of the IOP Publishing.)

for 3L- and 2L-MoS$_2$, respectively. Both VB and CB are mainly composed of Mo-$d_{x^2-y^2}$ and d_{z^2}. However, in the case of VB, there is some additional contribution from S-p_z orbital. This is clearly seen in the band-decomposed charge density (Figure 11(a)), where CBM and VBM mainly originate from around Mo-$d_{x^2-y^2}$ (in-plane lobes) and d_{z^2} (out-of-plane lobes). The contribution of S-p_z orbital to the VBM is also indicated by the out-of-plane lobes. With the application of NC strain, the LDOS corresponding to VB and CB moves towards the Fermi level resulting in the reduction of band gap. All the orbitals of Mo and S atoms contribute to the shift in CB, while in the case of VB the movement towards the Fermi level occurs mainly due to the contribution from Mo-d_{z^2} and S-p_z orbitals. The most dominant interaction happens between the Mo-d_{z^2} orbital of one layer and to the S-p_z orbital of another layer. Similar changes were observed for other layers and bulk MoS$_2$. Based on the LDOS and band-decomposed charge density analysis, one can conclude that the strong inter-layer interaction between Mo-d_{z^2} and S-p_z is the main cause for S–M transition under NC strain.

The change in LDOS is completely different in case of BC strain. The tail in LDOS of Mo-d_{z^2}, forming the VBM and CBM moves away from the Fermi level as the in-plane interaction increases caused by reduction in Mo-S bond lengths. Due to this movement of Mo-d_{z^2} orbitals, initially the band gap increases for lower strain values. With further increase in strain, LDOS of Mo-d_{z^2} crosses the tail of the LDOS of Mo-$d_{x^2-y^2}$, which now consists of both CBM and VBM. The in-plane orbitals begin to interact strongly, causing the shift in the position of VBM from Γ to K while the CBM moves from K to in between K and Γ. This can also be seen in the band-decomposed charge density plots where, the VBM changes to completely planar, while CBM has a slight out-of-plane contribution from Mo-d_{z^2} orbital (Figures 12(b) and 12(c)). With further increase in in-plane interaction, the Mo-$d_{x^2-y^2}$ and S-p_x orbitals hybridise strongly within the layer as seen in Figure 12(d). This strong intra-layer hybridisation leads to the closing of gap. This can also be seen in the band-decomposed charge density Figure 12(d), where the VBM and CBM show strong in-plane character near S–M transition.

In the case of BT strain, the VBM and the CBM has maximum contribution from Mo-d_{z^2} orbitals (Figure 12(e)). With the increase in BT strain, Mo-d_{z^2} and S-p_z orbitals start to hybridise weakly within the same layer. This weak hybridisation further leads to the delocalisation of orbitals thereby, causing S–M transition

(Figure 12(f)). Although in the case of both NC and BT strains, the S–M transition is dominated by Mo-d_{z^2} and S-p_z, in the case of NC strain the interaction happens between Mo and S atoms from different layers. However, for BT strain this interaction is within the same layer.

9.4.2 Effect of strain on transport properties of 2D MX$_2$

Application of strain/pressure modifies the dispersion of bands, hence, under strain the carrier effective masses can also be tuned. Hosseini *et al.* [92, 103], calculated the change in the carrier effective mass under tensile strain by Boltzmann transport calculations. For unstained cases WS$_2$ and WSe$_2$ exhibit smaller effective masses than that of the unstrained MoS$_2$ and MoSe$_2$. In all the cases, tensile strain decreases the effective mass of the K-valley while under compressive strain effective mass of Q valley decreases [92, 103]. Dubey *et al.* [100] also estimated the carrier effective mass for bilayer MoS$_2$ under the biaxial and uniaxial strain as shown in Figure 13. Their results show that as the biaxial or uniaxial strain varies from -6% to 6%, the effective mass change monotonically or remain constant, except for $m_e(K)$ which peaks at a biaxial compressive strain $\epsilon \leq 4\%$ [100]. This peak is associated with the relative variations in the profile of the conduction band at K-point. Similar tuning of carrier effective mass under strain was also proposed by Peelaers *et al.* [39]. Since the carrier mobility depends on the effective mass, a decrease in effective mass results in increase in mobility as shown by Esseni *et al.* [92, 103]. A relatively small tensile strain of 0.4% only weakly affects the effective mass of the K-valley (lowest valley), while it has a stronger effect on the shift of the valleys with respect to Fermi level. Therefore, this small strain has the largest (smallest) effect on the mobility of single-layer WSe$_2$ (MoS$_2$). For all the TMDs, application of tensile strain can result in many fold enhancement in the carrier mobility [92, 103].

TMDs have also been proposed as good material for thermo-electric applications [44, 101, 104–106]. Usually, the performance of a

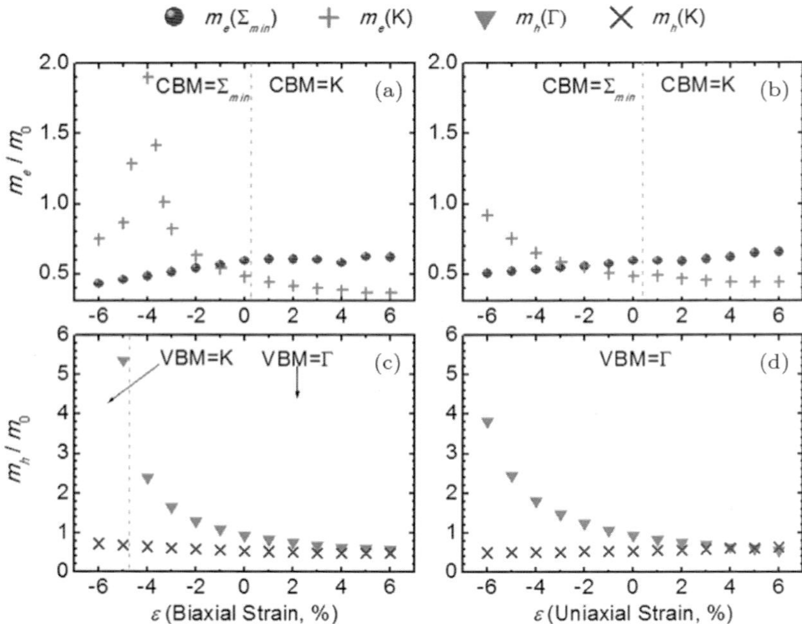

Figure 13. Variations of electron effective mass (m_e) under (a) biaxial strain, and (b) uniaxial strain, (c) and (d) present the variations of hole effective mass (m_h) under biaxial strain and uniaxial strain, respectively. (Reprinted from Ref. [100] with permission of the AIP Publishing.)

thermoelectric material is given by a dimensionless parameter (figure of merit) $ZT = S^2\sigma T/\kappa$, where σ, T, S, and κ are the electrical conductivity, the temperature, thermopower, and total thermal conductivity, respectively. Materials with a large thermopower and electrical conductivity and/or low thermal conductivity are expected to show good thermoelectric properties. Application of strain and hydrostatic pressure [44, 101] has been proposed as a route to enhance the thermopower and the electrical conductivity in MoS$_2$. As shown by Zhang *et al.* [101], due to enhanced electrical conductivities a significant value of the thermoelectric figure of merit over a wide temperature range for bulk MoS$_2$ under hydrostatic pressure can be achieved. In another study, Bhattacharyya and Singh [44] have investigated the effect of normal compressive, biaxial compressive, and biaxial tensile strain on thermoelectric properties of multilayers of MoS$_2$. Their results indicate that the maximum thermoelectric performance can be achieved under biaxial tensile strain. This enhancement in thermoelectric properties is mainly driven by an increase in electrical conductivity. In a recent study, similar enhancement in thermoelectric properties was obtained for phosphorene [107, 108]. These recent results suggest that strain engineering can act as an effective means to enhance the thermoelectric properties of layered materials.

9.4.3 Effect of strain on vibrational properties of 2D MX$_2$

Phonons, bosonic quanta of lattice vibrations, play an important role in understanding various macroscopic properties of materials — specific heats, thermal expansion and heat conduction. The absence of negative frequency in phonon dispersion can predict structural stability of materials under stress. Vibrational properties of materials can be studied by Raman spectroscopy, which is considered a very reliable experimental tool for attaining insight on the lattice vibrations and stability. MX$_2$ has two characteristic Raman active vibrational modes: E_{2G}^1 and A_{1G} mode. The in-plane E_{2G}^1 mode results from opposite vibration of two X atoms with respect to the M atom while the A_{1G} mode is associated with out-of-plane vibration of X atoms in opposite directions. The atomic displacements of these modes are shown in the inset of Figure 14(a). The frequencies of the E_{2G}^1 and A_{1G} modes are obtained by identifying the vibrational modes from the phonon dispersion curves at the Γ point. The frequencies of these Raman active modes change the number of layers of MX$_2$ and have been studied extensively in the literature [70, 109, 110]. With the increase in the number of layers the frequency of E_{2G}^1 mode decreases while that of the A_{1G} mode increases for all the semiconducting MX$_2$. With the increase in the number of layers, the dielectric screening of the long-range Coulomb interaction increases between the effective charges, causing a decrease in frequency of the E_{2G}^1 mode. The weak inter-layer interaction due to the additional layers is responsible for the increase in the frequency of the A_{1G} mode [70]. The occurrence of new peaks, which are

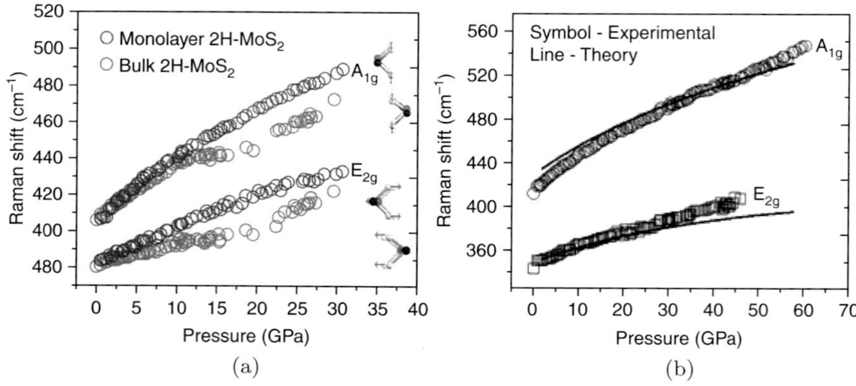

Figure 14. Raman shift of E_{2G} and A_{1G} mode as a function of hydrostatic pressure for (a) monolayer and bulk MoS$_2$, and (b) multilayer WS$_2$. Representative vibrations are shown in the inset of (a) for E_{2G} and A_{1G} mode. Part (a) and (b) are reprinted from Refs. [46] and [111], respectively, with permission of the American Chemical Society.

absent in bulk and monolayers has been reported in multilayers of MX$_2$ [109].

Application of strain shows change in the Raman frequencies [75, 112, 113] of bulk as well as 2D MX$_2$ as the phonon propagation is highly influenced by the M–X bond strength and the X–M–X bond angle. The change in phonon dispersion in monolayer MoS$_2$ is shown in Figure 15 under biaxial and uniaxial strain. The application of uniaxial strain introduces anisotropy in the M–X bond strengths and the X–M–X bond angles. This causes lifting of degeneracies of the in-plane E' and E'' modes at Γ. The average Raman shifts decrease with increasing strain for the two split modes. No such splitting occurs while applying biaxial strain as no anisotropy is introduced. The phonon stiffness changes isotropically because of the change in bond strengths. The phonon frequencies increase for compressive while decreases for tensile biaxial strain in MX$_2$ monolayers [75].

The effect of hydrostatic pressure on vibrational properties of 2D and bulk WS$_2$ and MoS$_2$ has also been investigated both experimentally and theoretically [38, 46, 111]. Frequencies of both E_{2G} and A_{1G} modes increase with increase in pressure for monolayer and bulk MoS$_2$ [46] as well as multilayer WS$_2$ [111] as shown in

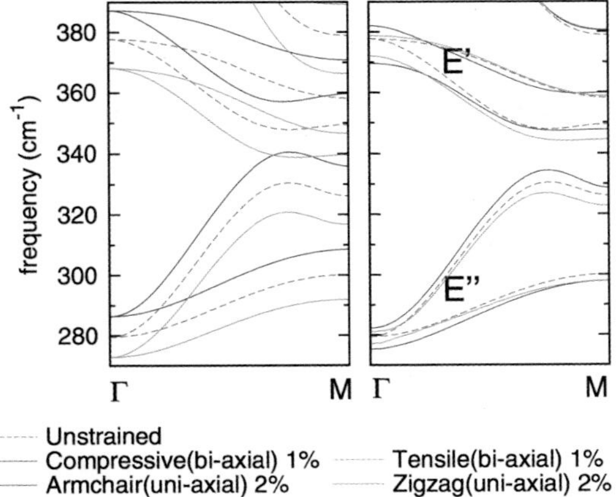

Unstrained
Compressive(bi-axial) 1%   Tensile(bi-axial) 1%
Armchair(uni-axial) 2%   Zigzag(uni-axial) 2%

Figure 15. Phonon dispersion of MoS$_2$ under biaxial compressive and tensile strain and uniaxial strain along armchair and zigzag direction. (Reprinted from Ref. [75] with permission of the American Physical Society.)

Figures 14(a) and 14(b), respectively. However, the rate of increase in frequency is high for the out of plane A_{1G} mode in all the plots. This variation can be explained based on the different vibrational pattern of the two modes. As mentioned before, under hydrostatic pressure the out-of-plane compression and hence the A_{1G} modes become more prominent than the in-plane compression. This causes faster transverse vibration of S–S atom as compared to the in-plane movement of Mo–S atom, which is the origin of the E_{2G} mode. This is the reason for the larger slope of the A_{1G} mode in the frequency vs. pressure plot [46]. For bulk MoS$_2$, a flat region is observed for the A_{1G} mode in Figure 14(a), where the Raman shift is nearly invariant with pressure between 10 GPa and 19 GPa. This region corresponds to an intermediate state, where inter-layer charge transfer takes place along with a possible occurrence of structural distortion before the material undergoes complete metallic transition [38].

Similar to 2H-phase Raman spectroscopy can also be used to characterise 1T′-phase [46, 69]. As discussed by Akinwande *et al.* [46] 1T′-phase of MoS$_2$ has three additional (J_1, J_2 and J_3) Raman active modes, which are not observed in the 2H-phase. The atomic vibration of these Raman modes was first studied by Calandra [69].

He showed that the J_1 mode of 1T'-MoS$_2$ involves two types of vibrations: an out-of-plane vibration of the Mo atoms forming the zig–zag chain and an in-plane vibration of one chain of atoms with respect to the other in the chain along shearing direction [69]. These two modes occur as a single peak in the experiments [46]. The J_2 Raman peak is originated from the relative motion of the two zigzag chains against each other [69]. The third Raman mode J_3 has a vibration such that such zigzag chain tends to break into two stripes with a slight contribution along the out-of-plane direction [69]. Under the application of hydrostatic pressure on 1T'-MoS$_2$, all the Raman active modes linearly increase with pressure, and the out-of-plane modes show the highest increase [46]. At a high pressure (above 27 GPa), the only dominant modes of this material is observed to be the J_1, A_{1G} and E_{2G} modes [46]. These above results indicate that the Raman spectroscopy can provide a wealth of information about changes in vibrational properties under strain/pressure. By careful monitoring of Raman modes the structural as well electronic phase transition can be detected [38, 114].

9.5 Experimental Progresses on Effect of Strain on Properties of 2D MX$_2$

The theoretical predictions of changing various properties of MX$_2$ as mentioned in the previous part of this chapter are promising if it can be achieved in experiments. Therefore, experimental verification is important specially in terms of the suitable methods to apply various strains in MX$_2$ and measuring properties at a particular strain. The real materials will have defects, which along with various experimental conditions such as temperature will have effect on the properties. Moreover, theoretical predictions are qualitative in nature and for exact quantitative description, one has to perform experiments. In this section, we will review some of the recent progresses in experimental work in this area of research.

There have been several experimental works to study structural, electronic, optical, vibrational and thermoelectric properties of bulk and 2D MX$_2$ under strain. Planar strain can be applied by growing or transferring 2D MX$_2$ layer on a flexible and stretchable substrate. The strain applied to the substrate will get transferred to the MX$_2$ layer. In this way, not only a planar strain but also a radial deformation (curvature) can be applied. Highly flexible and stretchable MoS$_2$ thin-film transistors have been fabricated on plastic substrates and no degradation in the electrical characteristics was indicated while applying uniaxial tensile strain [115] and bending the film to a large curvature [116]. This indicates high mechanical stability and a possibility to use MX$_2$ in flexible and stretchable devices. Bolotin *et al.* [117] performed Raman spectroscopy and photoluminescence (PL) measurement on MoS$_2$ monolayers and bilayers under uniaxial tensile strain of 0–2.2%. Phonon softening and breaking of degeneracy in the doubly degenerated in-plane E_{2G} Raman mode of both mono and bilayer MoS$_2$ was observed with increase in strain. However, the peak position of the out of plane A_{1G} mode remained unaffected by the planar strain. From the PL spectra, a linear redshift in the direct band to band peak (A peak) was observed for monolayer while red shift in A peak as well as I peak, originating from the transition across the indirect band gap was observed in bilayer MoS$_2$. A decrease in intensity of the A peak in monolayer indicates that there is transition from direct to indirect under uniaxial tensile strain. A similar red shift in PL peak and splitting of in-plane Raman mode under uniaxial tensile strain was also reported for mono and bilayer MoS$_2$ by other experimentalists [118, 119]. The result does not depend on the crystallographic direction of the applied strain i.e., along zigzag or armchair direction [118]. Similar effect on band gap has been reported for 2D WS$_2$ [112] and WSe$_2$ [113] under the application of uniaxial tensile strain. Javey *et al.* [113] have shown that an indirect to direct band gap transition can be achieved in bilayer WSe$_2$ under uniaxial tensile strain as shown in Figure 16.

Uniform and controllable biaxial compressive strain has been applied on trilayer MoS$_2$ in experiments using a piezoelectric substrate and PL and Raman spectra measurements were performed [120]. A large blue-shift in the direct band gap and an enhancement in the PL intensity was observed under biaxial

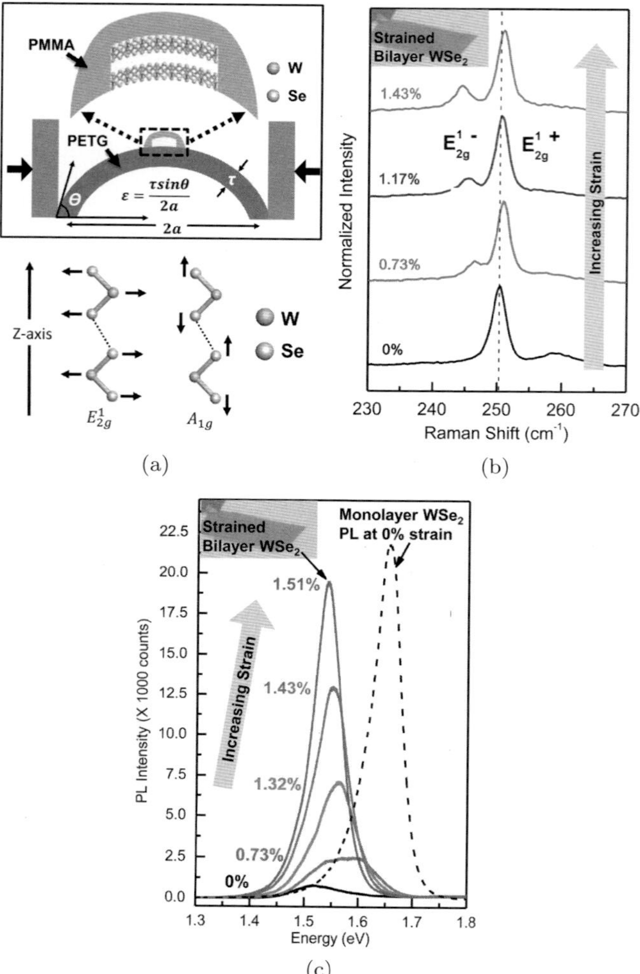

Figure 16. (a) Schematic of the two-point bending apparatus for generation of biaxial strain, (b) Raman spectra for bilayer WSe$_2$ at different strains, (c) PL spectra of bilayer WSe$_2$ at different strain. Sudden increase in PL intensity at 1.51% strain indicates indirect to direct band gap transition. (Reprinted from Ref. [113] with permission of the American Chemical Society.)

compressive strain as shown in Figure 17 [120]. Korn *et al.* [121] exploited the thermal expansion properties of a silicone-based substrate to uniformly apply biaxial tensile strain in MoS$_2$. Due to the large mismatch of the thermal expansion coefficients, the substrate

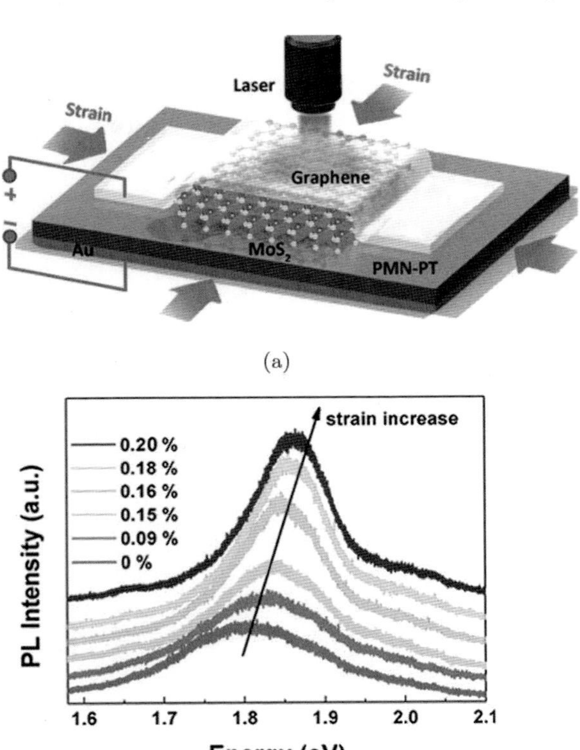

(a)

(b)

Figure 17. (a) Schematic diagram of Raman and PL measurements on MoS$_2$ that is sandwiched between a piezoelectric PMN-PT substrate and a graphene top electrode, (b) Photoluminescence spectra of trilayer MoS$_2$ under various strains. (Reprinted from Ref. [120] with permission of the American Chemical Society.)

was expanded more as compared to the MoS$_2$ layer, when heated, thus straining the layer uniformly [121]. The amount of planar strain applied in these experiments are not large enough to make the 2D MX$_2$ undergo semiconductor to metal transition as predicted by theory [113, 120, 121]. As mentioned before, due to the strong covalent nature of the bond between atoms within the layer, a large amount of force needs to be applied to obtain the required strain. Moreover, the strain is applied via the substrates and hence the method depends on the mechanical or piezoelectric properties of the substrate. Thus, there are experimental challenges to apply large strain along planar direction. An efficient method to apply normal compressive strain on 2D MX_2 has been developed by Frank *et al.* [91], where the pressure transmitting medium has been removed in a conventional anvil cell experiment. Their experiments showed direct to indirect transition in band gap at $\sim$0.5 GPa for a monolayer MoS_2.

Hydrostatic pressure has been applied on bulk MX_2 and reduction in the resistivity [45] was observed indicating possible advent of metallic phase at very high pressure. Recently, using a diamond anvil cell (DAC) reversible semiconductor to metallic transition has been shown experimentally for 2D MoS_2 and WS_2 under pressure [38, 111]. A six order decrease in resistivity, a two order decrease in mobility along with a four order increase in carrier concentration were observed after the advent of metallic phase in multilayer WS_2 at $\sim$22 GPa [111]. Pressure has been applied on monolayers also, though S–M transition was not yet achieved in experiments. This is because of the requirement of a comparatively high critical pressure (67.9 GPa) [46] to get metallic phase of monolayer MoS_2 than that for the multilayers ($\sim$19 GPa) [38]. The metallic properties of monolayer $1T'$-MoS_2, another metastable crystallographic phase of MoS_2 was not affected by the applied pressure [46]. The 2D MX_2 retained its crystal structure under high pressure as no major structural phase transition was observed in these works. However, some lattice distortions occurred before it becomes electronically metallic [38]. These are among the first experimental reports on achieving a complete metallic transition in 2D MX_2 under pressure. An unprecedented controlled tunability of the properties of 2D TMDs by strain offers a platform for both experimentalist and theorist to develop a completely new range of devices based on alternative physical principles.

9.6   Conclusion

Application of mechanical strain shows remarkable effects on various properties of 2D MX_2 layers. Owing to the structural anisotropy, different types of applied strains, such as uniaxial, biaxial, hydrostatic have distinct effects on the materials' properties. All the strains cause semiconductor to metal transition in semiconducting 2D MX$_2$ with different critical strains. Strain changes the inter-layer bonding type from weak vdW to strong covalent bonding without causing major structural changes. The vibrational properties also change significantly with applied strain. Development of experimental techniques to apply various types of strain in a controlled manner and excellent mechanical stability make 2D MX$_2$ an ideal material for straintronics devices. Some of the electronic and photonics devices, e.g., lasers, photodetectors and electromechanical devices require materials with varying band gap. MX$_2$ may find its prospective application in such devices as the required band gap may be obtained just by applying the corresponding strain on the same material in a reversible manner, thus, eliminating the use of many materials with varying band gap. The excellent mechanical stability under high strain and reversible change in properties make 2D MX$_2$, a suitable material for low dimensional sensors and stretchable devices. The strain sensitivity of various properties of MX$_2$ can be further exploited to develop many new micro and nanodimensional devices for everyday use to the sophisticated applications.

References

[1] J. M. Martin, C. Donnet, T. Le Mogne and T. Epicier, *Phys. Rev. B* **48**, 10583 (1993).

[2] O. Toshio, T. Ken-ichi and M. Koshiro, *J. Catal.* **48**, 229 (1977).

[3] S. Patel, *Cryst. Res. Technol.* **27**, 285 (1992).

[4] J. Bernede, J. Pouzet, E. Gourmelon and H. Hadouda, *Synt. Met.* **99**, 45 (1999).

[5] R. F. Frindt and A. D. Yoffe, *Proc. R. Soc. Lon. Ser.-A* **273**, 69 (1963).

[6] K. S. Novoselov, D. Jiang, F. Schedin, T. J. Booth, V. V. Khotkevich, S. V. Morozov and A. K. Geim, *Proc. Natl. Acad. Sci. USA* **102**, 10451 (2005).

[7] R. F. Frindt, *J. Appl. Phys.* **37**, 1928 (1966).

[8] P. Joensen, R. F. Frindt and S. R. Morrison, *Mater. Res. Bull.* **21**, 457 (1986).

[9] A. Schumacher, L. Scandella, N. Kruse and R. Prins, *Surf. Sci.* **289**, L595 (1993).

[10] J. N. Coleman, M. Lotya, A. ONeill, S. D. Bergin, P. J. King, U. Khan, K. Young, A. Gaucher, S. De, R. J. Smith, I. V. Shvets, S. K. Arora, G. Stanton, H.-Y. Kim, K. Lee, G. T. Kim, G. S. Duesberg, T. Hallam, J. J. Boland, J. J. Wang, J. F. Donegan, J. C. Grunlan, G. Moriarty, A. Shmeliov, R. J. Nicholls, J. M. Perkins, E. M. Grieveson, K. Theuwissen, D. W. McComb, P. D. Nellist and V. Nicolosi, *Science* **331**, 568 (2011).

[11] B. K. Miremadi and S. R. Morrison, *J. Appl. Phys.* **63**, 4970 (1988).

[12] D. Yang and R. Frindt, *J. Phys. Chem. Solids* **57**, 1113 (1996).

[13] H. S. S. Ramakrishna Matte, A. Gomathi, A. K. Manna, D. J. Late, R. Datta, S. K. Pati and C. N. R. Rao, *Angew. Chem. Int. Ed.* **49**, 4059 (2010).

[14] G. Eda, H. Yamaguchi, D. Voiry, T. Fujita, M. Chen and M. Chhowalla, *Nano Lett.* **11**, 5111 (2011).

[15] T. Korn, S. Heydrich, M. Hirmer, J. Schmutzler and C. Schuller, *Appl. Phys. Lett.* **99**, 102109 (2011).

[16] D. Yi, W. Yanli, N. Jun, S. Lin, S. Siqi and T. Weihua, *Physica B* **406**, 2254 (2011).

[17] A. Ayari, E. Cobas, O. Ogundadegbe and M. S. Fuhrer, *J. Appl. Phys.* **101**, 014507 (2007).

[18] A. M. Gabovich, A. I. Voitenko, J. F. Annett and M. Ausloos, *Supercond. Sci. Tech.* **14**, R1 (2001).

[19] E. Morosan, H. W. Zandbergen, B. S. Dennis, J. W. G. Bos, Y. Onose, T. Klimczuk, A. P. Ramirez, N. P. Ong and R. J. Cava, *Nat. Phys.* **2**, 544 (2006).

[20] K. Rossnagel, O. Seifarth, L. Kipp, M. Skibowski, D. Voß, P. Krüger, A. Mazur and J. Pollmann, *Phys. Rev. B* **64**, 235119 (2001).

[21] W. Z. Hu, G. Li, J. Yan, H. H. Wen, G. Wu, X. H. Chen and N. L. Wang, *Phys. Rev. B* **76**, 045103 (2007).

[22] A. F. Kusmartseva, A. Akrap, H. Berger, L. Forro and E. Tutis, *Nat. Mater.* **7**, 960 (2008).

[23] K. K. Kam and B. A. Parkinson, *J. Phys. Chem.* **86**, 463 (1982).

[24] S. W. Han, H. Kwon, S. K. Kim, S. Ryu, W. S. Yun, D. H. Kim, J. H. Hwang, J.-S. Kang, J. Baik, H. J. Shin and S. C. Hong, *Phys. Rev. B* **84**, 045409 (2011).

[25] S. Lebègue and O. Eriksson, *Phys. Rev. B* **79**, 115409 (2009).

[26] A. Splendiani, L. Sun, Y. Zhang, T. Li, J. Kim, C.-Y. Chim, G. Galli and F. Wang, *Nano Lett.* **10**, 1271 (2010).

[27] J. K. Ellis, M. J. Lucero and G. E. Scuseria, *Appl. Phys. Lett.* **99**, 261908 (2011).

[28] B. Radisavljevic, A. Radenovic, J. Brivio, V. Giacometti and A. Kis, *Nat. Nanotechnol.* **6**, 147 (2010).

[29] L. Liu, S. Kumar, Y. Ouyang and J. Guo, *IEEE T. Electron Dev.* **58**, 3042 (2011).

[30] Y. Yoon, K. Ganapathi and S. Salahuddin, *Nano Lett.* **11**, 3768 (2011).

[31] B. Radisavljevic, M. B. Whitwick and A. Kis, *ACS Nano.* **5**, 9934 (2011).

[32] D. Inosov, V. Zabolotnyy, D. Evtushinsky, A. Kordyuk, B. Büchner, R. Follath, H. Berger and S. Borisenko, *New J. Phys.* **10**, 125027 (2008).

[33] T. Kidd, D. Klein, T. Rash and L. Strauss, *Appl. Surf. Sci.* **257**, 3812 (2011).

[34] C. Andrei, G. Corneliu, M. Vadim, K. Leonid and F. Emeri, *J. Lumin,* **129**, 1945 (2009).

[35] A. Ramasubramaniam, D. Naveh and E. Towe, *Phys. Rev. B* **84**, 205325 (2011).

[36] Y. Jing, X. Tan, Z. Zhou and P. Shen, *J. Mater. Chem. A* **2**, 16892 (2014).

[37] S. Bhattacharyya and A. K. Singh, *Phys. Rev. B* **86**, 075454 (2012).

[38] A. P. Nayak, S. Bhattacharyya, J. Zhu, J. Liu, X. Wu, T. Pandey, C. Jin, A. K. Singh, D. Akinwande and J.-F. Lin, *Nat. Commun.* **5**, 3731 (2014).

[39] H. Peelaers and C. G. Van de Walle, *Phys. Rev. B* **86**, 241401 (2012).

[40] E. Scalise, M. Houssa, G. Pourtois, V. Afanas'ev and A. Stesmans, *Nano Res.* **5**, 43 (2012).

[41] W. S. Yun, S. Han, S. C. Hong, I. G. Kim and J. Lee, *Phys. Rev. B* **85**, 033305 (2012).

[42] Y. Qu, K. Jun, S. Zhengzheng, Z. Xueao, C. Shengli, W. Guang, Q. Shiqiao and L. Jingbo, *Phys. Lett. A* **376**, 1166 (2012).

[43] S. M. Tabatabaei, M. Noei, K. Khaliji, M. Pourfath and M. Fathipour, *J. Appl. Phys.* **113**, 163708 (2013).

[44] S. Bhattacharyya, T. Pandey and A. K. Singh, *Nanotechnology* **25**, 465701 (2014).

[45] M. Dave, R. Vaidya, S. G. Patel and A. R. Jani, *Bull. Mater. Sci.* **27**, 213 (2004).

[46] A. P. Nayak, T. Pandey, D. Voiry, J. Liu, S. T. Moran, A. Sharma, C. Tan, C.-H. Chen, L.-J. Li, M. Chhowalla, J.-F. Lin, A. K. Singh and D. Akinwande, *Nano Lett.* **15**, 346 (2015).

[47] P. E. Blöchl, *Phys. Rev. B* **50**, 17953 (1994).

[48] G. Kresse and D. Joubert, *Phys. Rev. B* **59**, 1758 (1999).

[49] J. P. Perdew, K. Burke and M. Ernzerhof, *Phys. Rev. Lett.* **77**, 3865 (1996).

[50] J. Heyd, G. E. Scuseria and M. Ernzerhof, *J. Chem. Phys.* **118**, 8207 (2003).

[51] J. Heyd, G. E. Scuseria and M. Ernzerhof, *J. Chem. Phys.* **124**, 219906 (2006).

[52] L. Hedin, *Phys. Rev.* **139**, A796 (1965).

[53] S. Grimme, *J. Comput. Chem.* **27**, 1787 (2006).

[54] M. Dion, H. Rydberg, E. Schröder, D. C. Langreth and B. I. Lundqvist, *Phys. Rev. Lett.* **92**, 246401 (2004).

[55] K. Lee, E. D. Murray, L. Kong, B. I. Lundqvist and D. C. Langreth, *Phys. Rev. B* **82**, 081101 (2010).

[56] J. M. Ziman, *Principles of the Theory of Solids*, Cambridge University Press, Cambridge (1964).

[57] N. W. Ashcroft and D. N. Mermin, *Solid State Physics*, Thomson Learning, Toronto (1976).

[58] L. L. Boyer, *Phys. Rev. B* **19**, 2824 (1979).

[59] G. K. Madsen and D. J. Singh, *Comput. Phys. Commun.* **175**, 67 (2006).

[60] X. Gonze and C. Lee, *Phys. Rev. B* **55**, 10355 (1997).

[61] G. Kresse and J. Hafner, *Phys. Rev. B* **47**, 558 (1993).

[62] A. Togo, F. Oba and I. Tanaka, *Phys. Rev. B* **78**, 134106 (2008).

[63] S. Baroni, S. De Gironcoli, A. Dal Corso and P. Giannozzi, *Rev. Mod. Phys.* **73**, 515 (2001).

[64] Y. Li, K.-A. N. Duerloo, K. Wauson and E. J. Reed, *Nat. Commun.* **7**, 10671 (2016).

[65] K.-A. N. Duerloo, Y. Li and E. J. Reed, *Nat. Commun.* **5**, 4214 (2014).

[66] J. Wilson and A. Yoffe, *Adv. Phys.* **18**, 193 (1969).

[67] D. Voiry, H. Yamaguchi, J. Li, R. Silva, D. C. Alves, T. Fujita, M. Chen, T. Asefa, V. B. Shenoy, G. Eda, *et al.*, *Nat. Mater.* **12**, 850 (2013).

[68] J. Heising and M. G. Kanatzidis, *J. Am. Chem. Soc.* **121**, 638 (1999).

[69] M. Calandra, *Phys. Rev. B* **88**, 245428 (2013).

[70] A. Molina-Sánchez and L. Wirtz, *Phys. Rev. B* **84**, 155413 (2011).

[71] R. Saito, G. Dresselhaus, M. S. Dresselhaus *et al.*, *Physical Properties of Carbon Nanotubes*, Vol. 35, World Scientific, Singapore (1998).

[72] B. Peng, H. Zhang, H. Shao, Y. Xu, X. Zhang and H. Zhu, *RSC Adv.* **6**, 5767 (2016).

[73] Y. Cai, J. Lan, G. Zhang and Y.-W. Zhang, *Phys. Rev. B* **89**, 035438 (2014).

[74] E. Cappelluti, R. Roldán, J. Silva-Guillén, P. Ordejón and F. Guinea, *Phys. Rev. B* **88**, 075409 (2013).

[75] C.-H. Chang, X. Fan, S.-H. Lin and J.-L. Kuo, *Phys. Rev. B* **88**, 195420 (2013).

[76] P. Johari and V. B. Shenoy, *ACS Nano.* **6**, 5449 (2012).

[77] H. Shi, H. Pan, Y.-W. Zhang and B. I. Yakobson, *Phys. Rev. B* **87**, 155304 (2013).

[78] K. Kośmider, J. González and J. Fernández-Rossier, *Phys. Rev. B* **88**, 245436 (2013).

[79] C. Ataca, H. Sahin and S. Ciraci, *J. Phys. Chem. C* **116**, 8983 (2012).

[80] D. Y. Qiu, H. Felipe and S. G. Louie, *Phys. Rev. Lett.* **111**, 216805 (2013).

[81] C. Espejo, T. Rangel, A. H. Romero, X. Gonze and G.-M. Rignanese, *Phys. Rev. B* **87**, 245114 (2013).

[82] T. Cheiwchanchamnangij and W. R. Lambrecht, *Phys. Rev. B* **85**, 205302 (2012).

[83] K. F. Mak, C. Lee, J. Hone, J. Shan and T. F. Heinz, *Phys. Rev. Lett.* **105**, 136805 (2010).

[84] M. Ghorbani-Asl, S. Borini, A. Kuc and T. Heine, *Phys. Rev. B* **87**, 235434 (2013).

[85] E. Scalise, M. Houssa, G. Pourtois, V. Afanas and A. Stesmans, *Physica E: Low-dimensional Systems and Nanostructures* **56**, 416 (2014).

[86] H. Pan and Y.-W. Zhang, *J. Phys. Chem. C* **116**, 11752 (2012).

[87] Q. Zhang, Y. Cheng, L.-Y. Gan and U. Schwingenschlögl, *Phys. Rev. B* **88**, 245447 (2013).

[88] D. M. Guzman and A. Strachan, *J. Appl. Phys.* **115**, 243701 (2014).

[89] J.-W. Jiang, *Nanoscale* **6**, 8326 (2014).

[90] S. Horzum, H. Sahin, S. Cahangirov, P. Cudazzo, A. Rubio, T. Serin and F. Peeters, *Phys. Rev. B* **87**, 125415 (2013).

[91] M. Peña-Álvarez, E. del Corro, A. Morales-Garca, L. Kavan, M. Kalbac and O. Frank, *Nano Lett.* **15**, 3139 (2015).

[92] M. Hosseini, M. Elahi, M. Pourfath and D. Esseni, *J. Phys. D: Appl. Phys.* **48**, 375104 (2015).

[93] S. Rhim, Y. S. Kim and A. Freeman, *Appl. Phys. Lett.* **107**, 241908 (2015).

[94] L. Wang, A. Kutana and B. I. Yakobson, *Ann. Phys.* **526**, L7 (2014).

[95] T. Li, *Phys. Rev. B* **85**, 235407 (2012).

[96] A. Ramasubramaniam, *Phys. Rev. B* **86**, 115409 (2012).

[97] J. Feng, X. Qian, C.-W. Huang and J. Li, *Nature Photon.* **6**, 866 (2012).

[98] P. Lu, X. Wu, W. Guo and X. C. Zeng, *Phys. Chem. Chem. Phys.* **14**, 13035 (2012).

[99] B. Amin, T. P. Kaloni and U. Schwingenschlögl, *RSC Adv.* **4**, 34561 (2014).

[100] L. Dong, A. M. Dongare, R. R. Namburu, T. P. O'Regan and M. Dubey, *Appl. Phys. Lett.* **104**, 053107 (2014).

[101] H. Guo, T. Yang, P. Tao, Y. Wang and Z. Zhang, *J. Appl. Phys.* **113**, 013709 (2013).

[102] G. Zhang and Y.-W. Zhang, *Mech. Mater.* **91**, 382 (2015).

[103] M. Hosseini, M. Elahi, M. Pourfath and D. Esseni, *IEEE T. Electron. Dev.* **62**, 3192 (2015).

[104] D. Wickramaratne, F. Zahid and R. K. Lake, *J. Chem. Phys.* **140**, 124710 (2014).

[105] D. Fan, H. Liu, L. Cheng, P. Jiang, J. Shi and X. Tang, *Appl. Phys. Lett.* **105**, 133113 (2014).

[106] G. Huai-Hong, Y. Teng, T. Peng and Z. Zhi-Dong, *Chin. Phys. B* **23**, 017201 (2013).

[107] H. Lv, W. Lu, D. Shao and Y. Sun, *Phys. Rev. B* **90**, 085433 (2014).

[108] B. Liao, J. Zhou, B. Qiu, M. S. Dresselhaus and G. Chen, *Phys. Rev. B* **91**, 235419 (2015).

[109] E. D. Terrones, H., Corro, S. Feng, J. M. Poumirol, D. Rhodes, D. Smirnov, N. R. Pradhan, Z. Lin, M. A. T. Nguyen, A. L. Elías, T. E. Mallouk, L. Balicas, M. A. Pimenta and M. Terrones, *Sci. Rep.* **4**, 4215 (2014).

[110] X. Luo, Y. Zhao, J. Zhang, Q. Xiong and S. Y. Quek, *Phys. Rev. B* **88**, 075320 (2013).

[111] A. P. Nayak, Z. Yuan, B. Cao, J. Liu, J. Wu, S. T. Moran, T. Li, D. Akinwande, C. Jin and J.-F. Lin, *ACS Nano.* **9**, 9117 (2015).

[112] Y. Wang, C. Cong, W. Yang, J. Shang, N. Peimyoo, Y. Chen, J. Kang, J. Wang, W. Huang and T. Yu, *Nano Res.* **8**, 2562 (2015).

[113] S. B. Desai, G. Seol, J. S. Kang, H. Fang, C. Battaglia, R. Kapadia, J. W. Ager, J. Guo and A. Javey, *Nano Lett.* **14**, 4592 (2014).

[114] Z.-H. Chi, X.-M. Zhao, H. Zhang, A. F. Goncharov, S. S. Lobanov, T. Kagayama, M. Sakata and X.-J. Chen, *Phys. Rev. Lett.* **113**, 036802 (2014).

[115] J. Pu, Y. Zhang, Y. Wada, J. Tse-Wei Wang, L.-J. Li, Y. Iwasa and T. Takenobu, *Appl. Phys. Lett.* **103**, (2013).

[116] J. Pu, Y. Yomogida, K.-K. Liu, L.-J. Li, Y. Iwasa and T. Takenobu, *Nano Lett.* **12**, 4013 (2012).

[117] H. J. Conley, B. Wang, J. I. Ziegler, J. Richard F. Haglund, S. T. Pantelides and K. I. Bolotin, *Nano Lett.* **13**, 3626 (2013).

[118] K. He, C. Poole, K. F. Mak and J. Shan, *Nano Lett.* **13**, 2931 (2013).

[119] C. R. Zhu, G. Wang, B. L. Liu, X. Marie, X. F. Qiao, X. Zhang, X. X. Wu, H. Fan, P. H. Tan, T. Amand and B. Urbaszek, *Phys. Rev. B* **88**, 121301 (2013).

[120] Y. Y. Hui, X. Liu, W. Jie, N. Y. Chan, J. Hao, Y.-T. Hsu, L.-J. Li, W. Guo and S. P. Lau, *ACS Nano.* **7**, 7126 (2013).

[121] G. Plechinger, A. Castellanos-Gomez, M. Buscema, H. S. J. van der Zant, G. A. Steele, A. Kuc, T. Heine, C. Schller and T. Korn, *2D Mater.* **2**, 015006 (2015).

Chapter 10

Point Defects, Grain Boundaries and Planar Faults in 2D h-BN and TMX$_2$: Theory and Simulations

Anjali Singh* and Umesh V. Waghmare[†]

*Theoretical Sciences Unit,
Jawaharlal Nehru Centre for Advanced Scientific
Research, Jakkur, Bangalore 560064, India
*anjalisingh@jncasr.ac.in
[†]waghmare@jncasr.ac.in*

"Crystals are like people: it is the defects in them which tend to make them interesting"

— *C. J. Humphreys*

Abstract. In contrast to graphene, two-dimensional (2D) materials such as h-BN and TMX$_2$ exhibit a non-zero band-gap which makes them suitable for applications in FETs. It is important to understand the nature of defect in these materials, and how they influence their properties. Here, we present analysis of point defects, interfaces and planar defects in these 2D materials and highlight the impact of these defects on electronic and vibrational properties. We discuss carbon doping in h-BN, which is interesting for engineering its electronic structure and band-gap through both compositional and configurational modification. We show that the planar structure h-BN with grain boundaries (GBs or interfaces) exhibits instability with respect to buckling (out-of-plane deformation), resulting in the formation of a wrinkle at the GB and long length-scale rippling of the structure. We present the electronic and vibrational signatures of these interfaces and STM images of the most stable interface, which should facilitate their experimental characterisation. Planar defects (faults) essentially give different polymorphs of materials such as TMX_2 compounds (M = (Mo, W); X = (S, Se)), whose monolayer consists of 2 or 3 atomic layers. Introduction of these planar faults leads to transition from insulating 1H structure to metallic c1T structure. We present the energy landscape associated with the fault or 1H to 1T phase transition, leading to prediction of novel 1T structures and intermediate structures associated with structural instabilities of the 1T structure. The metallic centrosymmetric 1T (c1T) structure of these compounds is shown to be unstable with respect to dimerisation or trimerisation of transition metal atoms, leading to a competing metallic $\sqrt{3} \times 1$ 1T form and ferroelectric semiconducting $\sqrt{3} \times \sqrt{3}$ 1T form respectively. While the former is a more stable 1T form of $MoSe_2$, WS_2 and WSe_2, the latter is a more stable 1T form of MoS_2 exhibiting rich ferroelectric dipolar domain structure. Due to vicinity to such metal-semiconductor transitions, their semiconducting forms are shown to exhibit anomalous response to electric field. To facilitate experimental verification of these subtle features of 1T forms of MX_2 monolayers, we present comparative analysis of their vibrational properties, and identify their Raman and Infra-Red spectroscopic signatures.

10.1   Introduction

Defects are inherently present in real materials and influence their properties such as electronic, optical, thermal properties, mechanical strength, electrical conductivity, corrosion and chemical reactivity. Structurally, it marks a region where the microscopic arrangement of ions differs drastically from that of a perfect crystal. Defects and imperfections appear during synthesis or treatment of a material [1–3]. For example, mechanical strength and ductility of metals are determined by defects, whereas electrical conductivity and thermal conductance of semiconductors are controlled by defects [4].

Defects are named planar, line, or point defects, depending on whether the imperfection is confined at the atomic scale in one, two or three dimensions. There are two types of defects (i) intrinsic and (ii) extrinsic defects [5]. Defects which appear due to thermodynamic effects and increase in number with increased temperature are known as intrinsic defects. Extrinsic defects are the ones which appear due to imperfection in stoichiometry and do not depend on temperature. Intrinsic defects include point defects (vacancies), line defects (dislocations), grain boundaries (GBs), stacking and growth faults. In the structures with reduced dimensionality, the types of defects are fewer, and their effects become more pronounced in two-dimensional (2D) materials. 2D materials are important to technological applications, and we present an analysis of GBs in hexagonal boron nitride (h-BN) and planar defects in Group-VI transition metal dichalcogenides (TMDCs) [6, 7]. We highlight the effects of these defects on electronic and vibrational properties of the materials. 2D carbon/boron-nitride alloys are attractive for engineering electronic structure and band gap via both compositional and configurational modification, which can make them suitable for optoelectronic and photocatalytic applications [8].

GBs or interfaces in 2D materials appear during growth of these materials [3]. At these interfaces, the atomistic structure deviates from the perfect crystal structure, resulting in changes in electronic and optical properties [3, 6]. GBs lead to out-of-plane deformation of flat 2D structures [6, 9], for example, h-BN has a tendency to form ripples in the presence of these interfaces. Planar defects are normally found in MX$_2$ type layered compounds whose single-layer consist of 2–3 atomic planes (e.g., MoS$_2$). A Planar defect or faults can be introduced by sliding one-layer (out of 2–3 layers) on relative to other layers, keeping the other layers fixed [7]. Such planar defects introduce a structural transformation, that is accompanied by semiconductor to metal phase transition and ferroelectric properties, which can be utilised in novel devices.

In this chapter, we focus on the effects of point defects (e.g., carbon doping in h–BN), interfaces (GBs) and planar defects in 2D materials [6–8].

Density functional theory (DFT) is a powerful tool for determination of the electronic ground state and structural properties of crystals, molecules and surfaces. A line defect in carbon nanotube was theoretically predicted using DFT [9], which later verified experimentally [10]. This shows how the theoretical prediction of defect structures facilitates experimental observation of these defects, and helps in understanding which properties of a material are affected by these defects. With these calculations, it is possible to focus on one type of defects at a time and study their interactions, which is not readily possible experimentally.

In this chapter, we present analysis of GBs in h-BN and planar defects in TMDCs and associated effects on properties (some of these are published in Refs. [6, 7]).

10.2 Methods

DFT has been remarkably successful in the exploration of materials by finding a solution to a fundamental equation that describes the quantum behaviour of electrons in atoms, molecules or extended system of electrons and nuclei. It is a theory to tackle quantum ground state of correlated many-fermion system, mapping it onto an effective independent particle method. DFT simulations have been used to make important contributions to a wide range of materials scientific questions, generating information that would be essentially impossible to determine from experiments, and predicting novel structures. The fundamental outcome of a DFT calculation is the non-empirical total energy of a collection of atoms, which captures the interatomic interaction and serves as a Hamiltonian governing atomic motion. Many properties of materials can be determined by taking first (e.g., stress, magnetisation and forces are first derivative of energy with respect to strain, magnetic fields and atomic positions) and second derivatives (e.g., elastic constants, interatomic force constants and magnetic susceptibility are the second derivative of energy with respect to strain, magnetic fields and atomic positions) of total energy. There are different computational methods based on DFT of which the most popular one is the one based on plane wave basis for periodic systems. Electronic wave function in a periodic structure is a product of a cell periodic part and a wavelike part [11],

$$\psi_{n,k}(r) = e^{ik.r}\phi_{n,k}(r), \tag{10.1}$$

k being Bloch vector and n being the band index. The cell-periodic part of the wave function ϕ can be expanded using a basis set consisting of a discrete set of plane waves with wave vectors given by the points of the reciprocal lattice.

$$\phi_{n,k}(r) = \sum_{k+G} c^n_{k,G}\, e^{iG.r}, \tag{10.2}$$

where G is reciprocal lattice vectors, which is defined as $G.R = 2\pi m$, where R is a real-space lattice vector of the crystal and m is an integer. The set of the G vectors is truncated by considering those plane waves whose kinetic energy is less than a fixed energy cutoff. Brillouin zone (BZ) integrations are sampled with k points on uniform meshes in BZ.

Bloch theorem cannot be applied to a system containing a single defect or in the direction perpendicular to a crystal surface due to broken translational symmetry. Thus, a simulation of defects or a surface is challenging and require careful analysis. In practice, one uses a large super cell with a defect or a surface in their simulations. Thus, the Bloch states and plane wave basis can be used effectively in their analysis within DFT.

Our first-principles calculations are based on density functional theory as implemented in Quantum ESPRESSO package [12], with ultrasoft pseudopotential [13] to model the interactions between valance electrons and ionic cores. We use a generalised gradient approximation (GGA) parametrised by Pardew *et al.* [14] to treat the exchange-correlation energy of the electrons. The energy cutoff of 30 Ry and 35 Ry were used to truncate the plane-wave basis used in representing Kohn–Sham wavefunctions and energy cutoff of 180 Ry and 280 Ry to represent charge density are used for h-BN and (Mo,W)(S, Se)$_2$, respectively. Structures are relaxed through minimisation of energy until the Hellman–Feynman force on each atom is less than 0.001 Ry/bohr. In the periodic supercell to model 2D materials, a vacuum of 15 Å is introduced along c-axis is used to ensure negligible interaction between the nearest sheets. The vibrational spectra determined using DFT linear response and a frozen phonon method with a finite difference formula and atomic displacements of ± 0.04 Å.

10.3   Defects in 2D h-BN

h-BN has a crystal structure analogs to graphite. A monolayer of h-BN with honeycomb structure is thus, analogs to graphene [15,16], where every alternative site of a hexagonal ring is decorated by B atom, and rest of the sites are occupied by N atoms. The electronic structure of h-BN monolayer and graphene differ remarkably from each other despite of their structural similarity: graphene is zero band-gap semiconductor whereas h-BN is a wide band-gap insulator with a gap $E_g > 5\,\text{eV}$ [17]. Although 2D h-BN is isostructural to graphene (small lattice mismatch of $\sim 1.8\%$), it exhibits rather different structural defects [18]. Vacancies, antisites (a B occupying N site and vice versa), GBs affect the properties of 2D h-BN vitally. These types of defects break the crystalline order and modify the properties of h-BN, and impact the performance of h-BN-based devices.

10.3.1   Point defects

At the vacancies or edges, which are lattice defects (atomic defect), individual Boron (B) and Nitrogen (N) atoms have been imaged and distinguished experimentally [1]. An aberration corrected high resolution transmission electron microscopy (HRTEM) and the exit-wave (EW) reconstruction [19] are used to distinguish between the boron and nitrogen atoms within an h-BN monolayer. The EW reconstruction method involving a series of through focus HRTEM images is effective in discriminating the boron and nitrogen atoms, since it is possible to enhance the high-spatial-frequency components and retrieve the phase image [1]. In HRTEM image (Figure 1(a)), the triangle-shaped holes (or vacancies) with various sizes are evident, which have been created mainly due to the knock on effect of the incident electron beam. However, it is not quite possible to deduce the exact atomic structure of these defects from such an image because the boron and nitrogen atoms cannot be distinguished. Nevertheless, one can notice the following points: (i) the triangle-shaped holes have discrete sizes, (ii) the smallest triangle is located at the corner of three bright spots (hexagons) and should correspond to a monovacancy, (iii) all of the defects have nearly a normal triangular shape, and (iv) all the triangles occur exactly in the same orientation.

A model has been presented by Jin *et al.* [1] to understand the type of vacancies introduced by electron irradiation. Mono-vacancy of boron (marked as V_B) consists of one missing boron atom due to

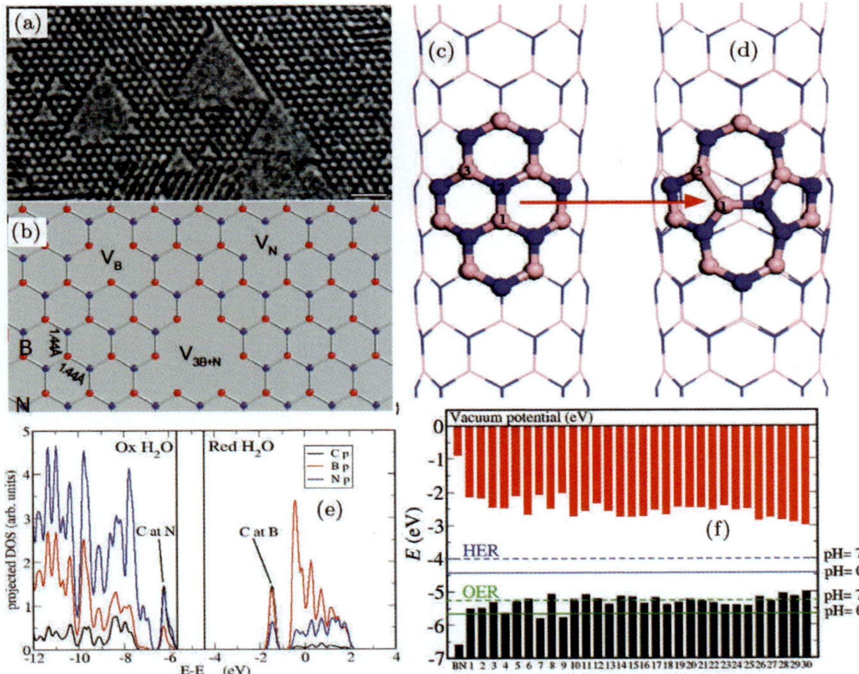

Figure 1. Zero-dimensional (point) defects in h-BN. (a) HRTEM image showing the lattice defect in h-BN, such as mono-vacancy and even larger vacancies which are triangular in shape and in the same direction. (b) Models for atomic defects in h-BN where V_B, V_N are mono-vacancies of B and N atoms. Note that V_B and V_N are in opposite orientation whereas V_B and V_{3B+N} (three B and one N are missing) in the same orientation. (part figures (a) and (b) reprinted with permission from Ref. [1]. Copyright 2009 American Physical Society). (c) The structure of boron nitride nanotube (BNNT). (d) The structure of Stone-Wales defect in BNNT which involve 90° rotation of BN bond. (part figures (c) and (d) reprinted with permission from Ref. [20]. Copyright 2008 American Chemical Society). (e) Electronic density of states of C-substituted h-BN [8] and (f) Band gap and band edge positions calculated for the symmetry inequivalent configurations of $C_{2x}B_{1-x}N_{1-x}$ (for $x = 0.22$) aligned with respect to vacuum level [8].

which three neighbouring atoms are the doubly coordinated nitrogen atoms. Another type of monovacancy is a nitrogen vacancy (V_N) which shows opposite configuration with B atoms. V_{3B+N} is a multi-vacancy defect, where three boron atoms and the centering nitrogen atom are missing, which is surrounded by six (doubly coordinated) nitrogen atoms. From the HRTEM image (Figure 1(a)), all the triangles are in the same orientation, from which one concludes that the V_B and V_N will not coexist because two types of vacancies should give rise to vacant triangles with different orientations (Figure 1(b)). (Note that V_B, V_{3B+N}, V_{6B+3N}, ... have the same orientation, whereas V_N, V_{B+3N}, ... have the opposite one). V_B's are dominant in an experimental sample and no V_N is found which can be connected with irradiation induced damage process. Since the energy carried by the incident electrons (120 keV) is larger than the threshold beam energies for a knock on of both atoms, the difference of the threshold energy cannot be the only reason for the dominance boron vacancies. The previous theoretical studies (and experimental investigations) claimed that the formation energy of a V_B is larger than that of V_N and that the formation of nitrogen monovacancy should be preferable [21]. These theoretical suggestions apparently contradict this experimental result [1]. However, it should be noted that most of these simulations represent the thermodynamic equilibrium conditions, which are different from the conditions in experiments [1]. The vacancies were created here mostly by the high energy charged electrons, and the defect structures cannot be exactly the same as the thermodynamically expected ones. No stable di-vacancy (V_{BN}) has been found so far, although one of the previous studies pointed out a possibility of a paired vacancy formation under the electron beam irradiation by considering the charge compensation of boron and nitrogen [22]. One infers that the V_{BN} should immediately transform to the V_{3B+N} due to the further removal of doubly coordinated boron atoms.

A Stone–Wales (SW) defect is a topological point defect well studied in graphene as a pair of bound dislocations [23, 24]. A SW defect involves an in-plane 90° rotation of a carbon–carbon bond with respect to the midpoint of the bond in graphene [24]. A similar rotation of a BN bond would result in a SW defect in 2D h-BN [20], resulting in a pair of pentagons and heptagons (Figure 1(c) and 1(d)). A theoretical report [20], suggested an SW defect in boron nitride nanotubes, though there is no experimental observation of this defect. A rotation of BN bond by 90° leads to the formation of homopolar B−B and N−N bonds (Figure 1(d)), which are energetically unfavourable, and hence an SW defect is not quite favourable in 2D h-BN.

The similarity in crystal structure (almost lattice matched!) of graphene and h-BN and contrast in electronic structure, suggest tunability of properties based on combination (solution) of the two materials. Shirodkar *et al.* [8] have reported electronic structure of carbon substituted 2D BN alloy, considering various distributions of ions at a give composition. With h-BN end-member of the solid solution as starting point, which already has a large gap, the gap was shown to be tunable with C substitution (Figure 1). In addition to the band gap, they determined the alignment of the valence and conduction band edges with respect to the vacuum: this alignment is important in understanding the electronic behaviour of these alloys, and also in assessing their potential for photochemical applications. The compositions they considered are $C_{2x}B_{1-x}N_{1-x}$, i.e., C replaces the same number of B and N atoms, keeping the stoichiometry with ratio B/N=1, and low values of $x(x < 1/3)$ [8]. These compositions give band gap values in a range that is useful for optoelectronic and photochemistry applications. The band gap decreases with increasing C concentration; more dispersed the distribution of the substituents is, the reduction in the gap is more pronounced. Configurations with only isolated C substitutions gives the smallest band gap (no C–C bond). However, larger concentrations of C-impurity substitutions make the defect bands wider. One would need to find a balance between gap opening and mobility while engineering the band gap of such materials [8]. Configurations consisting of C–C dimers in $(BN)_{1-x}C_x$ exhibit band alignment that is favourable for a single semiconductor photocatalysis of the water splitting reaction under conditions near to room temperature and neutral pH, though with too wide band gaps for effective solar energy absorption. On the other hand, configurations with only isolated C atoms have smaller band gaps. Their band edge positions do not straddle the oxygen evolution reaction (OER) level (Figure 1(f)), but satisfy the requirements for a cathode in a heterojunction photocatalyst (with respect to the hydrogen evolution reaction (HER) level) [8]. However, thermodynamic analysis shows that configurations with isolated C atoms would be rare in experimental samples due to the strong tendency of C substituents to cluster in the 2D h-BN lattice [8].

10.3.2   Grain boundaries

Towards development and engineering of 3D materials with nano-scale architecture based on h-BN and graphene [25], it is important to understand the interfaces or GBs in h-BN and graphene on the same footing. Liu *et al.* [26] have discussed structural aspect of GBs in 2D h-BN in terms of dislocations and GBs in monolayered h-BN, analysing square-octagon (4|8) and pentagon–heptagon (5|7) pair dislocations. They have studied different GBs by varying tilt angles from $0°$ to $240°$ [26]. The tilt angle is an angle between the orientation directions of two grains. Depending on the tilt angle, GBs can be either polar (B or N rich) composed of 5|7 dislocations or non-polar constituted by 4|8 dislocations. Compared to non-polar GB's, polar GBs have net charges and results in smaller band gaps suggesting interesting electronic and optical applications [26].

Though the Raman spectroscopic characterisation of defects in graphene has been quite effective, it lacks the specificity with respect to the kind of defects which requires the identification of specific vibrational signatures (if any). It is thus important to have a unified picture of structural stability and spectroscopic signatures of GBs in monolayer h-BN and graphene. We now present a detailed structural analysis of extended line defects or GBs in h-BN and graphene benefiting from the ideas of growth and stacking faults that are well established in metallurgy. This is expected to be useful in under-standing how these GB structures may form during growth of these materials in experiments. We discuss in detail about the structure and properties of GBs in h-BN in the subsequent sub-sections.

10.3.2.1   *Structural changes at the GBs in 2D h-BN*

We classify GBs into two types, though there are several ways to construct GBs. A stacking fault is defined as a misregistry between two semi-infinite halves of the 2D lattice, and a growth fault is associated with the removal of line of atoms in the 2D lattice (a layer of atoms is missed during the growth of the lattice). These GBs are essentially the inversion domain boundaries and 60° GBs. Growth fault has been well studied in carbon nanotubes and graphene [9].

(a) Stacking Fault

A perfect 2D lattice of h-BN with armchair edge is divided into two semi-infinite sheets by a slip line along $\vec{a}$ (Figure 2(a)). The

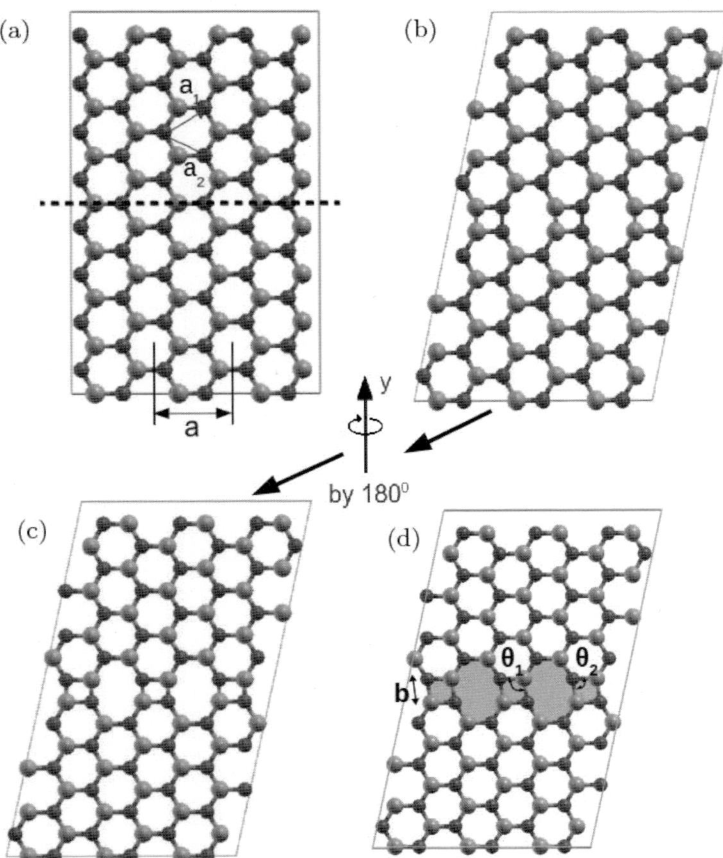

Figure 2. Construction of 4:8 GB in h-BN. (a) Pure h-BN lattice with the dashed line indicating the location where GB is introduced (b) h-BN lattice with stacking fault (c) h-BN with GB twisting, and (d) ground state structure of BN with 4:8 GB (grey shadow show GB with octagons and rhombii). Note that GB marked in shadow show structural modification after relaxation. B atoms are shown in light colour and N atoms are in dark colour [6].

upper half of the sheet is displaced with respect to the lower half by a misregistry vector, $\vec{v} = 1.5b\hat{x}$, where b is the B$-$N bond length. Across the GB due to introduction of this fault, homo-elemental bonds (B$-$B and N$-$N) are formed. These bonds are chemically as well as energetically unfavourable (Figure 2(b)). Hence, an 180° rotation of the upper half is performed around y-axis such that heteroelemental (B$-$N) bonds are formed, which are energetically more favourable [27] (Figure 2(c)). As a result, we get a pair of rectangular and octagonal rings, and this structure is named as the 4:8 GB. Due to periodic boundary conditions, two 4:8 GBs are generated, one at the centre of the periodic supercell and another at the edge. On relaxation, the structure of GBs evolve into interlinked rhombal and slightly irregular octagonal rings with equal opposite sides [6]. The B$-$N bonds perpendicular to GB are elongated by 0.02 Å with respect to these in bulk h-BN. The bond angles θ_1 and θ_2 of rhombii are 82° and 98°, respectively (Figure 2(d)). The energy of the 4:8 GB is 0.48 eV/Å for a defect concentration of 2 defects/nm. Introducing a 4:8 GB in graphene with a supercell with the same number of atoms as h-BN, we found that 4:8 GB is more stable in h-BN than in graphene due to heteroelemental sublattices facilitate the formation of stable bonds at the interface, in contrast to that in graphene [6].

(b) Growth Fault

To introduce a growth fault, we consider two grains of pure h-BN obtained by cutting the pure h-BN across the dashed line along the zigzag direction. These grains are labelled as 1 and 2 as shown in Figure 3(a). These grains are antisymmetric upon reflection across the dashed line (Figure 3(a)). In order to obtain symmetric grain facing each other with same polar edges, we replace B atom by N atom in grain 2 and vice versa (Figure 3(b)). To construct a growth fault, we (i) move grain 1 along x-direction, such that B atoms at the edges overlap along the dashed line and (ii) remove a line of B atoms on this line. As a result, we get a linear chain of rhombii at the interface with B atomic sites at the GB (Figure 3(c)). To maintain

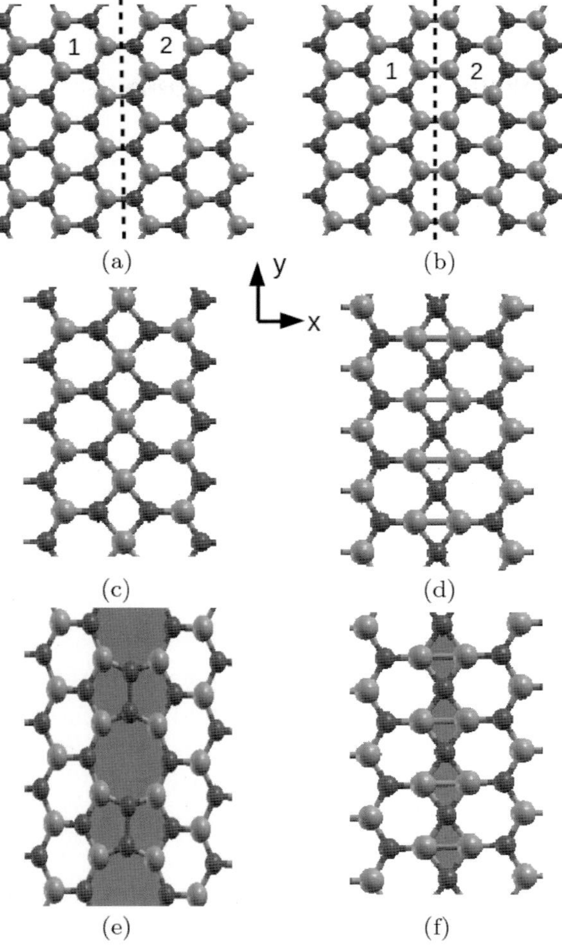

Figure 3. Construction of a growth fault in h-BN (a) h-BN lattice which is divided into two grains by the dashed line. Two grains are marked as 1 and 2. (b) Grain 2 is rotated by 180° (tilt) along z-directions with respect to grain 1 such that B−B bond is formed perpendicular to the dashed line (symmetric grain boundary with same polarity grains). Grain 1 is moved in x-direction, such that B atoms are placed perpendicular to dashed line overlap with each other and a layer of boron atoms are removed. This leads to formation of GB at the center with a chain of rhombii (c) Chain of rhombii at the edge (d) Note that in (c) B atoms occupying sites at GB whereas in (d) they are occupied by N atoms. GBs marked in shadow show structural modification after relaxation, (e) and (f) B atoms are shown in light colour and N atoms are in dark colour [6].

equal number of B and N atoms, we remove a line of N atoms from the edge of the supercell [6]. Due to periodic boundary conditions, another GB is generated at the edge of the supercell involving a chain of rhombii with N atoms occupying the sites at GB (Figure 3(d)). Structural relaxation of these structures leads to the formation of a pair of two slightly irregular pentagonal rings with apex angle $111°$ and an octagonal ring at the whose opposite sides are equal (Figure 3(e)). The resulting GB is named as 5:5:8 GB. The GBs with N−N and B−B bonds along the interfacial line are refereed as 5:5:8 N−N and 5:5:8 B−B GBs, respectively. GB at the edge is expected to transform into a 5:5:8 B−B GB but it retains in its initial structure of rhombii (Figure 3(f)), confirming its occurrence in h-BN at 0 K. We label this GB as 4:4 GB. This GB transforms to 5:5:8 B−B GB at 1000 K as discussed by Li *et al.* [28]. At 4:4 GB, N atom is coordinated with 4 B atoms saturating all its valence electrons which is energetically favourable, whereas the B atom is coordinated with 4 N atoms at the centre of the supercell (Figures 3(c) and 3(f)) [6]. Hence, the 4:4 GB with B atoms at the interfacial line transforms to 5:5:8 N−N GB. To determine the energies of these GBs, ribbons of h-BN containing these GBs at the central line of the supercell had to be simulated. The calculated energies of 5:5:8 B−B, 5:5:8 N−N and 4:4 GBs are 0.65 eV/Å, 0.77 eV/Å and 1.87 eV/Å, respectively (see Table 1). It is clear that 5:5:8 B−B is more stable than the other two (5:5:8 N−N and 4:4) GBs in h-BN [6].

To facilitate the direct comparison with energetics of GBs in graphene, simulations of these GBs in graphene were undertaken with the same supercell. We have two GBs, one at the centre of the supercell and another at the edge similar to h-BN [6]. However, they evolve quite differently from each other in graphene. A GB at the centre evolves into a GB with a pair of two irregular pentagons with apex angle $112°$ and irregular octagonal ring i.e., 5:5:8 GB (Figure 4(a)). The GB at the edge interestingly evolves into a new type of GB with two oppositely oriented pentagon-octagon pairs linked with a rectangular ring (Figure 4(b)). This GB, named as 6:5:8:4:8:5 GB [6], consists of an irregular pentagonal ring with apex angle $109°$ and irregular octagonal rings. The energies of 5:5:8 and

Table 1. Energies of different GBs (4:8, 5:5:8, 4:4 and 6:5:8:4:8:5 GBs) in h-BN and graphene [6].

Type of GBs	System	Energy (eV/Å)
4:8	h-BN	0.48
4:8	Graphene	0.74
5:5:8 N-N	h-BN	0.77
5:5:8 B-B	h-BN	0.65
4:4*	h-BN	1.87
6:5:8:4:8:5 N-N*	h-BN	1.10
6:5:8:4:8:5 B-B*	h-BN	0.79
5:5:8	Graphene	0.52
6:5:8:4:8:5*	Graphene	0.73

*Novel GBs.

6:5:8:4:8:5 GBs are 0.52 eV/Å and 0.83 eV/Å, respectively. Note that these estimates of energies include the energy cost associated with dangling bonds formed at the edge of the ribbon. We find that the 5:5:8 GB is lower in energy than the 6:5:8:4:8:5 GB, which indicates its greater stability in graphene. Our analysis also shows that the 5:5:8 GB is more stable in graphene than in h-BN (see Table 1). The reason for this is that extra energy is required for the formation of homo-elemental bonds at the GB in h-BN. We explored the feasibility of the 6:5:8:4:8:5 GB in h-BN and found two new types of GB structures [6] (6:5:8:4:8:5 B$-$B and 6:5:8:4:8:5 N$-$N) (Figures 4(c) and 4(d)). The structures of these GBs involve a pair of pentagons and an octagon oppositely oriented, connected by a rhombus and two sets of 5-8-4-8-5 polygons that are connected by a deformed hexagon with BB/NN bonds at the opposite sides [6]. The energies of 6:5:8:4:8:5 B$-$B and 6:5:8:4:8:5 N$-$N GBs are 0.79 eV/Å and 1.10 eV/Å, respectively, comparable to the energy of other GBs.

10.3.2.2 *Structural instability and the origin of wrinkles and ripples*

If a structure is a local minimum of energy with respect to all structural distortions, it is stable against structural distortions. To

404   *A. Singh and U. V. Waghmare*

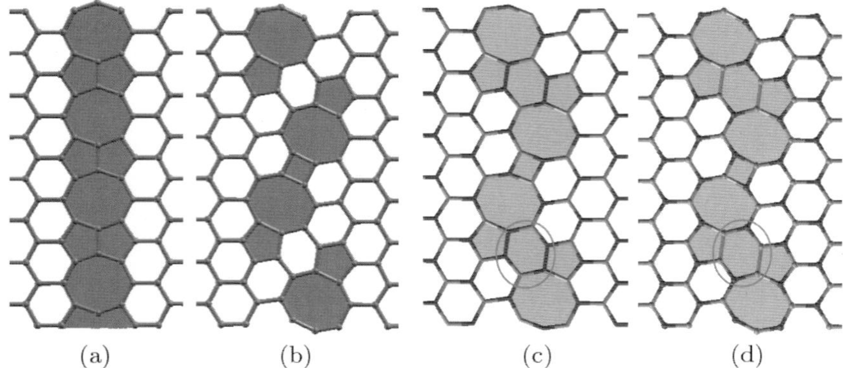

(a)   (b)   (c)   (d)

Figure 4. Structures of different GBs in graphene and h-BN. Structure of (a) 5:5:8 GB in graphene, (b) 6:5:8:4:8:5 GB in graphene, (c) 6:5:8:4:8:5 N−N GB in h-BN and (d) 6:5:8:4:8:5 B-B GB in h-BN. GBs structures are highlighted by shadows. Boron (B) and Nitrogen atoms are shown in light colour and dark colour, respectively [6].

assess that, long wavelength phonons of the supercell (at the Γ-point) are useful. The vibrational density of states (Figure 5) of h-BN with 4:8, 4:4 and 5:5:8 N−N GBs with planar structure reveals the presence of unstable modes (i.e., imaginary frequencies) [6]. In contrast, the 5:5:8 B−B GB does not have any unstable mode, showing that the structure of h-BN with the 5:5:8 B−B GB is stable, i.e., local minimum of energy (Figure 5). These unstable modes positively show that the structure is not a local minimum of energy, but is a saddle point [6]. Eigen-modes of phonons with imaginary frequencies ($\omega^2 < 0$) in the vibrational spectrum of h-BN with GBs involve atomic displacements, which lead to out-of-plane deformation along these interfaces (Figures 6(a) and 6(c)).

Freezing in small atomic displacements corresponding to eigen modes of the lowest energy (strongest instability) unstable modes distorts the structure (which are $53i\,\mathrm{cm}^{-1}$, $375i\ \mathrm{cm}^{-1}$ and $40i\,\mathrm{cm}^{-1}$ for 4:8, 4:4 and 5:5:8 N−N GBs, respectively). The unstable modes of 4:8 and 4:4 GBs have oscillatory deformation localised (like a sine or cosine wave) along the GB (Figures 6(a), 6(b) and Figures 7(a), 7(b)), whereas the unstable mode of 5:5:8 N−N has deformation wave running along the edge of the ribbon (Figures 6(c) and 6(d)) [6].

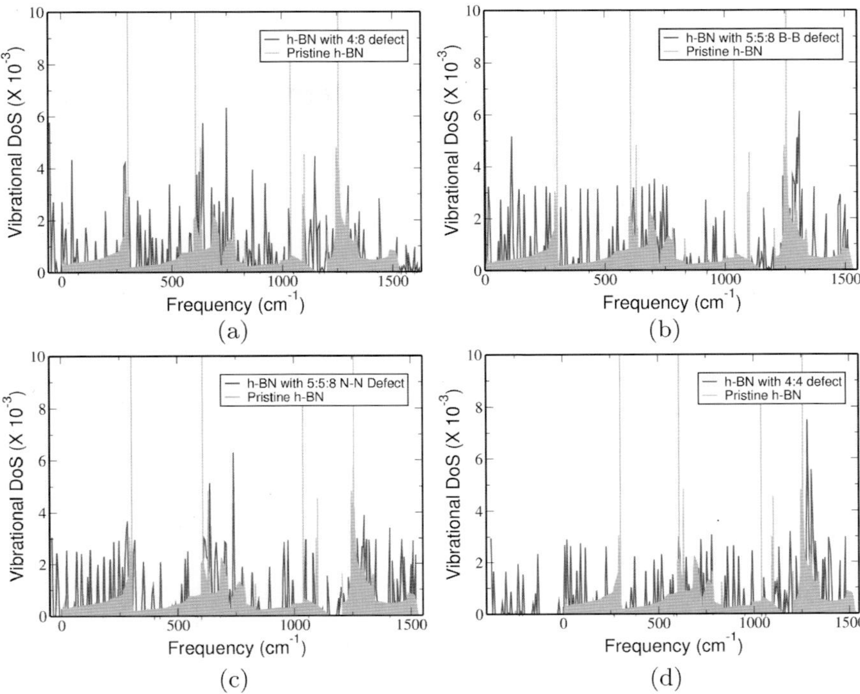

Figure 5. Vibrational density of states (VDoS) of pure h-BN and h-BN with GBs. (a) VDoS of h-BN with 4:8 GB (b) VDoS of h-BN with 5:5:8 B−B GB (c) VDoS of h-BN with 5:5:8 N−N GB and (d) VDoS of h-BN with 4:4 GB, superimposed on VDoS of pure h-BN [6].

The N atom is coordinated with four B atoms through sp^3 bonding. Therefore it naturally tries to deviate from the planar structure and change to the tetrahedral structure at the interface of 4:4 GB. A pair of two B atoms (at 4:4 GB) placed opposite to the N atom (see Figure 7(b)) attempt to move in the opposite direction to attain the tetrahedral structure. Thus, the h-BN sheet with these GBs buckles, i.e., becomes non-planar upon structural relaxation. Such buckling of the sheet leads to the reduction in energy of the 4:8 GB by 19.5%. For 4:8 GB, the rippling amplitude (which is defined as the difference in the displacements along the z-axis of the highest and the lowest atoms) is 0.35 Å, which is quite sizeable and should be readily observable in experiments. Due to buckling,

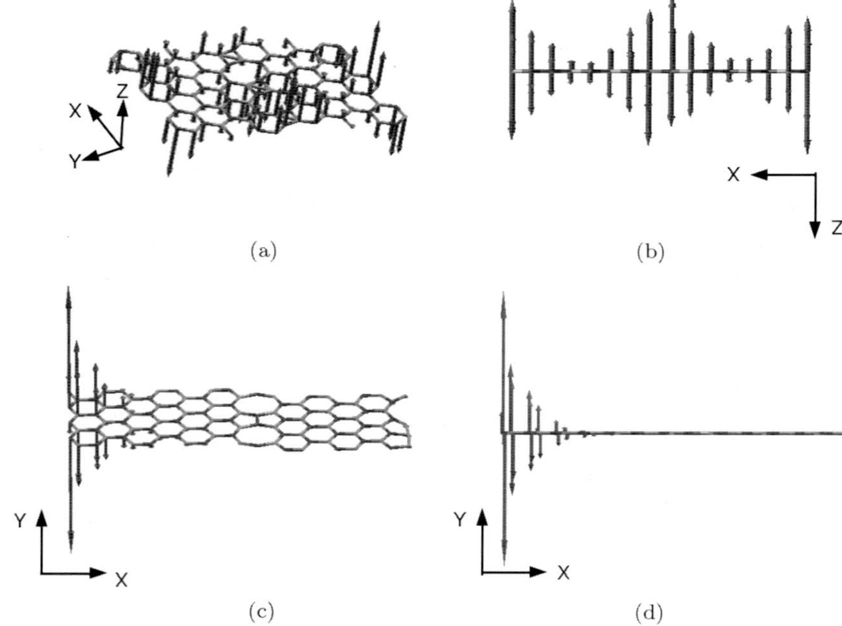

Figure 6. Visualisation of unstable phonon modes of h-BN with different GBs. (a) Perspective and (b) side views of atomic displacements of ($\omega \approx$) 53i cm^{-1} mode of h-BN with 4:8 GB. (c) Perspective and (d) side views of atomic displacements of ($\omega \approx$) 40i cm^{-1} mode of h-BN with 5:5:8 GB [6].

there is no considerable change in the energy of the 5:5:8 N$-$N GB even though its rippling amplitude is 0.42 Å, suggesting a rather flat energy landscape associated with rippling [6]. Freezing in the atomic displacements of the lowest energy unstable mode of h-BN with a 4:4 GB lowers its energy by 39.5% upon structural relaxation, with formation of a wrinkle (short length scale deformation) at the interface (Figure 7(c)). The chain of rhombii at the interface evolves to a completely different structure (a chain of two heptagons and a rhombus). The rippling amplitude of this GB (1.6 Å) is quite significant. Such a deformed structure involving large displacements of most of the atoms at the GB constitutes a wrinkle (Figure 7(c)). Freezing in the unstable mode of 4:4 GB with 375i cm^{-1} leads to quite a different structure, which evolve on relaxation to a pair of two heptagons sharing a B$-$B bond side connected with a rhombus (Figure 7(d)). Theoretically, 4:8 GB has been studied by Liu *et al.*

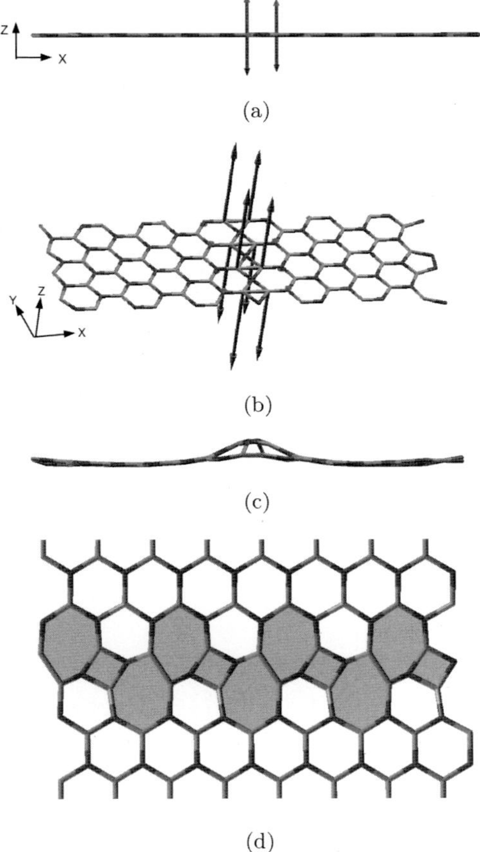

Figure 7. Phonon modes constituting structural instability of h-BN with 4:4 GB. (a) Side and (b) perspective views of atomic displacement of the mode with frequency $\sim$375i cm^{-1}. (c) Wrinkle formed at the centre of the sheet caused by freezing in the same mode (side view of the ribbon) and (d) Top view of 4:4 GB (which evolve on relaxation), which has two heptagons sharing a B−B bond side connected with rhombus. B atoms are shown in light colour and N atoms are in dark colour [6].

[26], but they missed the deviation from planarity of asym-0°(120° and 240°) [26] GBs (4:8 GB). In contrast, we find that h-BN with a 4:8 GB does show buckling and there is a noticeable reduction in energy (19.5%). Buckling or wrinkling thus arises inherently at the GBs, as a result of deviation from sp^2 hybridisation of atoms at the defect.

10.3.2.3 *Vibrational signatures of GBs*

The structural changes associated with these GBs in h-BN lead to characteristic changes in the vibrational spectra, and these changes can be identified in Raman and IR spectra. The unit cell of 2D h-BN consist of two atoms and belongs to point group D_{3h} and the space group $(P\bar{6}2m)$. The optical phonon modes of 2D h-BN are classified into $A_2' + 2E'$ symmetries, where mode A_2'' is IR active and an E' mode is active in both IR and Raman spectra. The E' is a doubly degenerate in-plane bond stretching mode. Calculated frequencies of A_2'' and E' modes are $791\,\mathrm{cm}^{-1}$ and $1382\,\mathrm{cm}^{-1}$, respectively. These calculated frequencies agree well (within 1%) with the observed values [29]. Due to structural modification at their GBs, the frequency and eigen displacement of A_2'' and E' modes are altered. To determine the frequencies of A_2'' and E' modes of the h-BN supercell with a 4:8 GB, we project its normal modes (e_ν) onto those of pristine h-BN (e_μ), by calculating an overlap matrix $(S_{\mu\nu})$,

$$S_{\mu\nu} = \langle e_\mu | e_\nu \rangle. \tag{10.3}$$

The A_2'' and E' modes of the defective h-BN are identified as those having the largest overlap $(|S_{\mu\nu}|)$ with the A_2'' and E' eigenvectors of pristine h-BN.

The symmetry equivalence between x- and y-directions in the hexagonal lattice is broken, and hence the degeneracy of E' mode is lifted in the presence of GBs. The frequencies of A_2'' mode reduces (mode softening) by $14\,\mathrm{cm}^{-1}$, $10\,\mathrm{cm}^{-1}$ and $14\,\mathrm{cm}^{-1}$ due to 4:8, 5:5:8 $N-N$ and 4:4 GBs, respectively (see Table 2) [6]. In the presence of the 5:5:8 $B-B$ GB, the A_2'' mode hardens by $38\,\mathrm{cm}^{-1}$. The E' mode softens by $21\ \mathrm{cm}^{-1}$ and $25\,\mathrm{cm}^{-1}$, and hardens by $5\,\mathrm{cm}^{-1}$ and $10\,\mathrm{cm}^{-1}$ in structures with 5:5:8 $N-N$, 5:5:8 $B-B$, 4:8 and 4:4 GBs, respectively [6]. The other in-plane E' mode softens by $26\,\mathrm{cm}^{-1}$ and $3\,\mathrm{cm}^{-1}$ (not significant) and hardens by $12\,\mathrm{cm}^{-1}$ and $106\,\mathrm{cm}^{-1}$ (significantly) in structures with 5:5:8 $N-N$, 5:5:8 $B-B$, 4:8 and 4:4 GBs, respectively, due to vibrations in the xy-plane involving displacements perpendicular to the GB (refer Table 2).

Table 2. Vibrational spectroscopic signatures in the Raman and IR active modes of h-BN with different GBs (4:8, 5:5:8, 4:4) and shifts in frequencies ($\Delta\omega$ in cm^{-1}) with respect to pristine h-BN [6].

Pristine (h-BN) cm^{-1})	h-BN with 4:8 GB (cm^{-1})	h-BN with 5:5:8 B-B GB (cm^{-1})	h-BN with 5:5:8 N-N GB (cm^{-1})	h-BN with 4:4 GB (cm^{-1})
791 (IR active)	777 ($\Delta\omega = -14$)	829 ($\Delta\omega = +38$)	781 ($\Delta\omega = -10$)	778 ($\Delta\omega = -14$)
1382 (Raman and IR active)	1389 ($\Delta\omega = +5$)	1392 ($\Delta\omega = +10$)	1361 ($\Delta\omega = -21$)	1357 ($\Delta\omega = -25$)
1382 (Raman and IR active)	1394 ($\Delta\omega = +12$)	1379 ($\Delta\omega = -3$)	1356 ($\Delta\omega = -26$)	1488 ($\Delta\omega = +106$)

While these shifts have been estimated at high density of GBs, they can be observed experimentally by IR and Raman spectroscopies. The softening (hardening) of these modes is readily understood in terms of stretching (contraction) of the in-plane (out-of-plane) bonds.

10.3.2.4 *Electronic signatures of GBs in h-BN*

We discuss the electronic signature of the lowest energy GB (4:8 GB) in h-BN through comparison with pristine h-BN. In electronic density of states (DoS), two extra peaks (Figure 8(a)) arise at the valence band maximum (VBM) and the conduction band minimum (CBM). These extra peaks constitute characteristic signatures of the 4:8 GB. It is clear from analysis of the charge density of these states as these electronic states are localised at GBs. A few extra peaks evident in the DoS of the non-planar structure highlight a redistribution of charge in the localised electronic states upon buckling, which is clear from the comparison of the electronic DoS of planar and non-planar GB structures (Figures 8(a) and 8(b)) [6].

Simulated scanning tunneling microscopy (STM) images to facilitate experimental observation and verification of GBs. They were obtained keeping the tip at a constant height of 1 Å above the

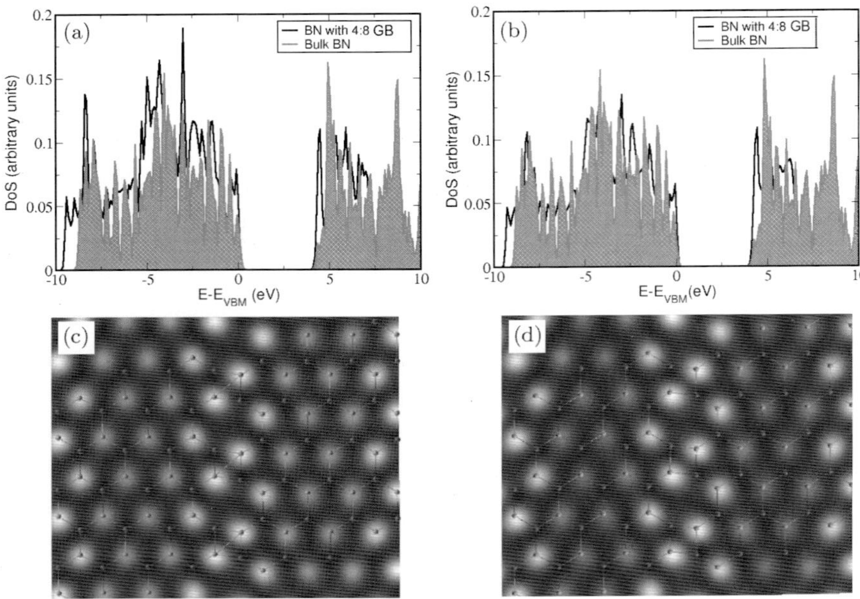

Figure 8. Electronic signature of h-BN with 4:8 grain boundary (GB). Density of states of (a) planar and (b) non-planar (buckled) structure compared with pure h-BN. Atomic structure juxtaposed on simulated STM images of (c) planar and (d) buckled (non-planar) h-BN with 4:8 GB [6].

surface of h-BN with a positive sample bias of 0.8 V with respect to the CBM. For a positive sample bias, induced charge accumulation is observed at B atoms whose orbitals constitute the conduction band. Simulated STM images (Figures 8(c) and 8(d)) highlight the difference between the electronic structure of the planar and non-planar structures of the GB in h-BN. We notice that the bright features in the bulk planar structure weaken upon buckling (clear from the STM image of the non-planar structure), revealing that charge accumulation remains weakly localised at the GB even after buckling [6].

10.4   Defects in 2D MX$_2$

Group VI transition metal (Mo, W) dichalcogenides (TMDCs) in their most common 2D 1H structure (2H-phase refers to the bulk crystal) (Figure 9) have a non-vanishing direct band gap, and hence

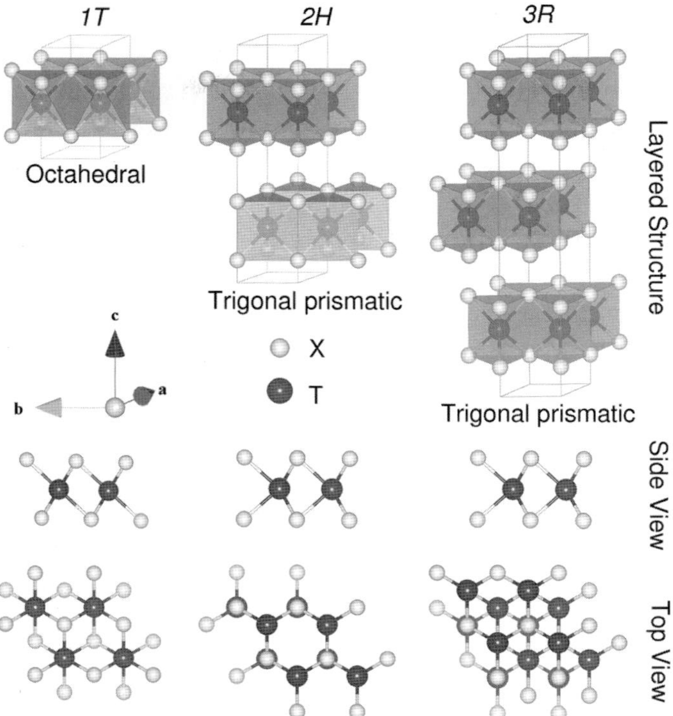

Figure 9. Structural polytypes of pristine TMDCs. Stacking of 1T, 2H and 3R layers are shown in bulk TMDCs. Side and top view of monolayer of TMDCs in 1T-, 1H- and 1R-phase [37]. (Reproduced by permission of The Royal Society of Chemistry).

are suitable for applications such as transistors (with high on–off ratios [30]), photodetector [31,32], electroluminescent devices [33,34], electrodes for Li-ion batteries [35], and supercapacitors [35]. The 1H structure consists of three $X-M-X$ stacking of triangular planar atomic layers, stacked in the *ABA* sequence. It effectively has a honeycomb lattice structure (in top view), with every alternate site occupied by metal atom M, and the rest of the sites are occupied by a pair of chalcogenide atoms X. The 1H form of 2D TMDCs is non-centrosymmetric and has a sizable band gap. The 1T structure is a polytype of MX_2 (M= (Mo, W); X= (S, Se)), with $X-M-X$ atomic lattices are stacked in the *ABC* sequence such that each M atom is octahedrally co-ordinated with six X atoms, and is at the centre of inversion. The centrosymmetric monolayered 1T polytype (refer to Figure 9) of TMDCs is metallic. In certain cases, 1T polytype is thermodynamically unstable and its structural distortion leads to other 1T forms. In a recent report, 1T structure was stabilised experimentally through Re substitution [36], and the atomistic mechanism of 1H to 1T transformation was uncovered through a careful electron microscopy. Additional polytype appear from variation in the stacking such 3R-phase (Figure 9) with rhombohedral structure, whose stacking is *AbACaCBcB*.

In this section, we focus on 1H and 1T derived polytypic forms of MX_2 compounds. We first present structural defects of different dimensionality and atomic structure. Defects in 2D TMDCs, can be classified as zero dimensional (point defects, dopants or non-hexagonal rings), one dimensional (GBs, edges and inter-plane heterojunction) and 2D planar defects associated with layer sliding.

10.4.1 Point defects

Vacancies are the most common defects in TMDCs, there are six types of intrinsic defects which are observed in monolayer MoS_2 grown with chemical vapour deposition (CVD) (i) mono-sulphur vacancies (V_S) (ii) disulphur vacancies (V_{S_2}) (iii) vacancy complex of Mo atom and three neighbouring sulphur atoms (in the same plane) (V_{MoS_3}), (iv) vacancy complex of Mo atom and neighbouring three disulphurs (V_{MoS_6}), and antisite defects with (v) Mo atom occupying S_2 column (Mo_{S2}) or (vi) a pair of S's column occupying the Mo atom site $S2_{Mo}$ [38] (Figure 10(a)). The Mo vacancies are rarer to find due to its tendency to form complexes with S's (experimentally also it is difficult to find Mo vacancy alone [38]). STM and atomic resolution annular dark field (STEM-ADF) imaging have the capability of imaging atom by atom which helps in unambiguously identifying the mono-sulphur vacancy from di-sulphur vacancy and antisite defects from the regular lattice sites. Most of the point defects, maintain 3-fold symmetry except Mo_{S2}. Stability of the defects is determined by defect formation energy, where formation energy depends on upon the chemical potential of Mo and S. It is observed that monovacancies

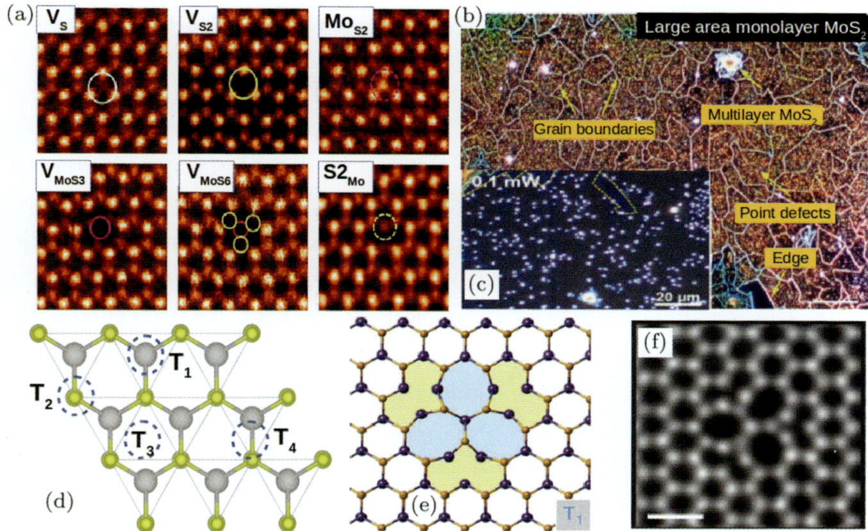

Figure 10. Zero dimension defects in monolayer MoS$_2$. (a) Atomic resolution ADF images of various intrinsic point defect present in monolayer MoS$_2$, including V_S, V_{S_2}, Mo$_{S2}$, V_{MoS_3}, V_{MoS_6} and S2$_{Mo}$. (Reprinted with permission from Ref. [38]. Copyright 2013 American Chemical Society). (b) Dark field optical image of CVD grown MoS$_2$ which clearly highlight point defect in the sample and (c) modulation of defect distribution with white light power for MoS$_2$ on TiO$_2$/Ag substrate. (part figures (b) and (c) are reprinted with permission from Ref. [41]. Copyright 2015 American Chemical Society). (d) A schematic diagram showing four sites available for adatom adsorption on 1H structure of TMDCs lattice. (e) The atomic model of trefoil defect in MX$_2$ compounds and (f) the magnified ADF image of WSe$_2$ observed at 500°. (Reprinted with permission from Ref. [42]). Copyright 2015 Creative Commons (visit http://creativecommons.org/licenses/by/4.0/).

are found frequently, whereas antisite defects are seen occasionally in experimental samples. Out of these point defects, V_S is the most stable one. The formation energy of disulphur vacancy is twice of mono-sulphur vacancy, which shows that mono-vacancies of sulphur do not have a tendency to combine. In contrast, in graphene di-vacancies are energetically more favourable than mono-vacancies [39, 40].

Atomic arrangement at vacancies can be obtained using transmission electron microscopy (TEM) or SEM [38, 43]. The distribution of such defects at macro-scale is determined using dark-field microscopy (Figure 10(b)), which have implications to electrical transport and thermal properties of a given material. These defects are visualised by selectively anchoring silver nanoparticle on defect sites of MoS_2 under light illumination [41] (Figure 10(c)). The distribution of point defects is much dependent on light power and exposure time. DFT calculations are performed in order to find out how Ag atom adsorbs at defect sites in MoS_2. Two models were considered (i) adsorption of Ag atom directly on the S vacancy site and (ii) adsorption of Ag atom on the oxygen atom saturated S vacancy. These calculations show that Ag adsorption on S vacancy is energetically not favourable, while Ag atoms adsorb readily on an oxygen atom saturating S vacancy site [41]. With the help of this technique, point defects can be analysed qualitatively and the distribution can be rationalised in terms of growth conditions.

When a foreign atom which can be absorbed at crystal surface rather than substituted in the lattice, is known as an adatom, constituting another type of point defect in 2D materials. These defects have special importance in 2D materials due to large surface area. There are four possible sites for an adsorbent (i) above Mo atom (T_1), (ii) above S atom (T_2), (iii) above or on the hexagonal void (T_3) and (iv) on the Mo$-$S bond (T_4) [44] (Figure 10(d)) Different adatoms defects have been studied as transition metal elements which can induce magnetisation in 1H MoS_2. Group IVA elements such as C, Si and Ge which can form stable, planar and buckled honeycomb structure (C was considered because it can be used to form graphene and MoS_2 complexes) adsorb such that Mo and S atoms exit in 1H MoS_2 as residue. MoS_2 in 1H structure appears to be a material suitable for functionalisation because adatoms can be adsorbed at different sites with significant binding energy. The most likely adatoms to be observed in MoS_2 are Mo and S [38]. It is predicted that S adatom is stable at T_2, whereas Mo adatom is stable at T_1 and T_3 [45]. MoS_2 attains a local magnetisation through adsorption of 3d transition metals, Si and Ge. C adatom shows significant electronic charge transfer from MoS_2 sheet to adatom, such excess surface charge at higher coverage of adatoms can improve the tribiological and catalytic properties of 1H-MoS_2.

Functionalisation of MoS$_2$ through adatom adsorption seems to be a promising way to extend their applications [44].

Stone–Wales defect, a well-known defect in graphene, involves a rotation of C−C bond by 90°. These defect cannot be seen in TMDCs due to the polar nature of chemical bonds. However, a 60° rotation of three bonds centred at metal atom will preserve the heteroatomic nature of bonding and trigonal lattice symmetry [42]. This 60° rotation of M−X bonds result in trefoil shape defect (Figure 10(e)). These trefoil defects are seen in experimental STEM images (Figure 10(f)), and are found normally in chalcogen deficient (these deficiencies arise from electron-beam irradiation) TMDCs. It is suggested that the actual transformation mechanism may not include only bond rotation, but also the migration and rearrangement of X (S, Se) atoms in the region of high X vacancy concentration. The trefoil defect with three divacancies of X is more favourable than trefoil defect with three single vacancies of X, based on first-principles energetics. This defect is observed only in MoSe$_2$, WS$_2$ and WSe$_2$, but not in MoS$_2$. These defects give rise to p-doping and local magnetic moments, and weakly affect the mechanical properties of TMDCs. The presence of these types of rotational defects provides a way to modify the local properties of TMDCs [42].

10.4.2 Grain boundaries

GBs have significant influence on mechanical, optical, electrical and thermal properties of materials. Certain type of GBs have been shown to weaken the mechanical strength [46] and degrade the electronic structure of graphene [47,48]. The 60° GBs are well studied in MoS$_2$ and are of two types, consisting of four-fold rings chains either edge-sharing (4|4E) or point-sharing (4|4P) [38]. Both the GBs behave electronically as 1D metallic wire in 2D semiconducting lattice of TMDCs. 4|4P structure is consist of 4 fold rings with point sharing at common S2 sites (Figure 11(a)). In this GB, Mo retains its 6 fold symmetry whereas symmetry of S atom at the GB changes from three-fold to four-fold. This 1D conducting stripe could provide new functionalities and form intrinsic electronic heterostructure in monolayer MoS$_2$. 4|4E GB are observed when two 60° rotated

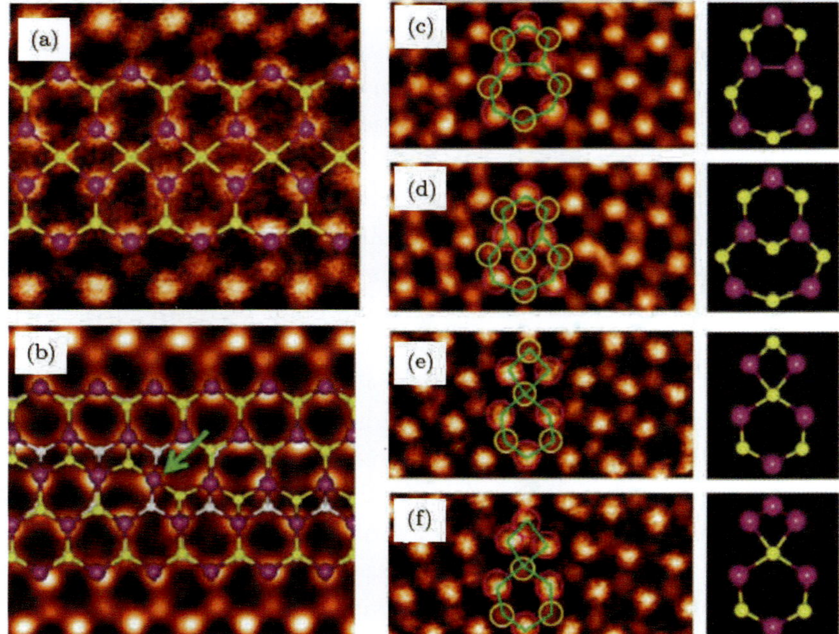

Figure 11. Atomic structure of 60° GBs and dislocations in monolayer MoS_2. (a) ADF image of the 4|4P 60° GB structures with the structural model overlapped. (b) ADF image and overlapped with structural model of a 4|4E type 60° grain boundary, and the grain boundary steps are linked by 4-fold coordinated Mo atoms as highlighted. (c), (d) STEM-ADF images of an 18.5° grain boundary consisting of dislocations with five- and seven-fold rings (5|7) and dislocations with six- and eight-fold rings (6|8). (e), (f) ADF images of a 17.5° grain boundary consisting of dislocations with four- and six-fold rings (4|6), either pristine (d) or with Mo-substitution (e). (Reprinted with permission from Ref. [38]. Copyright 2013 American Chemical Society).

GBs meet and consist of 4-fold rings with edge-sharing network (Figure 11(b)).

STEM-ADF images shows that S sites along GB are occupied by monosulphur atoms. The energy of 4|4E with 50% sulphur coverage is 0.1 eV/Å lower than the 100% sulphur coverage. The energy difference between 4|4E and 4|4P is 0.04 eV/Å [38]. The 4|4E GBs also serve as a perfect 1D metallic quantum wire embedded in the semiconducting MoS_2 sheet. Another type of GB is explored with a small tilt angle 18.5°. This GB is composed of 5|7 and 6|8 structures as shown in STEM-ADF image (Figures 11(c) and 11(d)). The structure of five and seven-fold rings (5|7) can be seen as basic dislocation core structure and addition of a di-sulphur or mono-sulphur into the Mo$-$Mo bonds generate 6|8 structure (Figures 11(d)). Such transitions from 5|7 to 6|8 dislocation are energetically favourable only under S rich conditions [38] suggesting high density of 6|8 structures. It is clear that grain boundary structure (Figures 11(e) and 11(f)) is more complex than that of graphene containing 4|4, 5|7, 6|8 and 4|6 dislocation core structures which extend three dimensionally (3D polyhedra of polyelemental compositions). To construct dislocation, one needs to remove half-lines (of arm-chair) in 2D materials and reconnecting the bonds [49]. The resulting topological mismatch is determined by arbitrary closed contour which is known as Burgers vector, b. A dislocation with $b = (1, 0)$ is generated by deleting an armchair half-line of atoms known as 5|7 dislocation (Figures 12(a) and 12(b)). Due to trigonal symmetry, there are three chemically equivalent possibilities with Burgers vectors $b = (1, 0)$, $(-1, 1)$, $(0, -1)$ (leads to Mo rich dislocations) with distinctive metal-metal

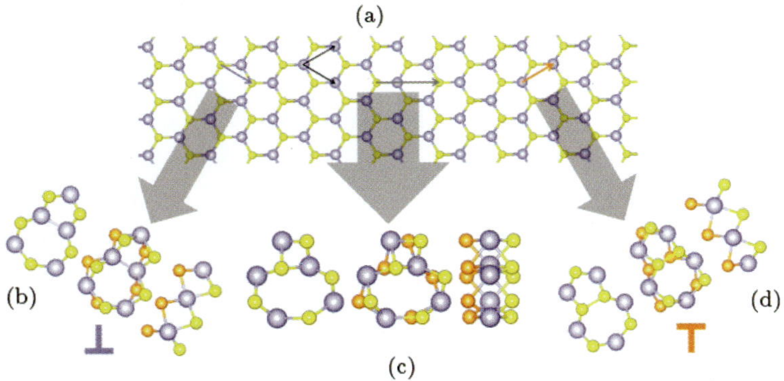

Figure 12. Basic edge dislocations formed by removal of shaded atoms from the lattice (a), with atomic structures of metal-polar $\perp$ shown in (b), sulphur-polar $\top$ in (d), and a 4|8 in (c). Thin black arrows show the basis vectors, while the Burgers vectors are coloured as blue (1, 0), orange (0, 1), and grey (1, 1). Each atomic structure is shown in front, isometric, and side (along the layer) views. Atoms coloured blue (metal, M), yellow (top layer sulphur) and orange (bottom layer sulphur). (Reprinted with permission from Ref [49]. Copyright 2013 American Chemical Society).

homoelemental bond. With $b = (-1, 0)$, $(1, -1)$, $(0, 1)$, sulphur–sulphur homoelemental bonds are obtained by deleting an inverted half-line, results in S-rich inverted dislocation (Figures 12(a) and 12(d)). Dislocations (4|8) of $\sqrt{3}$-times larger Burgers vector $b = \pm(1,1)$, $\pm(2,-1)$, $\pm(2,1)$, obtained by removing two parallel zigzag rows (Figures 12(a) and 12(b)), leads to formation of energetically favourable hetero-elemental bonds [49]. Linear array of such dislocations form a grain boundary between the mutually tilted crystalline domains. These GBs introduce the mid-gap electronic states and act as localised sinks for the carriers. These GBs are highly dependent on growth conditions [38] and do affect properties of the materials.

GBs in monolayer MoS_2 can play an important role in tunability of band gap [50]. Such tunability of band gap depends on the distance from the grain boundary and also on the grain misorientation angle. The band gap reduction can be as large as $0.85\,eV$ for the misorientation between two neighbouring grains of $18°$ (Figures 11(c) and 11(d)) and as small as $0.17\,eV$ for misorientation angle of $3°$. These type of GBs suggest possibility of flexible electronic and optoelectronic devices with tunable band gaps [50].

10.4.3 Planar defects and polytypes

Planar defects are found in materials whose monolayer is consist of 2 to 3 atomic layers, e.g., MX_2 (M = Mo, W; X = S, Se) compounds. These defects mainly occur due to change in stacking sequence and only reported for MX_2 compounds [7]. Details of this defect is discussed in subsequent sections.

10.4.3.1 *Structural changes at the planar defect: transition pathways from 1H to 1T structure*

A planar defect is introduced by displacing (or sliding) one of the X-atomic sublattice planes (e.g., top layer, X_{top}) with respect to the M-atom layer of the 1H structure. 1T form arises for a special value of such sliding. We consider different configurations for the energy landscape for a transformation between 1H and 1T polytypes. The slip or the glide vector $(\vec{d})$ in the plane of the 2D MX_2 is defined as:

$$\vec{d} = x\vec{a} + y\vec{b}, \qquad (10.4)$$

where $\vec{a}$ and $\vec{b}$ are the unit cell vectors. The atomic planes of metal and other X-atom are kept fixed. As x and $y \in [0,1)$ one considers configurations (x_i, y_j) points on a 6×6 mesh in the unit cell of real-space lattice in fractional coordinates. The chalcogen atoms (X) are allowed to relax only in the z-direction (perpendicular to sheet).

The DFT-based energies of configurations at (x_i, y_j) points on a 6×6 mesh are Fourier interpolated to construct the energy-surface. The energy surface (γ) is defined as;

$$\gamma(x,y) = \frac{E(x,y) - E(0,0)}{\mathcal{A}}$$

$$= \frac{1}{\sqrt{6}} \sum_{p=0}^{5} \frac{1}{\sqrt{6}} \sum_{q=0}^{5} A_{p,q} \exp\{i(2\pi k_p x + 2\pi k_q y)\}, \quad (10.5)$$

where $E(x,y)$ is the energy of a configuration $\vec{d}(x,y)$ relative to that of 1H structure ($E(0,0)$), and $\mathcal{A}$ is the area of the unit cell. $k_{p,q} \in \{0,1,2,3,-2,-1\}$ and $A_{p,q}$ are the coefficients obtained from discrete Fourier transform of energies of configurations $\vec{d}(x_i, y_j)$ corresponding to points on a 6×6 grid in real space [7]. The coefficients $A_{p,q}$ possess the symmetry of the crystal. We note that the γ-surface (see Figure 13) exhibits a reflection symmetry about $(\bar{1}, 1)$ direction (see Figure 13 and direction $\vec{p_3}$ in Figure 14), which is used to reduce the number of configurations simulated [7].

A local minimum of the γ-surface (denoted by C) located at the slip of $\left(\frac{2}{3}, \frac{1}{3}\right)$ corresponds to the 1T polytype. 1T polytype of MX$_2$ has ABC stacking of the X$-$M$-$X planes of atoms, and close packed network of MX$_6$ octahedra with M-atom at the site of inversion symmetry making it metallic (see Figure 13(c)) [7]. Our estimate of the energy of 1T structure of MX$_2$ compounds is in the range of 1.18–1.51 J/m^2 (0.7–0.8 eV/fu) (see Table 3, Figures 13 and 14), and comparable to the surface energy of metals like copper (≈ 1.73 J/m^2) and gold (≈ 1.48 J/m^2) [51]. We note that the relative energies of 1T polytypes of MSe$_2$ (with respect to the 1H form) are generally

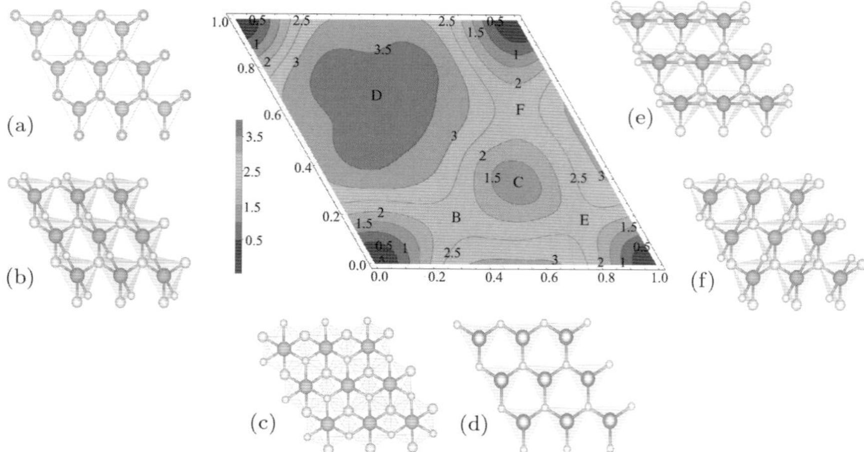

Figure 13. Contour plot of the γ-surface associated with glide of a Se-plane of MoSe$_2$ monolayer and top view of the structure at different configurations of γ-surface. Here, (a) is 1H polytype, (b), (e), (f) are symmetry equivalent transition states, (c) is the structure of c1T polytype and (d) is the structure at highest point at the energy surface (where Se-atom is sitting on the top of Mo-atom) of MoSe$_2$ structure. Se-atoms are denoted by light grey (small) spheres and Mo-atoms by grey (large) spheres. To distinguish between two Se planes, the top Se atoms are shown with circles of smaller radii than the bottom ones [7].

smaller than those of MS$_2$, and we expect that the 1T polytype is expected to be more stable in the selenides than in sulphides [7].

From Figure 13, we see that a MX$_2$ monolayer needs to overcome an energy barrier to transform from the 1H to the 1T structure. A configuration at the energy barrier i.e., the transition state is a point in γ-surface (denoted as $\gamma_B = \gamma_E = \gamma_F$) representing an unstable equilibrium, and occurs at slips of $(\frac{5}{6}, \frac{1}{6})$, $(\frac{5}{6}, \frac{2}{3})$ and $(\frac{1}{3}, \frac{1}{6})$. The estimated energies of the transition state of the four MX$_2$ compounds are in the range of 2.47–3.01 J/m^2 (1.5–1.7 eV/fu) (see Table 3 and Figure 13) [7]. In the structure (see (b), (d) and (f) in Figure 13), of transition state, the X-atom lies on top of the midpoint of the bond connecting two nearest M atoms in the hexagonal lattice. These three configurations are symmetry equivalent (hence of same energies), and have been indicated in the contour plot of γ-surface (see Figures 13(b), 13(e) and 13(f)). This implies that there are three equivalent paths (along $\vec{p_1}$, $\vec{p_2}$ and $\vec{p_3}$, where $\vec{p_1}$ and $\vec{p_2}$ are related by

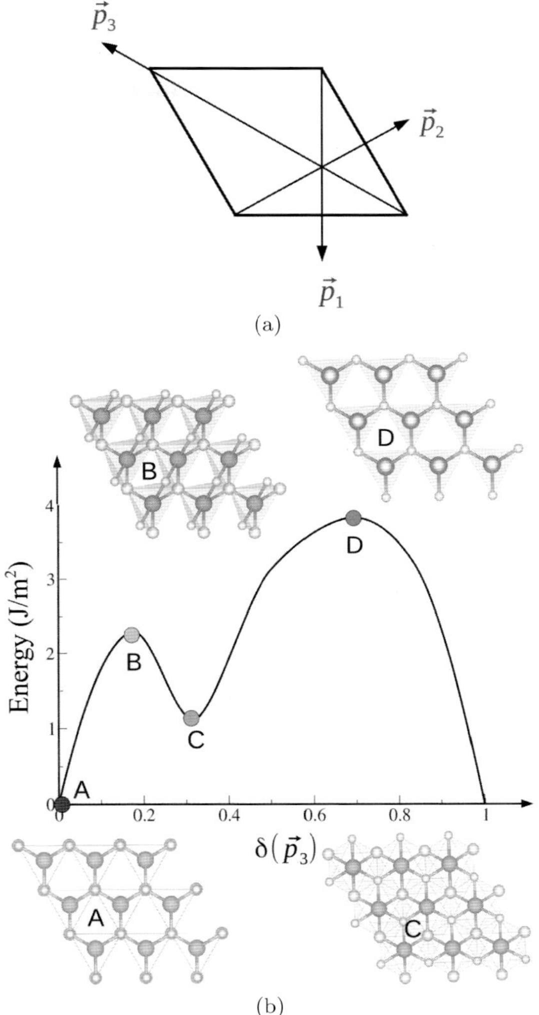

Figure 14. Unit cell with three symmetry directions forming the transition paths, section of the γ-surface along one of these directions and atomic structures of the extremum points along such a path. (a) Unit cell of MX$_2$ and the three directions denoted by arrows $\vec{p_1}$, $\vec{p_2}$ and $\vec{p_3}$ (where $\vec{p_1}$ and $\vec{p_2}$ are equivalent directions); (b) sections of γ-surface along the $\vec{p_3}$ direction. Top views of the structures of monolayer of MX$_2$ at extreme points of the γ-surface are shown in (b). X-atoms are denoted by light grey (small) spheres and M-atoms by grey spheres. To distinguish between two X planes, the top X-atoms are denoted by smaller radii than the bottom ones. [7]

Table 3. Energies of local minima (γ_C), minimum energy barrier (γ_B) and maximum energy barrier (γ_D) with respect to 1H structure, of energy surface for MX_2 (M = Mo, W and X = S, Se) [7].

Compound	γ_C, $(x, y = \frac{2}{3}, \frac{1}{3})$		γ_B, $(x, y = \frac{5}{6}, \frac{1}{6}; \frac{5}{6}, \frac{2}{3}; \frac{1}{3}, \frac{1}{6})$		γ_D, $(x, y = \frac{1}{3}, \frac{2}{3})$	
	e (J/m^2)	e (eV/fu)	e (J/m^2)	e (eV/fu)	e (J/m^2)	e (eV/fu)
MoS_2	1.51	0.83	2.80	1.54	4.37	2.40
$MoSe_2$	1.18	0.70	2.28	1.35	3.81	2.26
WS_2	1.61	0.88	3.01	1.65	4.51	2.47
WSe_2	1.29	0.77	2.47	1.47	3.94	2.35

reflection symmetry) of transformation of the ideal 1H structure to the 1T configuration, that involve the barriers of same energy (see Figure 14) [7].

The maximum energy on the γ-surface (denoted as γ_D) occurs at a slip of $\vec{d} = (\frac{1}{3}, \frac{2}{3})$, and it ranges from 3.81 to 4.51 J/m^2 (2.3–2.5 eV/fu) (see Table 3) for different MX_2 compounds. In this configuration, the top X sublattice plane is on the top of the M sublattice (Figure 13(a)), hence it is unstable with respect to slip along both x- and y-directions. The other set of calculations in which the top X-atom was allowed to relax in all the directions, showed that the energies of the local minima remain unchanged, and the energy of the transition state (i.e., the saddle point energy) reduces by $\approx 0.03\,J/m^2$.

10.4.3.2   *Electronic structure of planar defects in MX_2: sensitivity of gap*

We now examine the electronic band gap of MX_2 monolayers in structural configurations associated with glide on the uniform 6×6 mesh in the real space unit cell, and visualise its Fourier interpolation $E_g(x, y)$ with a contour plot (Figure 15). We find that the band gap vanishes in most of the region $[(x, y) \in [0, 1)]$ centred at the configuration of 1T structure, and is non-zero only in a tiny region centred at the 1H structure, where MoS_2, $MoSe_2$, WS_2 and WSe_2 monolayers exhibit direct band gaps of 1.68 eV, 1.46 eV, 1.80 eV and 1.52 eV, respectively [7]. We note that Mo compounds typically have

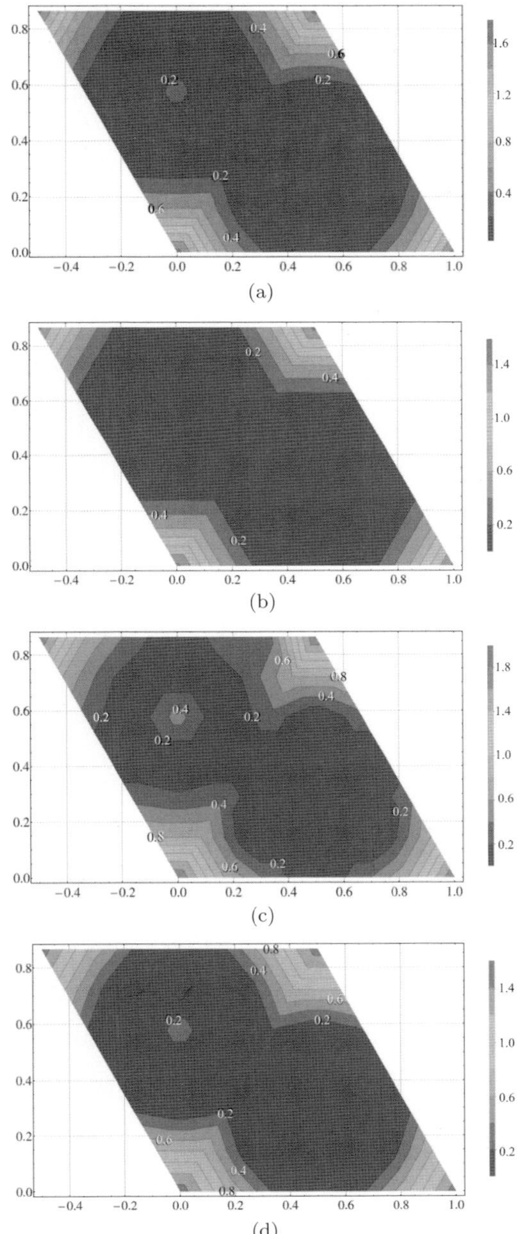

Figure 15. Contour plot of the band gap $E_g(x,y)$ of monolayer of MX$_2$ as a function of slip/glide for (a) MoS$_2$, (b) MoSe$_2$, (c) WS$_2$ and (d) WSe$_2$. Note the absence of islands with non-zero band gap in the region with metallic states in MoSe$_2$ [7].

a lower band gap than the respective W compounds and, amongst Mo/W TMDCs, compounds with Se have a lower band gap than ones with S in the respective metals. Lower band gaps indicate softer bonds [52], implying that bonding in $MoSe_2$ is the weakest amongst these compounds. In the contour plot of the band gaps (refer to Figure 15), we find that only $MoSe_2$ has a vanishing gap at its γ_D configuration, which is also related to the nature of bonding i.e., softer bonds in $MoSe_2$. We note that the pathway of the 1H to 1T polytype transition invariably passes through metallic states (see Figure 15).

10.4.3.3 *Vibrational properties and stability*

We now assess the structural stability of the 1H and 1T polytypes of MX_2 compounds through determination of their phonon spectra. If the phonon spectrum exhibits phonon modes with imaginary frequencies, the structure is locally unstable (i.e., it is a saddle point in the energy landscape), else it is stable. The eigen displacements of the unstable modes give energy lowering structural distortions. Lastly, phonons of the stable structural form not only predict the structural stability, but are also useful in Raman and infra-red (IR) characterisation of these structures.

1H Structure

Phonon dispersions of the 1H polytype of each of the four compounds exhibit no unstable modes (i.e., imaginary frequencies), establishing that the 1H structure is stable polytypic form of MX_2 (refer to Figure 16, left panel). Higher mass of Se results in lower frequencies of MSe_2 compounds than those of MS_2, and the bandwidth of phonon modes of 1H MSe_2 is typically smaller compared to those of 1H MS_2. Interestingly, the optical bands of WSe_2 with in-plane and out-of-plane atomic displacements are split, whereas they overlap in frequency, forming a single composite band of optical phonons in the other three compounds [7].

1T Structure

Our analysis of the vibrational spectra of the centrosymmetric 1T (c1T) polytype reveals that all four MX_2 compounds studied here

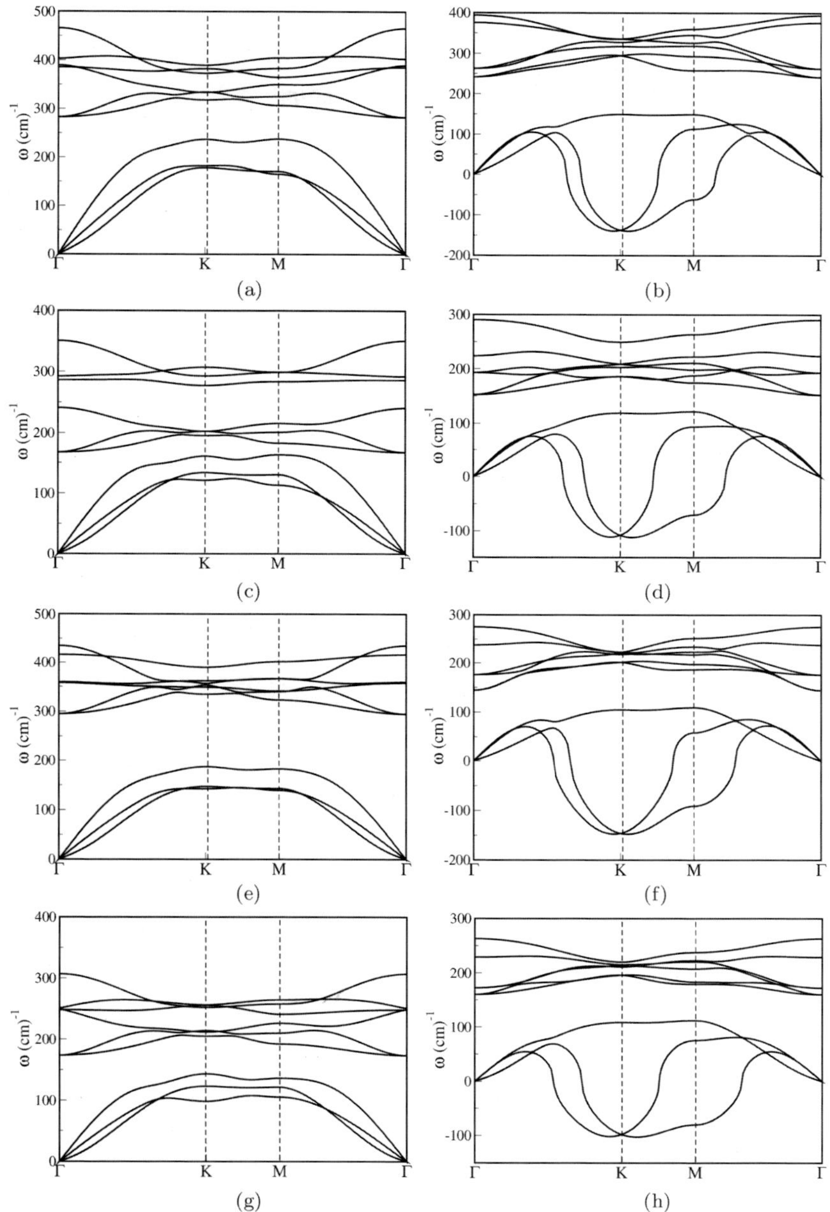

Figure 16. Phonon dispersion of monolayered MX$_2$ with 1H structure (left panel) and c1T structure (right panel), MoS$_2$ in (a) and (b), MoSe$_2$ in (c) and (d), WS$_2$ (e) and (f), and WSe$_2$ in (g) and (h). Note that the instability at the K-point in 1T MX$_2$ is doubly degenerate, and the degeneracy is split at points away from K. The instability at M-point is singly degenerate [7].

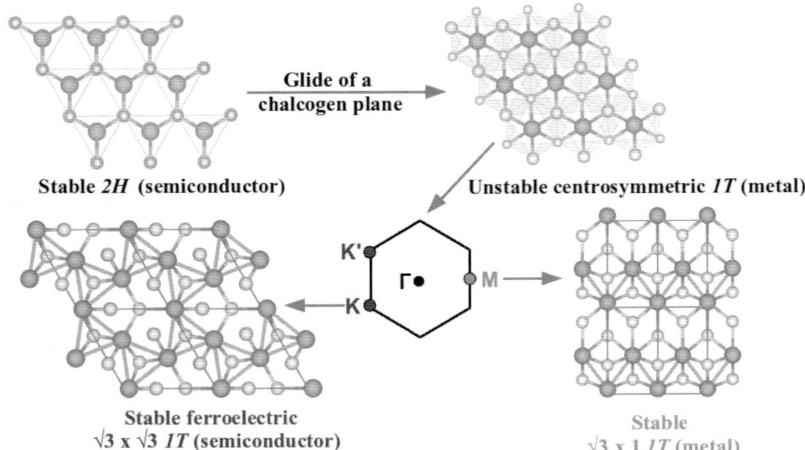

Figure 17. Top view of 1H, c1T, $\sqrt{3} \times \sqrt{3}$ 1T and $\sqrt{3} \times 1$ 1T structures. A schematic diagram showing systematic transition of 1T structure (unstable) to stable structures $\sqrt{3} \times \sqrt{3}$ 1T structure and $\sqrt{3} \times 1$ 1T structure which is led by K-point and M-point instabilities.

are structurally unstable, and exhibit structural instabilities with imaginary frequencies in the range of 100i to 200i cm^{-1} (refer to Figure 16, right panel). The unstable modes are doubly degenerate at the high-symmetry K-point of the BZ, and singly degenerate at the M-point of the BZ [7]. Structural distortions associated with these instabilities at K and M points transform the c1T structure into $\sqrt{3} \times \sqrt{3}$ and $\sqrt{3} \times 1$ superstructure (see Figure 17). We discuss the K-point instability and its consequences to the structure of the 1T form in section (a), followed by similar analysis of the M-point instability in section (b).

(a) K-point instability: $\sqrt{3} \times \sqrt{3}$ 1T structure

We first note that the degeneracy of the K-point instability is lifted at wavevector q in the neighbourhood of K (i.e., $q = K + \delta k$; refer to Figure 16 right panels). This evolves into two singly degenerate unstable modes, one of which constitutes a slightly stronger instability than that at the K-point. Since there are two equally unstable modes at the K-point that can lower the energy through distortion of the 1T polytype as opposed to a single strongly unstable mode at $K + \delta k$, we expect the lowest energy structure to arise from freezing of the K-point instability. The structural distortion associated with atomic displacements of the K-point mode with K$_3$ symmetry leads to a cell tripled ($\sqrt{3} \times \sqrt{3}$) superstructure with trimerisation of metal atoms [53] (Figure 17). The trimerisation of metal atoms with reduction in M$-$M bond lengths by 0.2 Å (refer to Table 4), which involves bonding through the d^2 electronic states of transition metals of these MX$_2$ compounds, lowers the energy of the 1T polytype by $\sim$0.28–0.32 eV/fu [7] (refer to Table 4) and is accompanied by the opening of a band gap of $\sim$0.6–0.8 eV marking a metal-semiconductor transition. Along with this, we perform full (atomic and cell) relaxation of the structure as stresses on the unit cell ($\sigma_{xx} = \sigma_{yy} \sim 12$ kbar to the 24 kbar) are a bit large due to trimerisation of the metal atoms. The lattice constants ($a = b$) of structure of MoS$_2$, MoSe$_2$, WS$_2$ and WSe$_2$ change by 1.8%, 1.2%, 2.2% and 1.5%, respectively, with respect to the lattice parameters of the c1T structure. Phonon dispersion of the $\sqrt{3} \times \sqrt{3}$ 1T structure of each of the four compounds exhibits no unstable modes, confirming its metastability [7] (Figure 18).

(b) M-point instability: $\sqrt{3} \times 1$ 1T structure

Freezing of the eigen-displacements of the unstable mode at M-point of the 1T polytype results in a superstructure (see Figure 17) with zigzag chains of metal atoms. Analysis of this superstructure shows that it is the same as the precursor (α phase) observed during the 1H to 1T-phase transformation by Lin *et al.* [36]. This distorted structure involving dimerisation of metal atoms (where the M$-$M bonds are contracted by 0.5 Å (refer to Table 4) is weakly metallic (see Figure 19(a)), and a better candidate for HER activity than 1T and 1H polytypes [54]. Along with this, a full relaxation of the structure (due to large stresses along the y-direction, $\sigma_{yy} \sim 30$ kbar -34 kbar) result in lattice constants b (a) of the structure of MoS$_2$, MoSe$_2$, WS$_2$ and WSe$_2$ vary by 3.7% (-0.2%), 4.0% (-1.1%), $+3.7$% (-0.4%) and 3.6% (-0.4%), respectively, with respect to the lattice parameters of the c1T structure [7]. Phonon dispersion of the $\sqrt{3} \times 1$ 1T

Table 4. Energetics, metal−metal bond lengths, polarisation and domain wall energies of 1T polytype of MX_2 compound.

Compound	$\Delta E'$ (eV/fu)	b_{c1T} (Å)	b_{d1T} ($\sqrt{3} \times 1$) (Å)	ΔE ($\sqrt{3} \times 1$) (eV/fu)	b_{d1T} ($\sqrt{3} \times \sqrt{3}$) (Å)	ΔE ($\sqrt{3} \times \sqrt{3}$) (eV/fu)	E_R (eV/fu)	P_z ($\sqrt{3} \times \sqrt{3}$) ($\mu C/cm^2$)	D_W ($\sqrt{3} \times \sqrt{3}$) (eV/Å)
MoS_2	0.82	3.2	2.7	−0.28	3.0	−0.28	0.004	0.27	0.004
$MoSe_2$	0.70	3.3	2.8	−0.34	3.1	−0.28	−0.072	0.26	−0.003
WS_2	0.89	3.2	2.7	−0.35	3.0	−0.31	−0.052	0.25	−0.002
WSe_2	0.76	3.3	2.8	−0.47	3.1	−0.32	−0.146	0.25	−0.02

Notes: $\Delta E' = E_{c1T} - E_{1H}$ and $\Delta E = E_{d1T} - E_{c1T}$ are given in eV per formula unit (distorted 1T (d1T) structures: $\sqrt{3} \times \sqrt{3}$ and $\sqrt{3} \times 1$ superstructure). b_{d1T} and b_{c1T} are the bond lengths of metal−metal bond of d1T and c1T structures, respectively. E_R is relative energy of $\sqrt{3} \times 1$ with respect to $\sqrt{3} \times \sqrt{3}$. P_z is the polarisation of 1T polytype of MX_2 in the direction perpendicular to sheet (z-direction). D_W is formation energy of the domain walls separating domains of opposite polarisation [7].

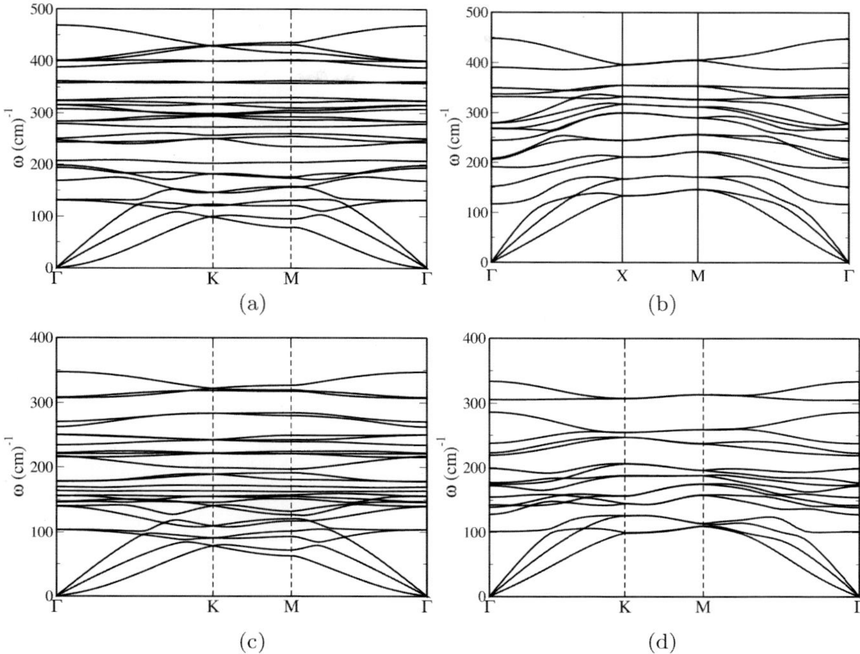

Figure 18. Phonon dispersion of monolayered MX$_2$ with $\sqrt{3} \times \sqrt{3}$ 1T structure (left panel) and $\sqrt{3} \times 1$ 1T structure (right panel), MoS$_2$ in (a) and (b), and MoSe$_2$ in (c) and (d) [7].

superstructure of each of the four compounds (see Figure 18) exhibits no unstable modes, thus confirming its metastability. We find that the $\sqrt{3} \times \sqrt{3}$ superstructure is energetically more stable than the $\sqrt{3} \times 1$ superstructure in MoS$_2$, which is in contrast to the other MX$_2$ compounds, where the $\sqrt{3} \times 1$ superstructure is more stable than $\sqrt{3} \times \sqrt{3}$ structure (refer to ΔE in Table 4) [7].

10.4.3.4 *Raman signatures of stable structural forms (1H, $\sqrt{3} \times \sqrt{3}$ 1T, $\sqrt{3} \times 1$ 1T)*

Raman spectroscopy is a robust technique in identification of a compound in a particular structure. Raman active modes of the stable structural forms of MX$_2$ compounds (Table 5) should facilitate experiments of characterisation of different structural forms of these compounds.

The Raman active A_{1g} modes of the 1H polytypes of MoS_2, WS_2 and $MoSe_2$, WSe_2 are at $388\,cm^{-1}$, $398\,cm^{-1}$, $285\,cm^{-1}$ and $392\,cm^{-1}$, respectively, whereas the doubly degenerate E_{2g}^1 Raman active mode is at $375\,cm^{-1}$, $344\,cm^{-1}$, $292\,cm^{-1}$ and $258\,cm^{-1}$ for MoS_2, WS_2, $MoSe_2$ and WSe_2, respectively [7] (Table 5). Note that the A_{1g} mode of $MoSe_2$ (285 cm^{-1}) is softer than the E_{2g}^1 mode, in contrast to the trend seen in other compounds, as was reported earlier by Sugai *et al.* [55]. This change in the trend in $MoSe_2$ can be traced back to its force constants relevant to shear and compressive deformations [55].

The signatures of the $\sqrt{3} \times \sqrt{3}$ 1T structure are evident seen in its Raman spectrum. The E_{1g} and A_{1g} Raman active modes of MoS_2 (WS_2) are at $292\,cm^{-1}$ ($292\,cm^{-1}$) and $402\,cm^{-1}$ ($411\,cm^{-1}$), respectively [7]. In comparison, the E_{1g} and A_{1g} modes of $MoSe_2$ (WSe_2) are at $148\,cm^{-1}$ ($152\,cm^{-1}$) and $227\,cm^{-1}$ ($257\,cm^{-1}$), respectively [7]. Along with E_{1g} and A_{1g} modes, there exist four additional Raman active modes (which we name J_1, J_2, J_3 and J_4) in all the MX_2 compounds. These modes (refer to Table 5) are the Raman signatures that can be used to clearly detect the $\sqrt{3} \times \sqrt{3}$ superstructure of these materials. We note that the Raman active modes of MoS_2 and

Table 5. Raman active modes of 1H and 1T ($\sqrt{3} \times \sqrt{3}$ and $\sqrt{3} \times 1$ superstructures) polytypes of MX_2 [7].

Structure	Raman Modes	MoS_2 (cm^{-1})	$MoSe_2$ (cm^{-1})	WS_2 (cm^{-1})	WSe_2 (cm^{-1})
1H	A_{1g}	388	285	398	392
	E_{2g}^1	375	292	344	258
1T ($\sqrt{3} \times \sqrt{3}$)	A_{1g}	402	227	411	257
	E_{1g}	292	148	292	152
	J_1	178	127	156	132
	J_2	203	159	150	128
	J_3	270	180	260	175
	J_4	376	319	360	273
1T ($\sqrt{3} \times 1$)	A_{1g}	391	222	391	254
	E_{1g}	268	139,143	258,268	141,147

WS$_2$ (refer to Table 5) lie in the same frequency range, and the same holds for MoSe$_2$ and WSe$_2$.

Spectroscopic signatures of the $\sqrt{3} \times 1$ superstructure can be identified in the E_{1g} and A_{1g} Raman active modes. Due to the lower symmetry of the dimerised structure, the degeneracy of the E_{1g} mode splits. We find that for MoS$_2$ (WS$_2$) the E_{1g} and A_{1g} modes are at $268\,\mathrm{cm}^{-1}$ ($258\,\mathrm{cm}^{-1}$ and $268\,\mathrm{cm}^{-1}$) and $391\,\mathrm{cm}^{-1}$ ($391\,\mathrm{cm}^{-1}$), respectively [7]. Similarly, the E_{1g} and A_{1g} modes of MoSe$_2$ (WSe$_2$) are at $139\,\mathrm{cm}^{-1}$ and $143\,\mathrm{cm}^{-1}$ ($141\,\mathrm{cm}^{-1}$ and $147\,\mathrm{cm}^{-1}$) and $222\,\mathrm{cm}^{-1}$ ($254\,\mathrm{cm}^{-1}$), respectively (see Table 5) [7]. In contrast, changes in frequencies of the A_{1g} and E_{1g} modes are softer in the $\sqrt{3} \times 1$ than in $\sqrt{3} \times \sqrt{3}$ superstructure. The E_{1g} modes of the $\sqrt{3} \times 1$ superstructure of MoS$_2$ (WS$_2$) and MoSe$_2$ (WSe$_2$) are softer by $25\,\mathrm{cm}^{-1}$ ($9\,\mathrm{cm}^{-1}$) and $34\,\mathrm{cm}^{-1}$ ($11\,\mathrm{cm}^{-1}$), respectively [7]. In contrast, the changes in frequencies of the A_{1g} modes are $11\,\mathrm{cm}^{-1}$ ($20\,\mathrm{cm}^{-1}$) and $5\,\mathrm{cm}^{-1}$ ($3\,\mathrm{cm}^{-1}$) for MoS$_2$ (WS$_2$) and MoSe$_2$ (WSe$_2$), respectively, in the two superstructures [7]. The change in the frequency of the A_{1g} mode is not as significant as the E_{1g} mode for all the compounds. Secondly, the shift in the frequency of the E_{1g} mode is more pronounced for the MS$_2$ compounds than in the MSe$_2$ compounds. Hence, the Raman signatures can be useful in distinguishing the $\sqrt{3} \times 1$ and $\sqrt{3} \times \sqrt{3}$ superstructures of MX$_2$ compounds experimentally.

10.4.3.5 *Transition pathways from $\sqrt{3} \times \sqrt{3}$ 1T to 1H, $\sqrt{3} \times 1$ 1T to 1H and $\sqrt{3} \times 1$ 1T to $\sqrt{3} \times \sqrt{3}$ 1T*

To understand the relationship between different structures, it is important to know energy barriers along the pathways of transitions from $\sqrt{3} \times \sqrt{3}$ to 1H, $\sqrt{3} \times 1$ to 1H and $\sqrt{3} \times 1$ 1T to $\sqrt{3} \times \sqrt{3}$ 1T structures. These can be determined the transition using the nudged elastic band method [56] as implemented in the Quantum ESPRESSO package [12]. Estimates of the energy barriers for the transitions from the $\sqrt{3} \times \sqrt{3}$ 1T to the 1H are $1.06\,\mathrm{eV/fu}$, $0.96\,\mathrm{eV/fu}$, $1.20\,\mathrm{eV/fu}$ and $1.10\,\mathrm{eV/fu}$ for MoS$_2$, MoSe$_2$, WS$_2$ and WSe$_2$, respectively (refer to Table 6) [7]. The energy barriers for a transition

Table 6. Energy barriers for the transition from (a) $\sqrt{3} \times \sqrt{3}$ to 1H (b) $\sqrt{3} \times 1$ to 1H and (c) $\sqrt{3} \times 1$ to $\sqrt{3} \times \sqrt{3}$ structure of MX_2 compounds [7].

Compound	E (eV/fu) ($\sqrt{3} \times \sqrt{3}$ to *1H*)	E (eV/fu) ($\sqrt{3} \times 1$ to *1H*)	E (eV/fu) ($\sqrt{3} \times 1$ to $\sqrt{3} \times \sqrt{3}$)
MoS_2	1.06	0.75	0.025
$MoSe_2$	0.96	0.70	0.085
WS_2	1.20	0.60	0.078
WSe_2	1.10	0.80	0.163

from the $\sqrt{3} \times 1$ 1T to the 1H structure are much smaller, i.e., 0.75 eV/fu, 0.70 eV/fu, 0.60 eV/fu and 0.80 eV/fu for MoS_2, $MoSe_2$, WS_2 and WSe_2, respectively. Finally, the energy barriers for a transition from the $\sqrt{3} \times 1$ 1T to the $\sqrt{3} \times \sqrt{3}$ 1T structure are even smaller, i.e., 0.025 eV/fu, 0.085 eV/fu, 0.078 eV/fu and 0.163 eV/fu for MoS_2, $MoSe_2$, WS_2 and WSe_2, respectively (refer to Table 6).

It is clear that the 1H to $\sqrt{3} \times 1$ structural transition is more readily possible than the one from the 1H to $\sqrt{3} \times \sqrt{3}$ structure in all four compounds. Hence, to stabilise the $\sqrt{3} \times \sqrt{3}$ structure in MoS_2, one needs to stabilise the $\sqrt{3} \times 1$ structure first [7]. The transition path calculations show that the $\sqrt{3} \times 1$ and $\sqrt{3} \times \sqrt{3}$ structures are quite comparable in energy and the barriers along the path between them are also quite small (refer to Table 6, Figure 19(b)). Notably, the intermediate structures (Figure 19(b)) have lower energy than the $\sqrt{3} \times \sqrt{3}$ structure in $MoSe_2$, WS_2 and WSe_2. A careful examination of the intermediate structures reveals that they are a essentially strained $\sqrt{3} \times 1$ structure. Hence, the $\sqrt{3} \times 1$ structure is more stable than the $\sqrt{3} \times \sqrt{3}$ structure in $MoSe_2$, WS_2 and WSe_2, whereas the $\sqrt{3} \times \sqrt{3}$ structure is more stable than the $\sqrt{3} \times 1$ structure in MoS_2 (Figure19(b)) [7]. Our analysis confirms, why the $\sqrt{3} \times 1$ structure is more commonly observed than the $\sqrt{3} \times \sqrt{3}$ structure.

10.4.3.6 *Ferroelectricity and domain walls in $\sqrt{3} \times \sqrt{3}$ 1T structure*

Through 2D materials (e.g., graphene and MoS_2) are suitable for high-speed and low-power nanoelectronic devices, incorporation of

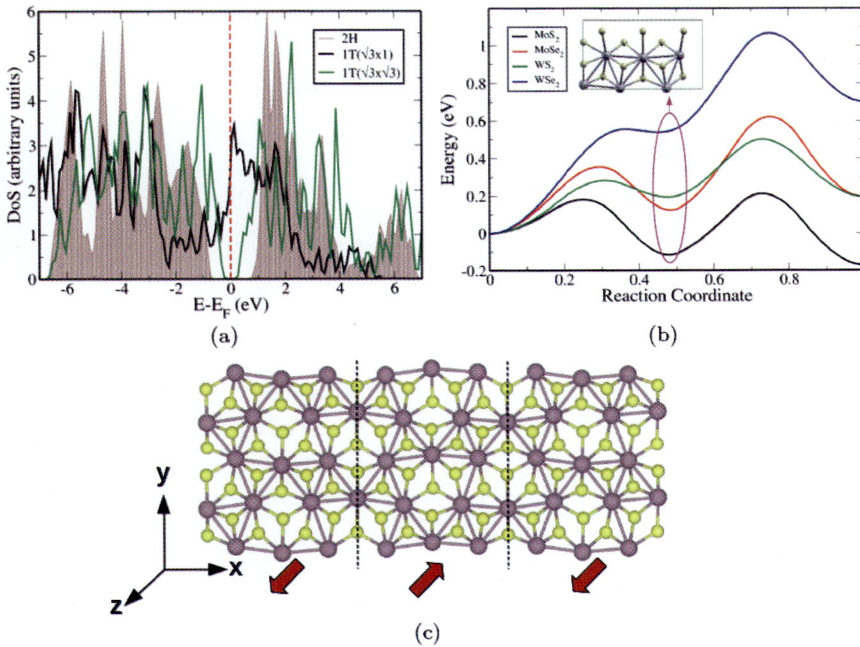

Figure 19. (a) Electronic density of states of MoS$_2$ in its 1H, $\sqrt{3} \times \sqrt{3}$ 1T and $\sqrt{3} \times 1$ 1T forms. (b) Transition pathways from the $\sqrt{3} \times \sqrt{3}$ 1T to $\sqrt{3} \times 1$ structure, respectively [7]. The structure of intermediated image is shown, where M atoms are denoted by grey spheres and X atoms are denoted by yellow sphere [7]. (c) Structure of the domain walls between up and down polarised states of $\sqrt{3} \times \sqrt{3}$ 1T. Black dashed lines in (c) show junction of up and down domain walls [7].

a smart functional property such as ferroelectricity can significantly enhance the range of their applications to sensors, actuators, and memories [57–59]. Ferroelectrics materials have a macroscopic electric polarisation arising from spontaneous ordering of electric dipoles which can be controlled by external electric and stress fields. These ferroelectric materials are typically insulators. Ferroelectric dipoles in ultrathin films perpendicular to the film surface are suppressed by their depolarising field and ferroelectricity disappears in the films with thickness below 24 Å in BaTiO$_3$ [60], 12 Å in PbTiO$_3$ [61], and 10 Å in polymer films [62]. However, 2D materials like graphene, h-BN and MoS$_2$ have not been explored for its existence, these materials are promising for (i) addressing the fundamental issue of

2D ferroelectricity and (ii) a possible combination of ferroelectricity and semiconducting transport properties relevant to applications. Among these, MoS_2 holds a special attraction for being a 2D ferroelectric semiconductor, as it exhibits polytypes with rich electronic structure [53] and a moderate band gap [63] and has been used effectively in a field effect transistor [64]. While a monolayer of $2H$-MoS_2 is non-centrosymmetric, its polarisation vanishes due to other symmetries of the structure. On the other hand, $1T$-MoS_2 is a promising candidate for ferroelectricity as its Mo site is at centre of inversion.

Ferroelectricity in c1T MoS_2 is predicted by Shirodkar *et al.* [65]. From the charge density difference of $\sqrt{3} \times \sqrt{3}$ 1T and c1T structures, it is clear that $\sqrt{3} \times \sqrt{3}$ 1T is non-centrosymmetric as change in charge density is localised only on one of the sulphur atoms. A spontaneous polarisation of $\sim 0.28\,\mu C/cm^2$ along the z direction is revealed, while the in-plane polarisation vanishes. The structural distortion of a K_3 mode involves a periodic array of dipole moments that average to a vanishing polarisation. Hence, a non-zero polarisation has to arise from a nonlinear coupling of the K_3 mode a polar mode [66]. Shirodkar *et al.* [65] have used symmetry analysis within a Landau theory to derive a precise form of the coupling responsible for ferroelectricity in the $\sqrt{3} \times \sqrt{3}$ 1T structure. Free energy is expressed as a symmetry invariant Taylor series in the relevant structure distortions called ordered parameters that connects c1T to $\sqrt{3} \times \sqrt{3}$ 1T structure taking c1T structure as the reference structure.

Symmetrised combinations of K_3 and K_3' modes form two sets of primary order parameters $S = (\eta_1, \eta_2, \eta_3, \eta_4)$ giving trimerisation of Mo. The polar mode Γ_2^- is the secondary order parameter η_5, which involves out-of-plane displacement of sulphur sublattices relative to the Mo sublattice inducing a polarisation along the z axis. Another secondary order parameter (η_6) is associated with changes in the effective thickness of the 1T monolayer, i.e., the Γ_1^+ mode with the full structural symmetry of c1T (for simplicity [65], the contribution of the Γ_1^+ mode is omitted in this analysis). Free energy as a symmetry-invariant Taylor expansion in order parameters is:

$$
\begin{aligned}
F = {} & g_{12}[(T - T_C)/T_C](\eta_1^2 + \eta_2^2 + \eta_3^2 + \eta_4^2) + g_{22}\eta_5^2 \\
& + g_{13}(\eta_1^3 - 3\eta_1\eta_2^2 + \eta_3^3 - 3\eta_3\eta_4^2) + g_{23}\eta_5(\eta_1^2 + \eta_2^2 - \eta_3^2 - \eta_4^2) \\
& + g_{14}[(\eta_1^2 + \eta_2^2)^2 + (\eta_2^2 + \eta_3^2)^2 + (\eta_3^2 + \eta_4^2)^2 + (\eta_4^2 + \eta_1^2)^2 \\
& + (\eta_3^2 + \eta_1^2)^2 + (\eta_4^2 + \eta_2^2)^2 - 2\eta_1^4 - 2\eta_2^4 - 2\eta_3^4 - 2\eta_4^4] \\
& + g_{24}[(\eta_1^2 + \eta_2^2)^2 + (\eta_3^2 + \eta_4^2)^2] + g_{34}\eta_5(\eta_1^3 - 3\eta_1\eta_2^2 \\
& + \eta_3^3 - 3\eta_3\eta_4^2) + g_{44}\eta_5^2(\eta_1^2 + \eta_2^2 + \eta_3^2 + \eta_4^2) + g_{54}\eta_5^4,
\end{aligned}
\tag{10.6}
$$

where T_C is the Curie temperature, and the $g_{\alpha\beta}$'s are coefficients that are determined from first-principles calculations (refer Ref. [65]). Minimisation of free energy in the subspace gives,

$$
\eta_5 = -\frac{g_{23}}{2g_{22}}\eta_1^2,
\tag{10.7}
$$

clearly showing that polarisation is induced as a quadratic function of trimerisation variable η_1, which becomes non-zero below the transition temperature.

The non-trivial geometry and symmetry of the four-dimensional (4D) structural subspace are essential to establish the existence of states with opposite polarisation [65] and switchability and, hence, the ferroelectricity of the d1T structure. In the η_1–η_2 plane, there are three minima of energy corresponding to symmetry-equivalent distorted $\sqrt{3} \times \sqrt{3}$ 1T (d1T) structures with polarisation of the same sign, $P_z \sim 0.22\,\mu\text{C/cm}^2$ (comparable to the estimate from first principles). These semiconducting states are separated from each other by semi-infinite lines corresponding to metallic states. Application of inversion symmetry transforms a d1T structure in the plane to that in the plane with reversed polarisation.

Based on the prediction of ferroelectricity in the $\sqrt{3} \times \sqrt{3}$ superstructure of MoS$_2$ [65], it is imperative to examine ferroelectric domains in the $\sqrt{3} \times \sqrt{3}$ superstructure of the MX$_2$ compounds. The $\sqrt{3} \times 1$ phase does not show ferroelectricity as it is metallic in nature. In the cell tripled ($\sqrt{3} \times \sqrt{3}$) structure, the inversion symmetry is broken due to the anharmonic coupling of the K-point unstable mode (K_3) with a stable polar mode (Γ_2^-) (as discussed above). This coupling leads to condensation of the polar mode in the structure, once the ordering driven by the primary instability K_3 sets in. Hence, the spontaneous polarisation perpendicular to the plane of the MX_2 sheet arises, ranging from 0.25 to 0.27 $\mu C/cm^2$ (refer to Table 4). While ferroelectricity in the $\sqrt{3} \times \sqrt{3}$ 1T structure results from trimerisation of M atoms that induce polar distortion of the c1T structure of MX_2, confirms ferroelectricity in $\sqrt{3} \times \sqrt{3}$ 1T form [67]. Thus, we confirm that the $\sqrt{3} \times \sqrt{3}$ 1T superstructure of $MoSe_2$, WS_2 and WSe_2 compounds also show ferroelectricity, in agreement with the predictions of Shirodkar *et al.* [65]. However, based on our results in previous section, these compounds will readily transform to $\sqrt{3} \times 1$ structure.

Ferroelectric application in memory devices depend on domains of different orientations of polarisation relevant to its switchability with an electric field. To switch polarisation through homogeneous nucleation of domains of polarisation with opposite orientation requires unrealistically large electric fields (coercive fields) [7]. The presence of heterogeneously distributed defects in MoS_2 leads to reduction of coercive fields [65], due to easier nucleation of ferroelectric domains near defects at lower electric fields. The dipolectronic devices requires MX_2 compounds to exhibit a stable ferroelectric dipolar domain structure. The stability of the domains is assessed with the domain wall energy as,

$$D_w = \frac{E(180°\text{domains}) - E_{FE}}{A}, \tag{10.8}$$

where, $E(180°$ domains) and E_{FE} are the energies of a configuration with 180° (up and down) polarised domains and bulk ferroelectric (with uniform polarisation), respectively. The area (A) is calculated as a product of length of domain wall and d (thickness of the 2D sheet of MX_2). A stable domain wall configuration indicates that a monolayer can support 180° domains of opposite polarisation (Figure 19(c)). Stable domain wall is necessary for polarisation switching which is a defining property of ferroelectrics [7]. Note that, the size of these domains is much larger than the periodicity of a typical antiferroelectric phase. From the minimum energy structure starting with two domain walls (per supercell) separating the domains with opposite polarisation (refer to Table 4) [7], it was found that only MoS$_2$ among all MX$_2$ compounds exhibits stable domain walls. Negative domain wall energy (refer to Table 4) of the other compounds show their natural instability and their tendency to transform to lower energy $\sqrt{3} \times 1$ 1T superstructure, upon structural relaxation with domain walls. Hence, the relative stability between the $\sqrt{3} \times 1$ and $\sqrt{3} \times \sqrt{3}$ superstructures of MoSe$_2$ and WX$_2$ is essentially reflected in their negative domain wall energies (refer to Table 4). The ferroelectric domain walls are thus the topological structural excitations that drive the $\sqrt{3} \times \sqrt{3}$ to $\sqrt{3} \times 1$ structural transformation in MoSe$_2$, WS$_2$ and WSe$_2$ due to the applied electric field.

While the ferroelectric semiconducting $\sqrt{3} \times \sqrt{3}$ 1T structure is locally stable in all MX$_2$ compounds, it is more stable than the $\sqrt{3} \times 1$ 1T structure only in MoS$_2$ [7]. Thus, only MoS$_2$ has the potential to be used as a semiconducting ferroelectric in the dipolectronic devices proposed in Ref. [65].

10.4.3.7 *Anomalous response to electric field ($\sqrt{3} \times \sqrt{3}$ 1T)*

A redistribution of electronic charge and ionic positions of an insulating single crystal in response to applied electric field determines the material's dielectric properties. The dielectric constant (ϵ) is a measure of the strength of screening of electric fields in a material. ϵ has been considered an important factor to capture the changes in optical properties which involve formation of excitons and trions [68] due to external electrostatic environment of 2D materials like MoS$_2$. In this subsection, we focus on ionic response of MX$_2$ monolayers to the electric field which is shown to be anomalous in nature. This anomaly shows that these monolayers are sensitive to external field and can easily undergo a metal-semiconductor phase transition (as discussed earlier section).

The dielectric properties of both 1H and $\sqrt{3} \times \sqrt{3}$ 1T forms (both of them are semiconductors) of MX$_2$ compounds are discussed in this subsection. The electronic contribution to dielectric constant ϵ^∞

Table 7. Born effective charges (Z^*) and electronic dielectric constants (ϵ^∞) for 1H and 1T ($\sqrt{3} \times \sqrt{3}$ superstructure) polytypes of MX_2 [7].

| | 1H | | | | | 1T($\sqrt{3} \times \sqrt{3}$) |
| | Z^*_{xx} | | Z^*_{yy} | | | |
Structure Compound	M	X	M	X	$\epsilon^\infty_{xx} = \epsilon^\infty_{yy}$	$\epsilon^\infty_{xx} = \epsilon^\infty_{yy}$
MoS_2	−1.0	0.5	−1.2	0.5	7.6	11.3
$MoSe_2$	−1.9	0.9	−2.0	0.9	8.5	11.5
WS_2	−0.4	0.2	−0.5	0.2	7.0	11.7
WSe_2	−1.2	0.5	−1.3	0.5	7.9	11.7

(optical dielectric constant) of $\sqrt{3} \times \sqrt{3}$ 1T form is significantly (by $\approx 50\%$) larger than that of the 1H form, due to its smaller band gap. $\epsilon^\infty_{xx} = \epsilon^\infty_{yy}$ is in the range of 7 to 8.5 for the 1H polytype whereas it ranges from 11 to 12 for the $\sqrt{3} \times \sqrt{3}$ 1T structure (refer to Table 7) [7]. The electronic, excitonic and trionic properties are dependent on dielectric constant of these materials.

The lower frequency (THz) dielectric response, which is directly proportional to squared mode (Born) effective charges and inversely proportional to the square of their frequencies [69] has contributions from phonons as well. Born effective charge ($Z^*_{i,\alpha,\beta}$) is defined as the force acting on an ion (i) in the 'α' direction when an electric field is applied in the 'β' direction (E_β). The tendency of a material to undergo a ferroelectric distortion or its vicinity to metallicity of the compound are reflected anomalous value of Z^* [70]. The measure of anomalous nature of Z^* measured by the deviation from the nominal charge, and it reflects metallicity or covalency. The dielectric constant or Born effective charges of $\sqrt{3} \times 1$ 1T structure of are physically meaningless due to strong screening of electric field by conduction electrons, it being metallic.

Born effective charges of 1H polytype of MX_2 compounds are expected to be closer to their nominal charges (i.e., +4 for M and −2 for X) due to its relatively wide band gap. However, we find that the values, and even the sign of Z^* are counter-intuitive. The Z^* of M atoms is in the range from −0.5 to −2 (refer to table) and that of the X atom is in that of $+0.2$ to $+0.9$ [7]. Due to strong covalency of these materials, the value of a Z^* is anomalous. The d-states of M atoms constituting the top most valence band just below the Fermi level are responsible for the opposite sign of Z^*. The anomaly is even stronger in $\sqrt{3} \times \sqrt{3}$ 1T form (see Table 7 [7]) and its Z^* is larger in magnitude by a factor of 2.5 than that of 1H polytype (due to smaller gap). We find that the in-plane Z^* of its M atoms are in the range of $+0.7$ to -5, and that of its X atom is in the range $+2$ to -2. These anomalies in Z^* are very likely to its vicinity primarily to a metal to semiconductor transition, and reveals tunability of its electron−phonon coupling with electric field [65].

10.5 Summary

It is clearly evident that defects in 2D materials, such as 2D h-BN and MX$_2$ compounds, greatly influence their structure and properties. GBs can intrinsically cause rippling and wrinkles in the h-BN sheet and affect the electronic structure greatly. It is challenging to identify these GBs experimentally, but with the help of shift in the Raman frequencies and high-resolution (HR) microscopy these GBs may be easily identified.

Planar defects are found only in materials such as MX$_2$ compounds (M = (Mo, W); X = (S, Se)) whose monolayers consist of 2 or 3 atomic layers. Upon the introduction of a planar defect, the system makes transition from insulator to metal because of increase in the covalency of M and X atoms. The metallic c1T polymorphic structure is unstable, exhibiting instabilities at valley point K and zone boundary point M. These structural instabilities of metallic structure lead to symmetry and energy distortion of the c1T structure and forms stable superstructures $\sqrt{3} \times \sqrt{3}$ and $\sqrt{3} \times 1$, respectively. The $\sqrt{3} \times 1$ structure associated with dimerisation of metal atoms remains metallic, while the $\sqrt{3} \times \sqrt{3}$ superstructure exhibits a non-zero electronic band gap arising from strong trimerisation of metal atoms.

The $\sqrt{3} \times \sqrt{3}$ structure of (Mo, W)(S, Se)$_2$ exhibits a spontaneous polarisation (0.25–$0.3\,\mu\mathrm{C/cm}^2$) along the direction perpendicular to the sheet, making them the thinnest known ferroelectrics. However, the $\sqrt{3} \times \sqrt{3}$ structure is (a) more stable than the $\sqrt{3} \times 1$ structure, and (b) exhibits a stable dipolar domain structure only in MoS_2. Thus, MoS_2 is the only suitable candidate for possible use in the dipolectronic devices. As a result of vicinity of these compounds to metal-semiconductor and ferroelectric transitions, all of them exhibit anomalous responses to an electric field.

In conclusion, the response of 2D materials is greatly altered by the presence defects. Any defect in a 2D material involving change in coordination number results in ripples and experimental observation of these defects is facilitated by theoretical predictions.

Acknowledgements

Anjali Singh is thankful to Jawaharlal Nehru Centre for Advanced Scientific Research, India, for a Research Fellowship. Umesh V. Waghmare acknowledges support from a J. C. Bose National Fellowship of the Department of Science and Technology, Govt. of India and US Air Force AOARD Grant No. FA 2386-15-1-0002.

References

[1] C. Jin, F. Lin, K. Suenaga and S. Iijima, *Phys. Rev. Lett.* **102**, 195505 (2009).
[2] J. Lahiri *et al.*, *Nat. Nanotechnol.* **5**, 326 (2010).
[3] P. Y. Huang *et al.*, *Nature* **469**, 389 (2011).
[4] J. M. Ziman, *Models of Disorder: The Theoretical Physics of Homogeneously Disordered Systems*, Cambrige University Press, Cambridge (1979).
[5] S. E. Dann *Reactions and Characterization of Solids*, The Royal Society of Chemistry, Cambridge (2000).
[6] A. Singh and U. V. Waghmare, *Phys. Chem. Chem. Phys.* **16**, 21664 (2014).
[7] A. Singh, S. N. Shirodkar and U. V. Waghmare, *2D Materials* **2**, 035013 (2015).
[8] S. N. Shirodkar, U. V. Waghmare, T. S. Fisher and R. Grau-Crespo, *Phys. Chem. Chem. Phys.* **17**, 13547 (2015).
[9] M. U. Kahaly, S. P. Singh and U. V. Waghmare, *Small* **4**, 2209 (2008).
[10] J. Lahiri *et al.*, *Nat. Nano.* **5**, 326 (2010).

[11] N. Ashcroft and N. Mermin, *Solid State Physics*, Saunders College, Philadelphia (1976).

[12] P. Giannozzi *et al.*, *J. Phys. Conden. Matter* **21**, 395502 (2009).

[13] D. Vanderbilt, *Phys. Rev. B* **41**, 7892 (1990).

[14] J. P. Perdew, K. Burke and M. Ernzerhof *Phys. Rev. Lett.* **77**, 3865 (1996).

[15] J. C. Meyer *et al.*, *Nano Lett.* **9**, 2683 (2009).

[16] A. Nag *et al.*, *ACS Nano* **4**, 1539 (2010).

[17] D. Golberg *et al.*, *ACS Nano.* **4**, 2979 (2010).

[18] L. S. Panchakarla *et al.*, *Adv. Mater.* **21**, 4726 (2009).

[19] W. Coene, G. Janssen, M. Op de Beeck and D. Van Dyck, *Phys. Rev. Lett.* **69**, 3743 (1992).

[20] Y. Li *et al.*, *J. Phys. Chem. C* **112**, 1365 (2008).

[21] I. Jimnez *et al.*, *Appl. Phys. Lett.* **68**, 2816 (1996).

[22] A. Zobelli *et al.*, *Nano Lett.* **6**, 1955 (2006).

[23] S. N. Shirodkar and U. V. Waghmare, *Phys. Rev. B* **86**, 165401 (2012).

[24] A. Stone and D. Wales, *Chem. Phys. Lett.* **128**, 501 (1986).

[25] D. P. Hashim *et al.*, *Scientific Rep.* **2**, 363 (2012).

[26] Y. Liu, X. Zou and B. I. Yakobson, *ACS Nano.* **6**, 7053 (2012).

[27] M. S. C. Mazzoni, R. W. Nunes, S. Azevedo and H. Chacham, *Phys. Rev. B* **73**, 073108 (2006).

[28] X. Li, X. Wu, X. C. Zeng and J. Yang, *ACS Nano.* **6**, 4104 (2012).

[29] R. Geick, C. H. Perry and G. Rupprecht, *Phys. Rev.* **146**, 543 (1966).

[30] B. Radisavljevic *et al.*, *Nat. Nanotechnol.* **6**, 147 (2011).

[31] Q. H. Wang *et al.*, *Nat. Nanotechnol.* **7**, 699 (2012).

[32] O. Lopez-Sanchez *et al.*, *Nat. Nanotechnol.* **8**, 497 (2013).

[33] R. Cheng *et al.*, *Nano Lett.* **14**, 5590 (2014).

[34] R. S. Sundaram *et al.*, *Nano Lett.* **13**, 1416 (2013).

[35] M. Chhowalla *et al.*, *Nat. Chem.* **5**, 263 (2013).

[36] Y.-C. Lin, D. O. Dumcenco, Y.-S. Huang and K. Suenaga, *Nat. Nanotechnol.* **9**, 391 (2014).

[37] A. Kuc, *Chemical Modelling: Volume 11*, The Royal Society of Chemistry, Cambridge, Vol. 11, pp. 1–29 (2015).

[38] W. Zhou *et al.*, *Nano Lett.* **13**, 2615 (2013).

[39] A. Krasheninnikov, P. Lehtinen, A. Foster and R. Nieminen, *Chem. Phys. Lett.* **418**, 132 (2006).

[40] G.-D. Lee *et al.*, *Phys. Rev. Lett.* **95**, 205501 (2005).

[41] H. Y. Jeong *et al.*, *ACS Nano.* **10**, 770 (2016).

[42] Y.-C. Lin *et al.*, *Nat. Commun.* **6**, 6736 (2015).

[43] A. M. van der Zande *et al.*, *Nat. Mater.* **12**, 554 (2013).

[44] C. Ataca and S. Ciraci, *J. Phys. Chem. C* **115**, 13303 (2011).

[45] H.-P. Komsa and A. V. Krasheninnikov, *Phys. Rev. B* **91**, 125304 (2015).

[46] Y. Wei *et al.*, *Nat. Mater.* **11**, 759 (2012).

[47] Q. Yu *et al.*, *Nat. Mater.* **10**, 443 (2011).

[48] O. V. Yazyev and S. G. Louie, *Nat. Mater.* **9**, 806 (2010).

[49] X. Zou, Y. Liu and B. I. Yakobson, *Nano Lett.* **13**, 253 (2013).

[50] Y. L. Huang *et al.*, *Nat. Commun.* **6**, 6298 (2015).

[51] M. J. Mehl, D. A. Papaconstantopoulos, N. Kioussis and M. Herbranson, *Phys. Rev. B* **61**, 4894 (2000).

[52] K. P. O'Donnell and X. Chen, *Appl. Phys. Lett.* **58**, 2924 (1991).

[53] C. Rovira and M. H. Whangbo, *Inorganic Chemistry* **32**, 4094 (1993).

[54] U. Gupta *et al.*, *APL Mat.* **2**, 092802 (2014).

[55] S. Sugai and T. Ueda, *Phys. Rev. B* **26**, 6554 (1982).

[56] G. Henkelman and H. Jónsson, *J. Chem. Phys.* **113**, 9978 (2000).

[57] J. F. Scott, *Ferroelectrics* **314**, 207 (2005).

[58] R. Guo *et al.*, *Nat. Commun.* **4**, 1990 (2013).

[59] G. Kirchmair *et al.*, *Nature* **460**, 494 (2009).

[60] J. Junquera and P. Ghosez, *Nature* **422**, 506 (2013).

[61] D. D. Fong *et al.*, *Science* **304**, 1650 (2004).

[62] A. V. Bune *et al.*, *Nature* **391**, 874 (1998).

[63] K. F. Mak *et al.*, *Phys. Rev. Lett.* **105**, 136805 (2010).

[64] R. B. *et al.*, *Nat Nano* **6**, 147 (2011).

[65] S. N. Shirodkar and U. V. Waghmare, *Phys. Rev. Lett.* **112**, 157601 (2014).

[66] C. J. Fennie and K. M. Rabe, *Phys. Rev. B* **72**, 100103 (2005).

[67] K. M. Rabe, *Functional Metal Oxides*, Wiley-VCH Verlag GmbH & Co. KGaA, Weinheim, pp. 221–244 (2013).

[68] Y. Lin *et al.*, *Nano Lett.* **14**, 5569 (2014).

[69] M. Born and K. Huang, *Dynamical Theory of Crystal Lattices*, Oxford University Press, Oxford (1954).

[70] U. V. Waghmare, N. A. Spaldin, H. C. Kandpal and R. Seshadri, *Phys. Rev. B* **67**, 125111 (2003).

Index